Corentin Briat

Linear Parameter-Varying and Time-Delay Systems

Analysis, Observation, Filtering & Control

Corentin Briat
Department of Biosystems Science
and Engineering
Swiss Federal Institute
of Technology—Zurich
Basel
Switzerland

ISSN 2197-117X ISSN 2197-1161 (electronic)
ISBN 978-3-662-44049-0 ISBN 978-3-662-44050-6 (eBook)
DOI 10.1007/978-3-662-44050-6

Library of Congress Control Number: 2014943506

Springer Heidelberg New York Dordrecht London

Printed on acid-free paper

Springer is part of Springer Science+Business Media (www.springer.com)

Advances in Delays and Dynamics

Volume 3

Series editor

Silviu-Iulian Niculescu, Laboratory of Signals and Systems, Gif-sur-Yvette, France
e-mail: Silviu.Niculescu@lss.supelec.fr

About this Series

Delay systems are largely encountered in modeling propagation and transportation phenomena, population dynamics, and representing interactions between interconnected dynamics through material, energy, and communication flows. Thought as an open library on delays and dynamics, this series is devoted to publish basic and advanced textbooks, explorative research monographs as well as proceedings volumes focusing on delays from modeling to analysis, optimization, control with a particular emphasis on applications spanning biology, ecology, economy, and engineering. Topics covering interactions between delays and modeling (from engineering to biology and economic sciences), control strategies (including also control structure and robustness issues), optimization, and computation (including also numerical approaches and related algorithms) by creating links and bridges between fields and areas in a delay setting are particularly encouraged.

More information about this series at http://www.springer.com/series/11914

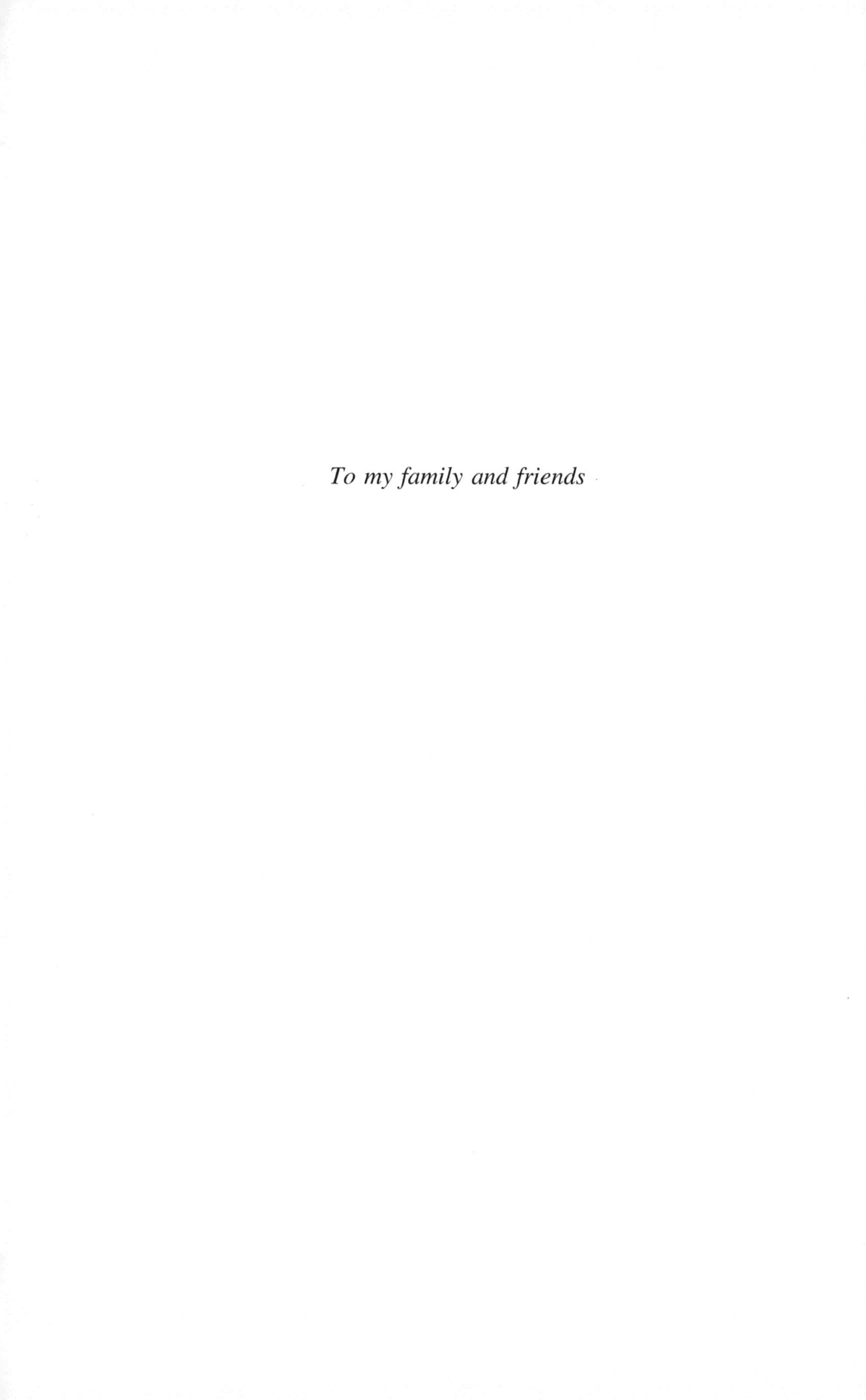

To my family and friends

Preface

The pay's not great, but the work is hard.
Bernard Black

This book provides an introduction to the analysis and control of linear parameter-varying systems, time-delay systems, and their interactions. The purpose is to give the readers some fundamental theoretical background on these topics and to give more insights into the possible applications of these theories. This monograph is intended to be self-contained and is written in an accessible way for readers ranging from undergraduate/Ph.D. students to engineers and researchers willing to know more about the fields of time-delay systems, parameter-varying systems, robust analysis, robust control, LPV gain-scheduling techniques, and LMI-based approaches. The only prerequisites are basic knowledge in linear algebra, ordinary differential equations, and (linear) dynamical systems. Most of the results are proved unless the proof is too complex or not necessary for a good understanding of the results. In the latter cases, suitable references are systematically provided.

This monograph is mostly adapted from my Ph.D. thesis [1] and subsequent works on LPV and time-delay systems. The title of my Ph.D. thesis was *Robust control and observation of linear parameter-varying time-delay systems*. It was supervised by Olivier Sename (Professor at the Grenoble Institute of Technology, Grenoble, France) and Jean-François Lafay (Professor at Ecole Centrale de Nantes, Nantes, France). My current feelings about my Ph.D. now is similar to what Andre Geim said about his Ph.D. thesis at his Nobel Prize lecture[1] in 2010: "It is as exciting as it sounds." But, retrospectively, it was to me an excellent learning research topic since it lies at the intersection of many fields such as time-delay systems, LPV systems, optimization, robust analysis, and control. I wrote the current monograph in the same spirit as I felt it at that time, that is, a book that covers these fields and their intersections, some of which being nowadays unavoidable when wandering in the fields of systems and control theory. After

[1] The title of his Ph.D. thesis was "Investigation of mechanisms of transport relaxation in metals by a helicon resonance method".

graduating, Olivier encouraged me to adapt my thesis into a book. But after 3 years of Ph.D. and very long months of tedious thesis writing, I was not as enthusiastic as he was. So I put this aside. In July 2012, I received a phone call from Silviu-Iulian Niculescu proposing me to write a book from it for the new Springer series *Advances in Delays and Dynamics*. And now, here we are!

This book is the first one I am writing. So, I tried to follow, as much as I could, the recommendations of the excellent textbook *Handbook of writing for the mathematical sciences* by Nicholas J. Higham. So, please be indulgent with me if this book is not a paragon of organization and exposition. I nevertheless hope that people will enjoy it.

Basel, Switzerland Corentin Briat

Reference

1. C. Briat, Robust Control and Observation of LPV Time-Delay Systems. Ph.D. thesis, Grenoble Institute of Technology (2008). http://www.briat.info/thesis/PhDThesis.pdf

Acknowledgments

Knowledge is in the end based on acknowledgement.
Ludwig Wittgenstein

I would like to thank first people who provided some help while writing this monograph. I am particularly grateful to Colette Bichsel who spent some of her precious time to produce some illustrations. I am also grateful to Charles Poussot-Vassal who provided some material for the examples on LPV systems. I also would like to thank Silviu-Iulian Niculescu for having given me the opportunity of writing this book and to Thomas Ditzinger, senior editor at Springer, for having guided me along the process. I also would like to thank my past and current coworkers with whom some of the results reported in this book have been developed.

I recommend to everyone willing to write a monograph the excellent *Handbook of Writing for the Mathematical Sciences* by Nicholas J. Higham. Lots of good advice in there. I am also very thankful to the people all across the world who developed all the different LaTeX packages I have used. Thanks to them, I saved a lot of time when preparing a satisfying version of this monograph. The list is unfortunately too long to be exhaustively enumerated here.

Also, a few words for some people who did not directly help in the process of writing this book, but did help at some point in the past, and therefore indirectly contributed to this book. My first thanks go to my parents and my family who helped me graduating from high-school, supporting my studies and also gave me the thirst for knowledge. I am also grateful to my professors in my engineering school (ESISAR) who taught me that mathematics (the real ones, not the ones in high school) were great, who made me discover automatic control and who instilled a quite complete knowledge in sciences and engineering in me (I suppose). They have greatly influenced my future since I initially intended to become an Electrical Engineer, so instead of writing these lines I would be probably designing antennas or microelectronic circuits, and perhaps even writing a book on these fascinating problems. I am also grateful to my Ph.D. supervisors Olivier Sename and Jean-François Lafay, who gave me the chance to do a Ph.D. under their supervision. Together with my supervisors, I would like to thank the people who accepted to be in my Ph.D. committee, namely Jean-Pierre Richard, Sophie Tarbouriech, Silviu-Iulian Niculescu, Andrea Garulli, and Erik Verriest, and who took some time to read and comment on my thesis. I also would like to thank Erik Verriest who made me discover that control could be applied to other things than electrical motors, water tanks, or any other captivating processes! Ecological models and impulsive systems have influenced my works since I am now interested in these fields. Some examples in this monograph are directly taken from our common works. I also would like to thank Karl-Henrik Johansson, Håkan Hjalmarsson, Henrik Sandberg, Gunnar Karlsson and, last but not the least, Ulf Jönsson for having given me the opportunity to work on the modeling and control of congestion in communication networks. Again some examples presented in this monograph are directly taken from this collaboration. Finally, I would like to thank Mustafa Khammash who accepted me in his "Control Theory and Systems Biology" group in ETH Zürich. I really appreciated that he had given me the chance to work again (and more seriously) on biological problems.

I am also grateful to all the people I met in my life and more especially those who have become friends because they made me what I have become, or what I think I have become. I am not writing a list of those people, but rather a list a places, hence: La Terrasse, Le Touvet, Crolles, Pontcharra, Chapareillan, Grenoble, Lans en Vercors, Valence, Lille, Altlanta, Stockholm, and Basel. Last but not the least, we also have Lion's Arch, Ozora, Idanha-a-Nova, Mulegns and Zernez. Finally, I am also thankful to all my liquorice, coffee and beef jerky, and to all my Sunday fellows.

Contents

Part I Linear Parameter-Varying Systems

Part III Linear Parameter-Varying Time-Delay Systems

Notations and Acronyms

Notations

$\mathbb{R}\,(\mathbb{C})$	Field of real (complex) numbers
$\mathbb{R}_{>0}\,(\mathbb{R}_{\geq 0})$	Set of positive (nonnegative) real numbers
$\mathbb{R}^n$	Space of $n \times 1$ real vectors
$\mathbb{S}^n$	Cone of $n \times n$ symmetric matrices
$\mathbb{S}^n_{\succ 0}(\mathbb{S}^n_{\succeq 0})$	Cone of $n \times n$ symmetric positive (semi)definite matrices
$\mathbb{C}_+\,(\mathbb{C}_-)$	Complex open right-half (left-half) plane
$\overline{\mathbb{C}}_+\,(\overline{\mathbb{C}}_-)$	Complex closed right-half (left-half) plane
$A \prec B$	$A - B$ is negative definite $(A, B \in \mathbb{S}^n)$
$A \preceq B$	$A - B$ is negative semidefinite $(A, B \in \mathbb{S}^n)$
j	Imaginary unit (i.e. $j^2 = -1$)
$\Re(z), \Im(z)$	Real and imaginary part of z
$\mathbf{co}\{U\}$	Convex hull of the set U
$\mathbf{vert}\{\Delta\}$	Set of vertices of the convex polyhedron Δ
$\mathrm{col}_i(x_i)$	Column vector with entries $x_1, x_2, \ldots$
I_n	Identity matrix of dimension n
$0_{n\times m}\,(0_n)$	Zero matrix of dimension $n \times m\,(n \times n)$
$\mathbb{1}_n$	Column vector of dimension n containing 1 entries
$A^{\mathrm{T}}\,(A^*)$	Transpose (conjugate transpose) of A
$\mathrm{He}\,[A]$	Hermitian operator defined as $\mathrm{He}[A] := A + A^*$
A^{-1}	Inverse of A
A^+	Moore-Penrose pseudoinverse of A
$\mathrm{trace}(A)$	Trace of A
$\det(A)$	Determinant of A
$\lambda(A)$	Set of eigenvalues of A
$\sigma(A)$	Set of singular values of A
$\lambda_{min}(A)$	Minimal eigenvalue of A (when it exists)
$\lambda_{max}(A)$	Maximal eigenvalue of A (when it exists)
$\bar{\sigma}(A)$	Maximal singular value of A

$\varrho(A)$	Spectral radius of A
ρ	Parameter in generic LPV systems
h	Delay in time-delay systems
$\boldsymbol{\Delta}_\rho$	Set of values of the parameter; i.e. $\rho \in \boldsymbol{\Delta}_\rho$
$\mathbf{V}_\rho$	Set of vertices of the set $\boldsymbol{\Delta}_\rho$, i.e. $\mathbf{V}_\rho = \mathbf{vert}\{\boldsymbol{\Delta}_\rho\}$
$\boldsymbol{\Delta}_\nu$	Set of values of the parameters derivative; i.e. $\dot{\rho} \in \boldsymbol{\Delta}_\nu$
$\mathbf{V}_\nu$	Set of vertices of the set $\boldsymbol{\Delta}_\nu$, i.e. $\mathbf{V}_\nu = \mathbf{vert}\{\boldsymbol{\Delta}_\nu\}$
$\mathscr{H}_{\mu,\bar{h}}$	Set of delay functions $h : \mathbb{R}_{\geq 0} \to \mathbb{R}_{\geq 0}$ defined as $\mathscr{H}_{\mu,\bar{h}} := \left\{h : \mathbb{R}_{\geq 0} \to \mathbb{R}_{\geq 0} : \ h(t) \leq \bar{h},\ \dot{h}(t) \leq \mu < 1\right\}$
$\mathscr{P}^\nu$	Set of parameter functions $\rho : \mathbb{R}_{\geq 0} \to \boldsymbol{\Delta}_\rho$ defined as $\mathscr{P}^\nu := \left\{\rho : \mathbb{R}_{\geq 0} \to \boldsymbol{\Delta}_\rho : \ \dot{\rho}(t) \in \boldsymbol{\Delta}_\nu,\ t \geq 0\right\}$
$\mathscr{P}^\infty$	$\mathscr{P}^\nu$ with $\boldsymbol{\Delta}_\nu = \mathbb{R} \times \cdots \times \mathbb{R}$
$\mathscr{P}_1^\nu$	$\mathscr{P}^\nu$ with $\boldsymbol{\Delta}_\rho = [-1,1] \times \cdots \times [-1,1]$
$\mathscr{P}_1^\infty$	$\mathscr{P}_1^\nu$ with $\boldsymbol{\Delta}_\nu = \mathbb{R} \times \cdots \times \mathbb{R}$
$\lvert\lvert w \rvert\rvert_q$	q-norm of the vector $w \in \mathbb{R}^n$ defined as $\lvert\lvert w \rvert\rvert_q := (\lvert w_1\rvert^q + \cdots + \lvert w_n\rvert^q)^{1/q}$
$\lvert\lvert M \rvert\rvert_q$	Induced q-norm of the matrix M defined as $\lvert\lvert M \rvert\rvert_q := \sup\limits_{\lvert\lvert w \rvert\rvert_q = 1} \{\lvert\lvert Mw \rvert\rvert_q\}$
$\lvert\lvert w \rvert\rvert_{L_q}$	L_q-norm of the function $w : \mathbb{R}_{\geq 0} \to \mathbb{R}^n$ defined as $\lvert\lvert w \rvert\rvert_{L_q} := \left(\int_0^\infty \lvert\lvert w(t) \rvert\rvert_q^q dt\right)^{1/q}$
L_p	Space of functions with finite L_p-norm
$\lvert\lvert T \rvert\rvert_{L_p - L_q}$	Induced-norm of the operator $T : L_p \to L_q$ defined as $\lvert\lvert T \rvert\rvert_{L_p - L_q} := \sup\limits_{\lvert\lvert w \rvert\rvert_{L_p} = 1} \{\lvert\lvert Tw \rvert\rvert_{L_q}\}$
$\lvert\lvert G \rvert\rvert_{H_\infty}$	H_∞-norm of the transfer function $G(s)$ defined as $\lvert\lvert G \rvert\rvert_{H_\infty} := \sup\limits_{s \in \overline{\mathbb{C}}_+} \bar{\sigma}(G(s))$

Acronyms

LTI	Linear Time-Invariant
LPV	Linear Parameter-Varying
LMI	Linear Matrix Inequality
LFT	Linear Fractional Transformation
LFR	Linear Fractional Representation
SOS	Sum-of-Squares
SDP	Semidefinite Program(ming)

Introduction

Tout le monde savait que c'était impossible.
Il est venu un imbécile qui ne le savait pas et qui l'a fait.[2]
Marcel Pagnol

The qualitative analysis of dynamical systems introduced by H. Poincaré at the end of the nineteenth century [1] gave birth to the fruitful field on dynamical systems theory, with all the profound implications and applications we have nowadays including, among others, systems and control theory. Before Poincaré, differential equations were mostly viewed as equations to be solved, similarly to as algebraic equations. Poincaré had the bright idea to try to study differential equations in a qualitative way, which essentially means that finding solutions is not the objective anymore, but instead, we focus on establishing certain properties of the solutions. This point of view is particularly relevant since many differential equations do not admit closed-form solutions and can only be solved numerically.

In the same vein of Poincaré's ideas, A.M. Lyapunov developed the theory of stability of dynamical systems during his Ph.D. thesis [2], which was supervised by P. Chebyshev. Stability is a fundamental property of dynamical systems having deep consequences in sciences and engineering. Stability essentially means that solutions of a dynamical system starting close to an equilibrium point (which is a resting point of the system), remain close to this equilibrium point. A typical example is the pendulum example. Pendulums with rigid rod admit two equilibrium points, one is when the rod is vertical and the mass down, the other is when the mass is up. Consider the first equilibrium point and assume that there is no friction. A small push from this resting position will result in sustained oscillations of bounded amplitude around it. This equilibrium point is therefore stable. An equilibrium point is, moreover, said to be asymptotically stable if it is stable and the trajectories starting nearby to it converge back to it. Taking again the pendulum example and adding friction to the problem will result in damped oscillations around the equilibrium point. Eventually, the pendulum will stop

[2] Everyone knew it was impossible. It came an imbecile who did not know and who did it.

oscillating and will return to its resting position. This equilibrium point is therefore asymptotically stable. Opposed to stable equilibrium points, unstable ones are resting positions from which arbitrarily small perturbations will be amplified, pushing then the dynamical system away from them. For instance, the second equilibrium point of the pendulum with friction, i.e., the one with the mass up, is unstable since when slightly pushed from its equilibrium position, it does not return there. Instead, it converges to the asymptotically stable equilibrium point.

A fundamental and appealing feature of Lyapunov's results is that, in the same spirit as Poincaré's ideas, the properties of the trajectories in a neighborhood of an equilibrium point can be assessed without even computing the solutions of the dynamical system. This can be actually performed using potential functions, now referred to as *Lyapunov functions*. These functions form the cornerstones of the powerful *Stability theory* also called *Lyapunov's theory of stability* or even *Lyapunov theory*.

This theory has been broadly accepted by systems and control theorists as a fundamental starting point for dealing with the analysis and control of dynamical systems. Whenever control systems are concerned, stability is one of the most important properties a control system should possess. Ensuring asymptotic stability of the closed-loop system is an efficient way for assessing that the controlled process behaves in the desired way, for instance, converges to a desired equilibrium point. Another striking point is the versatility of the approach which has been adapted, since then, to an immense variety of systems such as time-varying systems, discrete-time systems, hybrid systems, and infinite-dimensional systems. The dynamical systems we are interested in this monograph do not escape this rule, and Lyapunov theory will be shown to be an adequate tool for dealing with time-delay and linear parameter-varying systems. Whereas time-delay systems can be approached as a pure mathematical problem arising from a scientific field such as biology, ecology or physics, parameter-varying systems essentially come up from engineering problems such as filtering and control. In this regard, the field of linear parameter-varying time-delay systems is mostly of engineering interest only. In this respect, this monograph certainly fits better people having an engineering background than a background in mathematics. Some excellent monographs on the mathematics of delay systems moreover already exist; see for instance [3].

Structure of the Book

Part I is devoted to the representation, analysis and control of linear parameter-varying (LPV) systems. Parameter-varying systems are a large class of dynamical systems for which the future evolution of the state depends on the current state of the system plus some additional signals called parameters. These parameters act as inputs to the system and shape its internal structure. A typical example of parameter-dependent dynamics is vehicle dynamics. A vehicle, like a car, can indeed have dramatically different dynamics depending on the current speed and other inputs from the driver; the driver's inputs playing the role of parameters here. Parameters can also be internal to the system and be resulting from an approximation of a nonlinear system into a linear parameter-varying system. Considering these constant changes in the structure of LPV systems is an important information to consider in order to understand and characterize their behavior. LPV systems are analyzed in the same way as uncertain systems are, that is, using robust stability theory. Robust analysis and control has been initiated in the 1970s by some researchers, such that M. Athans [4] and coworkers, who started figuring out that uncertainties could lead to poor performance, drift in the controlled variables, instabilities, and so on. Robust analysis exactly addresses the analysis of systems perturbed by uncertainties, whereas robust control is concerned with the design of controllers for systems subject to uncertainties, that is, the design of robust controllers. This theory has been applied to an immense variety of systems over the last few decades.

Following the evolution of the parameters in view of capturing them, for instance, in a controller, offers an elegant way to adapt, in real-time, the controller structure to the current configuration of the system. This gives rise to the central concept of gain-scheduled controller. In the vehicle example previously mentioned, controllers can be adapted following the driver's inputs in order to obtain better performance, such as improved road-holding or comfort. LPV gain-scheduling can be seen as a direct extension of robust control techniques in which the controllers are time-invariant and do not adapt to the value of the uncertainties, which is consistent with the fact that we do not know uncertainties. The first gain-scheduling ideas in an LPV fashion have been proposed by J. Shamma in his Ph.D. thesis [5] in 1988 in the context of the gain-scheduled control of nonlinear systems. The major difficulty, at that time, was the lack of general theory for analyzing stability of LPV systems and for efficiently designing LPV gain-scheduled control laws. Modern robust control theory relying on convex optimization problems, notably involving linear matrix inequalities which appeared in the 1990s, have provided an appropriate framework for the design of gain-scheduled controllers in an LPV fashion

We consider linear parameter-varying systems in the first place as they are quite similar to linear time-invariant systems that we usually encounter in the first classes on control systems. In this respect, this monograph starts with something the reader is, in principle, familiar with. The first part indeed introduces the notion of uncertain system and different concepts of stability, in brief, the essential basics of robust analysis and robust control. Several stability results depending on the type of LPV systems are provided along with some discussions and examples. Finally, gain-scheduled design, which is the only rationale of the LPV formalism, is introduced and demonstrated as a way to go beyond robust design.

Part II is devoted to the representation and analysis of time-delay systems using time-domain techniques. Delays are indeed ubiquitous in our world. They can be encountered in various fields such as biology, physics, networks, or even economics, and may be used to represent several physical phenomena such as propagation and memory. As light propagates at a finite speed, observers looking at distant stars travel back in time. Propagation delay is therefore something astronomers face everyday. On the other hand, it is known that the behavior of many animals must be influenced by past experience in order to maximize the chances of survival. This means that, through memories, delays are involved in animal behavior. Propagation phenomena generally arise in networks, physics, epidemiology, and communication sciences whereas memory effects (or sometimes called hereditary effects) can be found in biology, ecology, and social sciences. Some of these dynamical systems can be described by delay-differential equations which are differential equations where the evolution of the current state depends on current and past state values, as opposed to only the current value for ordinary differential equations. On a mathematical level, delay systems have been around for quite a long time, with a particular uprise in the past few decades. Certainly the first person to have looked at delay-differential equations was Euler (how surprising!), but we can also trace back delay-differential equations in the works of Bernoulli, Lagrange, Laplace, Poisson and others, during the eighteenth century. Early in the twentieth century, it has been noticed that accurate descriptions of numerous problems in science and engineering were involving delays. Since then, much attention has been paid to delay problems and, in spite of this, it is still a very active research topic due to the inherent complexity and rich behavior these systems possess. On a qualitative level, delays indeed have, most of the time, a detrimental effect on the behavior of dynamical systems, at least as long as stability properties are concerned. They can, for instance, cause oscillations or divergent trajectories. They may, however, lead to important improvements in terms of behavior whenever complex systems are concerned and where stability is not a relevant nor meaningful concept anymore.

In this monograph, we will be mostly interested in the impact of delays on the stability of dynamical systems, and in the different ways for controlling and observing these systems. Time-delay systems are approached in the same spirit as LPV systems in the first part of this monograph, that is, by pinpointing differences between them and standard systems. We address first their representation, different stability notions specific to these systems are coming next and the possible associated designs are finally discussed. An extensive exposition of different stability results along with some comparative statements is notably provided, together with a particular emphasis on their conservativeness and the source of it—a discussion that is usually not clearly provided in the literature. The approaches considered in this monograph are exclusively time-domain approaches for the simple reason that the aim is to consider time-varying systems, either because the delay is time-varying or because the matrices of the time-delay system are time-varying. In this regard, frequency domain techniques are not discussed. Readers interested in frequency-domain techniques should, for instance, refer to [6, 7].

The last part finally addresses the merging of time-delay and LPV systems. This class is slightly richer since delays and parameters may interact with each other to yield parameter-dependent delays and delayed-parameters. Some examples are

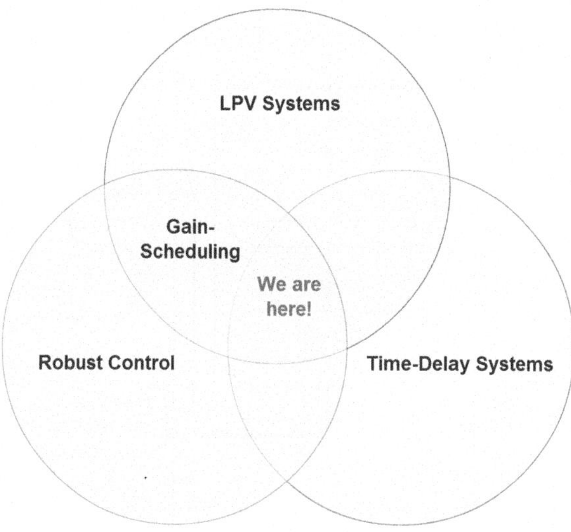

Fig. 1 Venn diagram of the topics treated in the monograph

first given for illustration and several methods for stability analysis of LPV time-delay systems are provided. Design methods for the filtering and observation of LPV time-delay systems as well as for control are also provided along with numerical examples.

The last topic, which is not present on the Venn diagram of Fig. 1 although extensively used in this monograph, is optimization. Most of the results provided in this book are formulated as linear matrix inequalities or, more generally, semidefinite programs. This type of optimization problems has been proven to be verifiable in polynomial-time using dedicated algorithms [8]. In this respect, they can be considered as a satisfying formulation for an answer to the stability or to design questions for time-delay and parameter-varying systems. Most of the basics about the manipulation of LMIs are presented in the main text of the monograph. Some advanced results can also be found in the main text and in the appendixes.

How to Read

This monograph is concerned with linear parameter-varying and time-delay systems as main topics and their combination as third topic. An additional topic is optimization, yet it is not intended to cover optimization in general but rather focus on matrix inequalities. In this respect, this monograph can be read in several different ways. The first way is the linear one, which is recommended to readers

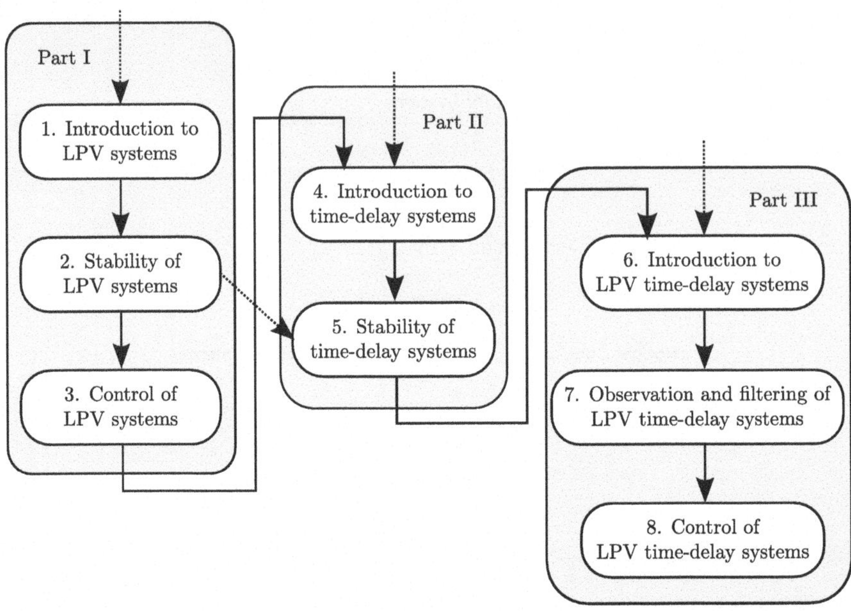

that are not familiar with any of the treated fields. A second way is to focus on one of the topics and, in this case, each part can be read almost independently of the others except, perhaps, for the last one where a basic knowledge of time-delay systems and LPV systems is preferable. Regarding the part on time-delay systems, some stability results are based on robust stability theory, so it may be interesting to have a look at Sect. 2.6 before attacking the analysis of time-delay systems.

The appendixes contain supplementary information on linear algebra, linear matrix inequalities, and numerous important results related to robust analysis and control. They complement the results included in the main text of the monograph and have been written to make them usable as a possible, yet not exhaustive, memo. The appendixes also contain unpublished results.

References

1. H. J. Poincaré, Sur le problème des trois corps et les équations de la dynamique. Acta Math. **13**, 1–270 (1890)
2. A. M. Lyapunov, *General Problem of the Stability of Motion* (Translated from Russian to French by E. Davaux and from French to English by A.T. Fuller). (Taylor & Francis, London, UK, 1992)
3. O. Diekmann, S. A. van Gils, S. M. Verduyn Lunel, H. O. Walther, *Delay Equations: Functional-, Complex-, and Nonlinear Analysis* (Springer-Verlag, New York, 1995)
4. M. Athans, Editorial on the LQG problem. IEEE Trans. Autom. Control. **16**(6), 528 (1971)

5. J. S. Shamma, Analysis and design of gain-scheduled control systems. Ph.D. thesis, Laboratory for Information and Decision Systems, Massachusetts Institute of Technology (1988)
6. S. I. Niculescu, *Delay Effects on Stability: a Robust Control Approach*, vol 269 (Springer-Verlag, Heidelberg, 2001)
7. K. Gu, V. L. Kharitonov, J. Chen, *Stability of Time-Delay Systems* (Birkhäuser, Boston, 2003)
8. Y. Nesterov, A. Nemirovskii, *Interior-Point Polynomial Algorithm in Convex Programming* (SIAM, Philadelphia, PA, 1994)

Part I
Linear Parameter-Varying Systems

Chapter 1
Introduction to LPV Systems

System Dynamics: Things today are the things of yesterday plus any changes. The changes are the result of the things of yesterday. Now extend this to tomorrow.

William S. Bonnell

Abstract The goal of this chapter is to introduce the main ways for representing linear parameter-varying systems and emphasize their ability to represent a wide class of dynamical systems. A classification of the types of parameters regarding their mathematical properties and their physical meaning is also given. Several real world examples of LPV systems are finally discussed in order to demonstrate the relevance of the LPV framework.

1.1 System Definition

Linear parameter-varying (LPV) systems are linear dynamical systems[1] whose mathematical description depends on parameters that change values over time. These parameters are generally considered as bounded and taking values inside a set $\mathbf{\Delta}_\rho$, often assumed to be a compact and convex polytope (e.g. a box). LPV systems are commonly described by equations of the form

$$\begin{aligned}\dot{x}(t) &= A(\rho(t))x(t) + E(\rho(t))w(t), \quad t \geq 0\\ z(t) &= C(\rho(t))x(t) + F(\rho(t))w(t)\\ x(0) &= x_0\end{aligned} \tag{1.1}$$

[1] For more details on dynamical systems, systems theory and related fundamental results, the reader should refer to [1–4]. Additional details on LPV systems can also be found in [5–7] and references therein.

C. Briat, *Linear Parameter-Varying and Time-Delay Systems*,
Advances in Delays and Dynamics 3, DOI 10.1007/978-3-662-44050-6_1

where x, w and z are the state, the input and the output of the system, respectively. The parameter vector ρ acts internally on the system by modifying its structure over time, and, consequently, modifying the overall input-output behavior of the system. It is assumed above that the matrices are continuous and bounded functions.

The class of LPV systems encompasses a wide variety of systems according to the type of trajectories of the parameters. For instance,

- LPV systems with arbitrarily fast varying parameters have parameters in the set

$$\mathscr{P}^{\infty} = \left\{\rho : \mathbb{R}_{\geq 0} \to \mathbf{\Delta}_{\rho}\right\}. \tag{1.2}$$

  For this class of parameters, we shall assume in what follows that the parameters behave sufficiently nicely so that mild solutions can be defined for all time, i.e. in the Carathéodory sense.
- LPV systems with slowly varying parameters have parameters in the set

$$\mathscr{P}^{\nu} = \left\{\rho : \mathbb{R}_{\geq 0} \to \mathbf{\Delta}_{\rho} : \dot{\rho} \in \mathbf{\Delta}_{\nu}\right\} \tag{1.3}$$

  where $\mathbf{\Delta}_{\nu}$ is a convex and compact polyhedron containing 0.
- LPV systems with piecewise constant parameters have parameters in the set

$$\mathscr{P}_{pc} = \left\{\rho : \mathbb{R}_{\geq 0} \to \mathbf{\Delta}_{\rho} : \rho_i \text{ piecewise constant}, i = 1, \ldots\right\}. \tag{1.4}$$

- Switched systems, see e.g. [8–13], with N modes can be represented as LPV systems with parameters in the set

$$\mathscr{P}_{ss} = \left\{\rho : \mathbb{R}_{\geq 0} \to \{0, 1\}^N : \sum_{i=1}^{N} \rho_i = 1\right\} \tag{1.5}$$

  where, again, some conditions have to be satisfied in order to have mild solutions at any time. In this case, we also have $A(\rho) = \sum_{i=1}^{N} A_i \rho_i$.
- Periodic systems, see e.g. [14–17], can be represented as LPV systems with parameters in the set of T-periodically varying parameters

$$\mathscr{P}_p = \left\{\rho : \mathbb{R}_{\geq 0} \to \mathbf{\Delta}_{\rho} : \rho(t) = \rho(t + T), \ t \geq 0\right\}. \tag{1.6}$$

The spirit of the LPV framework is the same as in robust analysis, and very few assumptions on the uncertainties/parameters are generally made. That is, only the sets $\mathscr{P}^{\nu}$ and $\mathscr{P}^{\infty}$ are generally considered in the LPV framework. This vagueness in the definition of the parameter trajectories makes the analysis quite difficult, and emphasizes the correspondence with the worst-case analysis point of view of robust analysis and robust control. This viewpoint is also the one considered in this monograph. When stronger assumptions are made of the parameter trajectories, i.e. by considering the sets (1.5) and (1.6), it is possible to adapt and specialize the tools to the considered class of parameters, ultimately leading to very specific and efficient approaches that will unfortunately not be treated here.

1.2 Types of Parameters

We put aside in this section the mathematical properties of parameter trajectories in LPV models to rather focus on their type and role. Parameters can indeed be used to approximate nonlinear dynamics, embed time-varying parts in a systematic fashion, or even introduce extra degrees of freedom that may be useful for a design perspective.

1.2.1 Approximating Nonlinear Systems: Quasi-LPV Systems

Whenever LPV systems are considered as approximations of nonlinear systems, scheduling parameters are functions of the state of the system. This particular type of LPV systems is referred to as *quasi-LPV systems*,[2] sometimes abbreviated *qLPV systems*; see for instance [18–26]. As an example, the following scalar nonlinear system

$$\dot{x}(t) = -x(t)^3 \tag{1.7}$$

can be represented as

$$\dot{x}(t) = -\rho(t)^2 x(t) \tag{1.8}$$

with $\rho(t) := x(t) \in \mathbb{R}$. While the above LPV representation is asymptotically stable for every parameter value $\rho \neq 0$, and therefore exactly characterizes the stability of the original nonlinear system, it is very important to keep in mind that, in general, an LPV approximation is not equivalent (in terms of stability, controllability or any other property) to the original nonlinear system. Moreover, an important additional difficulty is that nonlinear systems generally admit several LPV approximations/representations, and finding the most accurate one is not an easy task; see e.g. [22, 30, 31] for some approximation methods. To illustrate this, let us consider, for instance, the Van-der-Pol equation with reverse vector field considered in [30]:

$$\begin{aligned} \dot{x}_1(t) &= -x_2(t) \\ \dot{x}_2(t) &= x_1(t) - 0.3(1 - x_1(t)^2)x_2(t). \end{aligned} \tag{1.9}$$

[2] A similar framework is based on Takagi-Sugeno systems where nonlinear systems can be represented as a state-dependent convex combination of linear time-invariant systems (polytopic system); see e.g. [27–29].

The phase plot of the above nonlinear system is depicted in Fig. 1.1. The system (1.9) has an unstable limit cycle, i.e. every trajectory starting from inside the region delimited by the limit cycle converges to 0, while every trajectory starting from outside this region escapes to infinity. After a quick look at the model (1.9), the following LPV representation may be proposed

$$\begin{aligned}\dot{x}_1(t) &= -x_2(t)\\ \dot{x}_2(t) &= x_1(t) - 0.3(1-\rho(t)^2)x_2(t)\end{aligned} \tag{1.10}$$

where $\rho(t) = x_1(t)$. The above system can be shown to be quadratically asymptotically stable using the Lyapunov function $V(x) = x^{\mathrm{T}}Px$, $P \in \mathbb{S}^2_{\succ 0}$ (see Sect. 2.3.1) provided that $|\rho(t)| \leq 0.98$. The corresponding region of attraction then coincides with the level sets of the Lyapunov function V for which we have $x_1 \in [-0.98, 0.98]$. This region of attraction is depicted in Fig. 1.2 where we can see that the computed region is clearly a conservative region of attraction for the 0-equilibrium point. However, a more accurate nontrivial representation determined in [30] is given by

$$\dot{x}(t) = \begin{bmatrix} 0 & -1 \\ 1 + 0.24\rho_1(t)\rho_2(t) & -0.3 + 0.06\rho_1(t)^2 \end{bmatrix} x(t) \tag{1.11}$$

where $\rho_1(t) = x_1(t)$ and $\rho_2(t) = x_2(t)$. It can be proved that this system is asymptotically stable provided that $|\rho_1(t)| \leq 1.253$ and $|\rho_1(t)\rho_2(t)| \leq 0.85$. The corresponding region of attraction is depicted in Fig. 1.3 where we can see that this representation is better able to characterize the basin of attraction than system (1.10), but still remains conservative. Approximating nonlinear dynamics by LPV ones is still an open problem to date…

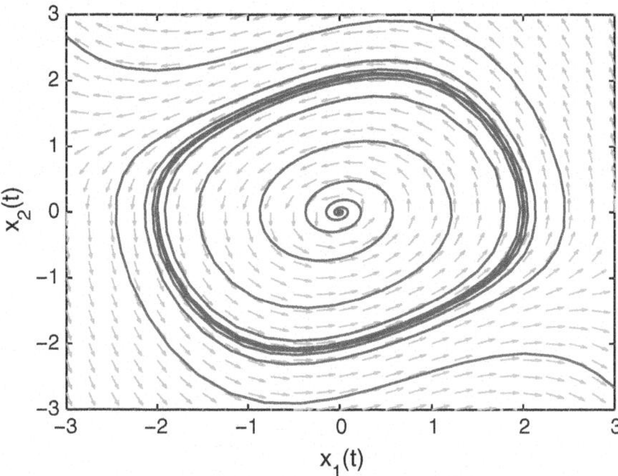

Fig. 1.1 Phase portrait of the Van-der-Pol equation (1.9) and some trajectories solutions of the system

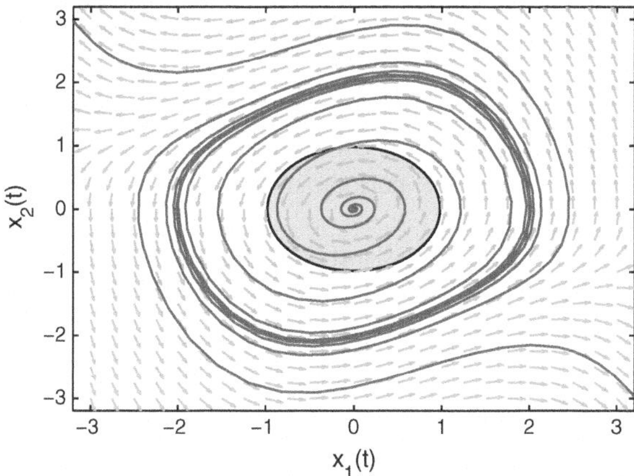

Fig. 1.2 Computed region of attraction (in *grey*, centered about the origin) using the LPV approximation (1.10) of the Van-der-Pol equation (1.9)

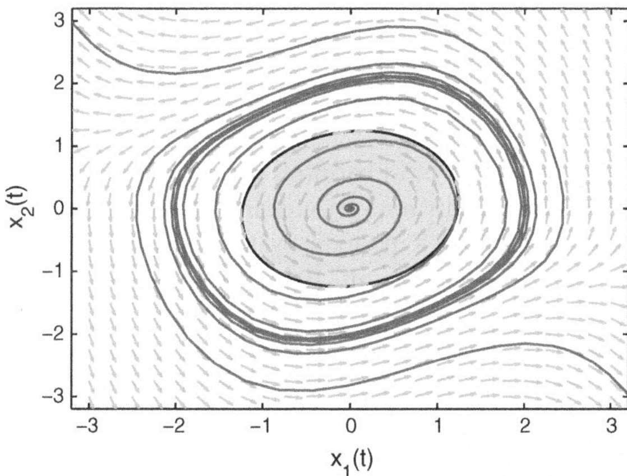

Fig. 1.3 Computed region of attraction (in *grey*, centered about the origin) using the LPV approximation (1.11) of the Van-der-Pol equation (1.9)

Another difficulty when dealing with the control of quasi-LPV systems obtained from the linearization of a nonlinear system lies in the presence of *hidden coupling terms* that may lead to an unstable closed-loop system even if the corresponding closed-loop LPV system is stable. For more information about these coupling terms see e.g. [32, 33].

1.2.2 Embedding Time-Varying Components: Intrinsic Parameters

Parameters can also be used to hide/embed time-varying components in order to use LPV gain-scheduling techniques for controlling the original system. For instance, the linear time-varying system (T-periodic in fact)

$$\dot{x}(t) = (-a + b\sin(\omega t))x(t),\ \ \omega = 2\pi/T \tag{1.12}$$

can be represented as

$$\dot{x}(t) = (-a + b\rho(t))x(t) \tag{1.13}$$

where $\rho(t) := \sin(\omega t) \in [-1, 1]$. As for qLPV approximations, stability of the original system is not equivalent to stability of the LPV approximation. From periodic systems theory, the periodic system (1.12) can be shown to be asymptotically stable if and only if $a > 0$. The LPV approximation (1.13) is, on the other hand, asymptotically stable if and only if $a > 0$ and $|b| < a$. This loss of equivalence stems from the fact that the LPV description embeds the actual periodic trajectory of the sine function into the more general set

$$\mathcal{E} = \left\{\rho : \mathbb{R}_{\geq 0} \to [-1, 1]\right\}$$

which includes the worst-case (most harmful) trajectories $\rho \equiv -1$ and $\rho \equiv 1$.

The above example is actually not very meaningful since stability analysis should have been performed according to periodic systems theory. It, nevertheless, has the merit to emphasize the conservatism of LPV descriptions that embed trajectories in wider sets.

Some other systems, however, naturally involve parameters representing some intrinsic time-varying components with a priori unknown trajectories. In this case, LPV representations may be exact. One such example is given by the system, see Fig. 1.4,

$$R(t)C(t)\frac{dV_{out}(t)}{dt} + \left(1 + R(t)\frac{dC(t)}{dt}\right)V_{out}(t) = V_{in}(t) \tag{1.14}$$

which represents an RC circuit with time-varying capacity $C(t)$ (assumed to be differentiable) and time-varying resistance $R(t)$.

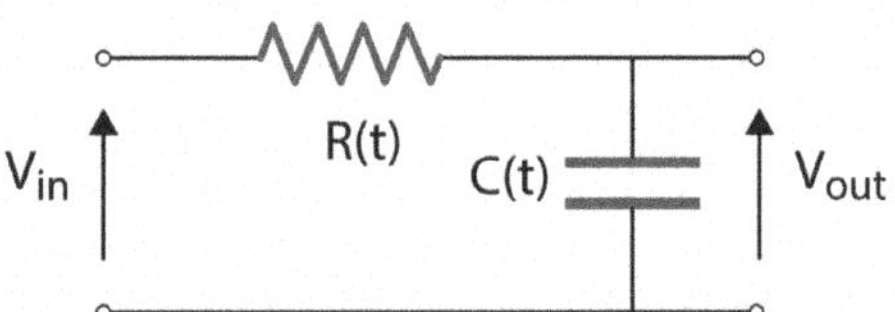

Fig. 1.4 RC circuit with time-varying resistance and capacity

1.2.3 Artificial/Extrinsic Parameters

Extrinsic parameters are mostly involved when design is the underlying objective, e.g. control design. These artificial parameters may then be used in the control law in order to shape its structure according to certain constraints or objectives. They may govern, for instance, different operating modes optimizing different criteria such as rate of convergence, H_∞-norm, etc.[3] In such scenarios, it is generally assumed that a high-level monitoring system adapts the values of the parameters according to some performance objectives and constraints.

To illustrate this, let us consider the following single-input single-output LTI system

$$\begin{aligned} \dot{x}(t) &= x(t) + u(t) \\ y(t) &= x(t) \end{aligned} \tag{1.15}$$

where $x \in \mathbb{R}$ and $u \in \mathbb{R}$ are the state and the control input, respectively. It is proposed to determine a control law such that

1. the output y tracks a differentiable reference signal r, and
2. the bandwidth of the closed-loop system can be adjusted in real-time.

The control law

$$u(t) = -(1 + \rho(t))x(t) + \rho(t)r(t), \quad \rho(t) > 0 \tag{1.16}$$

where ρ is an external parameter yields the closed-loop system

$$\dot{x}(t) = -\rho(t)(x(t) - r(t)). \tag{1.17}$$

It is immediate to see that the dynamics of the system is asymptotically stable, and that the bandwidth can be adjusted in real-time by playing with $\rho(t)$. A faster response is obtained with a larger parameter value.

1.3 Representation of LPV Systems

Now that we have introduced the different families of parameters, it is time to introduce the different LPV modeling paradigms that are omnipresent in the literature. Since the overall LPV framework is a direct descendent of robust analysis and control, readers familiar with the field of robust control will certainly understand this as a reinterpretation of uncertain systems into LPV systems.

---

[3] These ideas have been successfully applied in the context of switching controllers, see e.g. the references [9, 34–39].

Generic LPV systems are first presented in Sect. 1.3.1, polytopic LPV systems and LPV systems in LFT-form then follow in Sects. 1.3.2 and 1.3.3, respectively. Finally, LPV systems in input/output form are very briefly introduced in Sect. 1.3.4.

1.3.1 Generic LPV Systems

This formulation for LPV systems is the most natural one since LPV systems are directly taken as they are [5, 40] and represented as

$$\dot{x}(t) = A(\rho(t))x(t) \tag{1.18}$$

where $x \in \mathbb{R}^n$ is the state of the system and $\rho : \mathbb{R}_{\geq 0} \to \mathbf{\Delta}_\rho$ is the vector of time-varying parameters. No transformation nor preprocessing is applied to the system structure, and the parameter dependence is very general, e.g. polynomial, rational, exponential, etc. The only underlying assumptions are that the matrix function $A : \mathbf{\Delta}_\rho \to \mathbb{R}^{n\times n}$ be bounded, and that the parameters behave sufficiently well so that solutions to the differential equation are well-defined.

When the system depends polynomially on the parameters, the matrix $A(\rho)$ can be expressed as

$$A(\rho) = A_0 + \sum_i A_i \rho^{\alpha_i} \tag{1.19}$$

where ρ^{α_i} follows the multi-index notation.

A "less naive" formulation takes the form of an LPV system in descriptor form [41–44]:

$$\begin{aligned} \begin{bmatrix} I_n & 0 \\ 0 & 0 \end{bmatrix} \begin{bmatrix} \dot{x}(t) \\ \dot{y}(t) \end{bmatrix} &= \begin{bmatrix} \bar{A}_{11}(\rho) & \bar{A}_{12}(\rho) \\ \bar{A}_{21}(\rho) & \bar{A}_{22}(\rho) \end{bmatrix} \begin{bmatrix} x(t) \\ y(t) \end{bmatrix} \\ x(0) &= x_0 \\ y(0) &= -A_{22}(\rho(0))^{-1} A_{21}(\rho(0)) x_0 \end{aligned} \tag{1.20}$$

where the matrix functions $\bar{A}_{ij}(\rho)$ are continuous in $\rho \in \mathbf{\Delta}_\rho$, with the additional property for $\bar{A}_{22}(\rho)$ to be nonsingular[4] for all $\rho \in \mathbf{\Delta}_\rho$. The advantage of this formulation lies in the possibility for considering plants that depend rationally on the parameters and reformulating them as polynomial or affine descriptor LPV systems, provided that the state y is chosen adequately. Assume, indeed, that these matrices are affine or polynomial, then simple calculations show that the matrix $A(\rho)$ in (1.18) admits the following representation in terms of the matrices of the descriptor system (1.20):

$$A(\rho) = \bar{A}_{11}(\rho) - \bar{A}_{12}(\rho)\bar{A}_{22}^{-1}(\rho)\bar{A}_{21}(\rho). \tag{1.21}$$

[4] Nonsingularity of A_{22} automatically implies that the system is regular and impulse-free [45]. These conditions imply that for any compatible initial condition, there exists a unique continuous solution to (1.20).

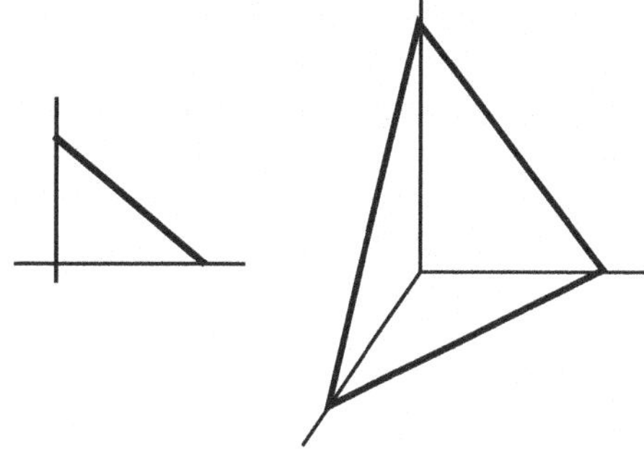

Fig. 1.5 N-unit simplex Λ_N. For $N = 2$ (*left*) the polytope is a simple segment while for $N = 3$ (*right*) the polytope consists of a triangular closed surface

This illustrates that rational and polynomial matrices $A(\rho)$ can be easily encoded in the descriptor representation (1.20). A similar idea is exploited in Sect. 1.3.3 where LPV systems in LFT-form are presented.

Example 1.3.1 The LPV system

$$\dot{x}(t) = \left(\frac{\rho(t)}{\rho(t)^2 + 1} - 3\right) x(t) \tag{1.22}$$

where $\rho \in [-1, 1]$ admits the following descriptor LPV representation:

$$\left[\begin{array}{c|ccc} 1 & 0 & 0 & 0 \\ \hline 0 & 0 & 0 & 0 \\ 0 & 0 & 0 & 0 \\ 0 & 0 & 0 & 0 \end{array}\right] \begin{bmatrix} \dot{x} \\ \hline \dot{y}_1 \\ \dot{y}_2 \\ \dot{y}_3 \end{bmatrix} = \left[\begin{array}{c|ccc} -3 & -1 & 0 & 1 \\ \hline \rho & 1 & 0 & 0 \\ 0 & -\rho & 1 & \rho \\ 0 & 0 & -\rho & 1 \end{array}\right] \begin{bmatrix} x \\ \hline y_1 \\ y_2 \\ y_3 \end{bmatrix}. \tag{1.23}$$

By applying the formula (1.21), the model (1.22) is retrieved.

1.3.2 Polytopic LPV Systems

The polytopic framework offers an elegant and convenient way for representing and analyzing LPV and uncertain systems; see, for instance, [46–50]. Unlike generic LPV systems of the previous section for which we, a priori, did not assume any particular dependence on the parameters, polytopic systems are, on the other hand, explicitly represented as a time-varying convex combination of LTI systems. This structural property can be exploited to obtain stability and stabilization results that can be easily verifiable using convex optimization techniques.

1.3.2.1 Convex Compact Polytopes and the N-Unit Simplex

Before explicitly characterizing polytopic LPV systems, it is important to mention some facts about the N-unit simplex and, more generally, about convex compact polytopes. Polytopes are generalizations of polyhedra to arbitrary dimensions. They are objects with 'flat sides': squares, cubes, triangles and tetrahedra are well-known examples. Below is the definition of the N-unit simplex, a very particular and useful polytope:

Definition 1.3.2 *(N-unit simplex)* The N-unit simplex, denoted by Λ_N, is defined as the set

$$\Lambda_N := \left\{ \chi \in \mathbb{R}^N_{\geq 0} : \sum_{i=1}^{N} \chi_i = 1 \right\}. \tag{1.24}$$

Note that the dimension of the N-unit simplex is equal to $N-1$ due to the rank-one relation between its components. Examples of unit simplices are depicted in Fig. 1.5.

From the above definition, it is easy to see that the N-unit simplex is a compact and convex polytope. As such, it can be alternatively and uniquely characterized by the set of its vertices:

$$V := \{v_1, \ldots, v_N\} \tag{1.25}$$

where

$$v_i = \begin{bmatrix} 0_{(i-1)\times 1} \\ 1 \\ 0_{(N-i)\times 1} \end{bmatrix} \tag{1.26}$$

and any point inside Λ_N can be uniquely written as a convex combination of the vertices v_i. The entire N-unit simplex can also be recovered by taking the *convex hull*[5] of the set of its vertices. We denote the operation of taking the convex hull by $\Lambda_N = \mathbf{co}\{V\}$ and the operation of taking the set of vertices by $V = \mathbf{vert}\{\Lambda_N\}$. The notion of convex hull is illustrated in Fig. 1.6. Note, however, that for this example the operation of taking the set of vertices will only return the points that lie on the boundary.

As a consequence, any compact convex polytope such as the box

$$\mathfrak{B} := [\alpha_1, \beta_1] \times \ldots [\alpha_\ell, \beta_\ell] \tag{1.27}$$

can be uniquely characterized in terms of the set of its vertices, and recovered by taking the convex hull of it. Indeed, for all $x \in \mathfrak{B}$, there exists $\lambda \in \Lambda_{2^\ell}$ such that

[5] The convex hull of a set S is the smallest convex set containing S.

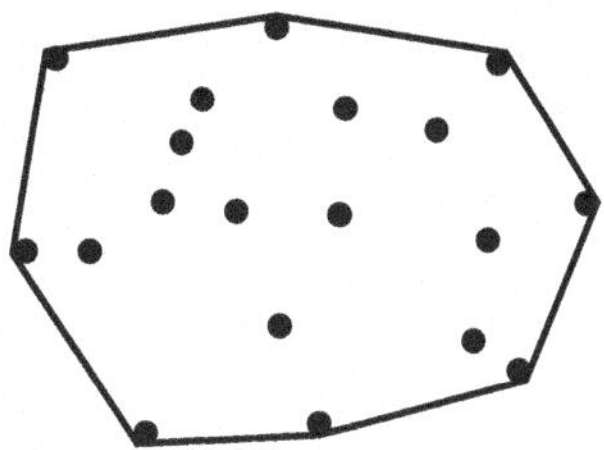

Fig. 1.6 Convex hull of a set of points in the plane

$$x = \sum_{i=1}^{2^\ell} \lambda_i b_i \tag{1.28}$$

where $\{b_1, \ldots, b_{2^\ell}\} = \mathbf{vert}[\mathfrak{B}] = \{\alpha_1, \beta_1\} \times \ldots \{\alpha_\ell, \beta_\ell\}$. It is interesting to note while that the dimension of the set $\mathfrak{B}$ is ℓ, the cardinal of $\mathbf{vert}[\mathfrak{B}]$ is 2^ℓ. This fact has deep implications on the tractability of certain problems involving polytopic LPV systems since the size of the set of parameters in the polytopic domain grows exponentially.

1.3.2.2 Polytopic LPV Systems

Based on the facts on polytopes described above, we are now ready to characterize polytopic systems in details. Such systems are represented in the following form

$$\dot{x}(t) = A(\lambda(t))x(t) \tag{1.29}$$

where $A(\lambda(t)) = \sum_{i=1}^{N} \lambda_i(t)A_i$, $A_i \in \mathbb{R}^{n\times n}$ and $\lambda(t) \in \Lambda_N$. Any LPV system can be represented, up to a certain degree of accuracy, by a polytopic LPV system. Systems that depend linearly on parameters taking values in a compact convex polyhedron, such as a box, can be exactly represented as polytopic systems. This is, most of the time, not true for systems having a more general parameter dependence, e.g. polynomial, or having parameters inside more general convex sets, e.g. a disc. Some of these facts are illustrated below.

Example 1.3.3 The LPV system

$$\dot{x}(t) = [A_1\rho_1(t) + A_2\rho_2(t)]x(t) \tag{1.30}$$

with $\rho(t) \in [-1, 1]^2$ admits an equivalent polytopic representation since the parameters take values in a convex compact polytope. The box $[-1, 1]^2$ can indeed be parametrized in terms of variables in the 4-unit simplex as shown below:

$$\dot{x}(t) = [A_1 f_1(\lambda(t)) + A_2 f_2(\lambda(t))]x(t) \tag{1.31}$$

where $\lambda(t) \in \Lambda_4$ and

$$\begin{aligned} f_1(\lambda) &= (\lambda_2 + \lambda_4) - (\lambda_1 + \lambda_3), \\ f_2(\lambda) &= (\lambda_3 + \lambda_4) - (\lambda_1 + \lambda_2). \end{aligned} \tag{1.32}$$

Note the increase of the number of parameters from 2 to 4.

Systems that depend polynomially on the parameters can, in general, not be exactly represented as polytopic LPV systems since it is not possible to represent, for instance, a univariate polynomial of degree 2 as a multivariate polynomial of degree 1, even on a compact set. This is illustrated in the following example.

Example 1.3.4 Let us consider the polynomially-dependent LPV system

$$\dot{x}(t) = (A_0 + A_1\rho(t) + A_2\rho(t)^2)x(t) \tag{1.33}$$

where $\rho(t) \in [-1, 1]$. Since there is no exact polytopic representation for 1.3.4, we then view the terms ρ and ρ^2 as distinct parameters. The corresponding set of values is therefore given by

$$S := \left\{(\chi, \chi^2) : \chi \in [-1, 1]\right\} \tag{1.34}$$

and is far from being convex. Therefore, we may consider instead the following convex covering

$$S \subset S_e := \mathbf{co}\left\{\begin{bmatrix}-1\\0\end{bmatrix}, \begin{bmatrix}1\\0\end{bmatrix}, \begin{bmatrix}-1\\1\end{bmatrix}, \begin{bmatrix}1\\1\end{bmatrix}\right\} \tag{1.35}$$

which is essentially the box $[-1, 1] \times [0, 1]$. Using then the set S_e, we get the polytopic description

$$\dot{x}(t) = \left[A_0 + A_1\sum_{i=1}^{4}(-1)^i\lambda_i(t) + A_2(\lambda_3(t) + \lambda_4(t))\right]x(t). \tag{1.36}$$

Since $S \subset S_e$, the polytopic representation above is not equivalent to the polynomial representation (1.33). Even more critically, all the points in S_e located

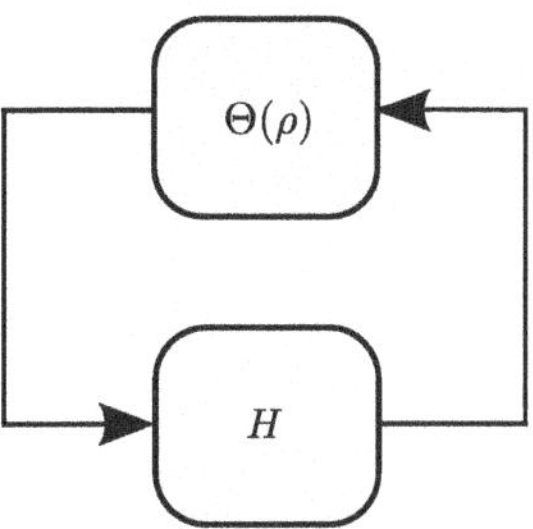

Fig. 1.7 Graphical representation of the LFT system (1.37) where H is the operator mapping w to z

off the parabola are very many, and these artefacts may seriously compromise the accuracy of the polytopic representation. The system may indeed be stable over S, but not over S_e.

It is also important to mention that many polytopic representations for (1.33) exist. Polytopic systems can be used to approximate arbitrarily well the parabola from below. However, approximating from above, by removing part of the epigraph, [a] is not possible without destroying convexity, a property lying at the core of the polytopic formulation. A possible solution to overcome this difficulty could be to consider a family of polytopic systems, each one of them approximating a small portion of the parabola.

[a] The epigraph of a function is the set of points lying on or above its graph.

Example 1.3.5 The closed-unit disc

$$\bar{\mathbf{D}} := \left\{ x \in \mathbb{R}^2 \;:\; ||x||_2 \leq 1 \right\}$$

is a convex semi-algebraic set which obviously cannot be alternatively represented as a polytope. However, it can be approximated as closely as desired by a convex polytope, e.g. a regular polygon.

Even if polytopic systems may result in inaccurate descriptions for some classes of LPV systems, they are still theoretically and computationally attractive since some properties of the considered LPV system may be connected to those of the subsystems, i.e. the extremal systems located at the vertices of the polytope. In some cases, stability of the polytopic LPV system can be related to the stability of the subsystems. This will be explained in more details in Sect. 2.5.

1.3.3 LPV Systems in LFT-Form

LPV systems in LFT-form are LPV systems expressed as interconnections of two subsystems, as shown in Fig. 1.7. The acronym LFT stands for *Linear Fractional Transformation* and is the reformulation procedure employed to express LPV/uncertain systems as two-block interconnections. This procedure is ubiquitous in robust analysis and control, for which it has been primarily developed. The interest for focusing on systems interconnections actually dates back to the mid 70s when researchers, puzzled by some unexplained and disappointing results on optimal control of aircrafts and submarines, started to suspect uncertainties to be responsible of these unsuccessful results. They then decided to investigate robustness issues in optimal control and laid the foundations of the to-be-successful robust control theory.[6]

The key idea behind LFT is to rewrite a complex system as an interconnection of a "simple and nice" part and a "complicated and annoying" part. The nice part should possess convenient properties such as linearity, time-invariance, etc. The annoying part, on the other hand, usually contains time-varying terms, nonlinearities, infinite-dimensional dynamics, etc. The overall system is then analyzed under this interconnected-systems paradigm for which many specific tools have been developed: the Popov criterion [3, 52, 53] for sector-bounded static nonlinearities (Lur'e problem [54]), gain concepts such as L_2-gain [55], singular-values concepts [56–58], dissipativity theory and its applications [59–63], topological separation [64, 65] and Integral Quadratic Constraints (IQCs) [66].

The interest of this formulation, in the LPV framework, has been emphasized in [60, 61] where a convex formulation of the design of gain-scheduled controllers with H_∞-performance has been described. Since then, numerous papers have been devoted to this problem; see e.g. [60–63, 67].

LPV systems in LFT-form, such as the one depicted in Fig. 1.7, are represented by

$$\begin{aligned} \dot{x}(t) &= Ax(t) + Bw(t) \\ z(t) &= Cx(t) + Dw(t) \\ w(t) &= \Theta(\rho(t))z(t). \end{aligned} \tag{1.37}$$

where x is the system state and w/z are loop-signals that describe the interconnection with the parameter dependent part $\Theta(\rho)$. It is immediate to see that the system consists of the interconnection of an LTI part (A, B, C, D), i.e. the "nice one", and a parameter-varying part $\Theta(\rho)$, i.e. the "annoying one". It is generally tacitly assumed that the interconnection is well-posed, i.e. the matrix $I - \Theta(\rho)D$ is invertible for all $\rho \in \mathbf{\Delta}_\rho$. Note that several possible LFT descriptions for a given LPV system exist, and finding the minimal one is desirable for reducing the complexity and conservatism of the approach [58, 68]. The analysis of such systems will be carried out in Sect. 2.6 .

[6] Readers interested in a short history of robust control theory should read M. G. Safonov's paper [51].

Proposition 1.3.6 *The LPV system in LFT-form* (1.37) *is equivalent to the LPV systems*

$$\dot{x} = (A - B(I - \Theta(\rho)D)^{-1}\Theta(\rho)C)x, \tag{1.38}$$

$$\dot{x} = (A - B\Theta(\rho)(I - D\Theta(\rho))^{-1}C)x, \tag{1.39}$$

and

$$\begin{bmatrix} I & 0 \\ 0 & 0 \end{bmatrix}\begin{bmatrix} \dot{x} \\ \dot{y} \end{bmatrix} = \begin{bmatrix} A & B \\ \Theta(\rho)C & I - \Theta(\rho)D \end{bmatrix}\begin{bmatrix} x \\ y \end{bmatrix}. \tag{1.40}$$

Proof The expression (1.38) is obtained by eliminating w from (1.37), and (1.39) can be obtained from (1.38) using the equality[7]

$$(I - \Theta(\rho)D)^{-1}\Theta(\rho) = \Theta(\rho)(I - D\Theta(\rho))^{-1}.$$

The representation (1.40) is obtained by using the identity

$$w = -\Theta(\rho)Cx + \Theta(\rho)Dw$$

and letting $y = w$.

In the light of the above proposition, we can clearly see that LPV systems in LFT-form can represent LPV systems with a rational dependence on the parameters. Note also that while the way to obtain the models (1.38) and (1.39) from the LPV system in LFT-form is immediate, finding a Linear Fractional Representation (LFR) corresponding to a system is much trickier. A systematic procedure for building an LFR from the initial LPV system is detailed in [68]. Several softwares can also be used to perform this in a safe and convenient way [69, 70].

Example 1.3.7 Let us take back the system of Example 1.3.1. The system (1.22) can be shown to admit the following (minimal) LFT representation:

$$\left[\begin{array}{c|c} A & B \\ \hline C & D \end{array}\right] = \left[\begin{array}{c|ccc} -3 & -1 & 0 & 1 \\ \hline 1 & 0 & 0 & 0 \\ 0 & 1 & 0 & -1 \\ 0 & 0 & 1 & 0 \end{array}\right] \quad \text{and} \quad \Theta(\rho) = \begin{bmatrix} \rho & 0 & 0 \\ 0 & \rho & 0 \\ 0 & 0 & \rho \end{bmatrix}. \tag{1.41}$$

To see this, just apply the result of Proposition 1.3.6.

[7] It is a particular case of the more general equality (A.3).

1.3.4 LPV Systems in Input/Output Form

For completeness, it seems important to mention the input/output description of LPV systems. This way of representation is much less spread than the three former ones and can be used for identification and control design; see e.g. [71–75] and references therein. LPV systems in input/output form are represented as

$$D(\sigma, \rho)y = N(\sigma, \rho)u \tag{1.42}$$

where u is the system input, y the system output. The operator σ may either be the time-derivative operator

$$(\sigma y)(t) = \dot{y}(t)$$

or the advance operator

$$(\sigma y)(t) = y(t+1)$$

depending on whether the system is in continuous-time or in discrete-time. It is important to stress here that despite sharing very similarities with the transfer function representation used for LTI systems, the framework is way different here since the system is time-varying. The polynomials N and D are polynomials in σ (operator) and ρ (time-varying parameters). Additionally, the time-varying nature of the system is captured through a non-commutativity property of the derivative operator (in the continuous-time setting) as

$$\sigma a = \dot{a} + a\sigma$$

for some differentiable function a. Noncommutative polynomials, also referred to as *Ore polynomials* or skew-polynomials [76, 77], have also been used to extend the concept of transfer function to nonlinear systems through the use of Ore algebra, see e.g. [78].

1.4 Examples

For a correct understanding of the rationale of LPV systems and LPV control, it seems important to address several examples. Old and recent applications will be considered, with a specific emphasis on the correspondence with the different types of parameters we have discussed in the previous sections.

The first example, treated in Sect. 1.4.1, pertains on the application of LPV control on an inverted pendulum and is taken from [79]. In this example, the initial nonlinear system is reformulated as a qLPV system in view of designing LPV gain-scheduled controllers.

The example of Sect. 1.4.2 is concerned with the LPV modeling of a web service system for combined quality of service and energy management control [80, 81]. This model is an example of LPV system with internal parameters which is completely determined using LPV identification techniques.

The third example, treated in Sect. 1.4.3 and inspired from [82], pertains on the LPV approximation of aperiodic sampled-data systems. Such systems can be indeed approximated as polynomial LPV systems in order to derive gain-scheduled controllers that adapt to sampling-period variations.

The fourth and last example, presented in Sect. 1.4.4, is taken from the PhD thesis [83] where global chassis control is addressed using LPV techniques. In this work, an artificial parameter is introduced in the design in order to penalize the control input when it enters a forbidden region.

Several other applications are quickly mentioned in Sect. 1.4.5 as an attempt to show the wide applicability and potential of the LPV framework.

1.4.1 Inverted Pendulum: Robust Control and Performance

This example is taken from [79] and is one the first experimental applications (if not the very first one) of gain-scheduling based on LPV techniques. The considered system is the inverted pendulum depicted in Fig. 1.8 consisting of two arms moving in the vertical plane.

As for most of the mechanical systems, the equations of motion are nonlinear. After some change of variables and several algebraic manipulations, the authors have been able to derive the following LPV representation:

$$\dot{x}(t) = A(\rho(t))x(t) + Bu(t) \tag{1.43}$$

where

$$A(\rho) = \begin{bmatrix} 0 & 1 & 0 & 0 \\ \alpha_2 & 0 & -\alpha_2 & 0 \\ 0 & 0 & 0 & \rho \\ 0 & 0 & 0 & -\alpha_1 \end{bmatrix}, \quad B = \begin{bmatrix} 0 \\ 0 \\ 0 \\ \alpha_3 \end{bmatrix}. \tag{1.44}$$

Above, $\alpha_1, \alpha_2, \alpha_3$ are constant terms depending on the constant of gravitation, the structure of the system and the actuator (a motor). The parameter ρ is given by $\rho = 2l_1 \sin(\varphi_1)$ where l_1 is the length of the first arm and φ_1 is the angle of the first arm. Gain-scheduled controllers have been designed in [79] using different techniques (polytopic and LFT formulations) and both led to performance improvements over robust control (μ-control).

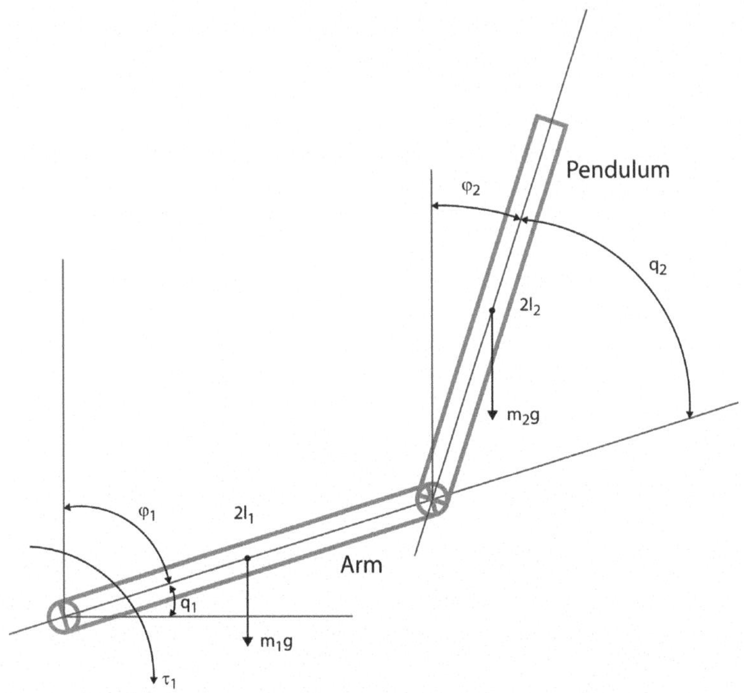

Fig. 1.8 Inverted pendulum considered in [79]

1.4.2 LPV Model for a Web Service System

The example discussed in this section pertains on the LPV modeling of a web service system and is taken from [81]. The model takes, as input variable, the admission probability (or rate), and both the request arrival rate and effective service time as scheduling parameters. Unlike mechanical systems, no theoretical model can be obtained from fundamental laws for such systems. It is therefore postulated that it admits the following representation

$$\begin{aligned} x_{k+1} &= Ax_k + (B_0 + B_1 s_k^f + B_2 s_k^f \lambda_k) p_k \\ y_k &= Cx_k + (D_0 + D_1 s_k^f + D_2 s_k^f \lambda_k) p_k + s_k^f \end{aligned} \tag{1.45}$$

where x_k is the state of the system, p_k is the probability that a request is admitted at time k and the output y_k is the server response time. The scheduling parameters are denoted by λ_k and s_k^f, where λ_k is the average requests arrival rate for the Web service application in the k-th time interval, and s_k^f is the effective service time in the k-th time interval.

In order to obtain numerical values for the matrices in the model (1.45), a least-squares identification procedure is performed in [81]. The considered identification

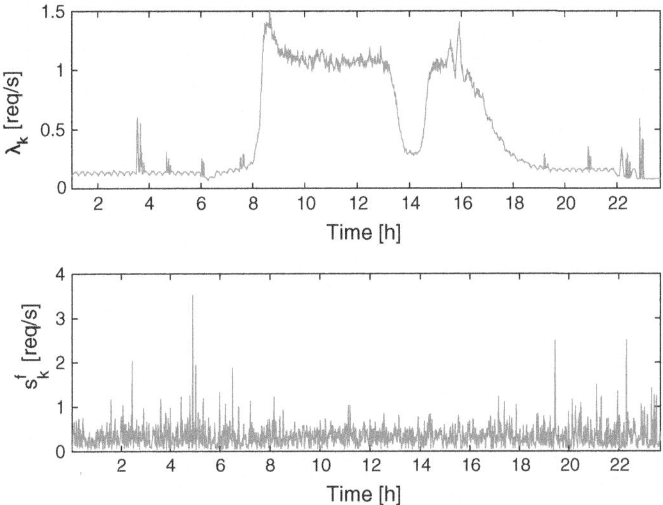

Fig. 1.9 Scheduling parameters: Average request arrival rate λ_k (*top*) and effective service time s_k^f (*bottom*)

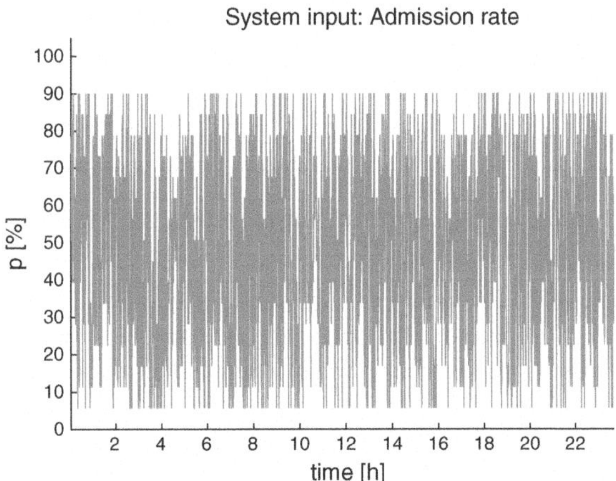

Fig. 1.10 System input: Admission rate

scenario can be seen in Figs. 1.9 and 1.10 where the scheduling parameters and input trajectories are shown.

The least-squares identification procedure returns the following numerical values:

$$A = \begin{bmatrix} 0.9445 & -0.0289 \\ -0.0280 & 0.9184 \end{bmatrix}, \; B_0 = \begin{bmatrix} 0.0424 \\ 0.1247 \end{bmatrix}, \; B_1 = \begin{bmatrix} -0.0602 \\ -0.0857 \end{bmatrix}, \; B_2 = \begin{bmatrix} 0.1926 \\ 0.2862 \end{bmatrix}$$

$$C = \begin{bmatrix} -0.3128 & 0.0086 \end{bmatrix}, \quad D_0 = -0.0678, \quad D_1 = -0.0542, \quad D_2 = 1.2212.$$

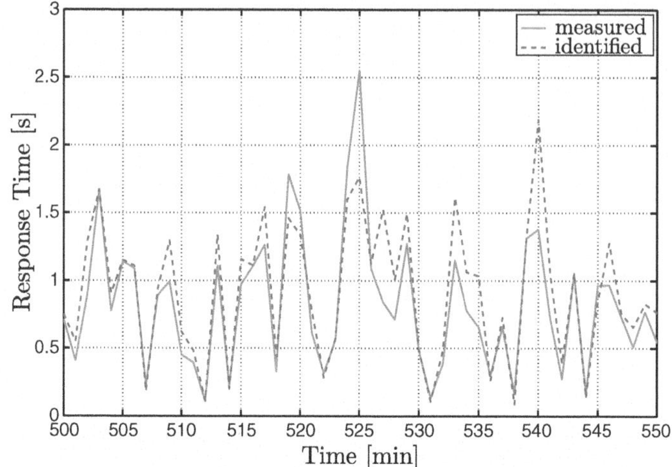

Fig. 1.11 Measured total service-time (*orange plain*) and simulated total service time (*green dashed*)

For validation, the measured output data is compared with the output obtained from the model. This comparison can be seen in Fig. 1.11 where we can see that the model matches reasonably well the real data.

1.4.3 LPV Models for Aperiodic Sampled-Data Systems

Aperiodic sampled-data systems can also be approximated by LPV systems in which the "varying sampling-period" plays the role of scheduling parameter. Gain-scheduling techniques can then be applied to derive controllers which adapt to the current sampling period value [82]. To illustrate this, let us consider the following aperiodic sampled-data system in discrete-time form

$$x(t_{k+1}) = A_d(T_k)x(t_k) + B_d(T_k)u(t_k) \tag{1.46}$$

where $x \in \mathbb{R}^n$, $u \in \mathbb{R}^m$ and $T_k := t_{k+1} - t_k \in \mathbb{R}_{>0}$ are the system state, the control input and the sampling-period at time k, respectively. The sampling-period-dependent matrices A_d and B_d are obtained from the matrices A and B of the initial continuous-time system as

$$A_d(T_k) = e^{AT_k} \text{ and } B_d(T_k) = \int_0^{T_k} e^{As} B\, ds. \tag{1.47}$$

Decomposing the matrix $A_d(T_k)$ as

$$A_d(T_k) = E_p(T_k) + R_p(T_k) \tag{1.48}$$

where

$$E_p(T_k) := I + AT_k + \frac{A^2 T_k^2}{2!} + \frac{A^3 T_k^3}{3!} + \ldots + \frac{A^p T_k^p}{p!}$$

is the Taylor expansion of order p of $A_d(T_k)$ and $R_p(T_k)$ is the remainder of the expansion. Note that a Padé approximation could have also been used to obtain a rational approximation of the exponential, rather than a polynomial one; see e.g. [84]. Based on the above decomposition, the system (1.46) can be approximated by the discrete-time LPV system

$$\tilde{x}_{k+1} = \left(E_p(T_k) + \Delta_1\right)\tilde{x}_k + \left[\left(\int_0^{T_k} E_p(s)ds\right) + \Delta_2\right] u_k \tag{1.49}$$

where Δ_1 and Δ_2 are uncertain matrices verifying $||\Delta_i||_2 \leq \alpha_i, i = 1, 2$ with

$$\begin{aligned} \alpha_1 &:= \max_{s \in [T_{min}, T_{max}]} ||R_p(s)||_2 \\ \alpha_2 &:= \max_{s \in [T_{min}, T_{max}]} \left|\left|\int_0^s R_p(\theta)d\theta\right|\right|_2, \end{aligned} \tag{1.50}$$

and where T_{min} and T_{max} are the minimal and maximal sampling period, respectively.

Note that the state has been changed to $\tilde{x}$ to emphasize the difference with the original system (1.46). Define now, for instance, the gain-scheduled state-feedback control law

$$u_k = K(T_k)\tilde{x}_k \tag{1.51}$$

where $K(T_k)$ is a sampling-period-dependent matrix. Since the approximated system is polynomial in T_k, it thus seems natural to choose the matrix $K(T_k)$ to be polynomial as well, and such that the (uncertain) closed-loop system (1.49)–(1.51) is stable. Due to their time-varying nature, such controllers are expected to be more efficient than their robust counterparts, i.e. K constant.

1.4.4 Automotive Suspension System

Another application of LPV control is *performance adaptation*. Parameters can indeed be introduced in loop-shaping weighting functions in H_∞/LPV synthesis, in order to adapt the characteristics of the closed-loop system in real time, e.g. the bandwidth of the closed-loop system, the weight on the control input, etc.

For instance, LPV control of semi-active suspensions is proposed in [85] in view of performing global chassis control. Semi-active suspensions are particular suspensions where the damping coefficient can be controlled in order to absorb energy in a desired way, see Fig. 1.12. The control input, i.e. the damping coefficient, can only take positive values since the suspension system is only able to absorb energy. A negative damping coefficient, on the other hand, would supply energy and is only available in active suspensions. The control input is then forced to lie in a certain region in the deflection speed/force plane, as depicted in Fig. 1.13. Ideally, the force produced by the suspension must be positive (negative) if the deflection speed is positive (negative).

An easy way to consider the constraint on the control input in the H_∞ framework is to use the following parameter dependent weighting-function acting on the control input

$$W_u(s,\rho) = \rho(u - v)\frac{1}{s/1000 + 1} \tag{1.52}$$

where u is the computed force and v is the achievable force which satisfies the quadrant constraints depicted in Fig. 1.13. The artificial parameter ρ is chosen to satisfy the following relation

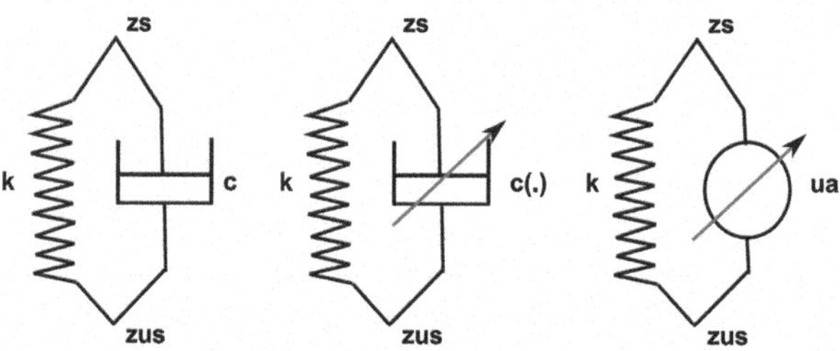

Fig. 1.12 Different types of suspensions, from *left* to *right*: passive, semi-active and active suspensions

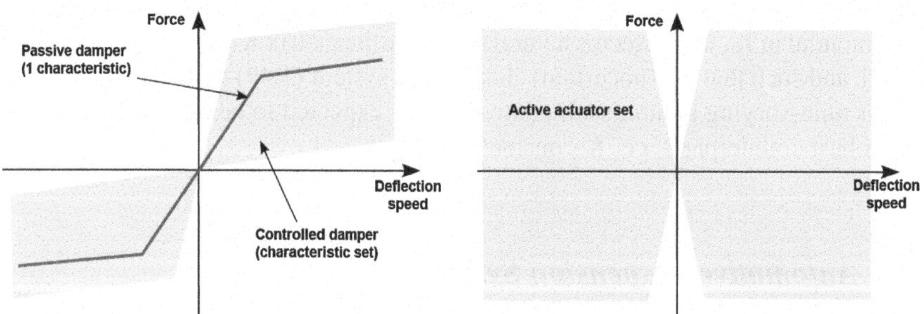

Fig. 1.13 Characteristics of passive, semi-active (*left*) and active (*right*) suspensions

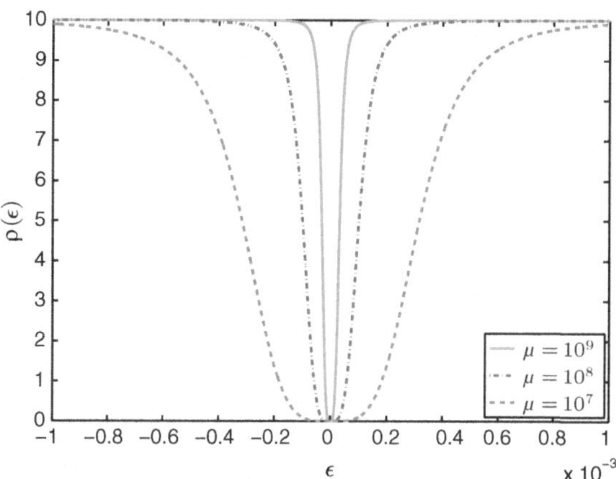

Fig. 1.14 Graph of the parameter ρ with respect to $\epsilon = u - v$

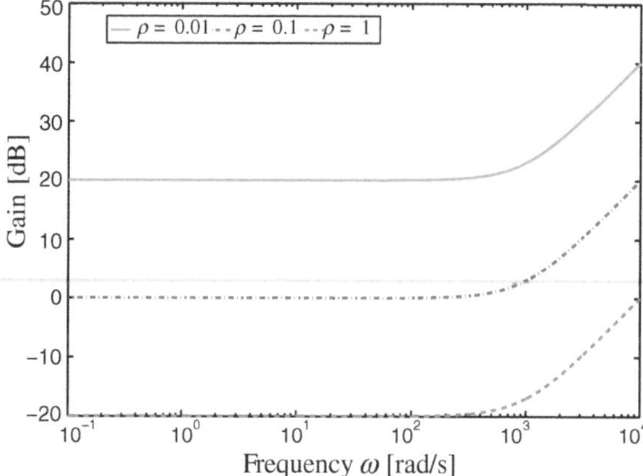

Fig. 1.15 Bode diagram of W_u^{-1} for different values of ρ

$$\rho(\varepsilon) = 10\frac{\mu\varepsilon^4}{\mu\varepsilon^4 + 1/\mu} \tag{1.53}$$

for a chosen large enough $\mu > 0$, e.g. $\mu = 10^8$. In this case, the parameter ρ ranges from 0 to 10 as shown in Fig. 1.14. By finally inspecting the Bode diagram of W_u^{-1}, see Fig. 1.15, we can notice that when ρ is large, i.e. when the computed force is far away from the achievable region, the control input will be attenuated down to 0, a value which is always achievable.

1.4.5 A Wide Range of Applications

We give here a non-exhaustive list of application of LPV modeling and control in the literature. In [25], the modeling and control of the air path system of diesel engines in view of reducing polluting gas is addressed. The control of elements in diesel engines is considered in [23, 24, 86–88] where the air flow, the fuel injection and/or the power unit are controlled. Turbocharged combustion engines are considered in [89]. In [90, 91], LPV systems are applied to modeling and control of turbofan engines. Electromagnetic actuators are considered in [92] whereas a robotic application is presented in [93]. In [94, 95], LPV control is applied to power system regulator. Wind turbines are considered in [96]. In the papers [21, 26, 97], LPV control is applied to the synthesis of missile autopilots. In [35], the attitude control of an F-16 Aircraft in response of the pilot orders for different angles of attack is addressed; aircrafts and spacecrafts are also considered in [98–101]. LPV vehicle suspensions modeling and control is presented in [102–107] while global chassis control (attitude control) is treated in [108–110]. Other automotive applications can be found in the papers [111, 112] and the recent book [113]. Fault detection and isolation using LPV techniques has been performed in [114] whereas LPV observers have been applied to estimate cell temperature in battery packs in [115]. Traffic control is considered in [116]. The control of asynchronous sampled-data systems is treated in an LPV fashion in [117] while LPV methods are applied to control time-delay systems in [118–121]. Finally, observer-based LPV control of nonlinear partial differential equations can be found in [122].

1.5 Control, Observation and Filtering of LPV Systems

To conclude on this introductory chapter on LPV systems, we briefly present in this section the different controller, observer and filter structures that can be considered in the context of LPV systems. The main difference with robust control theory lies in the possibility of adapting over time the structure of the controller/observer/filter according to the value of the parameters. This procedure is referred to as *gain-scheduling* and allows one to improve the performance of the designed process over non-scheduled ones.

Whereas design conditions will be given for controllers in Chap. 3, no design conditions will be explicitly given for observers and filters. The reason for this asymmetry lies in the fact that the design conditions derived in Chaps. 7 and 8 in the context of LPV time-delay systems, are general enough to be applied to non-delayed LPV systems. The LPV control problem is, however, very important to cover since very specific and important approaches (that are not covered in the chapter on control of LPV time-delay systems) deserve to be presented. Note that the design approaches for controllers can be applied to the design of observers and filters as well.

1.5.1 Observation and Filtering of LPV Systems

The goal of this section is to briefly introduce filters and observers for LPV systems of the form

$$\begin{aligned}\dot{x}(t) &= A(\rho(t))x(t) + B(\rho(t))u(t) + E(\rho(t))w(t)\\ z(t) &= C(\rho(t))x(t) + D(\rho(t))u(t) + F(\rho(t))w(t)\\ y(t) &= C_y(\rho(t))x(t) + F_y(\rho(t))w(t)\end{aligned} \tag{1.54}$$

where $x \in \mathbb{R}^n$, $u \in \mathbb{R}^m$, $w \in \mathbb{R}^p$, $z \in \mathbb{R}^q$ and $y \in \mathbb{R}^r$ are the state of the system, the known input, the disturbance output, the output to be estimated and the measured output, respectively. The parameters are assumed to take value in some compact set $\mathbf{\Delta}_\rho$ and to have reasonable trajectories ensuring that solutions to (1.54) exist.

1.5.1.1 Gain-Scheduled Filters for LPV Systems

A general filter for LPV systems is given by

$$\begin{aligned}\dot{x}_F(t) &= A_F(\rho(t))x_F(t) + B_{Fu}(\rho(t))u(t) + B_{Fy}(\rho(t))y(t)\\ z_F(t) &= C_F(\rho(t))x_F(t) + D_{Fu}(\rho(t))u(t) + D_{Fy}(\rho(t))y(t)\end{aligned} \tag{1.55}$$

where $x_F \in \mathbb{R}^{n_F}$ and $z_F \in \mathbb{R}^q$ are the state of the filter and the estimated output, respectively. We then design this filter such that it best estimates the signal z, i.e. we design the filter such that the gain of the transfer $w \to z - z_F$ is small, e.g. in the L_2-sense. When the dimension of the filter is smaller than the one of the system, the filter is said to be a *reduced-order filter* while when the dimension is the same, it is referred to as a *full-order filter*.

Filters have been designed in various settings; see e.g. [123–125] for the LFT-framework, [126] for the polytopic framework and [127] for the generic and affine frameworks. The case of inexact scheduling parameters has been considered [128, 129].

1.5.1.2 Gain-Scheduled Observers for LPV Systems

The goal of observers is to estimate the state of the system, that is, we usually design observers in such a way that the observation error is asymptotically stable and weakly affected by the disturbances.

Assuming that we have $z = Tx$ for some full rank matrix T and $y = C_y x$ in (1.54), the following observer can be used

$$\begin{aligned}\dot{\xi}(t) &= M(\rho(t))\xi(t) + N(\rho(t))y(t) + S(\rho(t))u(t)\\ \hat{z}(t) &= \xi(t) + Hy(t)\end{aligned} \tag{1.56}$$

where $\xi \in \mathbb{R}^q$ and $\hat{z} \in \mathbb{R}^q$ are the state of the system and the estimate of z, respectively. In such a case, the observation error is defined as $e(t) := z(t) - \hat{z}(t)$ and the observer should be designed such that the dynamical system governing $e(t)$ is asymptotically stable, and such that the gain of the transfer $w \to e$ is small, e.g. in the L_2-sense. The same terminology as for filters applies to observers. This type of observers has been considered, for instance, in [130].

More restrictive observers of the form

$$\dot{\hat{x}}(t) = A(\rho(t))\hat{x}(t) + B(\rho)u(t) + L(\rho)(y(t) - C_y(\rho(t))\hat{x}(t)) \tag{1.57}$$

with state $\hat{x} \in \mathbb{R}^n$ and gain $L(\rho)$ can be also used to estimate the full state x of the system (1.54) without any restriction on the structure of the system (1.54). In such a case, the observation error is defined as $e(t) := x(t) - \hat{x}(t)$. Such observers have been, for instance, considered in [131–133] in the polytopic setting and in [134] in the LFT framework. The observability problem for LPV systems has been addressed in [74, 135].

1.5.2 Control of LPV Systems

We consider, in this section, on the control of LPV systems, systems of the form:

$$\begin{aligned}
\dot{x}(t) &= A(\rho(t))x(t) + B(\rho(t))u(t) + E(\rho(t))w(t)\\
z(t) &= C(\rho(t))x(t) + D(\rho(t))u(t) + F(\rho(t))w(t)\\
y(t) &= C_y(\rho(t))x(t) + F_y(\rho(t))w(t)
\end{aligned} \tag{1.58}$$

where $x \in \mathbb{R}^n$, $u \in \mathbb{R}^m$, $w \in \mathbb{R}^p$, $z \in \mathbb{R}^q$ and $y \in \mathbb{R}^r$ are the state of the system, the control input, the disturbance output, the controlled output and the measured output, respectively. The parameters are assumed to take value in some compact set $\mathbf{\Delta}_\rho$ and to have reasonable trajectories ensuring that solutions to (1.54) exist.

1.5.2.1 Gain-Scheduled Static Controllers for LPV Systems

State-feedback and static output feedback controllers are part of the family of static controllers:

- Gain-scheduled static output-feedback controllers are given by

$$u(t) = K(\rho(t))y(t).$$

- Gain-scheduled state-feedback controllers are given by

$$u(t) = K(\rho(t)x(t).$$

State-feedback controllers have been proposed in [136]. Design conditions of state-feedback controllers in the generic and LFT frameworks are given in Chap. 3. The design of static-output feedback is much more complicated and will not be addressed in this monograph. This problem is well-known to be NP-hard in certain cases [137, 138] and several approaches have been developed to solve this challenging problem in the time-invariant setting; see e.g. [139–144]. The case of gain-scheduled static output-feedback controllers has been addressed in the discrete-time setting in [145].

1.5.2.2 Gain-Scheduled Dynamic Controllers for LPV Systems

Dynamic controllers may be classified in two main categories: observer-based output-feedback controllers and dynamic output-feedback controllers.

Observer-Based Controllers
As the name tells, this type of controllers consists of an observer part estimating the state of the system and a controller part that computes the control input from the estimated state. Observer-based output controllers take either of the following forms

$$\begin{aligned}\dot{\xi}(t) &= M(\rho(t))\xi(t) + N(\rho(t))y(t) + S(\rho(t))u(t)\\ \hat{x}(t) &= \xi(t) + Hy(t)\\ u(t) &= K(\rho(t))\hat{x}(t).\end{aligned}$$

or

$$\begin{aligned}\dot{\hat{x}}(t) &= A(\rho(t))\hat{x}(t) + B(\rho(t))u(t) + L(\rho)(y(t) - C_y(\rho(t))\hat{x}(t))\\ u(t) &= K(\rho(t))\hat{x}(t).\end{aligned}$$

where $\hat{x}, \xi \in \mathbb{R}^n$ are the states of the observers and M, N, S, L, H and K are matrices to be determined. Such controllers have been considered, for instance, in [136, 146, 147].

Dynamic Output-Feedback Controllers
Dynamic output feedback controllers have a similar structure to observer-based ones with the difference that the state is not aimed to be estimated. The controller is a one-block structure which simply computes a control input from the measured output, as seen below:

$$\begin{aligned}\dot{x}_c(t) &= A_c(\rho(t))x_c(t) + B_c(\rho(t))y(t)\\ u(t) &= C_c(\rho(t))x_c(t) + D_c(\rho(t))y(t)\end{aligned}$$

where $\mathbb{R}^{n_c}$ is the state of the controller. When the dimension of the controller is the same as the one of system, the controller is said to be of *full-order*, otherwise of *reduced-order* or *fixed-order*. The design of reduced-order controllers is a difficult problem, some instances of it being even known to be NP-hard [137, 138]. Yet some solutions exist; see e.g. [139, 142, 143].

Gain-scheduled dynamic output-feedback have been considered in [67, 136, 148–150] in the generic framework, in [61] in the polytopic framework and in [60–63] in the LFT framework. When the parameters are not exactly known, controllers that are resilient with respect to scheduling errors can also be designed; see e.g. [151, 152]. Several design conditions in the generic and LFT frameworks are given in Chap. 3.

References

1. G.D. Birkhoff, *Dynamical Systems* (American Mathematical Society, New York, 1927)
2. E. Sontag, *Mathematical Control Theory: Deterministic Finite Dimensional Systems* (Springer, New York, 1998)
3. H.K. Khalil, *Nonlinear Systems* (Prentice-Hall, USA, 2002)
4. S. Smale, *Differential Equations Dynamical Systems & an introduction to Chaos* (Academic Press, New York, 2004)
5. J.S. Shamma, Analysis and design of gain-scheduled control systems. PhD thesis, Laboratory for Information and decision systems—Massachusetts Institute of Technology, 1988
6. J.S. Shamma, An overview of LPV systems, in *Control of Linear Parameter Varying Systems with Applications*, ed. by J. Mohammadpour, C.W. Scherer (Springer, New York, 2012), pp. 3–26
7. J. Mohammadpour, C.W. Scherer (eds.), *Control of Linear Parameter Varying Systems with Applications* (Springer, New York, 2012)
8. J.P. Hespanha, A.S. Morse, Stability of switched systems with average dwell-time, in *38th Conference on Decision and Control*, Phoenix, Arizona, USA, 1999, pp. 2655–2660
9. D. Liberzon, *Switching in Systems and Control*. Birkhäuser, 2003
10. J.C. Geromel, P. Colaneri, Stability and stabilization of continuous-time switched linear systems. SIAM J. Control Optim. **45**(5), 1915–1930 (2006)
11. A.N. Michel, L. Hou, D. Liu, *Stability of Dynamical Systems—Continuous, Discontinuous and Discrete Systems* (Birkhäuser, Boston, 2008)
12. R. Goebel, R.G. Sanfelice, A.R. Teel, Hybrid dynamical systems. IEEE Control Syst. Mag. **29**(2), 28–93 (2009)
13. C. Briat, A. Seuret, Affine minimal and mode-dependent dwell-time characterization for uncertain switched linear systems. IEEE Trans. Autom. Control **58**(5), 1304–1310 (2013)
14. V.A. Yakubovich, V.M. Starzhinskii, *Linear Differential Equations with Periodic Coefficients* (Wiley, New York, 1975)
15. S. Bittanti, P. Colaneri, Periodic control systems, in *Proceedings of the IFAC Workshop*, Cernobbio-Como, Italy, Aug. 27–28, 2001, ed. by S. Bittanti, P. Colaneri
16. P. Colaneri, Theoretical aspects of continuous-time periodic systems. Annu. Rev. Control **29**, 205–215 (2005)
17. V.A. Yakubovich, A.L. Fradkov, D.J. Hill, A.V. Proskurnikov, Dissipativity of T-periodic linear systems. IEEE Trans. Autom. Control **52**(6), 1039–1047 (2007)
18. J. Shamma, M. Athans, Analysis of gain scheduled control for nonlinear plants. IEEE Trans. Autom. Control **35**(8), 898–907 (1990)
19. J.S. Shamma, M. Athans, Gain scheduling: potential hazards and possible remedies. IEEE Contr. Syst. Mag. **12**(3), 101–107 (1992)
20. D. Liberzon, J.P. Hespanha, A.S. Morse, Stability of switched systems: a Lie-algebraic condition. Syst. Control Lett. **37**, 117–122 (1999)
21. K. Tan, K.M. Grigoriadis, $\mathcal{L}_2 - \mathcal{L}_2$ and $\mathcal{L}_2 - \mathcal{L}_\infty$ output feedback control of time-delayed LPV systems, in *Conference on on Decision and Control*, 2000
22. J.Y. Shin, Analysis of linear parameter varying system models based on reachable sets, in *American Control Conference*, Anchorage, AK, USA, 2002
23. B. He, M. Yang, Robust LPV control of diesel auxiliary power unit for series hybrid electric vehicles. IEEE Trans. Power Electron. **21**(3), 791–798 (2006)

24. M. Jung, K. Glover, Calibrate linear parameter-varying control of a turbocharged diesel engine. IEEE Trans. Control Syst. Technol. **14**(1), 45–62 (2006)
25. X. Wei, L. del Re, Gain scheduled $\mathcal{H}_\infty$ control for air path systems of diesel engines using LPV techniques. IEEE Trans. Control Syst. Technol. **15**(3), 406–415 (2007)
26. B.A. White, L. Bruyere, A. Tsourdos, Missile autopilot design using quasi-LPV polynomial eigenstructure assignment. IEEE Trans. Aerosp. Electron. Syst. **43**(4), 1470–1483 (2007)
27. T. Takagi, M. Sugeno, Fuzzy identification of systems and its applications to modeling and control. IEEE Trans. Syst. Man Cybern. **15**(1), 116–132 (1985)
28. L.X. Wang, J.M. Mendel, Fuzzy basis functions, universal approximation and orthogonal least-squares. IEEE Trans. Neural Netw. **3**(5), 807–814 (1992)
29. K. Tanaka, H.O. Wang, *Fuzzy Control Systems Design and Analysis. A linear matrix inequality approach* (Wiley, New York, 2001)
30. F. Bruzelius, S. Petterssona, C. Breitholz, Linear parameter varying descriptions of nonlinear systems, in *American Control Conference*, Boston, Massachussetts, 2004
31. C.S. Mehendale, K.M. Grigoriadis, A new approach to LPV gain-scheduling design and implementation, in *43rd IEEE Conference on Decision and Control, Atlantis, Paradise Island, Bahamas*, 2004
32. R.A. Nichols, R.T. Reichert, W.J. Rugh, Gain scheduling for H-infinity controllers: a flight control example. IEEE Trans. Control Syst. Technol. **1**, 69–75 (1993)
33. D.A. Lawrence, W.J. Rugh, Gain scheduling dynamic linear controllers for a nonlinear plant. Automatica **31**(3), 381–390 (1995)
34. J.P. Hespanha, A.S. Morse, Switching between stabilizing controllers. Automatica **38**, 1905–1917 (2002)
35. B. Lu, F. Wu, S. Kim, Switching LPV control of an F-16 aircraft via controller state reset. IEEE Trans. Control Syst. Technol. **14**(2), 267–277 (2006)
36. P. Yan, H. Özbay, On switching H_∞ controllers for a class of linear parameter varying systems. Syst. Control Lett. **56**, 504–511 (2007)
37. W. Jiang, E. Fridman, A. Kruszewski, J.-P. Richard, Switching controller for stabilization of linear systems with switched time-varying delays, in *IEEE Conference on Decision and Control*, pp. 7923–7928, 2009
38. C. Prieur, A.R. Teel, Uniting local and global output feedback controllers. IEEE Trans. Autom. Control **56**(7), 1636–1649 (2011)
39. B. Demirel, C. Briat, M. Johansson, Supervisory control design for networked systems with time-varying communication delays, in *4th IFAC Conference on Analysis and Design of Hybrid Systems*, Eindovhen, The Netherlands, 2012, pp. 133–140
40. P. Apkarian, H. D. Tuan, Parametrized LMIs in control theory, in *Conference on Decision and Control*, Tampa, Florida, 1998
41. D.G. Luenberger, Dynamic equations in descriptor form. IEEE Trans. Autom. Control **22**, 312–321 (1977)
42. A. Rehm, F. Allgöwer, Selft-scheduled H_∞ output feedback control of descriptor systems. Comput. Chem. Eng. **24**, 279–284 (2000)
43. I. Masubuchi, T. Akiyama, M. Saeki, Synthesis of output feedback gain-scheduling controllers based on descriptor LPV system representation, in *42nd IEEE Conference on Decision and Control*, Maui, Hawaii, USA, 2003, pp. 6115–6120
44. I. Masubuchi, J. Kato, M. Saeki, A. Ohara, Gain-scheduled controller design based on descriptor representation of LPV systems: Application to flight vehicle control, in *43rd IEEE Conference on Decision and Control*, Atlantis, Paradise Island, Bahamas, USA, 2004, pp. 815–820
45. L. Dai, *Singular Control Systems* (Springer, Secaucus, 1989)
46. P. Apkarian, P. Gahinet, G. Becker, Self-scheduled control of linear parameter varying systems: a design example. Automatica **31**(9), 1251–1261 (1995)
47. J.C. Geromel, P. Colaneri, Robust stability of time-varying polytopic systems. Syst. Control Lett. **55**, 81–85 (2006)
48. R.A. Borges, P.L.D. Peres, $\mathcal{H}_\infty$ LPV filtering for linear systems with arbitrarily time-varying parameters in polytopic domains, in *Conference on Decision and Control*, San Diego, USA, 2006

49. R.C.L.F. Oliveira, V.F. Montagner, P.L.D. Peres, P. A. Bliman, LMI relaxations for $\mathcal{H}_\infty$ control of time-varying polytopic systems by means of parameter dependent quadratically stabilizing gains, in 3rd *IFAC Symposium on System, Structure and Control*, Foz do Iguassu, Brasil, 2007
50. G. Chesi, A. Garulli, A. Tesi, A. Vicino, Homogeneous polynomial forms for robustness analysis of uncertain systems, *Lecture Notes in Control and Information Sciences* (Springer, Berlin, 2009)
51. M.G. Safonov, Origins of robust control: early history and future speculations, in *7th IFAC Symposium on Robust Control Design*, 2012, pp. 1–8
52. V.M. Popov, Absolute stability of nonlinear systems of automatic control. Autom. Remote Control (Trans. Automatica i Telemekhanika) **22**(8), 857–875 (1961)
53. V.M. Popov, One problem in the theory of absolute stability of controlled systems. Autom. Remote Control **25**(9), 1129–1134 (1964)
54. A.I. Lur'e, *Some Nonlinear Problems in the Theory of Automatic Control (Nekotorye Nelineinye Zadachi Teorii Automaticheskogo Regulirovaniya)* (Gostekhizdat, Moscow, 1951)
55. G. Zames, On the input output stability of time-varying nonlinear feecback systems, part i : conditions using concepts of loop gain, conicity and positivity. IEEE Trans. Aut. Control **11**(2), 228–238 (1966)
56. M.G. Safonov, M. Athans, Gain and phase margin for multiloop LQG regulators. IEEE Trans. Autom. Control **22**(2), 173–178 (1977)
57. J.C. Doyle, Robustness of multiloop linear feedback systems, in *1978 IEEE Conference on Decision and Control*, 1978, pp. 12–18
58. K. Zhou, J.C. Doyle, K. Glover, *Robust and Optimal Control* (Prentice Hall, Upper Saddle River, 1996)
59. J.C. Willems, Dissipative dynamical systems i & ii. Ration. Mech. Anal. **45**(5), 321–393 (1972)
60. A. Packard, Gain scheduling via linear fractional transformations. Syst. Control Lett. **22**, 79–92 (1994)
61. P. Apkarian, P. Gahinet, A convex characterization of gain-scheduled $\mathcal{H}_\infty$ controllers. IEEE Trans. Autom. Control **5**, 853–864 (1995)
62. C.W. Scherer, LPV control and full-block multipliers. Automatica **37**, 361–375 (2001)
63. C.W. Scherer, I.E. Köse, Gain-scheduled control synthesis using dynamic D-scales. IEEE Trans. Autom. Control **57**(9), 2219–2234 (2012)
64. M.G. Safonov, M. Athans, On stability theory, in *17th Conference on Decision and Control*, San Diego, California, USA, 1978, pp. 301–314
65. T. Iwasaki, S. Hara, Well-posedness of feedback systems: insight into exact robustness analysis and approximate computations. IEEE Trans. Autom. Control **43**, 619–630 (1998)
66. A. Megretski, A. Rantzer, System analysis via integral quadratic constraints. IEEE Trans. Autom. Control **42**(6), 819–830 (1997)
67. P. Apkarian, R.J. Adams, Advanced gain-scheduling techniques for uncertain systems. IEEE Trans. Autom. Control **6**, 21–32 (1998)
68. C.W. Scherer, S. Weiland, Linear matrix inequalities in control. Lecture Notes, (2005), http://www.imng.uni-stuttgart.de/simtech/Scherer/lmi/
69. S. Hecker, A. Varga, J.F. Magni, Enhanced LFR-toolbox for Matlab. Aerosp. Sci. Technol. **9**, 173–180 (2005)
70. D. de Vito, A. Kron, J. de la Fontaine, M. Lovera, A Matlab toolbox for LMI-based analysis and synthesis of LPV/LFT self-scheduled H_∞ control systems, in *IEEE International Symposium on Computer-Aided Control System Design (CACSD'10)*, 2010, pp. 1397–1402
71. B. Bamieh, L. Giarré, Identification of linear parameter varying models. Int. J. Robust Nonlinear Control **12**, 841–853 (2002)
72. R. Tóth, F. Felici, P.S.C. Heuberger, P. M. J. van der Hof, Discrete time LPV I/O and state space representations—differences of behavior and pitfalls of interpolation, in *9th European Control Conference*, Kos, Greece, 2007, pp. 5418–5425
73. R. Tóth, Modeling and identification of linear parameter-varying systems—an orthonormal basis function approach, PhD thesis, Technische Universiteit Delft, 2008

74. R. Tóth, *Modeling and Identification of Linear Parameter-Varying Systems* (Springer, Germany, 2010)
75. V. Cerone, D. Piga, D. Reguto, R. Tóth, Fixed order LPV controller design for LPV models in input-output form, in *51st IEEE Conference on Decision and Control*, Maui, Hawaii, USA, 2012, pp. 6297–6302
76. O. Ore, Linear equations in non-commutative fields. Ann. Math. **32**, 463–477 (1931)
77. O. Ore, Theory of non-commutative polynomials. Ann. Math. **32**, 480–508 (1933)
78. M. Halás, An algebraic framework generalizing the concept of transfer functions to nonlinear systems. Automatica **44**, 1181–1190 (2008)
79. H. Kajiwara, P. Apkarian, P. Gahinet, LPV techniques for control of an inverted pendulum. IEEE Control Syst. Mag. **19**, 44–54 (1999)
80. C. Poussot-Vassal, M. Tanelli, M. Lovera, Linear parametrically varying MPC for combined quality of service and energy management in web service systems, in *American Control Conference*, 2010, pp. 3106–3111
81. C. Poussot-Vassal, M. Tanelli, M. Lovera, A control-theoretic approach for the combined management of quality-of-service and energy in service centers, in *Run-time Models for Self-managing Systems and Applications, Autonomic Systems*, ed. by D. Ardagna, L. Zhang (Springer, Basel, 2010), pp. 73–96
82. D. Robert, O. Sename, D. Simon, An $\mathcal{H}_\infty$ LPV design for sampling varying controllers: experimentation with a T-inverted pendulum. IEEE Trans. Control Syst. Technol. **18**(3), 741–749 (2010)
83. C. Poussot-Vassal, Global chassis control using a multivariable robust LPV control approach, PhD thesis, Grenoble-INP, 2008
84. C. Moler, C. van Loan, Nineteen dubious ways to compute the exponential of a matrix, Twenty-five years later. SIAM Rev. **45**(1), 3–49 (2003)
85. S.M. Savaresi, C. Poussot-Vassal, C. Spelta, O. Sename, L. Dugard, *Semi-active Suspension Control Design for Vehicles* (Butterworth Heinemann, Oxford, 2010)
86. C. Gauthier, O. Sename, L. Dugard, G. Meissonier, Modeling of a diesel engine common rail injection system, in *16th IFAC World Congress*, Prague, Czech Republic, 2005
87. C. Gauthier, O. Sename, L. Dugard, G. Meissonier, A $\mathcal{H}_\infty$ linear parameter varying (LPV) controller for a diesel engine common rail injection system, in *European Control Conference*, Kos, Greece, 2007
88. C. Gauthier, O. Sename, L. Dugard, G. Meissonnier, An LFT approach to $\mathcal{H}_\infty$ control design for diesel engine common rail injection system. Oil Gas Sci. and Technol. Rev. IFP **62**, 513–522 (2007)
89. A. Kominek, H. Werner, M. Garwon, M. Schultalbers, Identification of low-complexity LPV input-output models for control of a turbocharged combustion engine, in *Control of Linear Parameter Varying Systems with Applications*, ed. by J. Mohammadpour, C.W. Scherer (Springer, New York, 2012), pp. 445–460
90. L. Reberga, D. Henrion, F. Vary, LPV modeling of a turbofan engine, in *IFAC World Congress on Automatic Control*, Prague, Szech Republic, 2005
91. W. Gilbert, D. Henrion, J. Bernussou, Polynomial LPV synthesis applied to turbofan engines, in *IFAC Symposium on Automatic Control in Aerospace*, Toulouse, France, 2007
92. A. Forrai, T. Ueda, T. Yumura, Electromagnetic actuator control: a linear parameter-varying (LPV) approach. IEEE Trans. Ind. Electron. **54**(3), 1430–1441 (2007)
93. A. Kwiatkowski, H. Werner, LPV control of a 2-DOF robot using parameter reduction, in *European Control Conference*, Seville, Spain, 2005
94. Q. Liu, V. Vittal, N Elia, LPV supplementary damping controller design thyristor controlled series capacacitor (TCSC) device. IEEE Trans. Power Syst. **21**(3), 1242–1249 (2006)
95. Q. Liu, V. Vittal, N Elia, Expansion of system operating range by an interpolated LPV FACTS controller using multiple Lyapunov functions. IEEE Trans. Power Syst. **21**(3), 1311–1320 (2006)
96. F. Daher Adegas, C. Solth, J. Stoustrup, Structured linear parameter varying control of wind turbines, in *Control of Linear Parameter Varying Systems with Applications*, ed. by J. Mohammadpour, C.W. Scherer (Springer, Heidelberg, 2012), pp. 303–337

97. S. Lim, J. How, Application of improved $\mathcal{L}_2$ gain synthesis on LPV missile autopilot design, in *American Control Conference*, San Diego, CA, USA, 1999
98. J.M. Barker, G.J. Balas, Comparing linear parameter-varying gain-scheduled control techniques for active flutter suppression. J. Guidance, Control, Dyn. **23**(5), 948–955 (2000)
99. A. Corti, M. Lovera, Attitude regulation for spacecraft with magnetic actuators: an LPV approach, in *Control of Linear Parameter Varying Systems with Applications*, ed. by J. Mohammadpour, C.W. Scherer (Springer, Heidelberg, 2012), pp. 339–355
100. H.D. Hughes, F. Wu, LPV H_∞ control for flexible hypersonic vehicle, in *Control of Linear Parameter Varying Systems with Applications*, ed. by J. Mohammadpour, C.W. Scherer (Springer, Heidelberg, 2012), pp. 413–444
101. P. Seiler, G.J. Balas, A. Packard, Linear parameter-varying control for the X-53 active aeroelastic wing, in *Control of Linear Parameter Varying Systems with Applications*, ed. by J. Mohammadpour, C.W. Scherer (Springer, Heidelberg, 2012), pp. 483–512
102. P. Gaspard, I. Szaszi, J. Bokor, Active suspension design using LPV control, in *IFAC Symposium on Advances in Automotive Control*, University of Salerno, Italy, April 2004
103. C. Poussot-Vassal, O. Sename, L. Dugard, P. Gáspár, Z. Szabó, J. Bokor, Multi-objective qLPV $\mathcal{H}_\infty/\mathcal{H}_2$ control of a half vehicle, in *Proceedings of the 10th Mini-conference on Vehicle System Dynamics, Identification and Anomalies (VSDIA)*, Budapest, Hungary, November 2006
104. C. Poussot-Vassal, A. Drivet, O. Sename, L. Dugard, R. Ramirez-Mendoza, A self tuning suspension controller for multi-body quarter vehicle model, in *Proceedings of the 17th IFAC World Congress (WC)*, Seoul, South Korea, July 2008
105. C. Poussot-Vassal, O. Sename, L. Dugard, P. Gáspár, Z. Szabó, J. Bokor, New semi-active suspension control strategy through LPV technique. Control Eng. Pract. **16**, 1519–1534 (2008)
106. A.-L. Do, O. Sename, L. Dugard, LPV modeling and control of semi-active dampers in automotive systems, in *Control of Linear Parameter Varying Systems with Applications*, ed. by J. Mohammadpour, C.W. Scherer (Springer, Heidelberg, 2012), pp. 381–411
107. C. Poussot-Vassal, C. SPelta, O. Sename, S.M. Savaresi, L. Dugard, Survey and performance evaluation on some automotive semi-active suspension control methods—a comparative study on a single-corner model. Annu. Rev. Control **36**, 148–160 (2012)
108. P. Gáspár, Z. Szabó, J. Bokor, C. Poussot-Vassal, O. Sename, L. Dugard, Toward global chassis control by integrating the brake and suspension systems, in *Proceedings of the 5th IFAC Symposium on Advances in Automotive Control (AAC)*, Aptos, California, USA, August 2007
109. C. Poussot-Vassal, O. Sename, L. Dugard, P. Gáspár, Z. Szabó, J. Bokor, Attitude and handling improvements through gain-scheduled suspensions and brakes control, in *Proceedings of the 17th IFAC World Congress (WC)*, Seoul, South Korea, July 2008
110. Z. Szabó, P. Gáspár, J. Bokor, Design of integrated vehicle chassis control based on LPV methods, in *Control of Linear Parameter Varying Systems with Applications*, ed. by J. Mohammadpour, C.W. Scherer (Springer, Heidelberg, 2012), pp. 513–534
111. A. Kwiatkowski, J.P. Blath, H. Werner, M. Schultalbers, Application of LPV gain scheduling to charge control a SI engine, in *IEEE Conference on Computer Aided Control Systems Design*, Munich, Germany, 2006, pp. 2327–2331
112. A. Kwiatkowski, J. P. Blath, H. Werner, A. Ali, M. Schultalbers, Linear parameter varying PID controller design for charge control of a spark-ignited engine. Control Eng. Pract. **17**, 1307–1317 (2009)
113. O. Sename, P. Gáspár, J. Bokor (eds.), *Robust Control and Linear Parameter Varying Approaches—Application to Vehicle Dynamics* (Springer, Berlin, 2013)
114. M. Rodrigues, M. Sahnoun, D. Theilliol, J.-C. Ponsart, Sensor fault detection and isolation filter for polytopic lpv systems: a winding machine application. J. Process Control **23**, 805–816 (2013)
115. M. Debert, G. Colin, G. Bloch, Y. Chamaillard, An observer looks at the cell temperature in automotive battery packs. Control Eng. Pract. **21**, 1035–1042 (2013)
116. T. Luspay, T. Péni, B. Kulcsár, Constrained freeway traffic control via linear parameter varying paradigms, in *Control of Linear Parameter Varying Systems with Applications*, ed. by J. Mohammadpour, C.W. Scherer (Springer, Heidelberg, 2012), pp. 461–482

117. D. Robert, O. Sename, D. Simon, Synthesis of a sampling period dependent controller using LPV approach, in *5th IFAC Symposium on Robust Control Design ROCOND'06*, Toulouse, France, 2006
118. C. Briat, O. Sename, J.-F. Lafay, A LFT/$\mathcal{H}_\infty$ state-feedback design for linear parameter varying time delay systems, in *9th European Control Conference*, Kos, Greece, 2007, pp. 4882–4888
119. C. Briat, O. Sename, J.-F. Lafay, Delay-scheduled state-feedback design for time-delay systems with time-varying delays, in *17th IFAC World Congress*, Seoul, South Korea, 2008, pp. 1267–1272
120. C. Briat, O. Sename, J.-F. Lafay, Delay-scheduled state-feedback design for time-delay systems with time-varying delays—a LPV approach. Syst. Control Lett. **58**(9), 664–671 (2009)
121. C. Briat, O. Sename, J.-F. Lafay, $\mathcal{H}_\infty$ delay-scheduled control of linear systems with time-varying delays. IEEE Trans. Autom. Control **42**(8), 2255–2260 (2009)
122. S. M. Hashemi, H. Werner, Observer-based LPV control of a nonlinear PDE, in *50th IEEE Conference on Decision and Control*, Orlando, Florida, USA, 2011, pp. 2010–2015
123. H.D. Tuan, P. Apkarian, T.Q. Nguyen, Robust and reduced order filtering. IEEE Trans. Signal Process. **49**, 2875–2984 (2001)
124. H. D. Tuan, P. Apkarian, T. Q. Nguyen, Robust filtering for uncertain nonlinearly parametrized plants. IEEE Trans. Signal Process. **51**, 1806–1815 (2003)
125. N.T. Hoang, H.D. Tuan, P. Apkarian, S. Hosoe, Gain-scheduled filtering for time-varying discrete systems. IEEE Trans. Signal Process. **52**(9), 2464–2476 (2004)
126. R.A. Borges, V.F. Montagner, R.C.L.F. Oliveira, P.L.D. Peres, P.-A. Bliman, Parameter-dependent H_2 and H_∞ filter design for linear systems with arbitrarily time-varying parameters in poltopic domains. Signal Process. **88**, 1801–1816 (2008)
127. M. Sato, Filter design for LPV systems using quadratically parameter-dependent Lyapunov functions. Automatica **42**, 2017–2023 (2006)
128. M. Sato, Gain-scheduled H_2 filter synthesis via polynomially parameter-dependent Lyapunov functions with inexact scheduling parameters, in *19th International Symposium on Mathematical Theory of Networks and Systems*, Budapest, Hungary, 2010, pp. 2291–2298
129. M. Sato, Gain-scheduled H_∞ filters using inexactly measured scheduling parameters, in *American Control Conference*, Baltimore, Maryland, USA, 2010, pp. 3088–3093
130. G. Iulia Bara, J. Daafouz, F. Kratz, J. Ragot, Parameter-dependent state observer design for affine LPV systems. Int. J. control **74**(16), 1601–1611 (2001)
131. J. Daafouz, G. I. Bara, F. Katz, J. Ragot, State observers for discrete-time LPV systems: an interpolation based approach, in *39th IEEE Conference on Decision and Control*, Sydney, Australia, 2000, pp. 4571–4572
132. J. Daafouz, G. Milerioux, L. Rosier, Observer design with guaranteed bound for LPV systems, in *IFAC World Congress*, Prague, Czech Republic, 2005, pp. 961–966
133. F. Anstett, G. Millérioux, G. Bloch, Polytopic observer design for LPV systems based on minimal convex polytope finding. J Algorithms Comput. Technol. **3**(1), 23–43 (2009)
134. A.M. Gonzales, C. Hoffmann, C. Radisch, H. Werner, LPV observer design and damping control of container crane load swing, in *European Control Conference*, Zürich, Switzerland, 2013, pp. 1848–1853
135. G.J. Balas, J. Bokor, Z. Szabo, Invariant subspaces for LPV systems and their application. IEEE Trans. Autom. Contr. **48**, 2065–2069 (2003)
136. F. Wu, Control of linear parameter varying systems. PhD thesis, University of California Berkeley, 1995
137. V. Blondel, J.N. Tsitsiklis, NP-hardness of some linear control problems. SIAM J. Control Optim. **35**(6), 2118–2127 (1997)
138. M. Fu, Pole placement via static output feedback is NP-hard. IEEE Trans. Autom. Control **49**(5), 855–857 (2004)
139. L. El-Ghaoui, F. Oustry, M.A. Rami, A cone complementary linearization algorithm for static output-feedback and related problems. IEEE Trans. Autom. Control **42**, 1171–1176 (1997)

140. V.L. Syrmos, C.T. Abdallah, P. Dorato, K. Grigoriadis, Static output feedback—a survey. Automatica **33**(2), 125–137 (1997)
141. D. Peaucelle, D. Arzelier, Ellipsoidal sets for resilient and robust static output feedback. IEEE Trans. Autom. Control **50**, 899–904 (2005)
142. D. Henrion, LMI optimization for fixed-order H_∞ controller design, in *Positive Polynomials in Control, Lecture Notes in Control and Information Science*, ed. by D. Henrion, A. Garulli (Springer, Heidelberg, 2005), pp. 73–85
143. C.W.J. Hol, C.W. Scherer, A sum-of-squares approach to fixed-order H_∞ synthesis, in *Positive Polynomials in Control, Lecture Notes in Control and Information Science*, ed. by D. Henrion, A. Garulli (Springer, Heidelberg, 2005), pp. 45–71
144. D. Henrion, J.B. Lasserre, Convergent relaxations of polynomial matrix inequalities and static output feedback. IEEE Trans. Autom. Control **51**(2), 192–202 (2006)
145. J. de Caigny, J.F. Camino, R.C.L.F. Oliveira, P.L.D. Peres, J. Swevers, Gain-scheduled H_2 and H_∞ control of discrete-time polytopic time-varying systems. IET Control Theory Appl. **4**(3), 362–380 (2010)
146. B.-S. Hong, Observer-based paramterized LPV L2-gain control synthesis, in *American Control Conference*, Anchorage, Alaska, USA, 2002, pp. 4427–4432
147. W.P.M.H. Heemels, J. Daafouz, G. Millerioux, Observer-based control of discrete-time LPV systems with uncertain parameters. IEEE Trans. Autom. Control **55**(9), 2130–2135 (2010)
148. F. Wu, X.H. Yang, A. Packard, G. Becker, Induced $\mathcal{L}_2$-norm control for LPV systems with bounded parameter variation rates. Int. J. Robust Nonlinear Control **6**(9–10), 983–998 (1996)
149. F. Wu, A generalized LPV system analysis and control synthesis framework. Int. J. Control **74**, 745–759 (2001)
150. K. Tan, K.M. Grigoriadis, F. Wu, Output-feedback control of LPV sampled-data systems. Int. J. Control **75**(4), 252–264 (2002)
151. M. Sato, D. Peaucelle, Gain-scheduled output-feedback controllers using inexact scheduling parameters for continuous-ime LPV systems. Automatica **49**, 1019–1025 (2013)
152. J. Daafouz, J. Bernussou, J.C. Geromel, On inexact LPV control design of continuous-time polytopic systems. IEEE Trans. Autom. Control **53**(7), 1674–1978 (2008)

Chapter 2
Stability of LPV Systems

All stable processes we shall predict. All unstable processes we shall control.

John von Neumann

Abstract This chapter first presents the main stability and instability results for general dynamical systems. These results are then further adapted to the analysis of linear parameter-varying systems in the generic, polytopic and LFT frameworks. Notably, the notions of quadratic and robust stability using quadratic Lyapunov functions are introduced. Some other types of Lyapunov functions are also briefly discussed. The developed results rely on robust analysis techniques such as robust Lyapunov inequalities, the small-gain theorem, integral quadratic constraints and topological separation, and are expressed through linear matrix inequalities. A particular emphasis is made on the connections between the different approaches used to analyze LPV systems in LFT-form.

2.1 Chapter Outline

The analysis of LPV systems is mainly based on robust stability analysis approaches since an LPV system is nothing else but an uncertain system with time-varying parameters. The first section of this chapter therefore starts with some general definitions of stability of fixed points of general dynamical systems and then exposes very important stability theorems. Namely, the Lyapunov stability theorem, the Barbashin-Krasovskii theorem and the Chetaev's instability theorem. The linear system case is treated as a particular case of these results. Section 2.3 introduces the most common stability notions used to analyze uncertain and LPV systems, namely the notions of *quadratic stability* and *robust stability*, along with their respective class of Lyapunov functions. Several other types of Lyapunov functions, such as piecewise-quadratic or homogeneous ones are also briefly mentioned. In Sect. 2.4, results on generic LPV

C. Briat, *Linear Parameter-Varying and Time-Delay Systems*,
Advances in Delays and Dynamics 3, DOI 10.1007/978-3-662-44050-6_2

systems are obtained and expressed as parameter-dependent LMIs. Similar results pertaining on the stability analysis of polytopic LPV systems are derived in Sect. 2.5. The last section, Sect. 2.6, is about the analysis of LPV systems in LFT-form. Several approaches based on the notions of L_2-gain, small-gain theorems and scalings are first presented. Approaches relying on the full-block S-procedure, topological separation and integral quadratic constraints are then introduced. Throughout this section, all these approaches are related to each other on the basis of their corresponding stability criteria.

2.2 General Notions of Stability for Dynamical Systems

Before providing key results on stability of LPV systems, it is seems necessary to provide several general definitions and results about the stability of dynamical systems.

Definition 2.2.1 Let us consider the dynamical system

$$\begin{aligned} \dot{x}(t) &= f(x(t)),\ t \geq 0 \\ x(0) &= x_0 \end{aligned} \tag{2.1}$$

where f is a sufficiently nice function ensuring that the above dynamical system has a unique solution. Let us denote by $x(x_0, t)$ the solution to this dynamical system when the initial condition is x_0. Assume, for simplicity, that x^* is a fixed point of (2.1), i.e. $f(x^*) = 0$. Then, the equilibrium point x^* is said to be

- **stable** (in the sense of Lyapunov) if, for each $\varepsilon > 0$, there is $\delta = \delta(\varepsilon) > 0$ such that
$$||x^* - x_0|| \leq \delta \Rightarrow ||x^* - x(x_0, t)|| \leq \varepsilon \tag{2.2}$$
for all $t \geq 0$.
- **attractive** if there exists δ with the property that
$$||x^* - x_0|| \leq \delta \Rightarrow \lim_{t\to\infty} ||x^* - x(x_0, t)|| = 0. \tag{2.3}$$
- **asymptotically stable** (in the sense of Lyapunov) if it is both stable and attractive.
- **exponentially stable** if there exist $\delta, \alpha > 0$ and $\beta \geq 1$ such that
$$||x^* - x_0|| \leq \delta \Rightarrow ||x^* - x(x_0, t)|| \leq \beta e^{-\alpha t} ||x_0|| \tag{2.4}$$
for all $t \geq 0$.
- **unstable** if it is not stable in the sense of Lyapunov.

The *region of attraction* of an equilibrium point is defined as the set of initial states x_0 for which we have $x(x_0, t) \rightarrow x^*$ as t goes to infinity. If this region of attraction is the whole space, e.g. $\mathbb{R}^n$, then we say that the equilibrium point x^* is *globally attracting*. If the equilibrium point x^* is, furthermore, *globally stable*, then it is *globally asymptotically stable*. The definition of *global exponential stability* follows from the same idea.

It is important to mention that attractivity does not imply asymptotic stability. An equilibrium point can, indeed, be attractive but the trajectories of the system may not remain close to the equilibrium. For instance, the equilibrium point $x^* = 0$ of the nonlinear system

$$\begin{aligned}\dot{x}_1 &= x_1^2 - x_2^2\\ \dot{x}_2 &= 2x_1x_2\end{aligned} \tag{2.5}$$

is attractive but not asymptotically stable.

2.2.1 General Stability and Instability Results

Unlike LTI systems, stability of general systems cannot be inferred by looking at the explicit solutions since they are, most of the time, difficult or even impossible to compute. In the case of LPV systems, the solutions are even infinitely many, i.e. one solution per parameter trajectory. Additionally, the spectrum of the matrix $A(\rho)$ alone cannot be generally used to conclude on stability since time-variations must be taken into account. Lyapunov Theory allows us to overcome this difficulty by implicitly characterizing stability from the expression of the dynamical system, i.e. the matrix $A(\rho)$, through the use of a *Lyapunov function*.

Theorem 2.2.2 (Lyapunov's Stability Theorem [1, 2]) *Let us consider the general dynamical system*[a]

$$\begin{aligned}\dot{x}(t) &= f(x(t))\\ x(0) &= x_0\end{aligned} \tag{2.6}$$

having $x^* = 0$ as equilibrium point, i.e. $f(x^*) = 0$. Let $D \in \mathbb{R}^n$ be a domain containing $x^* = 0$ and $V : D \rightarrow \mathbb{R}$ be a continuously differentiable function such that

$$V(0) = 0 \text{ and } V(x) > 0 \text{ in } D - \{0\}, \tag{2.7}$$

$$\dot{V}(x) \leq 0 \text{ in } D. \tag{2.8}$$

Then, $x^* = 0$ *is a stable equilibrium point and* V *is called a Lyapunov function for* (2.6). *Moreover, if*

$$\dot{V}(x) < 0 \text{ in } D - \{0\}. \tag{2.9}$$

then $x^* = 0$ *is an asymptotically stable equilibrium.*

[a] It is tacitly assumed that the function f satisfies conditions for the existence of solutions for (2.6) for all $t \geq 0$.

Proof The proof is omitted but can be found in many textbooks on nonlinear systems such as [2]. ■

The above theorem states a local stability result only. When global stability is of interest, the following theorem should be considered instead:

Theorem 2.2.3 (Barbashin-Krasovskii Theorem [2, 3]) *Let* $x^* = 0$ *be an equilibrium point for* (2.6) *and let* $V : \mathbb{R}^n \to \mathbb{R}$ *be a continuously differentiable function such that*

1. $V(0) = 0$ *and* $V(x) > 0$ *for all* $x \neq 0$
2. $||x|| \to \infty \Rightarrow V(x) \to \infty$
3. $\dot{V}(x) < 0$ *for all* $x \neq 0$

then the equilibrium point $x^* = 0$ *is globally asymptotically stable.*

It is important to mention that, as stated, the above results are sufficient only. Converse theorems arise thus naturally. Converse results are very insightful since they provide information on the structure of the Lyapunov function to consider. Given a nonlinear dynamical system, finding a corresponding Lyapunov function can indeed be very fastidious [4]. The following converse result is due to Kurzweil:

Theorem 2.2.4 (Converse Lyapunov Theorem [5]) *Let* f *in* (2.6) *be a continuous function. If the system* (2.6) *is globally asymptotically stable at the origin, then there exists an infinitely differentiable Lyapunov function.*

For completeness, it is also important to mention that instability results also exist. The following one is due to Chetaev:

Theorem 2.2.5 (Chetaev's instability theorem[6–8]) *Let us consider system* (2.6). *Let* $D \in \mathbb{R}^n$ *be a domain containing* $x^* = 0$ *and* $V : D \to \mathbb{R}$ *be a* *continuously differentiable function and Ω be a subset containing $x^* = 0$, i.e. $0 \in D \cap \Omega$. If*

1. *$V(x) > 0$ and $\dot{V}(x) > 0$ for all $x \neq 0$ in D, and*
2. *$V(x) = 0$ for all x on the boundary of Ω*

then the system is unstable about the equilibrium $x^* = 0$.

2.2.2 The LTI System Case

Whenever LTI systems are considered, things turn to be much nicer since necessary and sufficient Lyapunov conditions for stability can be easily stated, as shown in the following result:

Theorem 2.2.6 *Let us consider the following n-dimensional LTI system*

$$\begin{aligned} \dot{x}(t) &= Ax(t),\ t \geq 0 \\ x(0) &= x_0. \end{aligned} \tag{2.10}$$

The following statements are equivalent:

1. *The system (2.10) is globally asymptotically stable.*
2. *The system (2.10) is globally exponentially stable.*
3. *The matrix A is Hurwitz, i.e. $\lambda(A) \subset \mathbb{C}_-$.*
4. *There exist matrices $P, Q \in \mathbb{S}^n_{\succ 0}$ such that the* ***Lyapunov equation***

$$A^T P + PA + Q = 0 \tag{2.11}$$

holds.

5. *There exists a matrix $P \in \mathbb{S}^n_{\succ 0}$ such that the* ***Lyapunov inequality***

$$A^T P + PA \prec 0 \tag{2.12}$$

holds.

Proof The equivalence of asymptotic stability and exponential stability is immediate in the LTI case, for instance by looking at the explicit solution $x(t) = e^{At}x_0$. The equivalence between 4. and 5. is also immediate. We give some short proofs for the other statements.

Proof of 4 ⇒ 1: Suppose (2.11) holds for some $P, Q \in \mathbb{S}^n_{\succ 0}$. Then defining $V(x) = x^{\mathrm{T}}Px$, $P \in \mathbb{S}^n_{\succ 0}$, we have that

$$\lambda_{min}(P)||x||_2^2 \leq V(x) \leq \lambda_{max}(P)||x||_2^2$$

and

$$\dot{V}(x) = x^{\mathrm{T}}(A^{\mathrm{T}}P + PA)x = -x^{\mathrm{T}}Qx \leq -\lambda_{min}(Q)||x||_2^2.$$

From Theorem 2.2.3, we can conclude on the global asymptotic stability of the system.

Proof of 3 ⇒ 4: We show here that by assuming $\Re[\lambda(A)] < 0$, it is possible to construct an explicit solution $P \in \mathbb{S}^n_{\succ 0}$ to Eq. (2.11) for any given $Q \in \mathbb{S}^n_{\succ 0}$. To find the solution, first pre- and post-multiply (2.11) by $e^{A^{\mathrm{T}}s}$ and e^{As}, respectively, to get

$$\frac{d}{ds}[e^{A^{\mathrm{T}}s}Pe^{As}] + e^{A^{\mathrm{T}}s}Qe^{As} = 0. \tag{2.13}$$

Since A has eigenvalues with negative real part, the integration of (2.13) from 0 to ∞ is well-defined and by doing so we finally get

$$P = \int_0^{\infty} e^{A^{\mathrm{T}}s}Qe^{As}ds. \tag{2.14}$$

This proves that under the assumption that the system is asymptotically stable, then for any given $Q \in \mathbb{S}^n_{\succ 0}$, there exists a matrix $P \in \mathbb{S}^n_{\succ 0}$ such that (2.11) holds.

Proof of 4 ⇒ 3: Let e_i be an eigenvector associated with the eigenvalue λ_i of the matrix A, for $i = 1, \ldots, p \leq n$. Pre and post-multiply (2.11) by e_i^* and e_i, we get

$$\begin{aligned} e_i^*\left(A^{\mathrm{T}}P + PA\right)e_i + e_i^*Qe_i &= 0 \\ 2\Re[\lambda_i]e_i^*Pe_i + e_i^*Qe_i &= 0 \end{aligned}$$

and thus

$$\Re[\lambda_i] < -\frac{e_i^*Qe_i}{2e_i^*Pe_i} < 0, \; i = 1, \ldots, p \tag{2.15}$$

since $e_i^*Qe_i > 0$ and $e_i^*Pe_i > 0$. The proof is complete. ■

In the theorem above, the inequality (2.12) is referred to as a *Linear Matrix Inequality* (LMI), where the inequality sign has to be understood as an inequality on the eigenvalues of the matrix on the left-hand side. Checking whether an LMI is feasible, i.e. admits a solution, is a convex feasibility problem for which efficient numerical tools exist [9, 10]. For more facts and results about LMIs, see Appendix B and references therein.

Below is an application of Theorem 2.2.6 in the LMI framework:

Example 2.2.7 Let us consider the LTI system (2.10) with matrix A given by

$$A = \begin{bmatrix} -1 & 0 \\ 1 & -1 \end{bmatrix}. \tag{2.16}$$

According to Theorem 2.2.6, a necessary and sufficient condition for asymptotic stability is the existence of a Lyapunov matrix $P \in \mathbb{S}^n_{\succ 0}$ such that the LMI (2.12) holds. Defining the matrix P as

$$P = \begin{bmatrix} p_1 & p_2 \\ p_2 & p_3 \end{bmatrix} \tag{2.17}$$

the LMI (2.12) then reads

$$A^{\mathrm{T}}P + PA = \begin{bmatrix} 2(p_2 - p_1) & -2p_2 + p_3 \\ \star & -2p_3 \end{bmatrix} \prec 0.$$

These LMI conditions are equivalent to the nonlinear inequalities[a]:

$$\begin{aligned} P \succ 0 &\Leftrightarrow \begin{cases} p_1 > 0 \\ p_1 p_3 - p_2^2 > 0 \end{cases} \\ A^{\mathrm{T}}P + PA \prec 0 &\Leftrightarrow \begin{cases} p_2 - p_1 < 0 \\ -4p_3(p_2 - p_1) - (p_3 - 2p_2)^2 < 0. \end{cases} \end{aligned} \tag{2.18}$$

A suitable choice is given, for instance, by $P = \begin{bmatrix} 3 & 2 \\ 2 & 2 \end{bmatrix} \succ 0$ and, in such a case, we have $A^{\mathrm{T}}P + PA = \begin{bmatrix} -2 & -2 \\ -2 & -4 \end{bmatrix} \prec 0$.

[a] We use here the fact that a matrix is negative (positive) definite if and only if all its principal minors are negative (positive).

2.3 Stability Notions for LPV and Uncertain Systems

Let us consider now linear systems with either time-varying or time-invariant parameters. Such systems can be generically represented as

$$\begin{aligned} \dot{x}(t) &= A(\rho(t))x(t), \ t \geq 0, \\ x(0) &= x_0 \end{aligned} \tag{2.19}$$

where $x \in \mathbb{R}^n$ is the state of the system and the parametric uncertainty vector $\rho(t)$ takes values in the compact set $\boldsymbol{\Delta}_\rho \subset \mathbb{R}^N$ where N is the number of parameters. Opposed to unperturbed LTI systems (i.e. not affected by uncertainties), different types of stability can be defined for uncertain and LPV systems. The most common ones are referred to as *quadratic stability* and *robust stability*. But, before defining them, it seems necessary to adapt the stability definitions in Definition 2.2.1. Note, however, that since LPV systems are linear systems, then all the stability properties are global. The term "global" is therefore implicitly meant in the definitions below.

Definition 2.3.1 Let us consider the LPV system

$$\begin{aligned} \dot{x}(t) &= A(\rho(t))x(t),\ t \geq 0 \\ x(0) &= x_0 \end{aligned} \tag{2.20}$$

where $\rho \in \mathscr{P}$ is the set of parameter trajectories, and let us denote by $x(x_0, \rho, t)$ the solution of this dynamical system given $\rho \in \mathscr{P}$ and $x_0 \in \mathbb{R}^n$. The system (the zero equilibrium point) is said to be

- **stable** if, for each $\varepsilon > 0$, there is $\delta = \delta(\varepsilon) > 0$ such that

$$||x_0|| \leq \delta \Rightarrow ||x(x_0, \rho, t)|| \leq \varepsilon \tag{2.21}$$

  for all $t \geq 0$ and all $\rho \in \mathscr{P}$.
- **attractive** if there exists δ with the property that

$$||x_0|| \leq \delta \Rightarrow \lim_{t\to\infty} ||x(x_0, \rho, t)|| = 0 \tag{2.22}$$

  for all $\rho \in \mathscr{P}$.
- **asymptotically stable** (in the sense of Lyapunov) if it is both stable and attractive.
- **exponentially stable** if there exist $\delta, \alpha > 0$ and $\beta \geq 1$ such that

$$||x_0|| \leq \delta \Rightarrow ||x(x_0, \rho, t)|| \leq \beta e^{-\alpha t}||x_0|| \tag{2.23}$$

  for all $t \geq 0$ and all $\rho \in \mathscr{P}$.
- **unstable** if it is not stable in the sense of Lyapunov.

2.3.1 *Quadratic Stability*

Quadratic stability is a straightforward extension of Theorem 2.2.6 since the same Lyapunov function is considered. This type of stability does not make any distinction between time-invariant parameters, slowly-varying parameters and parameters that vary arbitrarily fast. Therefore, quadratic stability may be very conservative.

Definition 2.3.2 (*Quadratic Stability*) System (2.19) is said to be quadratically stable if the positive definite quadratic form

$$V_q(x) = x^{\mathrm{T}} P_0 x, \quad P_0 \in \mathbb{S}^n_{\succ 0} \tag{2.24}$$

is a Lyapunov function for (2.19). Such a Lyapunov function is often referred to as a **common Lyapunov function** or a **parameter-independent Lyapunov function**.

Quadratic stability is only sufficient for asymptotic stability of an uncertain or LPV system. It is indeed possible to find systems that are asymptotically stable but not quadratically stable, as shown below.

Example 2.3.3 Let us consider the uncertain system (2.19) with matrix

$$A(\rho) = \begin{bmatrix} 1 & \rho \\ -4/\rho & -3 \end{bmatrix}$$

and time-invariant parameter $\rho \in [-1, -1/2] \cup [1/2, 1]$. The characteristic polynomial of this system given by

$$\det(sI - A(\rho)) = s^2 + 2s + 1$$

shows that it is asymptotically stable for all ρ in the uncertainty domain. We will show now that this system is not quadratically stable using a contradiction argument; see also [11] for a similar proof on a more complex system. To do so, let us assume that the system is quadratically stable. Then, there must exist a matrix $P = \begin{bmatrix} p_1 & p_2 \\ p_2 & p_3 \end{bmatrix} \succ 0$ such that the LMI

$$\begin{aligned} M_q(\rho) &:= A(\rho)^{\mathrm{T}} P + P A(\rho) \prec 0 \\ &= \begin{bmatrix} 2p_1 - \dfrac{8p_2}{\rho} & p_1\rho - 2p_2 - \dfrac{4p_3}{\rho} \\ \star & 2p_2\rho - 6p_3 \end{bmatrix} \prec 0 \end{aligned}$$

holds for all $\rho \in [-1, -1/2] \cup [1/2, 1]$. Therefore, for any $\rho_0 \in [1/2, 1]$ we have both $M_q(-\rho_0) \prec 0$ and $M_q(\rho_0) \prec 0$, and thus $M_q(-\rho_0) + M_q(\rho_0) \prec 0$. Computing this sum explicitly yields

$$\begin{aligned}M_q(-\rho_0) + M_q(\rho_0) &= [A(-\rho_0) + A(\rho_0)]^{\mathrm{T}} P + P[A(-\rho_0) + A(\rho_0)] \\ &= \begin{bmatrix} 4p_1 & -4p_2 \\ \star & -12p_3 \end{bmatrix}\end{aligned}$$

which cannot be negative definite due to the positive term in left-upper block; a contradiction. Consequently, the system is not quadratically stable.

It also seems important to relate the concept of quadratic stability to the spectrum of the matrix $A(\rho)$. It turns out that frozen stability[1] of $A(\rho)$ for all $\rho \in \boldsymbol{\Delta}_\rho$ is necessary for quadratic stability, regardless of the time-varying nature of the parameters.

Proposition 2.3.4 *If the system* (2.19) *is quadratically stable then the spectrum of* $A(\rho)$ *is contained in the open left-half plane for all* $\rho \in \boldsymbol{\Delta}_\rho$.

Proof The proof is similar to the one of Theorem 2.2.6. ∎

The converse is not true, as proved above in Example 2.3.3.

2.3.2 Robust Stability

To palliate some of the deficiencies of parameter-independent Lyapunov functions, such as the one illustrated in Example 2.3.3, a natural idea is to make the Lyapunov function parameter dependent. This leads us to the concept of robust stability. Unlike quadratic stability, this stability notion makes the distinction between constant and time-varying differentiable parameters. In the case of time-varying parameters, robust stability does indeed consider information on the rate of variation of the parameters.

Definition 2.3.5 (*Robust Stability*) System (2.19) is said to be robustly stable if the positive definite quadratic form

$$V_r(x, \rho) = x^{\mathrm{T}} P(\rho) x, \ \ P(\rho) \succ 0, \ \ \rho \in \boldsymbol{\Delta}_\rho \tag{2.25}$$

is a Lyapunov function for (2.19). Such a Lyapunov function is often referred to as a **parameter-dependent Lyapunov function**.

From the definitions above, it is clear that quadratic stability implies robust stability since quadratic stability is a particular case of robust stability where $P(\rho) = P_0$. The converse does not hold in general since it is easy to construct systems that are robustly stable, but not quadratically stable.

[1] Frozen stability of an LPV system with matrix $A(\rho)$ is defined as the stability of the matrix $A(\rho)$ for any fixed $\rho \in \boldsymbol{\Delta}_\rho$.

Example 2.3.6 Taking back the system of Example 2.3.3 and considering now a parameter-dependent quadratic form (2.25) with matrix

$$P(\rho) = P_0 + P_1\rho + P_2\rho^2 = \begin{bmatrix} p_1(\rho) & p_2(\rho) \\ \star & p_3(\rho) \end{bmatrix},$$

it is possible to show that

$$M_r(\rho_0) + M_r(-\rho_0) = \begin{bmatrix} 4(p_1^0 + p_1^1\rho^2) - 16p_2^1 & \star \\ \star & \star \end{bmatrix}$$

where $M_r(\rho) := A(\rho)^{\mathrm{T}}P(\rho) + P(\rho)A(\rho)$ and $p_i(\rho) := p_i^2\rho^2 + p_i^1\rho + p_i^0$. This LMI might be feasible since the (1,1) block may take negative values. To confirm this, we can solve numerically the stability conditions using some SDP solver to find the matrix function

$$P(\rho) = \begin{bmatrix} 50 + 6\rho^2 & 16\rho \\ \star & 1 + 7\rho^2 \end{bmatrix} \tag{2.26}$$

showing that the system is robustly stable. See Appendix B for more details on how to solve parameter dependent LMIs.

As for quadratic stability, it seems important to relate robust stability to spectrum properties of the matrix $A(\rho)$. We have the following results:

Proposition 2.3.7 *Assume that the uncertain parameter vector ρ is time-invariant. Then, the following statements are equivalent:*

1. *The system is robustly stable, i.e. there exists $P(\rho) \in \mathbb{S}^n_{\succ 0}$ for all $\rho \in \mathbf{\Delta}_\rho$ such that* (2.25) *is a Lyapunov function for system* (2.19).
2. *The spectrum of $A(\rho)$ in* (2.19) *is contained in the open left-half plane for all $\delta \in \mathbf{\Delta}_\rho$.*

Proof The proof is based on the straightforward extension of the proof of Theorem 2.2.6 to parameter dependent systems. Since the parameters are time-invariant, all the calculations still hold. ■

In the case of time-varying parameters, the picture is quite different. The rate of variation of the parameters indeed plays now a very important role.

Proposition 2.3.8 *Assume that the uncertain parameter vector ρ is time-varying. We have the following statements*

1. *Assume that the system is robustly stable, then the spectrum of $A(\rho)$ is bounded away from the imaginary axis for all $\rho \in \boldsymbol{\Delta}_\rho$. Moreover, the faster the parameters are, the farther are the eigenvalues from the imaginary axis.*
2. *If the spectrum of $A(\rho)$ is contained in the open left-half plane for all $\rho \in \boldsymbol{\Delta}_\rho$, then the system is robustly stable provided that the rate of variation of the parameters is sufficiently small.*

Proof Let us consider the Lyapunov function $V_r(x, \rho) = x^{\mathrm{T}} P(\rho) x$ as defined in Definition 2.3.5. Differentiating the function along the trajectories solution of the system (2.19), we get the expression

$$\dot{V}_r = x^{\mathrm{T}} \left[A(\rho)^{\mathrm{T}} P(\rho) + P(\rho) A(\rho) + \sum_{i=1}^{N} \dot{\rho}_i \frac{\partial P(\rho)}{\partial \rho_i} \right] x \tag{2.27}$$

where $(\rho, \dot{\rho}) \in \boldsymbol{\Delta}_\rho \times \boldsymbol{\Delta}_\nu$.

Proof of statement 1. Assume that the system is robustly stable, then there exists $P(\rho)$ such that (2.27) is negative definite for all $(\rho, \dot{\rho}) \in \boldsymbol{\Delta}_\rho \times \boldsymbol{\Delta}_\nu$. Let $e_i(\rho)$ be the eigenvector associated with eigenvalue $\lambda_i(\rho)$ of $A(\rho)$. Then, we have

$$e_i(\rho)^* \left[A(\rho)^{\mathrm{T}} P(\rho) + P(\rho) A(\rho) + \sum_{k=1}^{N} \dot{\rho}_k \frac{\partial P(\rho)}{\partial \rho_k} \right] e_i(\rho) < 0 \tag{2.28}$$

for all $(\rho, \dot{\rho}) \in \boldsymbol{\Delta}_\rho \times \boldsymbol{\Delta}_\nu$. This implies that

$$\Re[\lambda_i(\rho)] < -\sum_{k=1}^{N} \dot{\rho}_k \frac{y_i^k(\rho)}{2 z_i(\rho)} \tag{2.29}$$

where $z_i(\rho) := e_i(\rho)^* P(\rho) e_i(\rho) > 0$ and $y_i^k(\rho) = e_i(\rho)^* \dfrac{\partial P(\rho)}{\partial \rho_k} e_i(\rho)$. Since the parameters ρ are assumed to evolve non-monotonically, parameter derivatives can then take both positive and negative values over time. Therefore, the right-hand side of (2.29) is worst-case negative, pushing the eigenvalues of $A(\rho)$ away from the imaginary axis. Additionally, the faster the parameters are, the farther are the eigenvalues from the imaginary axis.

Proof of statement 2. Assume now that the system (2.19) is frozen stable. Therefore, for any $Q(\rho) \succ 0$, there exists $P(\rho)$ such that

$$A(\rho)^{\mathrm{T}} P(\rho) + P(\rho) A(\rho) = -Q(\rho). \tag{2.30}$$

Plugging this inside (2.27), we get that the system is robustly stable if

$$-Q(\rho) + \sum_{i=1}^{N} \dot{\rho}_i \frac{\partial P(\rho)}{\partial \rho_i} \prec 0 \tag{2.31}$$

for all $(\rho, \dot{\rho}) \in \boldsymbol{\Delta}_\rho \times \boldsymbol{\Delta}_\nu$. From the above expression, it is clear that when the parameters derivatives take too large values, the inequality is violated. Since this is true for any $Q(\rho)$, this shows that when the system is frozen stable, then it is also robustly stable provided that the parameters evolve sufficiently slowly. The proof is complete. ■

2.3.3 *Stability with Brief Instabilities*

In all the stability concepts and Lyapunov functions described above, it is most of the time assumed that the frozen LPV system (2.19) is asymptotically stable. This assumption has been actually relaxed in [12] where parameters are allowed to wander in the closed right-half plane for sufficiently short periods of time. In the same vein, a switched system approach is considered in [13] in order to analyze stability of LPV systems that are not frozen stable. In the latter case, notion of dwell-times arising from the analysis of hybrid systems [14], such as switched [15–17] or impulsive systems [18–20], are used.

2.3.4 *Other Types of Lyapunov Functions*

For completeness, it seems important to quickly mention other types of Lyapunov functions that can be used for analyzing uncertain and LPV systems.

2.3.4.1 Piecewise Quadratic Lyapunov Functions

An alternative to quadratic stability for systems having arbitrarily fast varying parameters relies on the use of piecewise (or composite) quadratic Lyapunov functions defined as [21–23]:

$$V(x) = \max_{i=1,\dots,m} \{x^{\mathrm{T}} P_i x\}$$

where $P_i \in \mathbb{S}^n_{\succ 0}$, $i = 1, \dots, m$. As shown in [21], piecewise quadratic Lyapunov functions improve over parameter-independent quadratic Lyapunov functions thanks to the use of multiple matrices P_i. The resulting conditions are, however, non convex and may be difficult to solve, especially when m is large.

2.3.4.2 Polyhedral Lyapunov Functions

Quadratic Lyapunov functions are intimately related to the 2-norm of the state vector. Polyhedral Lyapunov functions are, on the other hand, related to the ∞-norm of the state-vector as

$$V(x) = ||Q^{\mathrm{T}}x||_{\infty} \tag{2.32}$$

where $Q \in \mathbb{R}^{n\times m}$ is a full row rank matrix. Such Lyapunov functions have been shown to provide necessary and sufficient conditions for the stability of differential inclusions [24–28]. Note that differential inclusions can be used to represent, with some degree of accuracy, LPV and uncertain systems with arbitrarily fast-varying parameters.

2.3.4.3 Homogeneous Lyapunov Functions

Homogeneous Lyapunov functions are extensions of quadratic Lyapunov functions to higher order homogeneous polynomials with even degree. Such Lyapunov functions write

$$V(y) = y^{\mathrm{T}}Py \tag{2.33}$$

where $y \in \mathbb{R}^{d(n,m)}$ contains all the monomials of degree m, $P \in \mathbb{S}^{d(n,m)}_{\succ 0}$ and

$$d(n,m) = \frac{(n+m-1)!}{(n-1)!m!}.$$

When dealing with such Lyapunov functions, the system that has to be considered now is given by [29]:

$$\dot{y} = A^{\#}(\rho)y \tag{2.34}$$

where $A^{\#}(\rho)$ is the extended matrix of $A(\rho)$ in (2.19) given by

$$A^{\#}(\rho) := (K_m^{\mathrm{T}}K_m)^{-1}K_m^{\mathrm{T}}\left(\sum_{i=0}^{m-1} I_{n^{m-1-i}} \otimes A(\rho) \otimes I_{n^i}\right)K_m \tag{2.35}$$

where $K_m \in \mathbb{R}^{n^m \times d(n,m)}$ verifies $K_m y = \underbrace{x \otimes \ldots \otimes x}$. The symbol $\otimes$ denotes the Kronecker product.

Homogeneous Lyapunov functions have been shown to be very efficient for characterizing stability of linear uncertain systems, but not only; see e.g. [30–32] and references therein. The main drawback is the exponential increase of the computational complexity as the degree of the homogeneous polynomial increases.

2.4 Stability of Generic LPV Systems

We will consider, in this section, generic LPV systems of the form

$$\begin{aligned} \dot{x}(t) &= A(\rho(t))x(t) \\ x(0) &= x_0 \end{aligned} \tag{2.36}$$

$$\begin{bmatrix} I_n & 0 \\ 0 & 0 \end{bmatrix} \begin{bmatrix} \dot{x}(t) \\ \dot{y}(t) \end{bmatrix} = \bar{A}(\rho(t)) \begin{bmatrix} x(t) \\ y(t) \end{bmatrix} \\ x(0) = x_0 \tag{2.37}$$

where $x \in \mathbb{R}^n$, $y \in \mathbb{R}^{n_y}$ and $\rho \in \boldsymbol{\Delta}_\rho \subset \mathbb{R}^N$. The matrices of the above systems are assumed to be polynomial and to satisfy the assumptions stated in Sect. 1.3.1.

2.4.1 *Quadratic Stability*

2.4.1.1 General LPV Representation

Let us consider first the system (2.36). We have the following result:

Theorem 2.4.1 *The system* (2.36) *is quadratically stable if and only if there exists a matrix* $P \in \mathbb{S}^n_{\succ 0}$ *such that the LMI*

$$A(\rho)^T P + PA(\rho) \prec 0 \tag{2.38}$$

holds for all $\rho \in \boldsymbol{\Delta}_\rho$.

Proof The proof is an application of Definition 2.3.2 using the same developments as in the proof of Theorem 2.2.6. ■

The LMI (2.38) for quadratic stability is technically called a *semi-infinite dimensional LMI* due to the dependence on parameters. Whereas the LMI (2.12) for LTI systems defines the LMI constraint set

$$\mathcal{F} := \left\{ P \in \mathbb{S}^n_{\succ 0} : \; A^{\mathrm{T}} P + PA \prec 0 \right\}, \tag{2.39}$$

the LMI (2.38) actually defines infinitely many constraints sets:

$$\mathcal{F}_\rho := \left\{ P \in \mathbb{S}^n_{\succ 0} : \; A(\rho)^{\mathrm{T}} P + PA(\rho) \prec 0 \right\}, \; \rho \in \boldsymbol{\Delta}_\rho. \tag{2.40}$$

In the latter case, the overall feasibility problem is taken on the intersection of all the sets

$$\bar{\mathcal{F}} := \bigcap_{\rho \in \boldsymbol{\Delta}_\rho} \mathcal{F}_\rho. \tag{2.41}$$

Checking the feasibility of (2.38), or equivalently the non-emptiness of $\bar{\mathcal{F}}$, is, in general, a difficult problem which is, to date, still an active field of research, often referred to as *robust optimization*. Ways for solving such LMI problems are discussed in Appendix B.

In the special case where condition (2.38) is affine in ρ, an equivalent finite-dimensional representation to (2.38) exists:

Theorem 2.4.2 *Assume that $A(\rho)$ is affine in ρ. Then, the following statements are equivalent:*

1. *There exists $P \in \mathbb{S}^n_{\succ 0}$ such that the LMI*

$$A(\rho)^T P + PA(\rho) \prec 0 \tag{2.42}$$

holds for all $\rho \in \boldsymbol{\Delta}_\rho$.

2. *There exists $P \in \mathbb{S}^n_{\succ 0}$ such that the LMIs*

$$A(v)^T P + PA(v) \prec 0 \tag{2.43}$$

hold for all $v \in \mathbf{V}_\rho$.

Proof The proof exploits the convexity of the polytope $\boldsymbol{\Delta}_\rho$, from which we can state that for any $\rho \in \boldsymbol{\Delta}_\rho$, there exists $\lambda \in \Lambda_{2^N}$ such that

$$\rho = \sum_{i=1}^{2^N} \lambda_i v_i, \quad v_i \in \mathbf{V}_\rho. \tag{2.44}$$

Proof of 2 ⇒ 1: Assume that the LMIs (2.43) hold for all $v \in \mathbf{V}_\rho$. Then, considering $A(v_i)^\mathrm{T} P + PA(v_i)$, $v_i \in \mathbf{V}_\rho$, multiplying it by λ_i and summing over i yields

$$\sum_{i=1}^{2^N} \lambda_i \left[A(v_i)^\mathrm{T} P + PA(v_i) \right]. \tag{2.45}$$

Since by assumption the LMIs (2.43) hold for all $v \in \mathbf{V}_\rho$ and using the facts that (1) a sum of negative definite matrices is negative definite; and (2) for any $\rho \in \boldsymbol{\Delta}_\rho$, there is $\lambda \in \Lambda_{2^N}$ such that (2.44) holds, then we can conclude that (2.45) is negative definite for all $\lambda \in \Lambda_{2^N}$. Therefore, (2.42) holds for all $\rho \in \boldsymbol{\Delta}_\rho$.

Proof of 1 ⇒ 2: Assume that the LMI (2.42) holds for all $\rho \in \boldsymbol{\Delta}_\rho$, then it must also hold on the vertices of $\boldsymbol{\Delta}_\rho$, and therefore for all $v \in \mathbf{V}_\rho$. ■

The interest of this result lies in the fact that the equivalent conditions (2.43) are in *finite number*. We have thus been able to equivalently transform an infinite number of constraints into a finite number. Note, however, that this number grows exponentially in terms of the number of distinct parameters N since the number of vertices of $\boldsymbol{\Delta}_\rho$ is an exponential function of the dimension of $\boldsymbol{\Delta}_\rho$.

The use of the matrix cube theorem, see [33, 34] or Theorem B.3.2 in Appendix B.3.1, allows us to obtain a sufficient condition that is more appealing when the number of parameter is large. This yields the following result:

Theorem 2.4.3 *Assume that $A(\rho)$ is affine in ρ and can decomposed as*

$$A(\rho) = A_0 + \sum_{i=1}^{N} \rho_i A_i$$

and let the parameter vector take value in $[-1, 1]^N$. Assume further that there exist matrices $P \in \mathbb{S}^n_{\succ 0}$, $X_i \in \mathbb{S}^n$, $i = 1, \ldots, N$ such that the LMIs

$$-X_i \pm \left(A_i^T P + PA_i\right) \preceq 0 \tag{2.46}$$

and

$$A_0^T P + PA_0 + \sum_{i=1}^{N} X_i \prec 0 \tag{2.47}$$

hold. Then, the LMI (2.42) holds for all $\rho \in [-1, 1]^N$ and the corresponding system is quadratically stable.

Proof The proof is a simple application of the matrix cube theorem; see [33, 34] or Theorem B.3.2 in Appendix B.3.1. ■

2.4.1.2 Descriptor LPV Representation

When the system (2.37) is considered, the following result on quadratic stability holds:

Theorem 2.4.4 *The descriptor LPV system* (2.37) *is quadratically stable if and only if there exist* $P_1 \in \mathbb{S}^n_{\succ 0}$, $P_2 : \boldsymbol{\Delta}_\rho \to \mathbb{R}^{n_y \times n}$ *and* $P_3 : \boldsymbol{\Delta}_\rho \to \mathbb{R}^{n_y \times n_y}$ *such that the LMI*

$$\bar{A}(\rho)^T P(\rho) + P(\rho)^T \bar{A}(\rho) \prec 0 \tag{2.48}$$

holds for all $\rho \in \boldsymbol{\Delta}_\rho$ *where* $P(\rho) = \begin{bmatrix} P_1 & 0 \\ P_2(\rho) & P_3(\rho) \end{bmatrix}$.

Proof The proof is based on the use of the Lyapunov function

$$V(x, y) = \begin{bmatrix} x \\ y \end{bmatrix}^{\mathrm{T}} P(\rho)^{\mathrm{T}} \begin{bmatrix} I_n & 0 \\ 0 & 0 \end{bmatrix} \begin{bmatrix} x \\ y \end{bmatrix} \tag{2.49}$$

with satisfies the additional constraints $P(\rho)^{\mathrm{T}} \bar{E} = \bar{E}^{\mathrm{T}} P(\rho)$, $P_1 \succ 0$. The chosen structure for $P(\rho)$ fulfills these necessary constraints. Differentiating (2.49) yields the result. ■

Note that, in the result above, only the matrix P_1 needs to be positive definite since only the first state evolves according to a differential equation. The second state is indeed fully characterized by an algebraic equation and, therefore, the matrices P_2 and P_3 are arbitrary, i.e. no specific structure nor property. Note, moreover, that only P_1 needs to be parameter independent since this is the only matrix that is differentiated.

Note also that the LMI (2.48) involves the decision variable $P(\rho)$ which is a function of ρ. The feasibility problem involved in Theorem 2.4.4 is therefore meaningfully called an *infinite-dimensional feasibility problem*. This type of feasibility problem cannot be solved as such and must be approximated, we talk about *relaxation procedures*. Some relaxation schemes are discussed in Appendix B.

The result stated in Theorem 2.4.4 is illustrated in the example below:

Example 2.4.5 Let us consider back the system (1.22) of Example 1.3.1 on page 11. Quadratic stability can be assessed if there exists a constant P having the block-triangular structure defined in Theorem 2.4.4 and such that the LMIs

$$\begin{aligned} \bar{A}(-1)^{\mathrm{T}} P + P^{\mathrm{T}} \bar{A}(-1) \prec 0 \\ \bar{A}(1)^{\mathrm{T}} P + P^{\mathrm{T}} \bar{A}(1) \prec 0 \end{aligned} \tag{2.50}$$

hold. It turns out that the matrix P given by

$$P = \left[\begin{array}{c|ccc} 4 & 0 & 0 & 0 \\ \hline -5 & -7 & -1 & 1 \\ -6 & -1 & -7 & -1 \\ -4 & 1 & -1 & -7 \end{array}\right] \tag{2.51}$$

satisfies the LMIs (2.50), showing then that the system of Example 1.3.1 is quadratically stable.

2.4.2 Robust Stability

Let us consider first the system (2.36). We then have the following result:

Theorem 2.4.6 *The system* (2.36) *is robustly stable if there exists a differentiable matrix function* $P : \boldsymbol{\Delta}_\rho \to \mathbb{S}^n_{\succ 0}$ *such that the condition*

$$A(\rho)^T P(\rho) + P(\rho)A(\rho) + \sum_{i=1}^{N} \nu_i \frac{\partial P(\rho)}{\partial \rho_i} \prec 0 \tag{2.52}$$

holds for all $(\rho, \nu) \in \boldsymbol{\Delta}_\rho \times \mathbf{V}_\nu$.

Proof The proof is an application of Theorem 2.2.3 using the Lyapunov function of robust stability introduced in Definition 2.3.5. After differentiation, we get the LMI condition

$$A(\rho)^{\mathrm{T}} P(\rho) + P(\rho)A(\rho) + \sum_{i=1}^{N} \dot{\rho}_i \frac{\partial P(\rho)}{\partial \rho_i} \prec 0.$$

Noting that the dependence of the above inequality on $\dot{\rho}_i$ is affine, and that the values taken by ρ and $\dot{\rho}$ are (almost everywhere) independent, then the above LMI can be viewed as affine in $\dot{\rho}$. The same argument as in Theorem 2.4.2 can be applied and yields the result. ■

The following result is an extension of Theorem 2.4.4 to robust stability:

Theorem 2.4.7 *The descriptor LPV system* (2.37) *is robustly stable if there exist* $P_1 : \boldsymbol{\Delta}_\rho \to \mathbb{S}^n_{\succ 0}$, $P_2 : \boldsymbol{\Delta}_\rho \to \mathbb{R}^{n_y \times n}$ *and* $P_3 : \boldsymbol{\Delta}_\rho \to \mathbb{R}^{n_y \times n_y}$ *such that the LMI*

$$\tilde{A}(\rho)^T P(\rho) + P(\rho)^T \tilde{A}(\rho) + \sum_{i=1}^{N} \begin{bmatrix} \nu_i \dfrac{\partial P_1}{\partial \rho_i}(\rho) & 0 \\ 0 & 0 \end{bmatrix} \prec 0 \tag{2.53}$$

holds for all $(\rho, \nu) \in \boldsymbol{\Delta}_\rho \times \mathbf{V}_\nu$ *where* $P(\rho) = \begin{bmatrix} P_1(\rho) & 0 \\ P_2(\rho) & P_3(\rho) \end{bmatrix}$.

2.5 Stability of Polytopic LPV Systems

Polytopic LPV systems can be analyzed in the same way as general LPV systems discussed in the latter section. This is due to the fact that polytopic systems are affine LPV systems with parameters in the unit simplex. This property makes polytopic systems very convenient to work with.

This section will then be concerned with the analysis of the polytopic LPV system

$$\begin{aligned} \dot{x}(t) &= A(\lambda(t))x(t) \\ x(0) &= x_0 \end{aligned} \tag{2.54}$$

where $x \in \mathbb{R}^n$ is the system state, $A(\lambda) = \sum_{i=1}^{N} \lambda_i A_i$ and $\lambda \in \Lambda_N$.

2.5.1 *Quadratic Stability*

A necessary and sufficient condition for the quadratic stability of polytopic LPV systems is given below:

Theorem 2.5.1 *The polytopic LPV system* (2.54) *is quadratically stable in the sense of Definition 2.3.2 if and only if there exists a matrix* $P \in \mathbb{S}^n_{\succ 0}$ *such that the LMIs*

$$A_i^T P + P A_i \prec 0 \tag{2.55}$$

hold for all $i = 1, \ldots, N$.

Proof Using the Lyapunov function of Definition 2.3.2, we obtain the stability condition

$$\sum_{i=1}^{N} \lambda_i(t) \left[A_i^T P + P A_i \right] \prec 0 \tag{2.56}$$

where $\lambda(t) \in \Lambda_N$ for all $t \geq 0$.

Sufficiency: Assume that $A_i^{\mathrm{T}} P + P A_i \prec 0$ for all $i = 1, \ldots, N$. Then, following Theorem 2.4.2, we can conclude that (2.56) holds for all $\lambda(t) \in \Lambda_N$.

Necessity: Assume now that (2.56) holds for all $\lambda(t) \in \Lambda_N$, then necessarily the LMI must hold on the vertices of the set Λ_N, that is for any vector in $V = \mathbf{vert}\{\Lambda_N\}$, as defined in (1.25). ■

As in the case of affine LPV systems, a semi-infinite LMI condition has been turned into a finite set of LMIs. As before, quadratic stability results are limited in the sense that a common matrix P has to be found for all the subsystems A_i. The existence of a common Lyapunov function such that (2.55) holds for all $i = 1, \ldots, N$ is a difficult problem for which partial answers exist. An obvious necessary condition is that the matrices A_i all be Hurwitz. The following elegant result is taken from [35]:

Theorem 2.5.2 *Assume that there exists a nonsingular matrix $T \in \mathbb{R}^{n \times n}$ such that the matrices $T A_i T^{-1}$ are upper-triangular. Then, the following statements are equivalent:*

1. *The matrices A_i, $i = 1, \ldots, N$, are Hurwitz.*
2. *There exists a matrix $P \in \mathbb{S}^n_{\succ 0}$ such that the LMIs (2.55) hold for all $i = 1, \ldots, N$.*

Proof The proof that 2. ⇒ 1. follows from Theorem 2.2.6. The converse statement, however, is much more involved and needs Lie algebraic tools that are beyond the scope of this book. Interested readers should refer to [35] to get more insight on the proof. ■

As a corollary of the above result, when the Hurwitz matrices A_i commute with each other, there also exists a common quadratic Lyapunov function. More results of similar flavor can be found in [36–40] and references therein.

2.5.2 Robust Stability

For LPV systems, robust stability captures information on the rate of variation of the parameters. In the polytopic case, this rate of variation is contained in the term $\dot{\lambda}$. A difficulty here is that there is no immediate definition for a set that will contain the trajectories of $\dot{\lambda}$. To overcome this, we will assume here that the considered polytopic system approximates an LPV system with N_p parameters, denoted by ρ, for which derivative bounds are known. In this case, it is possible to define a polytope in which $\dot{\lambda}$ evolves within. This is stated below:

Proposition 2.5.3 *Assume that* $\dot{\rho} \in \boldsymbol{\Delta}_\nu = co\{\mathbf{V}_\nu\}$, $\mathbf{V}_\nu = \{d_1, \ldots, d_N\}$, $N = 2^{N_p}$ *and that the decomposition*

$$\rho(t) = \sum_{i=1}^{N} \lambda_i(t) v_i$$

holds with $\mathbf{V}_\rho = \{v_1, \ldots, v_N\}$. *Then, the set of all* $\dot{\lambda}$*'s is given by*

$$\dot{\Lambda}_N := \left\{ \begin{bmatrix} V \\ \mathbb{1}_N^T \\ 0 \end{bmatrix}^+ \begin{bmatrix} D \\ 0 \\ \mathbb{1}_N^T \end{bmatrix} \zeta : \eta \in \Lambda_N \right\} \tag{2.57}$$

where $\zeta(t) \in \Lambda_N$, $V = \begin{bmatrix} v_1 \ldots v_N \end{bmatrix}$ *and* $D = \begin{bmatrix} d_1 \ldots d_N \end{bmatrix}$. *Moreover, we have the identity*

$$\dot{\rho}(t) = \sum_{i=1}^{N} \zeta_i(t) d_i = \sum_{i=1}^{N} \dot{\lambda}_i(t) v_i. \tag{2.58}$$

Proof By differentiating $\rho(t) = \sum_{i=1}^{N} \lambda_i(t) v_i$ we get

$$\dot{\rho}(t) = \sum_{i=1}^{N} \dot{\lambda}_i(t) v_i.$$

The idea is now to look for a polytope that contains $\dot{\lambda}$. To this aim, we therefore impose that

$$\dot{\rho}(t) = \sum_{i=1}^{N} \zeta_i(t) d_i \tag{2.59}$$

where the d_i's are the vertices of the set $\boldsymbol{\Delta}_\nu$, i.e. the elements of $\mathbf{V}_\nu$. Equating these expressions, we get

$$\sum_{i=1}^{N} \dot{\lambda}_i(t) v_i = \sum_{i=1}^{N} \zeta_i(t) d_i$$

or, in a more compact form

$$V \dot{\lambda}(t) = D \zeta(t) \tag{2.60}$$

where $V = \begin{bmatrix} v_1 \ldots v_N \end{bmatrix}$ and $D = \begin{bmatrix} d_1 \ldots d_N \end{bmatrix}$. In addition to this equality, we have to also consider the relationships

$$\textstyle\sum_{i=1}^{N} \zeta_i(t) = 1 \quad \text{and} \quad \sum_{i=1}^{N} \dot{\lambda}_i(t) = 0.$$

All combined together, we obtain

$$\begin{bmatrix} V \\ \mathbb{1}_N^T \\ 0 \end{bmatrix} \dot{\lambda}(t) = \begin{bmatrix} D \\ 0 \\ \mathbb{1}_N^T \end{bmatrix} \zeta(t) - \begin{bmatrix} 0 \\ 0 \\ 1 \end{bmatrix}.$$

The remaining step is to solve the above equation for $\dot{\lambda}(t)$. Let us rewrite the above equation as $FX = E$ where $X = \dot{\lambda}$. It is known that there is a solution if and only if $E \in \text{span}[F]$. This is equivalent to the condition $\chi^{\mathrm{T}} E = 0$ with $F^{\mathrm{T}} \chi = 0$ and $\chi \neq 0$. It is immediate to observe that $\begin{bmatrix} V \\ \mathbb{1}_N^T \end{bmatrix}$ is full-row rank. Then, setting $\chi_0^{\mathrm{T}} = \begin{bmatrix} 0 & 0 & 1 \end{bmatrix}$ implies that $F^{\mathrm{T}} \chi_0 = 0$ and $\chi_0^{\mathrm{T}} E = 0$, showing the existence of a solution to the equation $FX = E$. The complete set of solutions is given by

$$\dot{\lambda}(t) = \begin{bmatrix} V \\ \mathbb{1}_N^T \\ 0 \end{bmatrix}^+ \left(\begin{bmatrix} D \\ 0 \\ \mathbb{1}_N^T \end{bmatrix} \zeta(t) - \begin{bmatrix} 0 \\ 0 \\ 1 \end{bmatrix} \right) + \left(I - \begin{bmatrix} V \\ \mathbb{1}_N^T \\ 0 \end{bmatrix}^+ \begin{bmatrix} V \\ \mathbb{1}_N^T \\ 0 \end{bmatrix} \right) z(t) \quad (2.61)$$

where $z(t) \in \mathbb{R}^N$ is an arbitrary vector and $\bar{V}^+$ is the Moore-Penrose pseudoinverse of $\bar{V}$. Since we only need a particular solution, we can set $z(t) \equiv 0$ without loss of generality, and the result finally follows from the fact that

$$\begin{bmatrix} V \\ \mathbb{1}_N^T \\ 0 \end{bmatrix}^+ \begin{bmatrix} 0 \\ 0 \\ 1 \end{bmatrix} = 0.$$

The proof is complete. ■

Example 2.5.4 Consider a two-parameter problem, i.e. $N_p = 2$, where

$$(\rho_1, \rho_2) \in [-1, 1] \times [-2, 3] \quad \text{and} \quad (\dot{\rho}_1, \dot{\rho}_2) \in [-2, 3] \times [-5, 6].$$

We then have the following matrices

$$V = \begin{bmatrix} -1 & -1 & 1 & 1 \\ -2 & 3 & -2 & 3 \end{bmatrix} \quad \text{and} \quad D = \begin{bmatrix} -2 & -2 & 3 & 3 \\ -5 & 6 & -5 & 6 \end{bmatrix}.$$

From Proposition 2.5.3, we have

$$\dot{\lambda}(t) = \begin{bmatrix} 1 & -0.1 & -0.25 & -1.35 \\ 0 & 1.1 & -1.25 & -0.15 \\ 0 & -1.1 & 1.25 & 0.15 \\ -1 & 0.1 & 0.25 & 1.35 \end{bmatrix} \zeta(t)$$

and it is easy to see that $\sum_{i=1}^{4} \dot{\lambda}_i(t) = 0$ for all $t \geq 0$ and that

$$\dot{\rho}(t) = V\dot{\lambda}(t) = D\zeta(t) \tag{2.62}$$

as desired.

Based on the above characterization of $\dot{\Lambda}_N$, we can state the following result:

Theorem 2.5.5 *The polytopic LPV system* (2.54) *is robustly stable if there exist matrices* $P_i \in \mathbb{S}^n_{\succ 0}$, $i = 1, \ldots, N$, *such that the parameter-dependent LMI*

$$A(\lambda)^T P(\lambda) + P(\lambda)A(\lambda) + P(\theta) \prec 0 \tag{2.63}$$

holds for all $(\lambda, \theta) \in \Lambda_N \times \textbf{vert}\{\dot{\Lambda}_N\}$ *where*

$$P(\lambda) = \sum_{i=1}^{N} \lambda_i P_i \textit{ and } A(\lambda) = \sum_{i=1}^{N} \lambda_i A_i.$$

Proof The proof is based on the parameter dependent Lyapunov function

$$V(x, \lambda) = x^{\mathrm{T}} P(\lambda)x$$

where $P(\lambda) = \sum_{i=1}^{N} \lambda_i P_i$, $P_i \in \mathbb{S}^n_{\succ 0}$, $i = 1, \ldots, N$. Positive definiteness of this quadratic form is equivalent to the constraints $P_i \succ 0$, $i = 1, \ldots, N$. The derivative of V along the trajectories solutions of system (2.54) is negative definite if and only if the LMI

$$A(\lambda)^{\mathrm{T}} P(\lambda) + P(\lambda)A(\lambda) + P(\dot{\lambda}) \prec 0 \tag{2.64}$$

holds for all $(\lambda, \dot{\lambda}) \in \Lambda_N \times \dot{\Lambda}_N$. The result immediately follows. ■

The main difficulty arising in the consideration of the LMI (2.63) lies in the fact that it is not affine in λ. Clearly, convexity is lost and, therefore, we cannot simply check the condition at the vertices of Λ_N as in the affine case. Some results have been obtained in the literature in order to deal directly with conditions that are quadratic or homogeneous in λ; see for instance [41–43]. Some other methods are also presented in Appendix B.

Below, an alternative result, relying on the use of *slack-variables*, is considered instead:

Theorem 2.5.6 *The polytopic LPV system (2.54) is robustly stable if there exist matrices $P_i \in \mathbb{S}^n_{>0}$, $i = 1, \ldots, N$, a matrix $X \in \mathbb{R}^{n\times n}$ and a sufficiently large scalar $\xi > 0$ such that the matrix inequalities*

$$\begin{bmatrix} -(X+X^T) & P_i + X^T A_i & X^T \\ \star & -\xi P_i + \sum_{j=1}^N P_j\theta_j & 0 \\ \star & \star & -P_i/\xi \end{bmatrix} \prec 0 \tag{2.65}$$

hold for all $i = 1, \ldots, N$ and all $\theta \in \mathbf{vert}\{\dot{\Lambda}_N\}$.

Proof The proof is made in two steps: the first step consists of showing that the feasibility of a certain matrix inequality implies the feasibility of (2.63). The second step uses convexity arguments to yield a finite set of LMIs.

Step 1: Let us consider the matrix inequality

$$\begin{bmatrix} -(X+X^{\mathrm{T}}) & P(\lambda) + X^{\mathrm{T}} A(\lambda) & X^{\mathrm{T}} \\ \star & -\xi P(\lambda) + P(\dot{\lambda}) & 0 \\ \star & \star & -P(\lambda)/\xi \end{bmatrix} \prec 0 \tag{2.66}$$

where $X \in \mathbb{R}^{n\times n}$ and $\xi > 0$. The goal is to show that the feasibility of (2.66) implies the feasibility of (2.63). To do so, first rewrite (2.66) as

$$\underbrace{\begin{bmatrix} 0 & P(\lambda) & 0 \\ \star & -\xi P(\lambda) + P(\dot{\lambda}) & 0 \\ \star & \star & -P(\lambda)/\xi \end{bmatrix}}_{\Psi} + \mathrm{He}\left(\begin{bmatrix} I \\ 0 \\ 0 \end{bmatrix} X^{\mathrm{T}} \begin{bmatrix} -I & A(\lambda) & I \end{bmatrix}\right) \prec 0.$$

The matrix X can be eliminated from (2.66) by projection (see the Projection Lemma in Appendix C.12) as follows. Let

$$U_1(\lambda) := \begin{bmatrix} -I & A(\lambda) & I \end{bmatrix}, \; U_2 := \begin{bmatrix} I & 0 & 0 \end{bmatrix},$$

and define the matrices $K_1(\lambda)$, K_2 to be bases of the null-spaces of $U_1(\lambda)$ and U_2, respectively. Their explicit expression is given by

$$K_1(\lambda) = \begin{bmatrix} A(\lambda) & I \\ I & 0 \\ 0 & I \end{bmatrix} \text{ and } K_2 = \begin{bmatrix} 0 & 0 \\ I & 0 \\ 0 & I \end{bmatrix}.$$

Projecting inequality (2.66) w.r.t. these null-spaces yields

$$\begin{aligned}\Psi_1 &:= K_1^{\mathrm{T}} \Psi K_1 \\ &= K_1^{\mathrm{T}} \begin{bmatrix} 0 & P(\lambda) & 0 \\ \star & -\xi P(\lambda) + P(\dot{\lambda}) & 0 \\ \star & \star & -P(\lambda)/\xi \end{bmatrix} K_1 \\ &= \begin{bmatrix} A(\lambda)^{\mathrm{T}} P(\lambda) + P(\lambda)A(\lambda) + \xi P(\lambda) + P(\dot{\lambda}) & P(\lambda) \\ \star & -P(\lambda)/\xi \end{bmatrix}\end{aligned}$$

and

$$\begin{aligned}\Psi_2 &:= K_2^{\mathrm{T}} \Psi K_2 \\ &= K_2^{\mathrm{T}} \begin{bmatrix} 0 & P(\lambda) & 0 \\ \star & -\xi P(\lambda) + P(\dot{\lambda}) & 0 \\ \star & \star & -P(\lambda)/\xi \end{bmatrix} K_2 \\ &= \begin{bmatrix} -\xi P(\lambda) + P(\dot{\lambda}) & 0 \\ \star & -P(\lambda)/\xi \end{bmatrix}.\end{aligned}$$

We have the immediate implication that if (2.66) holds, then both Ψ_1 and Ψ_2 are negative definite. A Schur complement on the inequality $\Psi_1 \prec 0$ gives

$$A(\lambda)^{\mathrm{T}} P(\lambda) + P(\lambda)A(\lambda) + P(\dot{\lambda}) \prec 0$$

which is identical to (2.64). The condition that $\Psi_2 \prec 0$ is clearly necessary to the feasibility of (2.66). Thus, we have proved that the feasibility of the matrix inequality (2.66) implies the feasibility of the LMI (2.64).

Step 2: Noting that the matrix inequality (2.66) is affine in λ and $\dot{\lambda}$, we can therefore convert the infinite set of LMIs into a finite set, and the result follows. ■

It is worth noting that condition (2.65) is not an LMI condition due to the presence of the decision variable $\xi > 0$. However, when ξ is fixed, the condition actually becomes an LMI. Finding a suitable $\xi > 0$ can be easily be performed using a line-search algorithm that keeps increasing the value of ξ until the problem becomes feasible, or stops when the value of ξ goes beyond a certain threshold value.

2.6 Stability of LPV Systems in LFT-Form

Let us consider in this section LPV systems in LFT-form given by

$$\begin{aligned}\dot{x}(t) &= Ax(t) + Bw(t) \\ z(t) &= Cx(t) + Dw(t) \\ w(t) &= \Theta(\rho(t))z(t).\end{aligned} \tag{2.67}$$

Above, the matrix $\Theta(\rho)$ is a, possibly structured, matrix containing the time-varying parameters. The LTI part can be viewed as a linear operator G mapping w to z having transfer function $\widehat{G}(s)$ defined as

$$\widehat{G}(s) := C(sI - A)^{-1}B + D. \tag{2.68}$$

The analysis of such systems can be carried out using many different techniques. First of all, the notions of *L_2-norm*, *L_2-gain* and *H_∞-norm* are introduced as a necessary background for stating the first important result that may be used to assess the stability of the system (2.67): the so-called *Small-Gain Theorem*. Using then the concepts of *scalings*, this result is refined into the *Scaled Small-Gain Theorem* and the *Dynamic Scaled Small-Gain Theorem*. Finally, results based on the *full-block S-procedure*, the concept of *topological separation* and *Integral Quadratic Constraints* are presented, together with some connections between them.

2.6.1 L_2-Norm and H_∞-Norm

Before stating the main results of this section, it is important to define the notions of L_2-norm, L_2-gain and H_∞-norm.

Definition 2.6.1 (*L_2-norm and L_2-space*) Let $w : \mathbb{R}_{\geq 0} \to \mathbb{C}^n$. The L_2-norm of w is defined as

$$||w||_{L_2} := \sqrt{\int_0^\infty w(s)^* w(s) ds}. \tag{2.69}$$

The space of signals mapping $\mathbb{R}_{\geq 0}$ to $\mathbb{C}^n$ with finite L_2-norm is denoted by $L_2(\mathbb{R}_{\geq 0}, \mathbb{C}^n)$. For simplicity, we use the shorthand L_2 for denoting any L_2-space.

Definition 2.6.2 (*L_2-gain*) Let Σ be a bounded[a] operator from L_2 to L_2. Its L_2-gain, denoted by $||\Sigma||_{L_2-L_2}$, is defined as

$$||\Sigma||_{L_2-L_2} = \sup_{||w||_{L_2}=1} \left\{ ||\Sigma w||_{L_2} \right\}.$$

[a] An operator from L_2 to L_2 is bounded if and only if $||\Sigma w||_{L_2}$ is finite for all $w \in L_2$.

Note that the L_2-gain is defined for any type of operators, e.g. linear, nonlinear, time-varying, etc.

Definition 2.6.3 (*H_∞-norm*) Let $\widehat{G} : \mathbb{C} \to \mathbb{C}^{m \times p}$ be a proper transfer function that is analytic in the closed right-half plane. The H_∞-norm of this transfer function, denoted by $||\widehat{G}||_{H_\infty}$, is given by

$$\begin{aligned} ||\widehat{G}||_{H_\infty} &= \sup_{s \in \bar{\mathbb{C}}_+} \bar{\sigma}(\widehat{G}(s)) \\ &= \sup_{\omega \in \mathbb{R}} \bar{\sigma}(\widehat{G}(j\omega)) \end{aligned}$$

where $\bar{\sigma}(\cdot)$ denotes the maximal singular-value; see Appendix A.3.

An important fact is that for an LTI operator G with transfer function $\widehat{G}(s)$, the L_2-gain of G coincides with the H_∞-norm of $\widehat{G}$, that is we have

$$||G||_{L_2 - L_2} = ||\widehat{G}||_{H_\infty}.$$

This leads to the following result:

Proposition 2.6.4 *The following statements are equivalent:*

1. *The LTI system G is asymptotically stable.*
2. *The system G has finite L_2-gain.*
3. *The transfer function $\widehat{G}$ has finite H_∞-norm.*

Proof The proof follows from the definition of the H_∞-norm. ■

Various methods can be applied to compute the H_∞-norm of a stable, rational and proper transfer function: bisection algorithms [44], Hamiltonian-matrix-based algorithms [45], Riccati equations and LMIs [46, 47]. Since LMIs are the most suitable tools for dealing with LPV systems, only the following LMI-based result is provided:

Lemma 2.6.5 (Bounded-Real Lemma [47, 48]**)** *Let us consider an LTI system with transfer function*

$$\widehat{G}(s) = C(sI - A)^{-1}B + D.$$

Then, the following statements are equivalent:

1. *The H_∞-norm of $\widehat{G}$ is smaller than $\gamma > 0$.*
2. *There exists $P \in \mathbb{S}^n_{\succ 0}$ such that the LMI*

$$\begin{bmatrix} A^T P + PA & PB & C^T \\ \star & -\gamma I & D^T \\ \star & \star & -\gamma I \end{bmatrix} \prec 0 \tag{2.70}$$

holds.

Proof Several proofs for this very important result exist. The proof based on the Kalman-Yakubovich-Popov Lemma (see Appendix C.1) is probably the shortest one. Another important one is based on dissipativity theory[2] and plays an important role in computing estimates for the H_∞-norm of time-delay systems, for instance.

A Schur complement on the LMI (2.70) yields

$$\begin{bmatrix} A^{\mathrm{T}}P + PA & PB \\ \star & 0 \end{bmatrix} + \begin{bmatrix} C & D \\ 0 & I \end{bmatrix}^{\mathrm{T}} \begin{bmatrix} \gamma^{-1}I & 0 \\ 0 & -\gamma I \end{bmatrix} \begin{bmatrix} C & D \\ 0 & I \end{bmatrix} \prec 0 \tag{2.71}$$

which is exactly of the form of the condition involved in the Kalman-Yakubovich-Popov Lemma. Note that since the conditions $P \succ 0$ and $A^{\mathrm{T}}P + PA \prec 0$ are equivalent to the stability of the matrix A, the matrix $j\omega I - A$ is therefore invertible for all $\omega \in \mathbb{R}$. This shows that the above LMI is equivalent to saying that

$$\begin{bmatrix} \widehat{G}(j\omega) \\ I \end{bmatrix}^{*} \begin{bmatrix} \gamma^{-1}I & 0 \\ 0 & -\gamma I \end{bmatrix} \begin{bmatrix} \widehat{G}(j\omega) \\ I \end{bmatrix} \prec 0 \tag{2.72}$$

holds for all $\omega \in \mathbb{R}$. Expanding the above inequality yields $\widehat{G}(j\omega)^{*}\widehat{G}(j\omega) \prec \gamma^{2}I$, or equivalently, $||\widehat{G}||_{H_\infty} < \gamma$. ■

The following proposition shows that the H_∞-norm of a system can be computed by solving a convex semidefinite problem (i.e. an LMI problem):

Proposition 2.6.6 *Assume that the LTI system with transfer function*

$$\widehat{G}(s) = C(sI - A)^{-1}B + D$$

has finite H_∞-norm given by γ^*. Then, we also have that

$$\begin{aligned} \gamma^* = \min_{P \in \mathbb{S}^n_{\succ 0}, \gamma > 0} \ & \gamma \\ \textit{s.t.} \ & (2.70) \textit{ holds.} \end{aligned} \tag{2.73}$$

Proof This is a simple implication of Lemma 2.6.5. ■

2.6.2 Small-Gain Theorem

Let us now consider the system interconnection depicted in Fig. 2.1 where $M(s)$ is a known transfer function and $\Delta(s)$ is an uncertain transfer function verifying

[2] For more details on dissipativity theory, see [49].

$||\Delta||_{H_\infty} \leq 1$. The transfer function (assumed to be causal) of the overall interconnection is given by

$$M_\Delta(s) = \frac{M(s)}{1 + M(s)\Delta(s)}.$$

The analysis of the stability of the interconnection, using for instance, the multivariable Nyquist criterion, is not easy due to the presence of the uncertain transfer function $\Delta(s)$. A convenient way for tackling this problem relies on the concepts of norms and gains previously defined. This leads us to the small-gain theorem:

Theorem 2.6.7 (Small-Gain Theorem [44, 50]) *Assume that* $||M||_{H_\infty} < 1$ *and* $||\Delta||_{H_\infty} \leq 1$, *then the interconnection of Fig. 2.1 is asymptotically stable.*

Proof Assuming that $||M\Delta||_{H_\infty} < 1$, then we have

$$||M_\Delta||_{H_\infty} \leq \frac{||M||_{H_\infty}}{1 - ||M\Delta||_{H_\infty}} < \infty. \tag{2.74}$$

Therefore, the closed-loop system is bounded in the H_∞-norm sense, which is equivalent to asymptotic stability. Using the submultiplicativity property[3] of the H_∞-norm, we have

$$\begin{aligned} ||M\Delta||_{H_\infty} &\leq ||M||_{H_\infty} ||\Delta||_{H_\infty} \\ &\leq ||M||_{H_\infty} \end{aligned} \tag{2.75}$$

where we have used the fact that $||\Delta||_{H_\infty} \leq 1$. Consequently, under the assumption that $||M||_{H_\infty} < 1$, we have $||M_\Delta||_{H_\infty} < \infty$ and the closed-loop system is asymptotically stable for any $||\Delta||_{H_\infty} \leq 1$. The proof is complete. ∎

In general, the small-gain is only a (strong) sufficient condition for stability, but turns out to be also necessary when $\Delta(s)$ is any unstructured uncertain transfer function matrix with H_∞-norm smaller or equal than 1. This is stated in the following proposition:

Proposition 2.6.8 ([44]) *The following statements are equivalent:*

1. *The interconnection of Fig. 2.1 is stable for all* $\Delta(s)$ *verifying* $||\Delta||_{H_\infty} \leq 1$.
2. *The transfer function* $M(s)$ *is such that* $||M||_{H_\infty} < 1$.

Proof The proof can be found in [44]. ∎

When the transfer functions $M(s)$ and $\Delta(s)$ are fixed, it is easy to construct them such that the interconnection is stable but the small-gain condition $||M\Delta||_{H_\infty} < 1$

---

[3] Any induced-norm satisfies the submultiplicativity property.

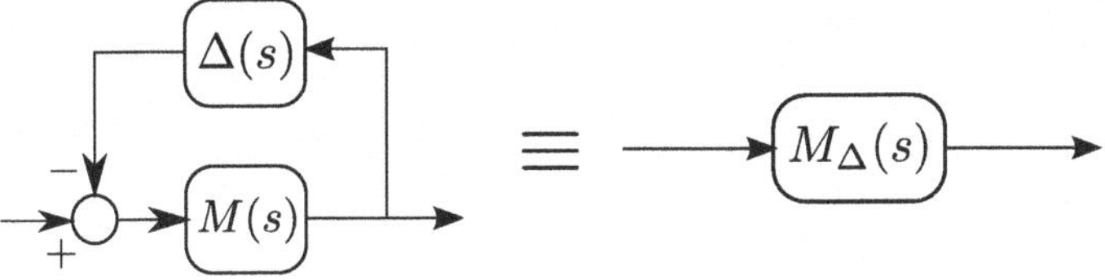

Fig. 2.1 Interconnection of two single-input single-output transfer functions

is violated. The following transfer functions

$$M(s) = \frac{10}{(s+1)(s+2)} \text{ and } \Delta(s) = \frac{10}{(s+3)(s+4)}$$

illustrate this fact.

As previously said, it is improper to talk about H_∞-norm when time-varying systems are considered since no transfer function in the Laplace domain can be defined for such systems. We therefore consider the L_2-gain instead. The small-gain argument then directly extends to LPV systems where the uncertain transfer function $\Delta(s)$ is now replaced by the parameter-varying matrix $\Theta(\rho)$, $\rho \in \boldsymbol{\Delta}_\rho$. Since the matrix depends on some parameters in a range of values, the worst-case L_2-gain must be considered.

To properly define it, let us consider the multiplication operator Θ_ρ defined as $\Theta_\rho(w)(t) = \Theta(\rho(t))w(t)$ for any signal $w \in L_2$ of appropriate dimensions.

Proposition 2.6.9 *The worst-case L_2-gain of the operator Θ_ρ is defined as*

$$\sup_{\rho \in \mathscr{P}} ||\Theta_\rho||_{L_2-L_2} \tag{2.76}$$

where $\mathscr{P}$ is the set of parameter trajectories and we have that

$$\sup_{\rho \in \mathscr{P}} ||\Theta_\rho||_{L_2-L_2} = \max_{\zeta \in \boldsymbol{\Delta}_\rho} ||\Theta(\zeta)||_2 \tag{2.77}$$

where $||\cdot||_2$ is the matrix induced 2-norm.

Proof Let $\eta := \Theta(\rho)\xi$, then we have

$$||\eta||_{L_2}^2 = \int_0^\infty \xi(s)^{\mathrm{T}} \Theta(\rho(s))^{\mathrm{T}} \Theta(\rho(s))\xi(s)ds$$

$$\leq \max_{\zeta \in \boldsymbol{\Delta}_\rho} \int_0^\infty \xi(s)^{\mathrm{T}} \Theta(\zeta)^{\mathrm{T}} \Theta(\zeta) \xi(s) ds$$

$$\leq \max_{\zeta \in \boldsymbol{\Delta}_\rho} \left\{ ||\Theta(\zeta)||_2^2 \right\} \int_0^\infty \xi(s)^{\mathrm{T}} \xi(s) ds$$

$$= \max_{\zeta \in \boldsymbol{\Delta}_\rho} \left\{ ||\Theta(\zeta)||_2^2 \right\} ||\xi||_{L_2}^2, \tag{2.78}$$

where we have used the fact that $\boldsymbol{\Delta}_\rho$ is compact. Note that equality holds for some constant parameter trajectories, i.e. those that maximize the 2-norm of $\Theta(\cdot)$. Taking finally the square-root yields the result. ■

Theorem 2.6.10 *Assume that* $||\Theta_\rho||_{L_2-L_2} \leq 1$ *for all* $\rho \in \mathscr{P}$, *then the LPV system* (2.67) *is asymptotically stable if*

$$||\widehat{G}||_{H_\infty} < 1. \tag{2.79}$$

The small-gain condition is a very simple stability test and is often very conservative in the LPV framework. Two main reasons for that:

- Unlike the Nyquist criterion, no information on the phase is taken into account, only gains are considered.
- No structural information on the interconnection is captured, e.g. $\Delta(s)$ or $\Theta(\rho)$ may have a specific structure, such as a block-diagonal structure, that can be exploited in the analysis conditions.

This leads us to the notion of scalings and the scaled small-gain theorem.

2.6.3 Constant D-Scalings and the Scaled-Small Gain Theorem

The rationale behind the scaled-small gain theorem is to improve the small-gain theorem by capturing information on the structure of the interconnection. In this section, we will assume that $\Theta(\rho)$ has the following structure

$$\Theta(\rho) = \mathrm{diag}(\rho_1 I_{\eta_1}, \ldots, \rho_p I_{\eta_p}) \tag{2.80}$$

where η_i is the number of occurrences of the parameter $\rho_i \in [-1, 1]$; hence $\Theta(\rho) \in \mathbb{R}^{\eta \times \eta}$, $\eta = \sum_{i=1}^p \eta_i$. Due to the structure of the matrix, it is easy to see that $||\Theta(\rho)||_2 \leq 1$ for all $\rho \in [-1, 1]^p$.

Definition 2.6.11 The set of D-scalings[a] associated with the matrix Θ is defined as

$$\mathcal{D}(\Theta) := \left\{ L \in \mathbb{S}^{\eta}_{\succ 0} : \Theta L^{1/2} = L^{1/2}\Theta \right\} \tag{2.81}$$

where $L^{1/2}$ denotes the positive definite symmetric square-root of L.

[a] Note that the definition of the set of D-scalings may slightly differ from one person to another. The definitions are nevertheless equivalent. For more details on the scalings the reader should refer to [51, 52].

According to the definition above, the role of the matrix L is to embed a structural information on the matrix Θ through a commutation property. We have the following properties for the set $\mathcal{D}(\Theta)$:

1. $\mathcal{D}(\Theta)$ is a convex subset of $\mathbb{S}^{\eta}_{\succ 0}$;
2. $I \in \mathcal{D}(\Theta)$;
3. $L \in \mathcal{D}(\Theta) \Longrightarrow L^{\mathrm{T}} \in \mathcal{D}(\Theta)$;
4. $L \in \mathcal{D}(\Theta) \Longrightarrow L^{-1} \in \mathcal{D}(\Theta)$;
5. $L_1, L_2 \in \mathcal{D}(\Theta) \Longrightarrow L_1 L_2 \Theta = \Theta L_1 L_2$.

By relying upon D-scalings, the small-gain theorem can be refined into the scaled small-gain theorem.

Lemma 2.6.12 (Scaled Bounded-Real Lemma [51–53]) *Assume that $L \in \mathcal{D}(\Theta)$. Then, the following statements are equivalent*

1. *The H_∞-norm of $\widehat{G}_L := L^{1/2}\widehat{G}L^{-1/2}$ is smaller than $\gamma > 0$.*
2. *There exists $P \in \mathbb{S}^{\eta}_{\succ 0}$ such that the LMI*

$$\begin{bmatrix} A^T P + PA & PBL & C^T L \\ \star & -\gamma L & D^T L \\ \star & \star & -\gamma L \end{bmatrix} \prec 0$$

holds.

Proof Since $L \in \mathcal{D}(\Theta)$, then we have $\Theta L^{1/2} = L^{1/2}\Theta$ and hence $\Theta = L^{-1/2}\Theta L^{1/2}$. Therefore, G can be equivalently substituted by $G_L := L^{1/2} G L^{-1/2}$ in the interconnection. Applying the Bounded-Real Lemma , i.e. Lemma 2.6.5, to the scaled-system G_L yields the LMI

$$\begin{bmatrix} A^{\mathrm{T}} P + PA & PBL^{-1/2} & C^{\mathrm{T}} L^{1/2} \\ \star & -\gamma I & L^{-1/2} D^{\mathrm{T}} L^{1/2} \\ \star & \star & -\gamma I \end{bmatrix} \prec 0.$$

A congruence transformation with respect to $\text{diag}(I, L^{1/2}, L^{1/2})$ finally gives the desired result. ■

It is immediate to see that the above result can be viewed as an extension of the bounded-real lemma. By indeed setting $L = I$, the bounded-real lemma is immediately retrieved. Based on the use of D-scalings and the scaled bounded real lemma, the scaled small-gain theorem can be stated:

Theorem 2.6.13 (Scaled Small-Gain Theorem) *Assume that $\Theta(\rho)$ has the structure* (2.80). *The LPV system* (2.67) *is asymptotically stable if there exists $P \in \mathbb{S}^n_{\succ 0}$ and $L \in \mathcal{D}(\Theta)$ such that the LMI*

$$\begin{bmatrix} A^TP + PA & PBL & C^TL \\ \star & -L & D^TL \\ \star & \star & -L \end{bmatrix} \prec 0$$

holds.

As for the small-gain theorem, a conservatism analysis can be carried out:

Proposition 2.6.14 (Conservatism – scaled small-gain theorem) *Assume that $\Theta(\rho)$ is block-diagonal and contains*

- *s repeated scalar blocks; and*
- *f unrepeated full blocks.*

Then, the scaled-small gain theorem is nonconservative if $2s + f \leq 3$. In all the other cases, the scaled-small gain theorem may be conservative.

Proof The proof is related to the exactness of the computation of the structured singular value using D-scalings, see [53]. ■

2.6.4 Frequency-Dependent D-Scalings

The extension of the scaled bounded-real lemma to frequency-dependent scalings is quite recent [54–58], even though frequency-dependent multipliers have been around for quite some time in IQC analysis [59] and multipliers theory [60]. Frequency-dependent D-scalings are expected to be less conservative than their constant counterparts due to their frequency-adapting characteristic. The goal of this section is simply to provide some introductory material on this approach which requires advanced techniques. Assume first that

$$\Theta = \text{diag}(\Theta_1(\rho), \ldots, \Theta_m(\rho))$$

where $||\Theta(\rho)||_2 \leq 1$ for all $\rho \in \boldsymbol{\Delta}_\rho$, and let $\mathcal{Q}$ be the set of matrices structured as

$$Q(s) = \text{diag}(q_1(s)I, \ldots, q_m(s)I) \tag{2.82}$$

in accordance with the structure of $\Theta(\rho)$. The components q_i are single-input single-output transfer functions that are real valued and bounded on the extended imaginary axis $j\mathbb{R}$. Stability of the LPV system is then guaranteed if there exists some multiplier $Q \in \mathcal{Q}$ for which

$$\begin{bmatrix} \widehat{G}(s) \\ I \end{bmatrix}^* \begin{bmatrix} Q(s) & 0 \\ 0 & -Q(s) \end{bmatrix} \begin{bmatrix} \widehat{G}(s) \\ I \end{bmatrix} \prec 0 \text{ and } Q(s) \succ 0 \text{ on } j\mathbb{R} \tag{2.83}$$

The key idea is to approximate any filter by a finite basis of elementary filters of the form

$$\begin{aligned} f_{1,\kappa}(s) &= \begin{bmatrix} 1 & f_1(s) & f_2(s) & \ldots & f_\kappa(s) \end{bmatrix} \\ f_{2,\kappa}(s) &= \begin{bmatrix} 1 & f_1(s)^* & f_2(s)^* & \ldots & f_\kappa(s)^* \end{bmatrix} \end{aligned}$$

where $f_{1,\kappa}(s)$ and $f_{2,\kappa}(s)$ are respectively stable and anti-stable[4] rows with $f(s)^* = f(-s)^{\mathrm{T}}$. Therefore, for any sufficiently large κ, any stable (anti-stable) filter can be uniformly approximated on $j\mathbb{R}$ by $f_{1,\kappa}(s)l_1$ ($f_{2,\kappa}(s)l_2$) using suitable real-valued columns vectors l_1 (l_2); see e.g. [61–63]. This implies that $Q(s)$ can be approximated by

$$\Psi_1(s)^* M \Psi_1(s) = \Psi_2(s)^* M \Psi_2(s) \tag{2.84}$$

where $\Psi_j := \text{diag}(I \otimes f_{j,\kappa}^{\mathrm{T}}, \ldots, I \otimes f_{j,\kappa}^{\mathrm{T}})$ and M is a real symmetric matrix such that $M := \text{diag}(I \otimes M_1, \ldots, I \otimes M_m)$ in which the M_i's are decision variables to be determined. Based on this characterization, we can now state the following result [54, 55]:

Theorem 2.6.15 *The following statements are equivalent:*

1. *The matrix A is Hurwitz and condition* (2.83) *holds for some Q of the form* (2.84).
2. *There exist real symmetric matrices X, Y and M of appropriate dimensions such that the LMIs*

$$\begin{bmatrix} I & 0 \\ A_p & B_p \\ C_p & D_p \end{bmatrix}^T \begin{bmatrix} 0 & X & 0 \\ \star & 0 & 0 \\ \star & \star & diag(M, -M) \end{bmatrix} \begin{bmatrix} I & 0 \\ A_p & B_p \\ C_p & D_p \end{bmatrix} \prec 0$$

[4] An anti-stable polynomial has all its roots in $\mathbb{C}_+$.

$$\begin{bmatrix} I & 0 \\ A_{\Psi_1} & B_{\Psi_1} \\ C_{\Psi_1} & D_{\Psi_1} \end{bmatrix}^T \begin{bmatrix} 0 & Y & 0 \\ \star & 0 & 0 \\ \star & \star & M \end{bmatrix} \begin{bmatrix} I & 0 \\ A_{\Psi_1} & B_{\Psi_1} \\ C_{\Psi_1} & D_{\Psi_1} \end{bmatrix} \succ 0$$

$$\begin{bmatrix} X_{11} - Y & X_{13} \\ \star & X_{33} \end{bmatrix} \succ 0$$

hold where $\left[\begin{array}{c|c} A_{\Psi_1} & B_{\Psi_1} \\ \hline C_{\Psi_1} & D_{\Psi_1} \end{array}\right]$ *is a minimal realization of* Ψ_1 *and*

$$\left[\begin{array}{c|c} A_p & B_p \\ \hline C_p & D_p \end{array}\right] := \left[\begin{array}{ccc|c} A_{\Psi_1} & 0 & B_{\Psi_1}C & D \\ 0 & A_{\Psi_2} & 0 & B_{\Psi_2} \\ 0 & 0 & A & B \\ \hline C_{\Psi_1} & 0 & D_{\Psi_1} & D_{\Psi_1}D \\ 0 & C_{\Psi_2} & 0 & D_{\Psi_2} \end{array}\right]$$

is a minimal realization of $\begin{bmatrix} \Psi_1 G \\ \Psi_2 \end{bmatrix}$. *Above the matrix* X *is partitioned according the matrix* A_p.

Proof The proof is quite long and technical. It is thus omitted here but can be found in [54–57]. ■

2.6.5 Full-Block S-Procedure

The full-block S-procedure is a quite general result that encompasses the small-gain and scaled small-gain results of the previous sections. The term *full-block* comes from the fact that the scalings involved are general matrices, as opposed to, for instance, block-diagonal scalings in the scaled small-gain theorem. The full-block S-procedure is therefore more general and is expected to cover a wider class of systems and uncertainties with a reduced conservatism. This framework has been successfully applied, among others, to the analysis of LPV systems [64–67], time-delay systems [68–70], interconnected systems [71], etc.

The results of this section can be obtained in many different ways. In the following, the results will be developed using a simple S-procedure argument, which is enough for proving an elementary analysis result for LPV systems.

To this aim, let us consider the system (2.67) where $\Theta(\rho) \in \mathbb{R}^{\eta\times\eta}$ is a possibly structured matrix. With the above facts in mind, we can state the following result:

Lemma 2.6.16 *The following statements are equivalent:*

1. *The system* (2.67) *is quadratically asymptotically stable for all* Θ *in the set*

$$\mathcal{F}_M := \left\{ \Theta : \begin{bmatrix} I \\ \Theta \end{bmatrix}^T M \begin{bmatrix} I \\ \Theta \end{bmatrix} \succeq 0 \right\} \tag{2.85}$$

for some given $M \in \mathbb{S}^{2\eta}$ *with* $M_{22} \prec 0$.

2. *There exist a matrix* $P \in \mathbb{S}^n_{\succ 0}$ *and a scalar* $\tau > 0$ *such that the LMI*

$$\begin{bmatrix} A^T P + PA & PB \\ \star & 0 \end{bmatrix} + \tau \begin{bmatrix} C & D \\ 0 & I \end{bmatrix}^T M \begin{bmatrix} C & D \\ 0 & I \end{bmatrix} \prec 0 \tag{2.86}$$

holds.

Proof A complete proof with meaningful discussions can be found in [65, 66]. We propose an alternative one based on the S-procedure. Consider first the quadratic form $V(x) = x^{\mathrm{T}} P x$, $P \in \mathbb{S}^n_{\succ 0}$, which will play the role of Lyapunov function. The key idea is to provide a decrease condition of the Lyapunov function for all $\Theta(\rho)$ in $\mathcal{F}_M$. Differentiating V along the trajectories solutions of the system, we get

$$\dot{V} = \begin{bmatrix} x \\ w \end{bmatrix}^{\mathrm{T}} \begin{bmatrix} A^{\mathrm{T}} P + PA & PB \\ \star & 0 \end{bmatrix} \begin{bmatrix} x \\ w \end{bmatrix} \tag{2.87}$$

where $w = \Theta(\rho)z$. The derivative must be negative definite for all pairs of signals (z, w) verifying

$$\begin{bmatrix} z \\ w \end{bmatrix}^{\mathrm{T}} M \begin{bmatrix} z \\ w \end{bmatrix} \geq 0 \tag{2.88}$$

where $z = Cx + Dw$. Invoking then the S-procedure, we get the condition (2.86). Using the fact that the S-procedure is lossless in the single quadratic constraint case, equivalence between the statements immediately follows. ■

Based on the above theorem, we obtain the following result:

Theorem 2.6.17 *The LPV system* (2.67) *is asymptotically stable if there exist matrices* $P \in \mathbb{S}^n_{\succ 0}$ *and* $M \in \mathbb{S}^{2\eta}$ *such that the LMIs*

$$\begin{bmatrix} A^TP + PA & PB \\ \star & 0 \end{bmatrix} + \begin{bmatrix} C & D \\ 0 & I \end{bmatrix}^T M \begin{bmatrix} C & D \\ 0 & I \end{bmatrix} \prec 0 \tag{2.89}$$

$$\begin{bmatrix} I \\ \Theta(\rho) \end{bmatrix}^T M \begin{bmatrix} I \\ \Theta(\rho) \end{bmatrix} \succeq 0 \tag{2.90}$$

hold for all $\rho \in \boldsymbol{\Delta}_\rho$.

Proof The proof is an application of Lemma 2.6.16. ■

The main difficulty in checking feasibility of the conditions in Theorem 2.6.17 lies in the presence of the semi-infinite condition (2.90). These type of robust feasibility problems can be solved using, for instance, sum of squares techniques as in [72, 73] where it is shown that the full-block S-procedure can be lossless in certain cases. The price to pay, however, is a high computational complexity.

The computational burden can be overcome, at the expense of conservatism, by considering a family $\mathcal{M}$ of matrices M for which the condition (2.90) is trivially satisfied for all matrices $\Theta(\rho)$, $\rho \in \boldsymbol{\Delta}_\rho$. Consequently, the semi-infinite dimensional LMI (2.90) can be removed from the problem, making it finite-dimensional and, therefore, more tractable.

The family of scalings $\mathcal{M}$ may differ according to the type of uncertain operators, say ∇, in some known class, say $\boldsymbol{\nabla}$. Before introducing popular scaling families, it is convenient to define the set $\mathcal{S}_{\boldsymbol{\nabla}}$ as the set of matrices that commute with all ∇'s in $\boldsymbol{\nabla}$, i.e.

$$\mathcal{S}_{\boldsymbol{\nabla}} := \left\{ M \in \mathbb{R}^{\eta\times\eta} : \nabla M = M\nabla,\ \nabla \in \boldsymbol{\nabla} \right\}.$$

2.6.5.1 Passive Uncertainties: Multipliers

When $\boldsymbol{\nabla}$ is the class of positive-real uncertainties, i.e.

$$\boldsymbol{\nabla} = \left\{ \nabla : \nabla + \nabla^* \geq 0 \right\} \subset \mathbb{C}^{\eta\times\eta} \tag{2.91}$$

then a suitable family $\mathcal{M}$ is given by

$$\mathcal{M} = \left\{ \begin{bmatrix} 0 & D \\ D & 0 \end{bmatrix} : D = D^* \succ 0,\ D \in \mathcal{S}_{\boldsymbol{\nabla}} \right\}. \tag{2.92}$$

2.6.5.2 Norm-Bounded Uncertainties: Constant D-Scalings

When $\boldsymbol{\nabla}$ is the class of norm-bounded uncertainties, i.e.

$$\boldsymbol{\nabla} = \left\{\nabla : \ ||\nabla||_{L_2-L_2} \leq 1\right\} \subset \mathbb{C}^{\eta\times\eta} \tag{2.93}$$

then a suitable family $\mathcal{M}$ is given by

$$\mathcal{M} = \left\{\begin{bmatrix} D & 0 \\ 0 & -D \end{bmatrix} : \ D = D^* \succ 0, \ D \in \mathcal{S}_{\nabla}\right\}. \tag{2.94}$$

2.6.5.3 Real Parametric Uncertainties

When $\boldsymbol{\nabla}$ is the class of norm-bounded real parametric uncertainties, i.e.

$$\boldsymbol{\nabla} = \left\{\mathrm{diag}(\delta_1 I_{\eta_1}, \ldots, \delta_p I_{\eta_m}) : \ \delta_i \in \mathbb{R}, \ |\delta_i| \leq 1\right\} \subset \mathbb{R}^{\eta\times\eta}$$

then suitable families $\mathcal{M}$ are given by

- D-G-scalings:

$$\mathcal{M} = \left\{\begin{bmatrix} D & G \\ G^* & -D \end{bmatrix} : \ D = D^* \succ 0, \ G + G^* = 0, \ D, G \in \mathcal{S}_{\nabla}\right\}. \tag{2.95}$$

- LFT-scalings:

$$\mathcal{M} = \left\{\begin{bmatrix} Q & S \\ S^{\mathrm{T}} & R \end{bmatrix} : \ R \prec 0, \ \begin{bmatrix} I_\eta \\ \nabla \end{bmatrix}^{\mathrm{T}} \begin{bmatrix} Q & S \\ S^{\mathrm{T}} & R \end{bmatrix} \begin{bmatrix} I_\eta \\ \nabla \end{bmatrix} \succeq 0, \ \nabla \in \mathbf{vert}\{\boldsymbol{\nabla}\}\right\}. \tag{2.96}$$

- Vertex separators:

$$\mathcal{M} = \left\{\begin{bmatrix} Q & S \\ S^{\mathrm{T}} & R \end{bmatrix} : \ R_i \preceq 0, \ \begin{bmatrix} I_\eta \\ \nabla \end{bmatrix}^{\mathrm{T}} \begin{bmatrix} Q & S \\ S^{\mathrm{T}} & R \end{bmatrix} \begin{bmatrix} I_\eta \\ \nabla \end{bmatrix} \succeq 0, \ \nabla \in \mathbf{vert}\{\boldsymbol{\nabla}\}\right\}. \tag{2.97}$$

  where the R_i's are the $\eta_i \times \eta_i$ diagonal blocks of R, i.e.

$$R = \begin{bmatrix} R_1 & * & \cdots & * \\ * & R_2 & \cdots & * \\ \vdots & & \ddots & \vdots \\ * & * & \cdots & R_m \end{bmatrix}.$$

We can clearly see that D-scalings are a particular case of D-G scalings, and are hence more conservative when bounded real parametric uncertainties are considered. The following result has been proved in [74]:

Proposition 2.6.18 *The D-G scaling is lossless if*

$$2(s+c)+f \leq 3 \tag{2.98}$$

where s, c and f are the numbers of repeated real scalar blocks, repeated complex scalar blocks and full complex blocks. When this condition does not hold, the D-G scalings may be conservative.

Regarding LFT-scalings, it is easily seen that they are less conservative than D-G scalings but more than vertex separators [75].

2.6.6 Topological and Quadratic Separation

Topological separation, introduced by M. Safonov in the seminal works [76, 77] is a fundamental result in the theory of interconnection of operators saying that[5]

> *...feedback system stability can be concluded if one can 'topologically' separate the infinite-dimensional function space containing the system's dynamical input-output relations into two regions, one region containing the dynamical input-output relation of the 'feedforward' element of the system and the other region containing the dynamical input-output relation of the 'feedback' element.*

The topological separation framework is very general and consists of a unified theory of stability where Lyapunov functions and contraction mappings (i.e. small-gain operators) are replaced by 'separating' functionals.

The idea of topological separation has been taken back later by T. Iwasaki in [75] where the results have been specialized to linear systems. In this case, separation can be made using *quadratic separators* leading to necessary and sufficient convex stability conditions. This framework has been later generalized, among others, to more general input-output relationships [78], LPV systems [79], robust and resilient control [80, 81], time-delay systems [82–85], sampled-data systems [86] and nonlinear systems [87].

2.6.6.1 Preliminary Results

Let us consider two finite-dimensional linear operators (matrices) $\Sigma_1 \in \mathbb{C}^{n_z \times n_w}$ and $\Sigma_2 \in \mathbb{C}^{n_w \times n_z}$ interconnected as in Fig. 2.2. The input and loop signals are related together via the system of linear equations

---

[5] Quoted from [76].

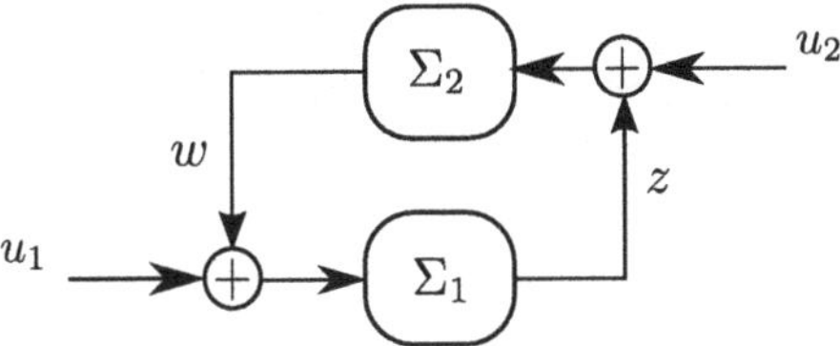

Fig. 2.2 General setup of the well-posedness framework

$$\begin{aligned} z &= \Sigma_1(w + u_1) \\ w &= \Sigma_2(z + u_2) \end{aligned} \tag{2.99}$$

where it is assumed that Σ_2 belongs to a compact family $\mathbf{\Sigma}_2$.

Before stating the key results, it is necessary to introduce the definition of well-posedness[6]:

Definition 2.6.19 (*Well-posedness of feedback systems*) The interconnection depicted in Fig. 2.2 is said to be well-posed if

1. For each pair $(u_1, u_2) \in \mathbb{C}^{n_w+n_z}$, there exists a unique pair $(z, w) \in \mathbb{C}^{n_z+n_w}$.
2. There exists $\gamma > 0$ such that

$$\left\| \begin{pmatrix} z \\ w \end{pmatrix} \right\|_2 \leq \gamma \left\| \begin{pmatrix} u_1 \\ u_2 \end{pmatrix} \right\|_2 \tag{2.100}$$

for all $(u, v) \in \mathbb{C}^{n_z+n_w}$.

As seen above, the well-posedness property concerns uniqueness and boundedness of solutions. This is particularly adapted to the analysis of dynamical systems where uniqueness and boundedness of solutions play very important roles in engineering and other fields. For constant matrices, it is immediate to obtain the following result:

Proposition 2.6.20 *The interconnection* (2.99) *is well-posed if and only if* $\det(\Sigma_2\Sigma_1 - I) \neq 0$ *for all* $\Sigma_2 \in \mathbf{\Sigma}_2$.

Proof Relationships (2.99) can be rewritten as

---

[6] This definition can be extended to a more general case where the operators depend on some parameters [75].

$$\underbrace{\begin{bmatrix} I & -\Sigma_1 \\ -\Sigma_2 & I \end{bmatrix}}_{\Omega} \begin{bmatrix} z \\ w \end{bmatrix} = \begin{bmatrix} \Sigma_1 & 0 \\ 0 & \Sigma_2 \end{bmatrix} \begin{bmatrix} u_1 \\ u_2 \end{bmatrix}. \tag{2.101}$$

Uniqueness and boundedness are immediately seen to be equivalent to the invertibility of Ω, i.e. $\det(\Omega) \neq 0$, for all $\Sigma_2 \in \mathbf{\Sigma}_2$. Using the Schur formula for determinants (see Appendix A.1), we get the condition $\det(\Sigma_2\Sigma_1 - I) \neq 0$ for all $\Sigma_2 \in \mathbf{\Sigma}_2$. The proof is complete. ■

Whereas the above condition is expressed in terms of standard linear algebra concepts, checking the invertibility of a class of matrices is not an easy task. The determinant condition, moreover, turns out to be inapplicable when more complex operators are considered. In [76], M. Safonov showed that well-posedness can be characterized in terms of graph separation. The theorem below is a simple version of this result:

Theorem 2.6.21 (Topological-separation theorem [76, 79]) *The interconnection* (2.99) *is well-posed if and only if the equality*

$$\mathcal{G}_1 \cap \mathcal{G}_2^- = \{0\}$$

holds for all $\Sigma_2 \in \mathbf{\Sigma}_2$ *where*

$$\mathcal{G}_1 := \left\{ \begin{pmatrix} z \\ w \end{pmatrix} : z = \Sigma_1 w \right\} \text{ and } \mathcal{G}_2^- := \left\{ \begin{pmatrix} z \\ w \end{pmatrix} : w = \Sigma_2 z,\ \Sigma_2 \in \mathbf{\Sigma}_2 \right\}.$$

The main difficulty is now to find a simple way for proving that the graph $\mathcal{G}_1$ does not intersect the inverse graph $\mathcal{G}_2^-$ except at 0. Note that the operators Σ_1 and Σ_2 are linear, the graphs are therefore convex sets, a very convenient property. We can then use separation ideas to get the following result:

Theorem 2.6.22 (Quadratic Separation Theorem [75]) *The following statements are equivalent:*

1. *The interconnection* (2.99) *is well-posed*
2. *There exist* $M \in \mathbb{S}^{n_w+n_z}$ *such that the conditions*

$$\begin{bmatrix} I & \Sigma_1 \end{bmatrix} M \begin{bmatrix} I \\ \Sigma_1^* \end{bmatrix} \prec 0 \tag{2.102a}$$

$$\begin{bmatrix} \Sigma_2 & I \end{bmatrix} M \begin{bmatrix} \Sigma_2^* \\ I \end{bmatrix} \succeq 0 \tag{2.102b}$$

hold for all $\Sigma_2 \in \mathbf{\Sigma}_2$.

Proof The proof can be found in [75]. It is recalled here for completeness.

Proof of 1 ⇒ 2: Assume that 1. holds. Then, from Proposition 2.6.20, we have that $\det(\Sigma_2\Sigma_1 - I) \neq 0$ for all $\Sigma_2 \in \mathbf{\Sigma}_2$. Let

$$M = \begin{bmatrix} \Sigma_1 \\ -I \end{bmatrix} \begin{bmatrix} \Sigma_1 \\ -I \end{bmatrix}^* - \epsilon I \tag{2.103}$$

then the left-hand side of (2.102a) becomes $-\epsilon(I + \Sigma_1\Sigma_1^*)$ which is obviously negative definite. The left-hand side of the condition (2.102b), however, becomes $(\Sigma_2\Sigma_1 - I)(\Sigma_2\Sigma_1 - I)^* - \epsilon(\Sigma_2\Sigma_2^* + I)$. Since the system is well-posed then $\det(I - \Sigma_2\Sigma_1) \neq 0$ and, thus, $(\Sigma_2\Sigma_1 - I)(\Sigma_2\Sigma_1 - I)^*$ is positive definite for all $\Sigma_2 \in \mathbf{\Sigma}_2$. Therefore, the left-hand side of (2.102b) can be made positive semidefinite through a suitable choice for $\epsilon > 0$.

Proof of 2 ⇒ 1: Assume that 2. holds. Then, from Finsler's lemma (see Appendix C.11) the condition (2.102a) is equivalent to the existence of $\xi > 0$ such that we have

$$M \prec \xi \begin{bmatrix} \Sigma_1 \\ -I \end{bmatrix} \begin{bmatrix} \Sigma_1 \\ -I \end{bmatrix}^*. \tag{2.104}$$

Pre- and post-multiplying the above inequality by $\begin{bmatrix} \Sigma_2 & I \end{bmatrix}$ and $\begin{bmatrix} \Sigma_2 & I \end{bmatrix}^*$, respectively, we get that

$$0 \preceq \begin{bmatrix} \Sigma_2 & I \end{bmatrix} M \begin{bmatrix} \Sigma_2^* \\ I \end{bmatrix} \prec \xi(\Sigma_2\Sigma_1 - I)(\Sigma_2\Sigma_1 - I)^*. \tag{2.105}$$

From (2.102b), we get that $(\Sigma_2\Sigma_1 - I)(\Sigma_2\Sigma_1 - I)^*$ is positive definite and hence that $\Sigma_2\Sigma_1 - I$ is invertible for all $\Sigma_2 \in \mathbf{\Sigma}_2$. The system (2.99) is therefore well-posed. ■

2.6.6.2 Application to LTI Systems

So far, no dynamical systems have been involved in the definitions and results. To illustrate how well-posedness can be applied in order to characterize stability of dynamical systems, let us consider the LTI system

$$\dot{x}(t) = Ax(t) + v(t) \tag{2.106}$$

where v is exogenous signal. This system can be equivalently represented as the interconnection in Fig. 2.3 where $\Sigma_1 = A$ and $\Sigma_2 = s^{-1}I$, s being the Laplace variable, and

$$\mathbf{\Sigma}_2 = \left\{\alpha I : \alpha \in \bar{\mathbb{C}}_+\right\}. \tag{2.107}$$

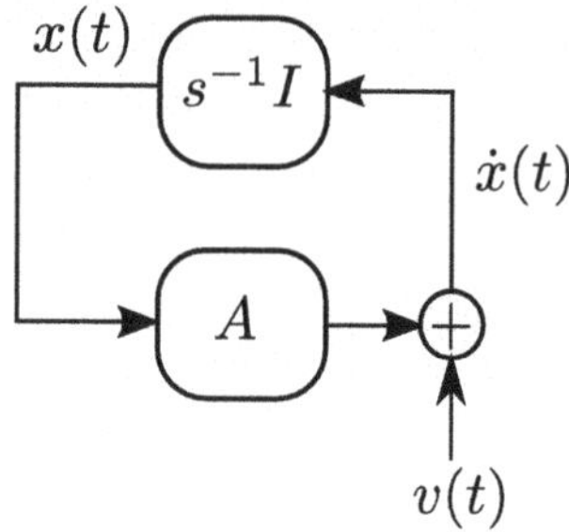

Fig. 2.3 Representation of a linear time invariant dynamical system in the well-posedness framework

We have the following proposition:

Proposition 2.6.23 *The following statements are equivalent:*

1. *The interconnection in Fig. 2.3 is well-posed for all* $s \in \bar{\mathbb{C}}_+$.
2. *The matrix* A *does not have any eigenvalue in the closed right-half plane.*
3. *The system* $\dot{x}(t) = Ax(t)$ *is asymptotically stable.*

Proof Assume that the interconnection is well-posed. Then, the loop signals x and $\dot{x}$ are uniquely defined for any input signal v. Taking the Laplace transform of the signals yields

$$s\widehat{x}(s) = A\widehat{x}(s) + \widehat{v}(s) \tag{2.108}$$

and therefore

$$(sI - A)\widehat{x}(s) = \widehat{v}(s)$$

where $\widehat{\cdot}$ denotes the Laplace transform of the corresponding signal. The signal $\widehat{x}$ is uniquely defined for all $s \in \bar{\mathbb{C}}_+$ if and only if s is not an eigenvalue of A. The proof of the converse statement is simply obtained by reversing the arguments. ■

The above proposition shows that the determinant condition of Proposition 2.6.20 corresponds to an eigenvalue criterion for stability analysis of LTI systems. The proposition below shows that the graph separation condition is equivalent to Lyapunov stability conditions:

Proposition 2.6.24 *The following statements are equivalent:*

1. *The interconnection in Fig. 2.3 is well-posed for all* $s \in \bar{\mathbb{C}}_+$.
2. *There exists a matrix* $X \in \mathbb{S}^n_{\succ 0}$ *such that the LMI*

$$AX + XA^T \prec 0 \tag{2.109}$$

holds.

3. *There exists a matrix* $P \in \mathbb{S}^n_{\succ 0}$ *such that the LMI*

$$A^T P + PA \prec 0 \tag{2.110}$$

holds.

Proof The result is immediate from Proposition 2.6.23 but it is interesting to show that Lyapunov conditions for LTI systems can be retrieved from Theorem 2.6.22. According to this theorem, the interconnection is well-posed for all $s \in \bar{\mathbb{C}}_+$ if and only if the LMIs

$$\begin{bmatrix} I & A \end{bmatrix} M \begin{bmatrix} I \\ A^{\mathrm{T}} \end{bmatrix} \prec 0 \quad \text{and} \quad \begin{bmatrix} s^{-1}I & I \end{bmatrix} M \begin{bmatrix} s^{-*}I \\ I \end{bmatrix} \succeq 0 \tag{2.111}$$

hold for all $s \in \bar{\mathbb{C}}_+$. The following alternative representation

$$\bar{\mathbb{C}}_+ = \left\{ s \in \mathbb{C} : s^{-1} + s^{-*} \geq 0 \right\} \tag{2.112}$$

indicates that we can use the class of multipliers for positive real uncertainties, sometimes referred to as *P-separators*.[7] Therefore, we pick

$$M = \begin{bmatrix} 0 & X \\ X & 0 \end{bmatrix} \tag{2.113}$$

where $X \in \mathbb{S}^n_{\succ 0}$. It is immediate to see that the second LMI in (2.111) is verified for this specific choice of separator, and we are left with the first LMI which is given, in this case, by

$$AX + XA^{\mathrm{T}} \prec 0. \tag{2.114}$$

Equivalence between statements 1. and 2. comes from the losslessness of the P-separator in the case of a single repeated scalar uncertainty.[8] The equivalence between statements 2. and 3. follows from a simple congruence transformation with respect to the matrix $P := X^{-1}$. The proof is complete. ■

2.6.6.3 Application to LPV Systems

We present here a basic quadratic stability analysis result for LPV systems using quadratic separation. Several conditions have been obtained in this framework, see e.g. [79, 88]. The goal of this part is to obtain a simple stability condition similar to the one of the full-block S-procedure, i.e. Lemma 2.6.16, in order to emphasize the correspondence between these two approaches. More advanced results can be

[7] See the list of separators/scalings in Sect. 2.6.5.

[8] The P-separator has the same nonconservativity properties as the D-scaling [53].

found in the aforementioned papers. It is crucial to note that Theorem 2.6.22 only holds for time-invariant operators Σ_1 and Σ_2. Therefore, the first step in view of generalizing this to more general operators such as time-varying, delay or integral operators, consists of extending Theorem 2.6.22. To this aim, let us consider the following interconnection

$$\begin{aligned} z(t) &= \Sigma_1(w(t) + u_1(t)) \\ w(t) &= \Sigma_2(z + u_2)(t) \end{aligned} \tag{2.115}$$

where Σ_1 is a real time-invariant matrix and Σ_2 is a time-varying causal linear operator from[9] L_{2e} to L_{2e}. We need the following extension of the definition of well-posedness:

Definition 2.6.25 The interconnection (2.115) is well-posed if the internal signals are uniquely defined in terms of $(u_1, u_2) \in L_{2e}$ and there exists $\gamma > 0$ such that

$$\left\| \begin{pmatrix} \mathbb{P}_T z \\ \mathbb{P}_T w \end{pmatrix} \right\|_{L_2} \le \gamma \left\| \begin{pmatrix} \mathbb{P}_T u_1 \\ \mathbb{P}_T u_2 \end{pmatrix} \right\|_{L_2} \tag{2.116}$$

hold for all $(u, v) \in L_{2e}$ and all $T \geq 0$.

With this in mind, we can state the following result which is a particular case of a more general statement proved in [89]:

Theorem 2.6.26 *The interconnection* (2.115) *is well-posed if there exists a Hermitian matrix M such that the conditions*

$$\begin{bmatrix} \Sigma_1 \\ I \end{bmatrix}^T M \begin{bmatrix} \Sigma_1 \\ I \end{bmatrix} \prec 0 \tag{2.117}$$

and

$$\int_0^T \begin{bmatrix} \vartheta(s) \\ \Sigma_2(\vartheta)(s) \end{bmatrix}^T M \begin{bmatrix} \vartheta(s) \\ \Sigma_2(\vartheta)(s) \end{bmatrix} \geq 0 \tag{2.118}$$

holds for all $\vartheta \in L_{2e}$ and all $T \geq 0$.

Proof The proof can be found in [89]. ■

Using the above result, we can state the following theorem which turns out to be equivalent to Lemma 2.6.16 obtained using the full-block S-procedure:

[9] The extended L_2-space, denoted L_{2e}, consists of the set of functions f such that all the projections $\mathbb{P}_T f$ onto $[0, T]$, $t \geq 0$, are in L_2.

Theorem 2.6.27 *The LPV system* (2.67) *with* Θ *in the set* (2.85) *is quadratically stable if there exist* $P \in \mathbb{S}^n_{\succ 0}$ *and matrices* $Q, R \in \mathbb{S}^{\eta}$, $S \in \mathbb{R}^{\eta\times\eta}$ *such that the LMIs*

$$\begin{bmatrix} PA + A^TP & PB \\ \star & 0 \end{bmatrix} + \begin{bmatrix} C & D \\ 0 & I \end{bmatrix}^T \begin{bmatrix} Q & S \\ \star & R \end{bmatrix} \begin{bmatrix} C & D \\ 0 & I \end{bmatrix} \prec 0 \tag{2.119}$$

$$\begin{bmatrix} I \\ \Theta(\rho) \end{bmatrix}^T \begin{bmatrix} Q & S \\ S^T & R \end{bmatrix} \begin{bmatrix} I \\ \Theta(\rho) \end{bmatrix} \succeq 0 \tag{2.120}$$

hold for all $\rho \in \boldsymbol{\Delta}_\rho$.

Proof First rewrite the LPV system (2.67) as

$$\begin{bmatrix} \dot{x}(t) \\ z(t) \end{bmatrix} = \underbrace{\begin{bmatrix} A & B \\ C & D \end{bmatrix}}_{\Sigma_1} \begin{bmatrix} x(t) \\ w(t) \end{bmatrix}, \quad \begin{bmatrix} x(t) \\ w(t) \end{bmatrix} = \underbrace{\begin{bmatrix} \mathcal{I} & 0 \\ 0 & \Theta(\rho(t)) \end{bmatrix}}_{\Sigma_2} \begin{bmatrix} \dot{x}(t) \\ z(t) \end{bmatrix}$$

where $\mathcal{I}$ is the integral operator defined as

$$\mathcal{I}(u)(t) = \int_0^t u(s)ds$$

where u is any integrable function.

We can clearly see that Σ_2 is a time-varying operator and thus that Theorem 2.6.26 must be used. We first show that a suitable multiplier M is defined as

$$M = \left[\begin{array}{cc|cc} 0 & 0 & P & 0 \\ 0 & Q & 0 & S \\ \hline P & 0 & 0 & 0 \\ 0 & S^{\mathrm{T}} & 0 & R \end{array}\right]. \tag{2.121}$$

Evaluating indeed (2.118), we get that

$$\begin{aligned} 2\int_0^T \vartheta_1(t)^{\mathrm{T}} P\mathcal{I}(\vartheta_1)(t)dt &= 2\int_0^T \left(\frac{d}{dt}\left[\mathcal{I}(\vartheta_1)(t)\right]\right)^{\mathrm{T}} P\mathcal{I}(\vartheta_1)(t)dt \\ &= \int_0^T \left[\mathcal{I}(\vartheta_1)(t)\right]^{\mathrm{T}} P\left[\mathcal{I}(\vartheta_1)(t)\right]dt \geq 0 \end{aligned} \tag{2.122}$$

for all $T \geq 0$ and that

$$\int_0^T \vartheta_2(s)^{\mathrm{T}} \begin{bmatrix} I \\ \Theta(\rho(s)) \end{bmatrix}^{\mathrm{T}} \begin{bmatrix} Q & S \\ S^{\mathrm{T}} & R \end{bmatrix} \begin{bmatrix} I \\ \Theta(\rho(s)) \end{bmatrix} \vartheta_2(s) ds \geq 0 \tag{2.123}$$

for all $\rho \in \boldsymbol{\Delta}_\rho$ whenever (2.120) holds for all $\rho \in \boldsymbol{\Delta}_\rho$. Therefore, with this specific choice for M, condition (2.118) is satisfied provided that (2.120) holds. Evaluating condition (2.117) finally gives the result. ■

2.6.7 Integral Quadratic Constraints Analysis

Integral Quadratic Constraints (IQCs) are important mathematical objects that can be used to implicitly characterize operators in an input/output framework, i.e. through relationships between their input and output signals, in the same vein as in previous input/output results such as small-gains, the full-block S-procedure and topological separation. IQCs have been initially introduced by V. A. Yakubovich in [90] in the context of stability analysis of nonlinear systems. However, the current IQC framework, as we understand it now, has been introduced by A. Megrestki in [59] and been obtained from an elegant blend of Eastern Europe ideas mixing IQCs, the Kalman-Yakubovich-Popov Lemma (KYP Lemma) [91–94] (see also Appendix C.1) and the S-procedure [95–98]. The combination of all these concepts and results makes the overall IQC framework widely applicable, since a lot of operators can be characterized in such a way and, thanks to the KYP-lemma, the resulting stability criteria may be expressed in terms of tractable optimization problems. On the top of that, it can also be shown that this framework encompasses several other stability criteria and techniques, such as Popov criterion, small-gain results, and can therefore be seen as a unifying theory of stability.

Many IQCs have been proposed to deal with a wide variety of operators such as constant real scalars, time-varying real scalars, slowly-varying real scalars [59, 99], delays [59, 100], odd nonlinearities in a sector [59], sampled-data systems [101, 102], distributed systems [103], networked systems [104] and even unstable systems [103] using an IQC interpretation of the ν-gap metric [105–107].

In the following, the main concepts and results of the IQC framework will be first presented in order to set up the main ideas. IQCs will then be applied to the stability analysis problem of LPV systems. It will be shown that results obtained using small-gain theorems, the full-block S-procedure and quadratic separation can be obtained and extended via the IQC framework.

2.6.7.1 Preliminaries

For completeness, it is important to introduce first the main ideas and concepts necessary to understand the IQC-framework. The first step is the definition of an IQC indeed:

Definition 2.6.28 Two signals $w \in L_2$ and $z \in L_2$ are said to **satisfy the IQC defined by** Π if the inequality

$$\int_{-\infty}^{+\infty} \begin{bmatrix} \widehat{v}(j\omega) \\ \widehat{w}(j\omega) \end{bmatrix}^* \Pi(j\omega) \begin{bmatrix} \widehat{v}(j\omega) \\ \widehat{w}(j\omega) \end{bmatrix} d\omega \geq 0 \tag{2.124}$$

holds where $\widehat{\cdot}$ denotes the Fourier transform of the corresponding signal.

It is tacitly assumed above that v and w are square integrable functions whereas Π can, on the other hand, be any Hermitian measurable function. Note, moreover, that if v and w satisfy several IQCs defined by $\Pi_1, \ldots, \Pi_n$, then they also satisfy the IQC defined by the positive linear combination $\Pi = z_1\Pi_1 + \ldots + z_n\Pi_n$ for any given $z_1, \ldots, z_n \geq 0$. This basically means that several IQCs can be used to characterize a given pair of signals (v, w).

By virtue of Plancherel's Theorem [108] or Parseval's Theorem [109, 110], the IQC (2.124) can be expressed in the time-domain as:

$$\int_0^{\infty} \sigma(x_\pi(t), w(t), v(t))dt \geq 0 \tag{2.125}$$

where σ is a quadratic form and x_π is defined as

$$\begin{aligned} \dot{x}_\pi(t) &= A_\pi x_\pi(t) + B_w w(t) + B_v v(t) \\ x_\pi(0) &= 0. \end{aligned} \tag{2.126}$$

For any bounded rational weighting function Π, the IQC (2.124) can be expressed in the form (2.125)–(2.126) by first factorizing Π as $\Pi(j\omega) = \Psi(j\omega)^* M \Psi(j\omega)$ where $\Pi(j\omega) = C_\psi(j\omega I - A_\pi)^{-1}\begin{bmatrix} B_w & B_v \end{bmatrix} + D_\psi$, and by then defining σ from C_ψ, D_ψ and M. This will be illustrated later in this section when stability analysis of LPV systems will be addressed, and in Sect. 5.9 in the context of delay systems.

2.6.7.2 Stability of Interconnections

Let us consider now the interconnection

$$\begin{aligned} v &= Gw + f \\ w &= \Delta(v) + e \end{aligned} \tag{2.127}$$

where e, f are the input signals and v, w are the looped signals. Above, we assume that the operator G is a stable linear time-invariant operator with transfer function $\widehat{G}(s)$ and Δ is a bounded causal operator.

Before providing the main result on stability analysis of interconnections of the form (2.127), we need first to extend the definition of IQC to operators:

Definition 2.6.29 We say that a bounded operator $\Delta : L_{2e} \to L_{2e}$ satisfies the IQC defined by Π if (2.124) holds for all $w = \Delta(v)$, $v \in L_2$.

We then have the following result taken from [59]:

Theorem 2.6.30 *Let $\widehat{G}(s)$ be the transfer function the linear system G and Δ be defined as in* (2.127), *and assume further that*

1. *for every $\tau \in [0, 1]$, the interconnection of G and $\tau\Delta$ is well-posed;*[a]
2. *for every $\tau \in [0, 1]$, the IQC defined by Π is satisfied by $\tau\Delta$;*
3. *there exists $\epsilon > 0$ such that the inequality*

$$\begin{bmatrix} \widehat{G}(j\omega) \\ I \end{bmatrix}^* \Pi(j\omega) \begin{bmatrix} \widehat{G}(j\omega) \\ I \end{bmatrix} \preceq -\epsilon I \tag{2.128}$$

holds for all $\omega \in \mathbb{R}$.

Then, the feedback interconnection (2.127) *of G and Δ is stable.* [a] The interconnection is well-posed if the map $(v, w) \mapsto (e, f)$ defined by (2.127) has a causal inverse on L_{2e}. The interconnection is stable if, in addition, the inverse is bounded. When G is linear, well-posedness means that $I - G\Delta$ is causally invertible.

The rationale for using τ-dependent conditions in the two first statements is to resolve an inherent difficulty of the IQC formalism. The IQC (2.124) is, indeed, only defined for square summable signals, but, if the interconnection is not stable, the signals v and w may not be square summable. We therefore end up with a circular argument since we need to assume stability (so that we have square summability of the signals) in order to assess it. The idea to resolve this is to consider an additional parameter τ such that stability is immediate for $\tau = 0$, whereas $\tau = 1$ gives the system to be analyzed. Then, the IQCs are used to show that as the parameter τ increases from zero to one, there can be no transition from stability to instability.

Proposition 2.6.31 *Assume that Δ satisfies IQCs defined by $\Pi_1, \ldots, \Pi_N$ and that $\Pi_q(s)$ and $\widehat{G}(s)$ are proper transfer function with no poles on the imaginary axis. Then, there exist a Hurwitz matrix $A \in \mathbb{R}^{n\times n}$ and matrices $B \in \mathbb{R}^{n\times m}$, $M_q \in \mathbb{R}^{(n+m)\times(n+m)}$, $q = 1, \ldots, N$, such that the equality*

$$\begin{bmatrix} \widehat{G}(j\omega) \\ I \end{bmatrix}^* \Pi_i(j\omega) \begin{bmatrix} \widehat{G}(j\omega) \\ I \end{bmatrix} = \begin{bmatrix} (j\omega I - A)^{-1}B \\ I \end{bmatrix}^* M_i \begin{bmatrix} (j\omega I - A)^{-1}B \\ I \end{bmatrix} \tag{2.129}$$

holds for all $i = 1, \ldots, N$ and for all $\omega \in \mathbb{R}$.

Invoking now the Kalman-Yakubovich-Popov Lemma, we get the following result [59]:

Lemma 2.6.32 *The following statements are equivalent:*

1. *There exist scalars* $z_1, \ldots, z_N \geq 0$ *such that*

$$\begin{bmatrix} \widehat{G}(j\omega) \\ I \end{bmatrix}^* \left(\sum_{i=1}^{N} z_i \Pi_i(j\omega) \right) \begin{bmatrix} \widehat{G}(j\omega) \\ I \end{bmatrix} \preceq -\epsilon I \tag{2.130}$$

   holds for all $\omega \in \mathbb{R}$.
2. *There exist scalars* $z_1, \ldots, z_N \geq 0$ *and matrix* $P \in \mathbb{S}^n$ *such that the matrix inequality*

$$\begin{bmatrix} A^T P + PA & PB \\ \star & 0 \end{bmatrix} + \sum_{i=1}^{N} z_i M_i \prec 0 \tag{2.131}$$

   holds.

Whenever the matrices M_i's are fixed, the condition (2.131) is an LMI in the variables $P, z_1, \ldots, z_N$ and can be easily checked.

2.6.7.3 Stability Analysis of LPV Systems

Let us consider back the LPV system in LFT-form (2.67) where $\Theta(\rho) \in \mathbb{R}^{\eta \times \eta}$ and $\rho \in \boldsymbol{\Delta}_\rho$. We have the following result:

Theorem 2.6.33 *Assume that* Θ *satisfies the IQC defined by*

$$\Pi(j\omega) = \begin{bmatrix} Q & S \\ S^T & R \end{bmatrix} \tag{2.132}$$

where $Q, R \in \mathbb{S}^\eta$ *and* $S \in \mathbb{R}^{\eta \times \eta}$ *are real matrices, i.e. defined such that the inequality*

$$Q + S\Theta(\rho) + \Theta(\rho)^T S + \Theta(\rho)^T R \Theta(\rho) \succeq 0 \tag{2.133}$$

holds for all $\rho \in \boldsymbol{\Delta}_\rho$. *Then, the system* (2.67) *is asymptotically stable if there exists a matrix* $P \in \mathbb{S}^n_{\succ 0}$ *such that the LMI*

$$\begin{bmatrix} A^T P + PA & PE \\ \star & 0 \end{bmatrix} + \begin{bmatrix} C & F \\ 0 & I \end{bmatrix}^T \begin{bmatrix} Q & S \\ \star & R \end{bmatrix} \begin{bmatrix} C & F \\ 0 & I \end{bmatrix} \prec 0 \tag{2.134}$$

holds.

It is immediate to recognize the results that have been obtained using the full-block S-procedure and in the quadratic separation framework in Sects. 2.6.5 and 2.6.6, respectively. Note also that small-gain results are also included since they correspond to particular values for the matrices Q, S and R.

References

1. A.M. Lyapunov, *General Problem of the Stability of Motion* (Translated from Russian to French by E. Davaux and from French to English by A. T. Fuller). Taylor and Francis, London, (1992)
2. H.K. Khalil, *Nonlinear Systems* (Prentice-Hall, Upper Saddle River, 2002)
3. E.A. Barbašin, N.N. Krasovskiĭ, On stability of motion in the large (in Russian). Dokl. Akad. Nauk SSSR **86**, 453–456 (1934)
4. M. Malisoff, F. Mazenc, *Constructions of Strict Lyapunov Functions* (Springer, London, 2009)
5. J. Kurzweil, Transl. Am. Math. Soc. On the inversion of Liapunov's second theorem. **24**, 19–77 (1963). (originally appeared in Czechoslovak Mathematical Journal **81**, 217–259 1956
6. N.G. Chetaev, A theorem on instability (in Russian). Dokl. Akad. Nauk SSSR **1**, 529–531 (1934)
7. N.G. Chetaev, *The Stability of Motion (Translated from Russian)* (Pergamon Press, New York, 1961)
8. N.N. Krasovskiĭ, *Stability of motion. Applications of Lyapunov's second Method to Differential Systems and Equations with Delay (Translated from Russian)* (Stanford University Press, Stanford, 1963)
9. S. Boyd, L. El-Ghaoui, E. Feron, V. Balakrishnan, *Linear Matrix Inequalities in Systems and Control Theory* (SIAM, Philadelphia, 1994)
10. Y. Nesterov, A. Nemirovskii, *Interior-Point Polynomial Algorithm in Convex Programming* (SIAM, Philadelphia, 1994)
11. F. Wu, X.H. Yang, A. Packard, G. Becker, Induced $\mathcal{L}_2$-norm control for LPV systems with bounded parameter variation rates. Int. J. Robust Nonlinear Control **6**(9–10), 983–998 (1996)
12. D.J. Stilwell, W.J. Rugh, Stability and $\mathscr{L}_2$ gain properties of LPV systems. Automatica **38**, 1601–1606 (2002)
13. J.P. Hespanha, O. Yakimenko, I. Kaminer, A. Pascoal, *Linear parametrically varying systems with brief instabilities: an application to integrated vision/IMU navigation*, 40th Conference on Decision and Control, vol. 3, Orlando, FL, (2001), pp. 2361–2371
14. R. Goebel, R.G. Sanfelice, A.R. Teel, Hybrid dynamical systems. IEEE Control Syst. Mag. **29**(2), 28–93 (2009)
15. A.S. Morse, Supervisory control of families of linear set-point controllers - Part 1: Exact matching. IEEE Trans. Autom. Control **41**(10), 1413–1431 (1996)

16. J. P. Hespanha and A. S. Morse, *Stability of switched systems with average dwell-time*, in 38th Conference on Decision and Control, Phoenix, Arizona, USA, (1999), pp. 2655–2660
17. C. Briat, A. Seuret, Affine minimal and mode-dependent dwell-time characterization for uncertain switched linear systems. IEEE Trans. Autom. Control **58**(5), 1304–1310 (2013)
18. J.P. Hespanha, D. Liberzon, A.R. Teel, Lyapunov conditions for input-to-state stability of impulsive systems. Automatica **44**(11), 2735–2744 (2008)
19. C. Briat, A. Seuret, A looped-functional approach for robust stability analysis of linear impulsive systems. Syst. Control Lett. **61**, 980–988 (2012)
20. C. Briat, A. Seuret, Convex dwell-time characterizations for uncertain linear impulsive systems. IEEE Trans. Autom. Control **57**(12), 3241–3246 (2012)
21. L. Xie, S. Shishkin, M. Fu, Piecewise Lyapunov functions for robust stability of linear time-varying systems. Syst. Control Lett. **31**(3), 165–171 (1997)
22. M. Johansson, A. Rantzer, Computation of piecewise quadratic Lyapunov functions for hybrid systems. IEEE Trans. Autom. Control **43**(4), 555–559 (1998)
23. T. Hu, Z. Lin, Composite quadratic Lyapunov functions for constrained control systems. IEEE Trans. Autom. Control **48**(3), 440–450 (2003)
24. A.P. Molchanov, S. Ye Pyatnitskiy, Criteria of asymptotic stability of differential and difference inclusions encountered in control theory. Syst. Control Lett. **13**, 59–64 (1989)
25. F. Blanchini, Nonquadratic Lyapunov functions for robust control. Automatica **31**(3), 451–461 (1995)
26. F. Blanchini, S. Miani, A new class of universal Lyapunov functions for the control of uncertain linear systems. IEEE Trans. Autom. Control **44**(3), 641–647 (1999)
27. R. Ambrosino, M. Ariola, F. Amato, Stability and instability conditions using polyhedral Lyapunov functions. IET Control Theory Appl. **4**(7), 1179–1187 (2008)
28. T. Hu, F. Blanchini, Non-conservative matrix inequality conditions for stability/stabilization of linear differential inclusions. Automatica **46**, 190–195 (2010)
29. G. Chesi, A. Garulli, A. Tesi, A. Vicino, *Homogeneous Polynomial Forms for Robustness Analysis of Uncertain Systems*, Lecture Notes in Control and Information Sciences (Springer, Berlin, 2009)
30. F. Wirth, A converse Lyapunov theorem for linear parameter-varying and linear switching systems. SIAM J. Control Optim. **44**(1), 210–239 (2005)
31. G. Chesi, LMI techniques for optimization over polynomials in control: a survey. IEEE Trans. Autom. Control **55**(11), 2500–2510 (2010)
32. G. Chesi, P. Colaneri, J.C. Geromel, R. Middleton, R. Shorten, A nonconservative LMI condition for stability of switched systems with guaranteed dwell-time. IEEE Trans. Autom. Control **57**(5), 1297–1302 (2012)
33. A. Ben-Tal, A. Nemirovski, On tractable approximations of uncertain linear matrix inequalities affected by interval uncertainty. SIAM J. Optim. **12**(3), 811–833 (2002)
34. A. Ben-Tal, A. Nemirovski, C. Roos, Extended matrix cube theorems with applications to μ-theory in control. Math. Oper. Res. **28**(3), 497–523 (2003)
35. D. Liberzon, J.P. Hespanha, A.S. Morse, Stability of switched systems: a Lie-algebraic condition. Syst. Control Lett. **37**, 117–122 (1999)
36. A.A. Agrachev, D. Liberzon, Lie-algebraic stability criteria for switched systems. SIAM J. Control Optim. **40**(1), 253–269 (2001)
37. R.K. Yedavalli, A. Sparks, *Conditions for the existence of a common quadratic Lyapunov function via stability analysis of matrix families*, in American Control Conference, Anchorage, Alaska, USA, (2002), pp. 1296–1301
38. R.K. Yedavalli, A noncombinatorial necessary and sufficient 'vertex' solution for checking robust stability of polytopes of matrices induced by interval parameters, in 42nd IEEE Conference on Decision and Control, Denver, Colorado, USA, (2003), pp. 3846–3851
39. P. Mason, R. Shorten, The geometry of convex cones associated with the Lyapunov inequality and the common Lyapunov function problem. Electron. J. Linear Algebra **412**, 42–63 (2005)
40. P. Mason, U. Boscain, Y. Chitour, Common polynomial Lyapunov functions for linear switched systems. SIAM J. Control Optim. **45**(1), 226–245 (2006)

41. P. Apkarian, H.D. Tuan, Parameterized LMIs in control theory. SIAM J. Control Optim. **38**(4), 1241–1264 (2000)
42. R.C.L.F. Oliveira, P.L.D. Peres, LMI conditions for robust stability analysis based on polynomially parameter-dependent Lyapunov functions. Syst. Control Lett. **55**, 52–61 (2006)
43. R.C.L.F. Oliveira, P.L.D. Peres, Parameter-dependent LMIs in robust analysis: characterization of homogeneous polynomially parameter-dependent solutions via LMI relaxations. IEEE Trans. Autom. Control **57**(7), 1334–1340 (2002)
44. K. Zhou, J.C. Doyle, K. Glover, *Robust and Optimal Control* (Prentice Hall, Upper Saddle River, 1996)
45. J.C. Doyle, B. Francis, A. Tanenbaum, *Feedback Control Theory* (Macmillan, London, 1990)
46. J.C. Doyle, K. Glover, P.P. Khargonekar, B.A. Francis, State space solutions to standard $\mathcal{H}_2$ and $\mathcal{H}_\infty$ control problems. IEEE Trans. Autom. Control **34**, 831–847 (1989)
47. C.W. Scherer, *The Riccati inequality and state-space H_∞-optimal control*, Ph.D. thesis, University of Würzburg, Germany, (1990)
48. B.D.O. Anderson, S. Vongpanitlerd, *Network Analysis and Synthesis - A Modern Systems Approach* (Prentice-Hall, Englewood Cliffs, 1973)
49. J.C. Willems, Dissipative dynamical systems i and ii. Ration. Mech. Anal. **45**(5), 321–393 (1972)
50. C.A. Desoer, M. Vidyasagar, *Feedback Systems: Input-Output Properties* (Academic Press, New York, 1975)
51. A. Packard, Gain scheduling via Linear Fractional Transformations. Syst. Control Lett. **22**, 79–92 (1994)
52. P. Apkarian, P. Gahinet, A convex characterization of gain-scheduled $\mathcal{H}_\infty$ controllers. IEEE Trans. Autom. Control **5**, 853–864 (1995)
53. A. Packard, J.C. Doyle, The complex structured singular value. Automatica **29**, 71–109 (1993)
54. C.W. Scherer, I.E. Köse, *On robust controller synthesis with dynamic D-scalings*, 46th Conference on Decision and Control, New Orleans LA, USA, (2007)
55. C.W. Scherer, I.E. Köse, Gain-scheduling with dynamic D-scalings, in 46th IEEE Conference on Decision and Control, New Orleans, LA, USA, (2007), pp. 6316–6321
56. C.W. Scherer, I.E. Köse, Robustness with dynamic IQCs: an exact state-space characterization of nominal stability with applications to robust estimation. Automatica **44**, 1666–1675 (2008)
57. C.W. Scherer, I.E. Köse, Gain-scheduled control synthesis using dynamic D-scales. IEEE Trans. Autom. Control **57**(9), 2219–2234 (2012)
58. C.W. Scherer, Gain-scheduled synthesis with dynamic positive real multipliers, in 51st IEEE Conference on Decision and Control, Maui, Hawaii, USA, (2012), pp. 6641–6646
59. A. Megretski, A. Rantzer, System analysis via integral quadratic constraints. IEEE Trans. Autom. Control **42**(6), 819–830 (1997)
60. G. Zames, P. Falb, Stability conditions for systems with monotone and slope-restricted nonlinearities. SIAM J. Control Optim. **6**, 89–108 (1968)
61. A. Pinkus, n-widths in apprximmation theory, *In Volume 7 of Ergebnisse der Mathematik und ihrer Grenzgebiete* (Springer, Berlin, 1985)
62. S.P. Boyd, G.H. Barrat, *Linear controller design - Limits of Performance* (Prentice-Hall, Englewood Cliffs, 1991)
63. C.W. Scherer, Multiobjective H_2/H_∞ control. IEEE Trans. Autom. Control **40**, 1054–1962 (1995)
64. C.W. Scherer, *Robust generalized $\mathcal{H}_2$ control for uncertain and LPV systems with general scalings*, in Conference on Decision and Control, (1996), pp. 3970–3975
65. C.W. Scherer, A full-block $\mathcal{S}$-procedure with applications, in 36th IEEE Conference on Decision and Control, (1997), pp. 2602–2607
66. C.W. Scherer, LPV control and full-block multipliers. Automatica **37**, 361–375 (2001)
67. C.W. Scherer, C.W.J. Hol, Matrix sum-of-squares relaxations for robust semi-definite programs, Mathematical Programming Series B, *107*:189–211, (2006)
68. F. Wu, Robust quadratic performance for time-delayed uncertain linear systems. Int. J. Robust Nonlinear Control **13**, 153–172 (2003)

69. U. Münz, J.M. Rieber, F. Allgöwer, Robust stabilization and H_∞ control of uncertain distributed delay systems, *Topics in Time-Delay Systems*, volume 388 of Lecture Notes in Control and Information Sciences, Springer, Berlin Heidelberg, (2009), pp. 221–231
70. C. Briat, O. Sename, J.-F. Lafay, A full-block $\mathcal{S}$-procedure application to delay-dependent $\mathcal{H}_\infty$ state-feedback control of uncertain time-delay systems, in 17th IFAC world Congress, Seoul, South Korea, (2008), pp. 12342–12347
71. C. Langbort, R.S. Chandra, R. D'Andrea, Distributed control design for systems interconnected over an arbitrary graph. IEEE Trans. Autom. Control **49**(9), 1502–1519 (2004)
72. C.W. Scherer, Relaxations for robust linear matrix inequality problems with verifications for exactness. SIAM J. Matrix Anal. Appl. **27**(2), 365–395 (2005)
73. C.W. Scherer, C.W.J. Hol, Matrix sum-of-squares relaxations for robust semi-definite programs. Math. Program. Series B **107**, 189–211 (2006)
74. G. Meinsma, Y. Shrivastava, M. Fu, A dual formulation of mixed μ and the losslessness of (D,G) scaling. IEEE Trans. Autom. Control **42**(7), 1032–1036 (1997)
75. T. Iwasaki, S. Hara, Well-posedness of feedback systems: insight into exact robustness analysis and approximate computations. IEEE Trans. Autom. Control **43**, 619–630 (1998)
76. M.G. Safonov, M. Athans, *On stability theory*, in 17th Conference on Decision and Control, San Diego, California, USA, (1978), pp 301–314
77. M.G. Safonov, *Stability, Robustness of Multivariable Feedback Systems* (MIT Press, Cambridge, 1980)
78. D. Peaucelle, D. Arzelier, D. Henrion, F. Gouaisbaut, Quadratic separation for feedback connection of an uncertain matrix and an implicit linear transformation. Automatica **43**, 795–804 (2007)
79. T. Iwasaki, LPV system analysis with quadratic separator. in 37th IEEE Conference on Decision and Control, (1998), pp. 3021–3026
80. T. Iwasaki, Chapter 14 of advances on LMI methods in control, in *Control Synthesis for Well-Posedness of Feedback Systems*, ed. by L. El-Ghaoui, S.-I. Niculescu (SIAM, Natick, 2000)
81. D. Peaucelle, D. Arzelier, Ellipsoidal sets for resilient and robust static output feedback. IEEE Trans. Autom. Control **50**, 899–904 (2005)
82. F. Gouaisbaut, D. Peaucelle, Stability of time-delay systems with non-small delay, in Conference on Decision and Control, San Diego, California, USA, (2006), pp. 840–845
83. F. Gouaisbaut, D. Peaucelle, Robust stability of time-delay systems with interval delays, in 46th IEEE Conference on Decision and Control, LA, USA, New Orleans, (2007), p. 2007
84. Y. Ariba, F. Gouaisbaut, Input-output framework for robust stability of time-varying delay systems, in 48th Conference on Decision and Control, Shanghai, China, (2009), pp. 274–279
85. Y. Ariba, F. Gouaisbaut, K.H. Johansson, *Stability interval for time-varying delay systems*, in 49th Conference on Decision and Control, Atlanta, Georgia, USA, (2010), pp. 1017–1022
86. Y. Ariba, C. Briat, K.H. Johansson, *Simple conditions for L_2 stability and stabilization of networked control systems*. in 18th IFAC World Congress, Milano, Italy, (2011), pp. 96–101
87. A.R. Teel, On graphs, conic relations, and input-output stability of nonlinear feedback systems. IEEE Trans. Autom. Control **41**(5), 702–709 (1996)
88. T. Iwasaki, G. Shibata, LPV system analysis via quadratic separator for uncertain implicit systems. IEEE Trans. Autom. Control **46**, 1195–1208 (2001)
89. Y. Ariba, F. Gouaisbaut, D. Peaucelle, *Stability analysis of time-varying delay systems in quadratic separation framework*, in International Conference on Mathematical Problems in Engineering, Aerospace and Sciences (ICNPAA'08), Genoa, Italy, 25–27 June 2008
90. V.A. Yakubovich, S-procedure in nonlinear control theory, Vestnik Leningrad University, pp. 62–77, (1971), English translation in Vestnik Leningrad University Math., vol. 4, pp. 73–93, (1977)
91. V.M. Popov, Absolute stability of nonlinear systems of automatic control. Autom. Remote Control (Translated from Automatica i Telemekhanika) **22**(8), 857–875 (1961)
92. V.A. Yakubovitch, The solution to certain matrix inequalities in automatic control. Soviet Math Dokl. **3**, 620–623 (1962)

93. J.P. LaSalle, Complete stability of a nonlinear control system. Proc. Natl. Acad. Sci. U.S.A. **48**, 600–603 (1962)
94. R.E. Kalman, Lyapunov functions for the problem of Lur'e in automatic control. Proc. Natl. Acad. Sci. U.S.A. **49**, 201–205 (1963)
95. V.A. Yakubovich, S-procedure in nonlinear control theory (in russian). Vestn. Leningrad University **1**, 62–77 (1971)
96. V.A. Yakubovich, S-procedure in nonlinear control theory (english translation). Vestn. Leningrad University **4**, 73–93 (1977)
97. U.T. Jönsson, *A lecture on the S-procedure*, Technical report, Division of Optimization and Systems Theory, Royal Institute of Technology, Stockholm, Sweden, (2001)
98. I. Pólik, T. Terlaky, A survey of the s-lemma. SIAM Rev. **49**(3), 371–418 (2007)
99. U.T. Jönsson, A. Rantzer, Systems with uncertain parameters - time-variations with bounded derivatives. Int. J. Robust Nonlinear Control **6**, 969–982 (1996)
100. C.Y. Kao, A. Rantzer, Stability analysis of systems with uncertain time-varying delays. Automatica **43**, 959–970 (2007)
101. H. Fujioka, Stability analysis of systems with aperiodic sample-and-hold devices. Automatica **45**, 771–775 (2009)
102. C.-Y. Kao, H. Fujioka, On stability of systems with aperiodic sampling devices (to appear). IEEE Trans. Autom. Control **58**(3), 2085–2090 (2013)
103. M. Cantoni, U.T. Jönsson, C.-Y. Kao, Robustness analysis for feedback interconnections of distributed systems via integral quadratic constraints. IEEE Trans. Autom. Control **57**(2), 302–317 (2012)
104. U.T. Jönsson, C.-Y. Kao, H. Fujioka, A popov criterion for networked systems. Syst. Control Lett. **56**(9–10), 603–610 (2007)
105. G. Vinnicombe, Frequency domain uncertainty and the graph topology. IEEE Trans. Autom. Control **38**(9), 1371–1383 (1993)
106. G. Vinnicombe, The robustness of feedback systems with bounded complexity controllers. IEEE Trans. Autom. Control **41**(6), 795–803 (1996)
107. G. Vinnicombe, *Uncertainty and Feedback - $\mathscr{H}_\infty$ loop-shaping and ν-gap metric* Imperial College Press, London, (2000)
108. M. Plancherel, Contribution a l'étude de la représentation d'une fonction arbitraire par les intégrales définies. Rend. del Circolo Matematico di Palermo **30**, 298–335 (1910)
109. M.-A. Parseval des Chênes, Mémoire sur les séries et sur l'intégration complète d'une équation aux différences partielle linéaire du second ordre, à coefficiens constans (in French). Sci. Math. et Physiques (Savans étrangers) **1**, 638–648 (1806)
110. W.D. Kammler, *A first Course in Fourier Analysis* (Cambridge University Press, Cambridge, 2007)

Chapter 3
Control of LPV Systems

The basic tool for the manipulation of reality is the manipulation of words. If you can control the meaning of words, you can control the people who must use the words.
Philip K. Dick

Abstract This chapter is devoted to the development of control-laws for linear parameter-varying systems. The LPV framework allows for an easy design of gain-scheduled controllers, that is, controllers whose state-space structure depends on the value of the parameters thereby used as scheduling variables. Ways for designing gain-scheduled state feedback and dynamic output feedback controllers are presented in the generic, polytopic and LFT settings. The goal of this chapter is not to detail the most advanced results obtained to date but rather to give a clear picture of what can be done. Suitable references pointing towards more recent and efficient results are given for completeness.

3.1 Gain-Scheduling: The LPV Way

Whereas tools for analyzing LPV systems are directly inherited from robust analysis and robust control theory, the full flavor of LPV theory is only revealed when design is the main purpose. The main difference with robust control lies in the fact that the parameters are assumed to be known or measurable in the LPV framework, whereas they are unknown, by assumption, in robust control theory. The parameters can be, therefore, used in the control law in a scheduling fashion, giving rise to *LPV gain-scheduled controllers* (Fig. 3.1). It is, however, important to stress that gain-scheduling existed before the LPV-way, see e.g. [1, 2]. LPV gain-scheduling techniques treat the problem in a direct and global way whereas, in preceding gain-scheduling techniques, the controller is constructed from a family of local linear controllers designed using linear time-invariant methods.

The first gain-scheduled ideas in an LPV fashion have been proposed by J. Shamma in his Ph.D. thesis [3] in 1988, and subsequent papers [4–6]. The major difficulty, at that time, was the lack of general theory for analyzing stability of LPV

C. Briat, *Linear Parameter-Varying and Time-Delay Systems*,
Advances in Delays and Dynamics 3, DOI 10.1007/978-3-662-44050-6_3

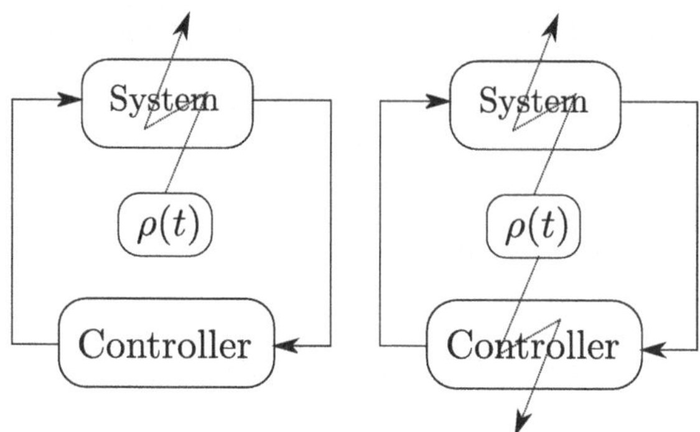

Fig. 3.1 Comparison of the robust (*left*) and LPV (*right*) paradigms.

systems and designing LPV-based gain-scheduled control laws. Indeed, as prophetically stated in [6]:

> The limitations of gain scheduling may be summarized as follows. If the possibility of fast parameter variations is not addressed in the design process, then guaranteed properties of the overall gain scheduled design cannot be established. The examples presented here demonstrate that this limitation is a consequence of fundamental control concepts.
>
> In order to remove these limitations, the development of a theory for LPV systems is needed. This development would involve the modification of robust control design methodologies such as H_∞ and μ-synthesis in order to explicitly address time-variations. It is stressed that such a modification is not simply a "time-varying" version of current practice. This difference is due to the absence of knowledge of future time-variations. The availability of such information is a critical assumption in the current literature on time-varying robust control.

As predicted, a suitable framework for analyzing and controlling LPV systems emerged from robust control ideas, such as H_∞-control, and the use of LMIs. Modern robust optimization techniques considerably strengthened this framework by providing a rigourous way for dealing with parameter-dependent LMIs; see Appendix B. The surveys [2, 7] give insights on the state-of-the-art in the year 2000 whereas the monograph [8] provides more recent developments of the theory of LPV systems and control.

3.2 Types of Controllers

Let us consider the following generic LPV system

$$\begin{aligned}\dot{x}(t) &= A(\rho(t))x(t) + B(\rho(t))u(t)\\ y(t) &= C(\rho(t))x(t)\end{aligned} \tag{3.1}$$

where x, u and y are the state of the system, the control input and the measured output. Two main classes of parameters will be considered. The first family is the class of bounded parameters with bounded derivatives

$$\mathscr{P}^{\nu} := \left\{\rho : \mathbb{R}_{\geq 0} \to \boldsymbol{\Delta}_{\rho},\ \dot{\rho}(t) \in \boldsymbol{\Delta}_{\nu},\ t \geq 0\right\} \tag{3.2}$$

whereas the second class of parameters is the class of systems with arbitrary variation rates

$$\mathscr{P}^{\infty} := \left\{\rho : \mathbb{R}_{\geq 0} \to \boldsymbol{\Delta}_{\rho}\right\} \tag{3.3}$$

where we restrict ourselves, in the latter case, to parameter trajectories for which solutions to (3.1) can be actually defined. As in the previous section, the sets $\boldsymbol{\Delta}_{\rho}$ and $\boldsymbol{\Delta}_{\nu}$ are assumed to be compact and convex polyhedra; e.g. boxes.

As it will be shortly shown, gain-scheduled controllers can be defined as extrapolations of their time-invariant counterparts. They are, however, not equivalent in terms of possibilities and ease of design. The control laws described below assume perfect knowledge of the parameters, i.e. the parameters in the controller and the system are equal to each other. Several works address the case of inexact scheduling parameters; see e.g. [9–11].

3.2.1 Gain-Scheduled State-Feedback

The most simple control-law that can be designed is the gain-scheduled state-feedback which writes

$$u(t) = K(\rho(t))x(t). \tag{3.4}$$

This is an immediate extension of the LTI state-feedback to the LPV setting: the controller matrix is now a function of the parameters. This controller is the easiest to design but it requires the knowledge of the full state of the system to be implemented.

3.2.2 Gain-Scheduled Static-Output-Feedback

A, seemingly simple, gain-scheduled control law is the gain-scheduled static-output feedback control law which takes the form

$$u(t) = K(\rho(t))y(t). \tag{3.5}$$

This class of controllers is very easy to implement since the control-input is directly computed from the measured output. The main difficulty, however, lies in the difficulty to obtain tractable conditions for the design of such control laws. Note that some instances of this problem are known to be NP-hard, see e.g. [12–14]. Yet some methods exist to design them, at least, in the robust setting, see e.g. [15–18], some of which being extendable to the LPV case.

3.2.3 Gain-Scheduled Dynamic-Output-Feedback

A very important class of controllers is the class of dynamic-output feedback controllers:

$$\begin{aligned}\dot{x}_c(t) &= A_c(\rho(t))x_c(t) + B_c(\rho(t))y(t)\\ u(t) &= C_c(\rho(t))x_c(t) + D_c(\rho(t))y(t)\end{aligned} \tag{3.6}$$

where x_c is the state of the controller. This class of controllers has been studied, for instance, in [19–22]. Interestingly, when the order of the controller is equal to the order of the process, i.e. $\dim(x) = \dim(x_c)$, the design problem turns out to admit convex solutions in several setups. When the controller is of reduced-order, i.e. $\dim(n_c) < \dim(x)$, the problem is known to be NP-hard due to the presence of a rank-constraint (nonconvex) in the stabilization conditions; see e.g. [23].

Observer-based control laws can be understood as a particular case of dynamic output-feedback. However, the structure of the observer may not always allow for the derivation of synthesis conditions that are exact and convex. This is, for instance, the case when both the observer and controller gains are aimed to be determined at the same time, through a single LMI condition. Separate design, however, always results in convex synthesis conditions; see e.g. [10]. In this chapter, only general dynamic output-feedback controllers of the form (3.6) will be considered.

3.3 Generic Parameter Dependent Systems

In this section, we will consider LPV systems of the form

$$\begin{aligned}\dot{x}(t) &= A(\rho(t))x(t) + B(\rho(t))u(t) + E(\rho(t))w(t)\\ z(t) &= C(\rho(t))x(t) + D(\rho(t))u(t) + F(\rho(t))w(t)\\ y(t) &= C_y(\rho(t))x(t) + F_y(\rho(t))w(t)\\ x(0) &= x_0\end{aligned} \tag{3.7}$$

where $x \in \mathbb{R}^n$ is the state of the system, $u \in \mathbb{R}^m$ is the control input, $w \in \mathbb{R}^p$ is the exogenous input, $z \in \mathbb{R}^q$ is the controlled output, $y \in \mathbb{R}^r$ is the measured output and $x_0 \in \mathbb{R}^n$ is the initial condition. The parameter trajectories are either in $\mathscr{P}^\nu$ or $\mathscr{P}^\infty$. Singular LPV systems of the form

$$\begin{aligned}
\begin{bmatrix} I & 0 \\ 0 & 0 \end{bmatrix} \begin{bmatrix} \dot{x}(t) \\ \dot{x}_a(t) \end{bmatrix} &= \tilde{A}(\rho(t)) \begin{bmatrix} x(t) \\ x_a(t) \end{bmatrix} + \tilde{B}(\rho(t))u(t) + \tilde{E}(\rho(t))w(t), \\
z(t) &= \tilde{C}(\rho(t)) \begin{bmatrix} x(t) \\ x_a(t) \end{bmatrix} + \tilde{D}(\rho(t))u(t) + \tilde{F}(\rho(t))w(t), \ x_a(t) \in \mathbb{R}^{\eta} \\
y(t) &= \tilde{C}_y(\rho(t)) \begin{bmatrix} x(t) \\ x_a(t) \end{bmatrix} + \tilde{F}_y(\rho(t)) \\
x(0) &= x_0
\end{aligned} \tag{3.8}$$

where

$$\tilde{A}(\rho) = \begin{bmatrix} A_{11}(\rho) & A_{12}(\rho) \\ A_{21}(\rho) & A_{22}(\rho) \end{bmatrix}, \ \tilde{B}(\rho) = \begin{bmatrix} B_1(\rho) \\ B_2(\rho) \end{bmatrix} \text{ and } \tilde{E}(\rho) = \begin{bmatrix} E_1(\rho) \\ 0 \end{bmatrix} \tag{3.9}$$

will also be considered.

As explained in Sect. 1.3.1, the latter formulation is beneficial when dealing with LPV systems of the form (3.7) having rational dependence on the parameters.

In the following, state-feedback and dynamic output feedback results will be provided, both in the quadratic and robust stabilization settings. In order to design judicious controllers, the considered stability conditions will also characterize the L_2-gain of the transfer from w to z. Using such a result, the controllers can be determined such that a certain L_2-gain is ensured for the closed-loop system.

3.3.1 Quadratic Stabilization by State-Feedback

Let us consider first quadratic stabilization by state-feedback with L_2-gain performance constraint. As stated in Sect. 2.3.1, quadratic stability and stabilization address the case of arbitrarily varying parameters, i.e.$\rho \in \mathscr{P}^{\infty}$.

Theorem 3.3.1 *The LPV system* (3.7) *is quadratically stabilizable by a state-feedback of the form* (3.4) *if and only if there exist a matrix* $X \in \mathbb{S}^n_{\succ 0}$ *and a matrix function* $Y : \boldsymbol{\Delta}_\rho \to \mathbb{R}^{m \times n}$ *such that the LMI*

$$\begin{bmatrix} \mathrm{He}\left[A(\rho)X + B(\rho)Y(\rho)\right] & E(\rho) & [C(\rho)X + D(\rho)Y(\rho)]^{\mathrm{T}} \\ \star & -\gamma I_p & F(\rho)^{\mathrm{T}} \\ \star & \star & -\gamma I_q \end{bmatrix} \prec 0 \tag{3.10}$$

holds for all $\rho \in \boldsymbol{\Delta}_\rho$. *Moreover, the state-feedback controllaw given by*

$$u = Y(\rho)X^{-1}x \tag{3.11}$$

ensures that we have $||z||_{L_2} \leq \gamma||w||_{L_2} + \left(\gamma x_0^{\mathrm{T}} X^{-1} x_0\right)^{1/2}$ *for all* $w \in L_2$ *and all parameter trajectories in* $\mathscr{P}^{\infty}$.

Proof The closed-loop system is given by

$$\begin{aligned}\dot{x}(t) &= [A(\rho(t)) + B(\rho(t))K(\rho(t))]x(t) + E(\rho(t))w(t)\\ z(t) &= [C(\rho(t)) + D(\rho(t))K(\rho(t))]x(t) + F(\rho(t))w(t).\end{aligned} \tag{3.12}$$

Substituting this model into the Bounded Real Lemma, i.e. Lemma 2.6.5, with constant matrix P yields the LMI

$$\begin{bmatrix} \mathrm{He}\,[PA(\rho) + PB(\rho)K(\rho)] & PE(\rho) & [C(\rho) + D(\rho)K(\rho)]^{\mathrm{T}} \\ \star & -\gamma I_p & F(\rho)^{\mathrm{T}} \\ \star & \star & -\gamma I_q \end{bmatrix} \prec 0 \tag{3.13}$$

for all $\rho \in \boldsymbol{\Delta}_{\rho}$. A congruence transformation with respect to the matrix $\mathrm{diag}(X, I, I)$, $X := P^{-1}$, and the change of variables $Y(\rho) = K(\rho)X$ yield the result. To prove the bound on the L_2-norm of z, just note that (3.13) is equivalent to saying that

$$\dot{V}(x(t)) - \gamma w(t)^{\mathrm{T}} w(t) + \gamma^{-1} z(t)^{\mathrm{T}} z(t) < 0 \tag{3.14}$$

holds for all $\mathrm{col}(x(t), w(t)) \neq 0$ and where $V(x) = x^{\mathrm{T}} P x$. Integrating from 0 to ∞, we get that

$$-V(x_0) \leq \gamma||w||_{L_2}^2 - \gamma^{-1}||z||_{L_2}^2 \tag{3.15}$$

where we have used the fact that the system is asymptotically stable, i.e. $\lim_{s\to\infty} V(x(s)) = 0$. Reorganizing the terms yields

$$||z||_{L_2}^2 \leq \gamma^2||w||_{L_2}^2 + \gamma V(x_0) \tag{3.16}$$

and the result follows. ■

The following result concerns quadratic stabilization of singular LPV systems:

Theorem 3.3.2 *Assume for simplicity that* $B_2(\rho)$ *is full-column rank. The LPV system* (3.8) *is quadratically stabilizable by a state-feedback of the form* (3.4) *if and only if there exist a constant matrix* $X_1 \in \mathbb{S}_{\succ 0}^n$, *matrix functions*

$X_2 : \boldsymbol{\Delta}_\rho \to \mathbb{R}^{\eta\times n}$, $X_3 : \boldsymbol{\Delta}_\rho \to \mathbb{R}^{\eta\times\eta}$, $Y_1 : \boldsymbol{\Delta}_\rho \to \mathbb{R}^{m\times n}$ and $Y_2 : \boldsymbol{\Delta}_\rho \to \mathbb{R}^{m\times\eta}$ such that the LMI

$$\begin{bmatrix} \mathrm{He}\left[\tilde{A}(\rho)X(\rho)+\tilde{B}(\rho)Y(\rho)\right] & \tilde{E}(\rho) & \left[\tilde{C}X(\rho)+\tilde{D}Y(\rho)\right]^{\mathrm{T}} \\ \star & -\gamma I_p & \tilde{F}(\rho)^{\mathrm{T}} \\ \star & \star & -\gamma I_q \end{bmatrix} \prec 0 \quad (3.17)$$

holds for all $\rho \in \boldsymbol{\Delta}_\rho$ with

$$X(\rho) = \begin{bmatrix} X_1 & 0 \\ X_2(\rho) & X_3(\rho) \end{bmatrix} \text{ and } Y(\rho) = \begin{bmatrix} Y_1(\rho) & Y_2(\rho) \end{bmatrix}. \quad (3.18)$$

Moreover, the state-feedback control law given by

$$u = (I + K_2(\rho)A_{22}(\rho)^{-1}B_2(\rho))^{-1}\left(K_1(\rho) - K_2(\rho)A_{22}(\rho)^{-1}A_{21}(\rho)\right)x \quad (3.19)$$

where[a]

$$\begin{aligned} K_1(\rho) &= (Y_1(\rho) - Y_2(\rho)X_3(\rho)^{-1}X_2(\rho))X_1^{-1} \\ K_2(\rho) &= Y_2(\rho)X_3(\rho)^{-1} \end{aligned} \quad (3.20)$$

ensures that the L_2-gain of the transfer $w \to z$ is smaller than $\gamma > 0$ for all parameter trajectories in $\mathscr{P}^\infty$.

[a] Perturb X_1 and X_3 in the case where they are not invertible.

Proof Substituting the closed-loop system in the conditions of quadratic stability with L_2 performance yields

$$\begin{bmatrix} \mathrm{He}\left[P(\rho)(\tilde{A}(\rho)+\tilde{B}(\rho)K(\rho))\right] & P(\rho)\tilde{E}(\rho) & (\tilde{C}(\rho)+\tilde{D}(\rho)K(\rho))^{\mathrm{T}} \\ \star & -\gamma I_p & \tilde{F}(\rho)^{\mathrm{T}} \\ \star & \star & -\gamma I_q \end{bmatrix} \prec 0 \quad (3.21)$$

which must hold for all $\rho \in \boldsymbol{\Delta}_\rho$ where

$$P(\rho) = \begin{bmatrix} P_1 & 0 \\ P_2(\rho) & P_3(\rho) \end{bmatrix}.$$

Simple calculations show that

$$X(\rho) := P(\rho)^{-1} = \begin{bmatrix} X_1 & 0 \\ X_2(\rho) & X_3(\rho) \end{bmatrix} = \begin{bmatrix} P_1^{-1} & 0 \\ -P_3(\rho)^{-1}P_2(\rho)P_1^{-1} & P_3(\rho)^{-1} \end{bmatrix}.$$

A congruence transformation with respect to the matrix $\text{diag}(X(\rho), I, I)$ and the changes of variables

$$\begin{aligned} Y_1(\rho) &= K_1(\rho)X_1 + K_2(\rho)X_2(\rho) \\ Y_2(\rho) &= K_2(\rho)X_3(\rho) \end{aligned} \tag{3.22}$$

yield the result. To finally obtain the closed-form for the controller we note that $u = K_1(\rho)x + K_2(\rho)x_a$ but we have that $x_a = -A_{22}(\rho)^{-1}A_{21}(\rho)x - A_{22}(\rho)^{-1}B_2(\rho)u$ and thus we obtain (3.19). It remains to prove that the matrix $I+K_2(\rho)A_{22}(\rho)^{-1}B_2(\rho)$ is indeed invertible. First note that admissibility of the closed-loop system is equivalent to the feasibility of (3.17). Therefore, this implies that $A_{22}(\rho) + B_2(\rho)K_2(\rho)$ is invertible. Since, $A_{22}(\rho)$ is invertible by assumption, this is equivalent to say that

$$I + B_2(\rho)K_2(\rho)A_{22}(\rho)^{-1}$$

is invertible as well and, after multiplying on the right by $B_2(\rho)$, which is full-column rank, we get that the matrix

$$B_2(\rho)[I + K_2(\rho)A_{22}(\rho)^{-1}B_2(\rho)]$$

is also full column-rank. This finally implies that the matrix $I+K_2(\rho)A_{22}(\rho)^{-1}B_2(\rho)$ is full-rank, hence invertible. The proof is complete. ■

We have the following corollary when $K_2 = 0$:

Corollary 3.3.3 *The LPV system* (3.8) *is quadratically stabilizable by a state-feedback of the form* (3.4) *if there exist a constant matrix* $X_1 \in \mathbb{S}^n_{\succeq 0}$, *matrix functions* $X_2 : \boldsymbol{\Delta}_\rho \to \mathbb{R}^{\eta\times n}$, $X_3 : \boldsymbol{\Delta}_\rho \to \mathbb{R}^{\eta\times\eta}$ *and* $Y : \boldsymbol{\Delta}_\rho \to \mathbb{R}^{m\times n}$ *such that the LMI*

$$\begin{bmatrix} \text{He}\left[\tilde{A}(\rho)X(\rho) + \tilde{B}(\rho)\tilde{Y}(\rho)\right] & \tilde{E}(\rho) & \left[\tilde{C}X(\rho) + \tilde{D}\tilde{Y}(\rho)\right]^{\mathrm{T}} \\ \star & -\gamma I_p & \tilde{F}(\rho)^{\mathrm{T}} \\ \star & \star & -\gamma I_q \end{bmatrix} \prec 0 \tag{3.23}$$

holds for all $\rho \in \boldsymbol{\Delta}_\rho$ *where*

$$X = \begin{bmatrix} X_1 & 0 \\ X_2(\rho) & X_3(\rho) \end{bmatrix} \text{ and } \tilde{Y}(\rho) = \begin{bmatrix} Y(\rho) & 0 \end{bmatrix}. \tag{3.24}$$

Moreover, the state-feedback control law given by

$$u = Y(\rho)X_1^{-1}x \tag{3.25}$$

ensures that the L_2-gain of the transfer $w \to z$ is smaller than γ for all parameter trajectories in $\mathscr{P}^{\infty}$.

3.3.2 Quadratic Stabilization by Dynamic-Output Feedback

Let us consider now the dynamic control law (3.6) and the system (3.7). We have the following result:

Theorem 3.3.4 *There exists a gain-scheduled dynamic output feedback control law (3.6) of order n that quadratically stabilizes (3.7) and ensures that the L_2-gain of the transfer $w \to z$ is less than $\gamma > 0$ if and only if there exist matrices $X_1, Y_1 \in \mathbb{S}^n_{\succ 0}$ such that the LMIs*

$$N_Y^{\mathrm{T}} \begin{bmatrix} A(\rho)Y_1 + Y_1 A(\rho)^{\mathrm{T}} & Y_1 C(\rho)^{\mathrm{T}} & E(\rho) \\ \star & -\gamma I_q & F(\rho) \\ \star & \star & -\gamma I_p \end{bmatrix} N_Y \prec 0 \tag{3.26}$$

$$N_X^{\mathrm{T}} \begin{bmatrix} X_1 A(\rho) + A(\rho)^{\mathrm{T}} X_1 & X_1 E(\rho) & C(\rho)^{\mathrm{T}} \\ \star & -\gamma I_p & F(\rho)^{\mathrm{T}} \\ \star & \star & -\gamma I_q \end{bmatrix} N_X \prec 0 \tag{3.27}$$

and

$$\begin{bmatrix} X_1 & I \\ \star & Y_1 \end{bmatrix} \succ 0 \tag{3.28}$$

hold for all $\rho \in \boldsymbol{\Delta}_\rho$ and for full-rank matrices $N_X(\rho)$, $N_Y(\rho)$ defined as

$$\begin{bmatrix} C_y(\rho) & F_y(\rho) & 0_{r\times q} \end{bmatrix} N_X(\rho) = 0 \text{ and } N_Y(\rho)^{\mathrm{T}} \begin{bmatrix} B(\rho) \\ D(\rho) \\ 0_{p\times m} \end{bmatrix} = 0. \tag{3.29}$$

Proof The closed-loop system obtained from the interconnection of (3.6) and (3.7) is given by

$$\begin{bmatrix} \dot{x} \\ \dot{x}_c \\ \hline z \end{bmatrix} = \left(\left[\begin{array}{c|c} \bar{A}(\rho) & \bar{E}(\rho) \\ \hline \bar{C}(\rho) & F(\rho) \end{array} \right] + \left[\begin{array}{c} \bar{B}(\rho) \\ \hline \bar{D}(\rho) \end{array} \right] \bar{K}(\rho) \left[\bar{C}_y(\rho) \,\middle|\, \bar{F}_y(\rho) \right] \right) \begin{bmatrix} x \\ x_c \\ \hline w \end{bmatrix} \tag{3.30}$$

where $\bar{C}(\rho) = \begin{bmatrix} C(\rho) & 0 \end{bmatrix}$, $\bar{D}(\rho) = \begin{bmatrix} 0 & D(\rho) \end{bmatrix}$ and

$$\begin{aligned} \bar{A}(\rho) &= \begin{bmatrix} A(\rho) & 0 \\ 0 & 0 \end{bmatrix}, & \bar{E}(\rho) &= \begin{bmatrix} E(\rho) \\ 0 \end{bmatrix}, & \bar{B}(\rho) &= \begin{bmatrix} 0 & B(\rho) \\ I & 0 \end{bmatrix}, \\ \bar{C}_y(\rho) &= \begin{bmatrix} 0 & I \\ C_y(\rho) & 0 \end{bmatrix}, & \bar{F}_y(\rho) &= \begin{bmatrix} 0 \\ F_y(\rho) \end{bmatrix}, & \bar{K}(\rho) &= \begin{bmatrix} A_c(\rho) & B_c(\rho) \\ C_c(\rho) & D_c(\rho) \end{bmatrix}. \end{aligned} \tag{3.31}$$

Furthermore, let

$$\begin{aligned} A_{cl}(\rho) &= \bar{A}(\rho) + \bar{B}(\rho)\bar{K}(\rho)\bar{C}_y(\rho), \\ C_{cl}(\rho) &= \bar{C}(\rho) + \bar{D}(\rho)\bar{K}(\rho)\bar{F}_y(\rho). \end{aligned} \tag{3.32}$$

The closed-loop system is quadratically stable if and only if the LMI

$$\begin{bmatrix} \mathrm{He}\,[XA_{cl}(\rho)] & X\bar{E}(\rho) & C_{cl}(\rho)^{\mathrm{T}} \\ \star & -\gamma I_p & \bar{F}_y(\rho) \\ \star & \star & -\gamma I_q \end{bmatrix} \prec 0 \tag{3.33}$$

holds for all $\rho \in \boldsymbol{\Delta}_\rho$, for some $X \in \mathbb{S}^{2n}_{\succ 0}$ and some $\bar{K} : \boldsymbol{\Delta}_\rho \to \mathbb{R}^{(n+m)\times(n+p)}$. Letting

$$X = \begin{bmatrix} X_1 & X_2 \\ \star & X_3 \end{bmatrix}, \quad Y := X^{-1} = \begin{bmatrix} Y_1 & Y_2 \\ \star & Y_3 \end{bmatrix} \tag{3.34}$$

and applying the Projection Lemma (see Appendix C.12) yield the result. Moreover, by virtue of the Completion Lemma (see Appendix C.13) the condition (3.28) is equivalent to the existence of positive definite matrices (3.34). ■

Remark 3.1 The conditions of Theorem 3.3.4 can be modified in order to specify existence conditions for a dynamic output feedback of reduced order, say $n_c < n$. In this case, the rank constraint

$$\operatorname{rank} \begin{bmatrix} X_1 & I \\ \star & Y_1 \end{bmatrix} \leq n + n_c \tag{3.35}$$

must be added to Theorem 3.3.4 and (3.28) becomes

$$\begin{bmatrix} X_1 & I \\ \star & Y_1 \end{bmatrix} \succeq 0. \tag{3.36}$$

Due to the presence of the rank-constraint, the problem is non-convex (NP-hard). Several approaches have been proposed to solve such problems; see e.g. [15, 24].

So far, the conditions are only existence conditions, no explicit formula has been provided for constructing the controller matrices. The following result addresses this point [25, 26]:

Proposition 3.3.5 *Assume that D and F_y are full-column and full-row rank, respectively. Then, the controller can be constructed by using the following procedure:*

1. *Compute $D_c(\rho)$ solution of*

$$\bar{\sigma}\left(F(\rho)+D(\rho)D_c(\rho)F_y(\rho)\right)<\gamma \tag{3.37}$$

and set $D_{cl}(\rho):=F(\rho)+D(\rho)D_c(\rho)F_y(\rho)$.

2. *Solve for $\hat{B}_c(\rho)$ and $\hat{C}_c(\rho)$ in*

$$\begin{bmatrix} 0 & F_y(\rho) & 0 \\ F_y(\rho)^{\mathrm{T}} & -\gamma I_p & D_{cl}(\rho)^{\mathrm{T}} \\ 0 & D_{cl}(\rho) & -\gamma I_q \end{bmatrix}\begin{bmatrix} \hat{B}_c(\rho)^{\mathrm{T}} \\ \star \\ \star \end{bmatrix} = -\begin{bmatrix} C_y(\rho) \\ E(\rho)^{\mathrm{T}}X_1 \\ C(\rho)+D(\rho)D_c(\rho)C_y(\rho) \end{bmatrix}$$

$$\begin{bmatrix} 0 & D(\rho)^{\mathrm{T}} & 0 \\ D(\rho) & -\gamma I_q & D_{cl}(\rho) \\ 0 & D_{cl}(\rho)^{\mathrm{T}} & -\gamma I_p \end{bmatrix}\begin{bmatrix} \hat{C}_c(\rho) \\ \star \\ \star \end{bmatrix} = -\begin{bmatrix} B(\rho)^{\mathrm{T}} \\ C(\rho)^{\mathrm{T}}Y_1 \\ \left(E(\rho)+B(\rho)D_c(\rho)F_y(\rho)\right]^{\mathrm{T}} \end{bmatrix}.$$

3. *Compute*

$$\begin{aligned}\hat{A}_c(\rho) = &-\left[A(\rho)+B(\rho)D_c(\rho)C_y(\rho)\right]^{\mathrm{T}} \\ &+\begin{bmatrix} \left(XE(\rho)+\hat{B}_c(\rho)F_y(\rho)\right)^{\mathrm{T}} \\ C(\rho)+D(\rho)D_c(\rho)C_y(\rho) \end{bmatrix}^{\mathrm{T}} M \begin{bmatrix} (E(\rho)+B(\rho)D_c(\rho)F_y(\rho))^{\mathrm{T}} \\ C(\rho)Y_1+D(\rho)\hat{C}_c(\rho) \end{bmatrix}\end{aligned}$$

where

$$M=\begin{bmatrix} -\gamma I_p & D_{cl}(\rho)^{\mathrm{T}} \\ \star & -\gamma I_q \end{bmatrix}^{-1}.$$

4. *Solve for X_2 and Y_2 in the factorization problem*

$$X_2Y_2^{\mathrm{T}}=I-X_1Y_1 \tag{3.38}$$

using singular value decomposition.

5. *Solve for the controller matrices*

$$\begin{aligned} A_c(\rho) = &X_2^{-1}\left(\hat{A}_c(\rho)-X_1(A(\rho)-B(\rho)D_c(\rho)C_y(\rho))Y_1\right. \\ &\left.-\hat{B}_c(\rho)C_y(\rho)Y_1-X_1B(\rho)\hat{C}_c(\rho)\right)Y_2^{-\mathrm{T}} \end{aligned} \tag{3.39}$$

$$B_c(\rho)=X_2^{-1}\left(\hat{B}_c(\rho)-X_1B(\rho)D_c(\rho)\right) \tag{3.40}$$

$$C_c(\rho)=\left[\hat{C}_c(\rho)-D_c(\rho)C_y(\rho)Y_1\right]Y_2^{-\mathrm{T}}. \tag{3.41}$$

The first step can be addressed numerically by solving for $D_c(\rho)$ in the LMI

$$\begin{bmatrix} -\gamma I & F(\rho) + D(\rho)D_c(\rho)F_y(\rho) \\ \star & -\gamma I \end{bmatrix} \prec 0. \tag{3.42}$$

If an analytical solution is preferred, then we can use the results of [27] to obtain

$$D_c(\rho) = -(D(\rho)^{\mathrm{T}}\Phi(\rho)D(\rho))^{-1}D(\rho)^{\mathrm{T}}\Phi(\rho)F(\rho)F_y(\rho)^{\mathrm{T}}R_c(\rho) \tag{3.43}$$

where $R_c(\rho) = (F_y(\rho)F_y(\rho)^{\mathrm{T}})^{-1}$ and

$$\Phi(\rho) = \left(-\gamma^2 I - F(\rho)F(\rho)^{\mathrm{T}} + F(\rho)F_y(\rho)^{\mathrm{T}}R_c(\rho)F_y(\rho)F(\rho)^{\mathrm{T}}\right)^{-1}. \tag{3.44}$$

3.3.3 Robust Stabilization by State-Feedback

Let us address now the case of robust stabilization. This type of results is concerned with the case of parameter trajectories in $\mathscr{P}^\nu$.

Theorem 3.3.6 *The LPV system (3.7) is robustly stabilizable by a state-feedback of the form (3.4) if there exist a differentiable matrix function $X : \boldsymbol{\Delta}_\rho \to \mathbb{S}^n_{\succ 0}$ and a matrix function $Y : \boldsymbol{\Delta}_\rho \to \mathbb{R}^{m\times n}$ such that the LMI*

$$\begin{bmatrix} \Xi(\rho,\nu) & E(\rho) & [C(\rho)X(\rho) + D(\rho)Y(\rho)]^{\mathrm{T}} \\ \star & -\gamma I_p & F(\rho)^{\mathrm{T}} \\ \star & \star & -\gamma I_q \end{bmatrix} \prec 0 \tag{3.45}$$

holds for all $(\rho,\nu) \in \boldsymbol{\Delta}_\rho \times \mathbf{V}_\nu$ where

$$\Xi(\rho,\nu) := \mathrm{He}\left[A(\rho)X(\rho) + B(\rho)Y(\rho)\right] - \sum_{i=1}^{N} \nu_i \frac{\partial X}{\partial \rho_i}(\rho). \tag{3.46}$$

Moreover, the state-feedback control lawgiven by

$$u = Y(\rho)X(\rho)^{-1}x \tag{3.47}$$

ensures that the L_2-gain of the transfer $w \to z$ of the closed-loop system is less than γ for all parameter trajectories in $\mathscr{P}^\nu$.

Proof The closed-loop system is given by

$$\dot{x} = (A(\rho) + B(\rho)K(\rho))x + E(\rho)w$$
$$z = (C(\rho) + D(\rho)K(\rho))x + F(\rho)w.$$

Substituting in the conditions in the Bounded Real Lemma yields the inequality

$$\begin{bmatrix} \mathcal{M}(\rho) + \partial P(\rho) & P(\rho)E(\rho) & [C(\rho) + D(\rho)K(\rho)]^{\mathrm{T}} \\ \star & -\gamma I_p & F(\rho)^{\mathrm{T}} \\ \star & \star & -\gamma I_q \end{bmatrix} \prec 0$$

where $\mathcal{M}(\rho) = \mathrm{He}\,[P(\rho)A(\rho) + P(\rho)B(\rho)K(\rho)]$, $\partial P(\rho) = \sum_{i=1}^{N} \nu_i \dfrac{\partial P}{\partial \rho_i}(\rho)$ and $(\rho, \dot{\rho}) \in \boldsymbol{\Delta}_\rho \times \boldsymbol{\Delta}_\nu$. Performing first a congruence transformation with respect to the matrix $\mathrm{diag}(X(\rho), I, I)$ where $X(\rho) := P(\rho)^{-1}$, and using the fact that a nonsingular and differentiable matrix $Q(t)$ verifies[1]

$$Q^{-1}(t)\frac{dQ(t)}{dt}Q^{-1}(t) = -\frac{d}{dt}\left[Q(t)^{-1}\right] \tag{3.48}$$

we get the final result where $Y(\rho) = K(\rho)X(\rho)$. ■

3.3.4 Robust Stabilization by Dynamic-Output Feedback

Theorem 3.3.7 *There exist a gain-scheduled dynamic output feedback control law (3.6) of order n that robustly stabilizes (3.7) if and only if there exist differentiable matrix functions $X_1, Y_1 : \boldsymbol{\Delta} \to \mathbb{S}^n_{\succ 0}$ such that the LMIs*

$$N_Y(\rho)^{\mathrm{T}} \begin{bmatrix} \mathrm{He}[A(\rho)Y_1(\rho)] - \sum_i \nu_i \dfrac{\partial Y_1(\rho)}{\partial \rho_i} & Y_1(\rho)C(\rho)^{\mathrm{T}} & E(\rho) \\ \star & -\gamma I_q & F(\rho) \\ \star & \star & -\gamma I_p \end{bmatrix} N_Y(\rho) \prec 0 \tag{3.49}$$

$$N_X(\rho)^{\mathrm{T}} \begin{bmatrix} \mathrm{He}[X_1(\rho)A(\rho)] + \sum_i \nu_i \dfrac{\partial X_1(\rho)}{\partial \rho_i} & X_1(\rho)E(\rho) & C(\rho)^{\mathrm{T}} \\ \star & -\gamma I_p & F(\rho)^{\mathrm{T}} \\ \star & \star & -\gamma I_q \end{bmatrix} N_X(\rho) \prec 0 \tag{3.50}$$

and

$$\begin{bmatrix} X_1(\rho) & I \\ \star & Y_1(\rho) \end{bmatrix} \succ 0 \tag{3.51}$$

[1] To show this, use the relation $Q(t)Q(t)^{-1} = I$ and differentiate it.

hold for all $(\rho, \nu) \in \boldsymbol{\Delta}_\rho \times \mathbf{V}_\nu$ *and for full-rank matrices* $N_X(\rho)$, $N_Y(\rho)$ *defined as*

$$\begin{bmatrix} C_y(\rho) & F_y(\rho) & 0_{r\times q} \end{bmatrix} N_X(\rho) = 0 \text{ and } N_Y(\rho)^{\mathrm{T}} \begin{bmatrix} B(\rho) \\ D(\rho) \\ 0_{p\times m} \end{bmatrix} = 0.$$

Moreover, in such a case, the controller also ensures that the L_2*-gain of the transfer* $w \to z$ *of the closed-loop system is less than* γ *for all parameter trajectories in* $\mathscr{P}^\nu$.

Proposition 3.3.8 *The construction is very similar to as in the quadratic stabilization case:*

1. *Compute* $\hat{A}_c(\rho)$, $\hat{B}_c(\rho)$, $\hat{C}_c(\rho)$ *and* $\hat{D}_c(\rho)$ *as in Proposition 3.3.5.*
2. *Compute the controller matrices using the formulas*

$$\begin{aligned} A_c(\rho,\dot{\rho}) =& X_2(\rho)^{-1}\Big[X_1(\rho)Y_1'(\rho,\dot{\rho}) + X_2(\rho)Y_2'(\rho,\dot{\rho})^{\mathrm{T}} && (3.52)\\ &+ \hat{A}_c(\rho) - X_1(\rho)(A(\rho) - B(\rho)D_c(\rho)C_y(\rho))Y_1(\rho) && (3.53)\\ &- \hat{B}_c(\rho)C_y(\rho)Y_1(\rho) - X_1(\rho)B(\rho)\hat{C}_c(\rho)\Big]Y_2(\rho)^{-\mathrm{T}} && (3.54)\\ B_c(\rho) =& X_2(\rho)^{-1}\Big(\hat{B}_c(\rho) - X_1(\rho)B(\rho)D_c(\rho)\Big) && (3.55)\\ C_c(\rho) =& \Big[\hat{C}_c(\rho) - D_c(\rho)C_y(\rho)Y_1(\rho)\Big]Y_2(\rho)^{-\mathrm{T}} && (3.56)\end{aligned}$$

where

$$Y_j'(\rho,\dot{\rho}) = \sum_i \dot{\rho}_i \frac{\partial Y_j(\rho)}{\partial \rho_i}, \quad j = 1, 2.$$

An important difference with respect to the result on quadratic stability lies in the presence of the terms $Y_j'(\rho, \dot{\rho})$ that make the controller matrices dependent on $\dot{\rho}$. In general, derivatives of parameters are difficult to measure or to estimate, e.g. due to the presence of noise, and this makes the controllers defined above difficult to implement. This fact motivated the definition of *practically valid controllers* which refers to controllers that *do not depend* on the parameter derivatives [26].

In order to overcome this difficulty, several approaches have been provided in the literature. In [26], controllers are made practically valid by assigning a specific structure to the variables X_1, Y_1, X_2 and Y_2. In [28], controllers are scheduled with a filtered version of the parameters, the derivative of which is known. A different method which does not rely on filtering nor imposing the matrices a specific structure has been recently proposed in [29].

3.4 Polytopic LPV Systems

This section will be concerned with the control of polytopic LPV systems of the form

$$\begin{aligned}\dot{x}(t) &= A(\lambda(t))x(t) + Bu(t) + E(\lambda(t))w(t)\\ z(t) &= C(\lambda(t))x(t) + Du(t) + F(\lambda(t))w(t)\\ x(0) &= x_0\end{aligned} \tag{3.57}$$

where $x \in \mathbb{R}^n$ is the system state, $u \in \mathbb{R}^m$ is the control input, $w \in \mathbb{R}^p$ is the exogenous input and $z \in \mathbb{R}^q$ is the controlled output. The λ-dependent matrices are defined as

$$\begin{aligned}A(\lambda) &= \sum_{i=1}^{N} \lambda_i A_i, \quad E(\lambda) = \sum_{i=1}^{N} \lambda_i E_i\\ C(\lambda) &= \sum_{i=1}^{N} \lambda_i C_i, \quad F(\lambda) = \sum_{i=1}^{N} \lambda_i F_i\end{aligned}$$

where $\lambda \in \Lambda_N$. As in the previous sections, we assume that the trajectories of the parameters are such that solutions to (3.57) are well-defined. Note that the matrices B and D do not depend on λ for some technical reasons that will be explained later. It must be, however, pointed out that the system (3.57) is non-restrictive in the sense that any polytopic LPV system, i.e. with $B(\lambda)$ and $D(\lambda)$ matrices, can be turned into a system of the form (3.57) by suitably filtering the input by a low-pass filter [20] and augmenting the system with the state of the filter.

In this section on polytopic LPV systems, we will consider a polytopic version of the state-feedback (3.4) having the form

$$u(t) = \left(\sum_{i=1}^{N} \lambda_i(t) K_i\right) x(t) \tag{3.58}$$

where the K_i's are the gains to be determined. The case of dynamic output feedback is not treated since this is very similar to the case of generic LPV systems; see e.g. [30].

3.4.1 *Quadratic Stabilization by State-Feedback*

Let us start with the case of quadratic stabilization. We have the following result:

Theorem 3.4.1 *The LPV system* (3.57) *is quadratically stabilizable using a state-feedback of the form* (3.58) *if there exist a matrix* $X \in \mathbb{S}^n_{\succ 0}$*, matrices* $Y_i \in \mathbb{R}^{m\times n}$, $i = 1, \ldots, N$, *and a scalar* $\gamma > 0$ *such that the LMIs*

$$\begin{bmatrix} \mathrm{He}\,[A_iX + BY_i] & E_i & (C_iX + DY_i)^{\mathrm{T}} \\ \star & -\gamma I_p & F_i^{\mathrm{T}} \\ \star & \star & -\gamma I_q \end{bmatrix} \prec 0 \tag{3.59}$$

hold for all $i = 1, \ldots, N$. *Moreover, the state-feedback control law given by* (3.58) *with the matrices* $K_i = Y_iX^{-1}$ *ensures that the* L_2*-gain of the transfer* $w \to z$ *is smaller than* $\gamma > 0$ *for all* $\lambda : \mathbb{R}_{\geq 0} \to \Lambda_N$.

Proof Substituting the closed-loop system into the Bounded-Real Lemma yields the LMI condition

$$\sum_{i=1}^{N} \lambda_i \begin{bmatrix} \mathrm{He}[P(A_i + BK_i)] & PE_i & (C_i + DK_i)^{\mathrm{T}} \\ \star & -\gamma I_p & F_i^{\mathrm{T}} \\ \star & \star & -\gamma I_q \end{bmatrix} \prec 0 \tag{3.60}$$

that must hold for all $\lambda \in \Lambda_N$. Performing a congruence transformation with respect to $\mathrm{diag}(X, I_p, I_q)$, $X = P^{-1}$, and exploiting the convexity (in λ) of the LMI condition yield the result where we have set $Y_i := K_iX$. ■

It is important to stress that if B or D were depending on λ, we would have obtained quadratic terms in λ in the LMI condition (3.60). These terms are more difficult to handle since convexity is usually lost. Several methods can be used to deal with such a case; see e.g. [31–34].

3.4.2 Robust Stabilization by State-Feedback

Let us continue with the robust stabilization case:

Theorem 3.4.2 *The LPV system* (3.57) *is robustly stabilizable using a state-feedback of the form* (3.58) *if there exist matrices* $Q_i \in \mathbb{S}^n_{\succ 0}$, $i = 1, \ldots, N$, *a matrix* $W \in \mathbb{R}^{n\times n}$ *and a sufficiently large scalar* $\xi > 0$ *such that the matrix* *inequalities*

$$\begin{bmatrix} -\mathrm{He}[W] & Q_i + A_i W + BY_i & W & E_i & (C_i W + DY_i)^{\mathrm{T}} \\ \star & -\xi Q_i + \sum_{j=1}^{N} Q_j\theta_j & 0 & 0 & 0 \\ \star & \star & -Q_i/\xi & 0 & 0 \\ \star & \star & \star & -\gamma I_q & F_i^{\mathrm{T}} \\ \star & \star & \star & \star & -\gamma I_q \end{bmatrix} \prec 0 \tag{3.61}$$

hold for all $i = 1, \ldots, N$, *and all* $\theta \in$ ***vert***$\{\dot{\Lambda}_N\}$. *Moreover, the state-feedback control law given by* (3.58) *with matrices* $K_i = Y_i W^{-1}$ *ensures that the* L_2*-gain of the transfer* $w \to z$ *is smaller than* $\gamma > 0$ *for all* $\lambda : \mathbb{R}_{\geq 0} \to \Lambda_N$, $\dot{\lambda} \in \dot{\Lambda}_N$.

Proof The proof is similar as for quadratic stabilization with the difference that we use Theorem 2.5.6. ■

3.5 LPV Systems in LFT-Form

Only quadratic stabilization by state-feedback and dynamic output feedback using constant D-scalings [19, 20] will be addressed in this section. The rationale for restricting us to this case is for simplicity of exposure in introducing recent ideas and tools behind LPV control design in LFT-form. The recent results based on the full-block S-procedure [21] and frequency-dependent D-scales [22] are obviously more accurate but way more complicated to expose. The purpose of this section is to give the readers a familiarity with the ideas and tools involved in LPV control in LFT-form but practitioners are encouraged to use more recent results, such as the one in [22].

The following class of LPV systems in LFT-form are considered in this section

$$\begin{aligned} \begin{bmatrix} \dot{x}(t) \\ z_0(t) \\ z_1(t) \\ \hline y(t) \end{bmatrix} &= \left[\begin{array}{ccc|c} A & E_0 & E_1 & B \\ C_0 & F_{00} & F_{01} & D_0 \\ C_1 & F_{10} & F_{11} & D_1 \\ \hline C_y & F_{y0} & F_{y1} & 0 \end{array}\right] \begin{bmatrix} x(t) \\ w_0(t) \\ w_1(t) \\ \hline u(t) \end{bmatrix} \\ w_0(t) &= \Theta(\rho(t)) z_0(t) \\ x(0) &= x_0 \end{aligned} \tag{3.62}$$

where $x \in \mathbb{R}^n$, $w_1 \in \mathbb{R}^p$, $z_1 \in \mathbb{R}^q$, $u \in \mathbb{R}^m$, $y \in \mathbb{R}^r$ are the state of the system, the exogenous inputs, the controlled output the control input and the measured output, respectively. The channel $z_0, w_0 \in \mathbb{R}^{n_0}$ is the scheduling channel through which the parameters act on the system. The matrix $\Theta(\rho)$ is assumed to be diagonal with affine entries in ρ and such that

$$||\Theta(\rho)||_2 \leq 1 \tag{3.63}$$

for all $\rho \in \boldsymbol{\Delta}_\rho := [-1, 1]^N$ where $N > 0$ is the number of distinct parameters. We also need the assumption that the system (3.62) is well-posed, i.e. the matrix $I - F_{00}\Theta(\rho)$ is invertible for all $\rho \in \boldsymbol{\Delta}_\rho$.

The following result, proved in [21, 35], will play a central role in the derivation of the next results:

Lemma 3.5.1 (Dualijection Lemma [21, 35]) *Let $M \in \mathbb{S}^n$ be a nonsingular symmetric matrix having n^+ positive eigenvalues and $n^- := n - n^+$ negative eigenvalues, and let $S \in \mathbb{R}^{n \times n^-}$ be a full column rank matrix which can be written as*

$$S = \begin{bmatrix} I_{n^-} \\ S_1 + S_2 X S_3 \end{bmatrix} \tag{3.64}$$

for some given matrices S_1, S_2 and S_3.

Then, the following statements are equivalent:

1. *There exists X such that*

$$\begin{bmatrix} I_{n^-} \\ S_1 + S_2 X S_3 \end{bmatrix}^{\mathrm{T}} M \begin{bmatrix} I_{n^-} \\ S_1 + S_2 X S_3 \end{bmatrix} \prec 0 \tag{3.65}$$

holds.

2. *The matrix inequalities*

$$\mathcal{N}_3^{\mathrm{T}} \begin{bmatrix} I_{n^-} \\ S_1 \end{bmatrix}^{\mathrm{T}} M \begin{bmatrix} I_{n^-} \\ S_1 \end{bmatrix} \mathcal{N}_3 \prec 0 \tag{3.66}$$

and

$$\mathcal{N}_2^{\mathrm{T}} \begin{bmatrix} S_1^{\mathrm{T}} \\ -I_{n^+} \end{bmatrix}^{\mathrm{T}} M^{-1} \begin{bmatrix} S_1^{\mathrm{T}} \\ -I_{n^+} \end{bmatrix} \mathcal{N}_2 \succ 0 \tag{3.67}$$

hold for full-rank matrices $\mathcal{N}_2, \mathcal{N}_3$ defined as $S_3\mathcal{N}_3 = 0$ and $S_2^{\mathrm{T}}\mathcal{N}_2 = 0$.

This result is an elegant mix of the Dualization Lemma (see Appendix C.9) and the Projection Lemma (see Appendix C.12), whence the name *Dualijection Lemma.* It provides a compact and generic way for solving control problems as long as the hypothesis on the matrices M and S are satisfied. It can be straightforwardly generalized to the case where the upper-block of the matrix S is not the identity matrix anymore. In this case, we have the following result:

Lemma 3.5.2 (Generalized Dualijection Lemma) *Let $M \in \mathbb{S}^n$ be a nonsingular symmetric matrix having n^+ positive eigenvalues and $n^- := n - n^+$ negative eigenvalues, $S \in \mathbb{R}^{n\times n^-}$ be a full column rank matrix which can be written as*

$$S = \begin{bmatrix} J \\ S_1 + S_2XS_3 \end{bmatrix} \tag{3.68}$$

for some given matrices S_1, S_2, S_3 and J where J is nonsingular.

Then, the following statements are equivalent:

1. *There exists X such that*

$$\begin{bmatrix} J \\ S_1 + S_2XS_3 \end{bmatrix}^{\mathrm{T}} M \begin{bmatrix} J \\ S_1 + S_2XS_3 \end{bmatrix} \prec 0 \tag{3.69}$$

holds.

2. *The matrix inequalities*

$$\mathcal{N}_3^{\mathrm{T}} \begin{bmatrix} J \\ S_1 \end{bmatrix}^{\mathrm{T}} M \begin{bmatrix} J \\ S_1 \end{bmatrix} \mathcal{N}_3 \prec 0 \tag{3.70}$$

and

$$\mathcal{N}_2^{\mathrm{T}} \begin{bmatrix} J^{-T} S_1^{\mathrm{T}} \\ -I_{n^+} \end{bmatrix}^{\mathrm{T}} M^{-1} \begin{bmatrix} J^{-T} S_1^{\mathrm{T}} \\ -I_{n^+} \end{bmatrix} \mathcal{N}_2 \succ 0 \tag{3.71}$$

hold for full-rank matrices $\mathcal{N}_2$, $\mathcal{N}_3$ defined as $S_3\mathcal{N}_3 = 0$ and $S_2^{\mathrm{T}}\mathcal{N}_2 = 0$.

3.5.1 Quadratic Stabilization by State-Feedback

Let us consider first in this section gain-scheduled state-feedback controllers of the form

$$\begin{aligned} \begin{bmatrix} u(t) \\ z_c(t) \end{bmatrix} &= \begin{bmatrix} K_{ux} & K_{uc} \\ K_{cx} & K_{cc} \end{bmatrix} \begin{bmatrix} x(t) \\ w_c(t) \end{bmatrix} \\ w_c(t) &= \Theta(\rho(t)) z_c(t) \end{aligned} \tag{3.72}$$

where z_c/w_c is the controller scheduling channel. Note that the system and the controller are scheduled exactly in the same way, i.e. in an LFT-fashion and with respect to the same matrix $\Theta(\rho)$. Note that scheduling may not be necessarily performed using the same parameter matrix as, for instance, in the approach based on the full-block S-procedure where the scheduling matrix is constructed in order to satisfy a certain condition.

The next result makes use of constant D-scalings. For convenience, the set of D-scalings associated with a matrix $\Theta(\rho) \in \mathbb{R}^{n_0 \times n_0}$ is recalled below

$$\mathcal{D}(\Theta) = \left\{ L \in \mathbb{S}^{n_0}_{\succ 0} : L^{1/2}\Theta = \Theta L^{1/2} \right\} \tag{3.73}$$

where $L^{1/2}$ is the positive square-root of L. We then have the following result:

Theorem 3.5.3 *Assume that there exist matrices $X \in \mathbb{S}^{n}_{\succ 0}$, $L_1, J_1 \in \mathcal{D}(\Theta)$ and a scalar $\gamma > 0$ such that the LMIs*

$$\begin{bmatrix} L_1 & I \\ \star & \tilde{L}_1 \end{bmatrix} \succ 0 \tag{3.74}$$

$$\begin{bmatrix} F_{00} & F_{01} \\ F_{10} & F_{11} \end{bmatrix}^{\mathrm{T}} \begin{bmatrix} L_1 & 0 \\ \star & \gamma^{-1} I_q \end{bmatrix} \begin{bmatrix} F_{00} & F_{01} \\ F_{10} & F_{11} \end{bmatrix} - \begin{bmatrix} L_1 & 0 \\ 0 & \gamma I_p \end{bmatrix} \prec 0 \tag{3.75}$$

and

$$\mathcal{N}^{\mathrm{T}} \left(\begin{bmatrix} \mathrm{He}[AX] & \star & \star \\ C_0 X & -\tilde{L}_1 & \star \\ C_1 X & 0 & -\gamma I_q \end{bmatrix} + \begin{bmatrix} E_0 & E_1 \\ F_{00} & F_{01} \\ F_{10} & F_{11} \end{bmatrix} \tilde{M} \begin{bmatrix} E_0 & E_1 \\ F_{00} & F_{01} \\ F_{10} & F_{11} \end{bmatrix}^{\mathrm{T}} \right) \mathcal{N} \prec 0 \tag{3.76}$$

hold where $\mathcal{N}$ is defined as a full-rank matrix satisfying

$$\begin{bmatrix} B^{\mathrm{T}} & D_0^{\mathrm{T}} & D_1^{\mathrm{T}} \end{bmatrix} \mathcal{N} = 0$$

and

$$\tilde{M} := \begin{bmatrix} \tilde{L}_1 & 0 \\ 0 & \gamma^{-1} I_p \end{bmatrix}. \tag{3.77}$$

Then, there exists a well-posed gain-scheduled state-feedback of the form (3.72) that stabilizes the system (3.62) and ensures that the L_2-gain of the transfer $w_1 \to z_1$ is less than $\gamma > 0$ for all $\rho \in \mathscr{P}_1^{\infty}$.

Proof The closed-loop system can be written as

$$\left[\begin{array}{c}\dot{x}(t)\\ \hline z_0(t)\\ z_c(t)\\ \hline z_1(t)\end{array}\right] = \left(\bar{A} + \bar{B}\bar{K}\bar{C}\right)\left[\begin{array}{c}x(t)\\ \hline w_0(t)\\ w_c(t)\\ \hline w_1(t)\end{array}\right]$$

$$\begin{bmatrix} w_0(t)\\ w_c(t)\end{bmatrix} = \Theta_a(\rho(t))\begin{bmatrix} z_0(t)\\ z_c(t)\end{bmatrix} \tag{3.78}$$

where $\Theta_a(\rho) := \mathrm{diag}(\Theta(\rho), \Theta(\rho))$ and

$$\bar{A} = \left[\begin{array}{c|cc|c} A & E_0 & 0 & E_1\\ \hline C_0 & F_{00} & 0 & F_{01}\\ 0 & 0 & 0 & 0\\ \hline C_1 & F_{10} & 0 & F_{11}\end{array}\right], \quad \bar{B} = \left[\begin{array}{cc} B & 0\\ \hline D_0 & 0\\ 0 & I\\ \hline D_1 & 0\end{array}\right],$$

$$\bar{C} = \left[\begin{array}{c|cc|c} I & 0 & 0 & 0\\ 0 & 0 & I & 0\end{array}\right], \quad \bar{K} = \begin{bmatrix} K_{ux} & K_{uc}\\ K_{cx} & K_{cc}\end{bmatrix}. \tag{3.79}$$

Substituting this system in the Scaled-Bounded Real Lemma (i.e. Lemma 2.6.12) yields the LMI (after factorization)

$$\mathcal{S}^{\mathrm{T}}\left[\begin{array}{cccc|cccc} 0 & 0 & 0 & 0 & P & 0 & 0 & 0\\ \star & -L_1 & -L_2 & 0 & 0 & 0 & 0 & 0\\ \star & \star & -L_3 & 0 & 0 & 0 & 0 & 0\\ \star & \star & \star & -\gamma I_p & 0 & 0 & 0 & 0\\ \hline \star & \star & \star & \star & 0 & 0 & 0 & 0\\ \star & \star & \star & \star & \star & L_1 & L_2 & 0\\ \star & \star & \star & \star & \star & \star & L_3 & 0\\ \star & \star & \star & \star & \star & \star & \star & \gamma^{-1}I_q\end{array}\right]\mathcal{S} \prec 0 \tag{3.80}$$

where $L = \begin{bmatrix} L_1 & L_2\\ \star & L_3\end{bmatrix} \in \mathcal{D}(\Theta_a)$ and

$$\mathcal{S} = \left[\begin{array}{cccc} I & 0 & 0 & 0\\ 0 & I & 0 & 0\\ 0 & 0 & I & 0\\ 0 & 0 & 0 & I\\ \hline A & E_0 & 0 & E_1\\ C_0 & F_{00} & 0 & F_{01}\\ 0 & 0 & 0 & 0\\ C_1 & F_{10} & 0 & F_{11}\end{array}\right] + \left[\begin{array}{cc} 0 & 0\\ 0 & 0\\ 0 & 0\\ 0 & 0\\ \hline B & 0\\ D_0 & 0\\ 0 & I\\ D_1 & 0\end{array}\right]\bar{K}\begin{bmatrix} I & 0 & 0 & 0\\ 0 & 0 & I & 0\end{bmatrix}. \tag{3.81}$$

Note then that the LMI (3.80) can be rewritten as

$$\begin{bmatrix} I\\ \bar{A} + \bar{B}\bar{K}\bar{C}\end{bmatrix}^{\mathrm{T}} M \begin{bmatrix} I\\ \bar{A} + \bar{B}\bar{K}\bar{C}\end{bmatrix} \prec 0 \tag{3.82}$$

where M is the central matrix of (3.80). From the structure of M, it is clear that it has $n + 2n_0 + p$ negative eigenvalues and $n + q + 2n_0$ positive eigenvalues. Hence, the matrix is invertible. The rank of $\mathcal{S}$ is equal to $n + 2n_0 + p$. Lemma 3.5.1 therefore applies and we get the LMIs of Theorem 3.5.3 where $\tilde{L} = \begin{bmatrix} \tilde{L}_1 & \tilde{L}_2 \\ \star & \tilde{L}_3 \end{bmatrix} := L^{-1}$.

It remains to prove that the controller is well-posed in the sense that the control input u is causally and uniquely defined by the state x. First note that, upon feasibility of the conditions of the scaled bounded real lemma, the overall closed-loop system is well-posed. This means that the matrix

$$\begin{bmatrix} I & 0 \\ 0 & I \end{bmatrix} - \begin{bmatrix} F_{00} & D_0 K_{uc} \\ 0 & K_{cc} \end{bmatrix} \begin{bmatrix} \Theta(\rho) & 0 \\ 0 & \Theta(\rho) \end{bmatrix} \tag{3.83}$$

is invertible for all $\rho \in [-1, 1]^N$. Since the open-loop system is well-posed by assumption, i.e. $I - F_{00}\Theta(\rho)$ invertible for all $\rho \in [-1, 1]^N$, then the invertibility condition above is equivalent to the invertibility of $I - K_{cc}\Theta(\rho)$, which is equivalent, in turn, to the well-posedness of the controller. Therefore, the controller is well-posed. The proof is complete. ∎

Proposition 3.5.4 *The controller matrices can be constructed using the following procedure [20]:*

1. *Compute the matrices L_2 and L_3 such that*

$$L := \begin{bmatrix} L_1 & L_2 \\ \star & L_3 \end{bmatrix} \succ 0, \ L^{-1} = \begin{bmatrix} \tilde{L}_1 & \tilde{L}_2 \\ \star & \tilde{L}_3 \end{bmatrix} \tag{3.84}$$

by first using a singular value decomposition on each diagonal block of $I - L_1\tilde{L}_1$ to determine the product $L_2\tilde{L}_2^{\mathrm{T}}$ and then determining L by solving the equation:

$$\begin{bmatrix} \tilde{L}_1 & \tilde{L}_2 \\ I & 0 \end{bmatrix} L = \begin{bmatrix} I & 0 \\ L_1 & L_2 \end{bmatrix}. \tag{3.85}$$

2. *Solve for $\bar{K}$ in the LMI*

$$\Psi + \mathcal{U}\bar{K}\mathcal{V}^{\mathrm{T}} + \mathcal{V}\bar{K}^{\mathrm{T}}\mathcal{U}^{\mathrm{T}} \prec 0 \tag{3.86}$$

where

$$\Psi = \begin{bmatrix} A^{\mathrm{T}}P + PA & PE_0 & 0 & PE_1 & C_0^{\mathrm{T}} & 0 & C_1^{\mathrm{T}} \\ \star & -L_1 & -L_2 & 0 & F_{00}^{\mathrm{T}} & 0 & F_{10}^{\mathrm{T}} \\ \star & \star & -L_3 & 0 & 0 & 0 & 0 \\ \star & \star & \star & -\gamma I_p & F_{01}^{\mathrm{T}} & 0 & F_{11}^{\mathrm{T}} \\ \star & \star & \star & \star & -\tilde{L}_1 & -\tilde{L}_2 & 0 \\ \star & \star & \star & \star & \star & -\tilde{L}_3 & 0 \\ \star & \star & \star & \star & \star & \star & -\gamma I_q \end{bmatrix}$$

$$\mathcal{U} = \begin{bmatrix} PB & 0 \\ 0 & 0 \\ 0 & 0 \\ 0 & 0 \\ D_0 & 0 \\ 0 & I \\ D_1 & 0 \end{bmatrix} \text{ and } \mathcal{V}^{\mathrm{T}} = \begin{bmatrix} I & 0 & 0 & 0 & 0 & 0 & 0 \\ 0 & 0 & I & 0 & 0 & 0 & 0 \end{bmatrix}.$$

Proof The first step follows from the application of the Completion Lemma; see Appendix C.13. The second statement follows from the substitution of the closed-loop system into the Scaled-Bounded Real Lemma (Lemma 2.6.12) to obtain the inequality (3.86). Note that since P and L are known, then the inequality (3.86) is an LMI. ■

3.5.2 *Quadratic Stabilization by Dynamic Output-Feedback*

In this section, we consider gain-scheduled dynamic output-feedback of the form

$$\begin{bmatrix} \dot{x}_c(t) \\ z_c(t) \\ u(t) \end{bmatrix} = \begin{bmatrix} A_c & B_c & B_y \\ C_c & D_{cc} & D_{cy} \\ C_u & D_{uc} & D_{uy} \end{bmatrix} \begin{bmatrix} x_c(t) \\ w_c(t) \\ y(t) \end{bmatrix}$$
$$w_c(t) = \Theta(\rho(t))z_c(t) \tag{3.87}$$

where z_c/w_c is the scheduling channel. The design of such controllers using constant D-scalings has been solved independently in [19, 20]. In what follows, however, a proof that differs from the ones of the above references will be provided. The main reason for proposing this alternative proof is to show the convenience of Lemma 3.5.1 for solving gain-scheduled controller design problems in more complicated frameworks such as the one based on the full-block S-procedure and the use of frequency-dependent D-scales. The complexity of the proofs are, on the other hand, very similar.

The following result states sufficient conditions for the existence of gain-scheduled controllers of the form (3.87):

Theorem 3.5.5 *Assume that there exist matrices* $P_1, \tilde{P}_1 \in \mathbb{S}^n_{\succ 0}$, $L_1, \tilde{L}_1 \in \mathcal{D}(\Theta)$ *and a scalar* $\gamma > 0$ *such that the LMIs*

$$\begin{bmatrix} P_1 & I \\ \star & \tilde{P}_1 \end{bmatrix} \succ 0 \quad \begin{bmatrix} L_1 & I \\ \star & \tilde{L}_1 \end{bmatrix} \succ 0 \tag{3.88}$$

$$\mathcal{N}_C^{\mathrm{T}} \left(\begin{bmatrix} \mathrm{He}[P_1 A] & P_1 E_0 & P_1 E_1 \\ \star & -L_1 & 0 \\ \star & \star & -\gamma I_p \end{bmatrix} + \begin{bmatrix} C_0^{\mathrm{T}} & C_1^{\mathrm{T}} \\ F_{00}^{\mathrm{T}} & F_{10}^{\mathrm{T}} \\ F_{01}^{\mathrm{T}} & F_{11}^{\mathrm{T}} \end{bmatrix} M_1 \begin{bmatrix} C_0^{\mathrm{T}} & C_1^{\mathrm{T}} \\ F_{00}^{\mathrm{T}} & F_{10}^{\mathrm{T}} \\ F_{01}^{\mathrm{T}} & F_{11}^{\mathrm{T}} \end{bmatrix}^{\mathrm{T}} \right) \mathcal{N}_C \prec 0 \tag{3.89}$$

and

$$\mathcal{N}_B^{\mathrm{T}} \left(\begin{bmatrix} \mathrm{He}[A\tilde{P}_1] & \tilde{P}_1 C_0^{\mathrm{T}} & \tilde{P}_1 C_1^{\mathrm{T}} \\ \star & -\tilde{L}_1 & 0 \\ \star & \star & -\gamma I_q \end{bmatrix} + \begin{bmatrix} E_0 & E_1 \\ F_{00} & F_{01} \\ F_{10} & F_{11} \end{bmatrix} \tilde{M}_1 \begin{bmatrix} E_0 & E_1 \\ F_{00} & F_{01} \\ F_{10} & F_{11} \end{bmatrix}^{\mathrm{T}} \right) \mathcal{N}_B \prec 0 \tag{3.90}$$

hold with

$$M_1 = \begin{bmatrix} L_1 & 0 \\ \star & \gamma^{-1} I_q \end{bmatrix}, \quad \tilde{M}_1 = \begin{bmatrix} \tilde{L}_1 & 0 \\ \star & \gamma^{-1} I_p \end{bmatrix} \tag{3.91}$$

and where $\mathcal{N}_C$ *and* $\mathcal{N}_B$ *are full-rank matrices satisfying*

$$\begin{bmatrix} C_y & F_{y0} & F_{y1} \end{bmatrix} \mathcal{N}_C = 0 \text{ and } \begin{bmatrix} B^{\mathrm{T}} & D_0^{\mathrm{T}} & D_1^{\mathrm{T}} \end{bmatrix} \mathcal{N}_B = 0. \tag{3.92}$$

In such a case, there exists a gain-scheduled dynamic output-feedback of order n *of the form (3.87) that stabilizes system (3.62) and ensures that the* L_2*-gain of the transfer* $w_1 \to z_1$ *is less than* γ *for all* $\rho \in \mathscr{P}_1^{\infty}$.

Proof The closed-loop system is given by

$$\begin{bmatrix} \dot{x}(t) \\ \dot{x}_c(t) \\ \hline z_0(t) \\ z_c(t) \\ \hline z_1(t) \end{bmatrix} = (\bar{A} + \bar{B}\Omega\bar{C}) \begin{bmatrix} x(t) \\ x_c(t) \\ \hline w_0(t) \\ w_c(t) \\ \hline w_1(t) \end{bmatrix} \tag{3.93}$$

where

$$\bar{A} = \left[\begin{array}{cc|cc|c} A & 0 & E_0 & 0 & E_1 \\ 0 & 0 & 0 & 0 & 0 \\ \hline C_0 & 0 & F_{00} & 0 & F_{01} \\ 0 & 0 & 0 & 0 & 0 \\ \hline C_1 & 0 & F_{10} & 0 & F_{11} \end{array}\right], \quad \bar{B} = \left[\begin{array}{ccc} 0 & 0 & B \\ I & 0 & 0 \\ \hline 0 & 0 & D_0 \\ 0 & I & 0 \\ \hline 0 & 0 & D_1 \end{array}\right]$$

$$\Omega = \begin{bmatrix} A_c & B_c & B_y \\ C_c & D_{cc} & D_{cy} \\ C_u & D_{uc} & D_{uy} \end{bmatrix} \text{ and } \bar{C} = \left[\begin{array}{cc|cc|c} 0 & I & 0 & 0 & 0 \\ 0 & 0 & 0 & I & 0 \\ C_y & 0 & F_{y0} & 0 & F_{y1} \end{array}\right]. \tag{3.94}$$

The rest of the procedure follows from the same lines as for state-feedback design, i.e. we apply Lemma 3.5.1 to eliminate the controller matrices and obtain the LMI conditions. ■

Proposition 3.5.6 *The controller matrix can be constructed using the following procedure [20]:*

1. *Compute the matrices $L \in \mathcal{D}(\Theta_a)$ from L_1 and $\tilde{L}_1$ as in Proposition 3.5.4.*
2. *Compute the matrix $P \succ 0$ from the matrices P_1 and $\tilde{P}_1$ using singular value decomposition.*
3. *Solve for $\bar{K}$ in the LMI*

$$\Psi + \mathcal{U}\bar{K}\mathcal{V}^{\mathrm{T}} + \mathcal{V}\bar{K}^{\mathrm{T}}\mathcal{U}^{\mathrm{T}} \prec 0 \tag{3.95}$$

where

$$\Psi = \begin{bmatrix} \mathrm{He}[P_1 A] & A^{\mathrm{T}} P_2 & P E_0 & 0 & P E_1 & C_0^{\mathrm{T}} & 0 & C_1^{\mathrm{T}} \\ \star & 0 & 0 & 0 & 0 & 0 & 0 & 0 \\ \star & \star & -L_1 & -L_2 & 0 & F_{00}^{\mathrm{T}} & 0 & F_{10}^{\mathrm{T}} \\ \star & \star & \star & -L_3 & 0 & 0 & 0 & 0 \\ \star & \star & \star & \star & -\gamma I_p & F_{01}^{\mathrm{T}} & 0 & F_{11}^{\mathrm{T}} \\ \star & \star & \star & \star & \star & -\tilde{L}_1 & -\tilde{L}_2 & 0 \\ \star & \star & \star & \star & \star & \star & -\tilde{L}_3 & 0 \\ \star & \star & \star & \star & \star & \star & \star & -\gamma I_q \end{bmatrix},$$

$$\mathcal{U} = \begin{bmatrix} P_2 & 0 & P_1 B \\ P_3 & 0 & P_2^{\mathrm{T}} B \\ 0 & 0 & 0 \\ 0 & 0 & 0 \\ 0 & 0 & 0 \\ 0 & 0 & D_0 \\ 0 & I & 0 \\ 0 & 0 & D_1 \end{bmatrix}, \mathcal{V}^{\mathrm{T}} = \begin{bmatrix} 0 & I & 0 & 0 & 0 & 0 & 0 & 0 \\ 0 & 0 & 0 & I & 0 & 0 & 0 & 0 \\ C_y & 0 & F_{y0} & 0 & F_{y1} & 0 & 0 & 0 \end{bmatrix}.$$

Remark 3.2 Unlike for the state-feedback case, the obtained controller may not be well-posed since the scaled-bounded real lemma only ensures that the matrix

$$I - \begin{bmatrix} D_{cc} & D_{cy} F_{y0} \\ D_0 D_{uc} & F_{00} + D_0 D_{uy} F_{y0} \end{bmatrix} \begin{bmatrix} \Theta(\rho) & 0 \\ 0 & \Theta(\rho) \end{bmatrix} \tag{3.96}$$

is invertible for all $\rho \in \boldsymbol{\Delta}_\rho$ and this does not imply, in general, that $I - D_{cc}\Theta(\rho)$ is invertible for all $\rho \in \boldsymbol{\Delta}_\rho$.

When $D_0 = 0$ or $F_{y0} = 0$, well-posedness of the interconnection is then equivalent to well-posedness of the system and the controller, separately. These matrices can be made parameter independent by filtering the input or the output of the system by low-pass filters. Another procedure is proposed in [20] to generate well-posed controllers.

References

1. T. Takagi, M. Sugeno, Fuzzy identification of systems and its applications to modeling and control. IEEE Trans. Syst. Man Cybern. B Cybern. **15**(1), 116–132 (1985)
2. D.J. Leith, W.E. Leithead, Survey of gain-scheduling analysis and design. Int. J. Control **73**(11), 1001–1025 (2000)
3. J.S. Shamma, Analysis and design of gain-scheduled control systems. PhD thesis, Laboratory for Information and decision systems—Massachusetts Institute of Technology, 1988
4. J. Shamma, M. Athans, Analysis of gain scheduled control for nonlinear plants. IEEE Trans. Autom. Control **35**(8), 898–907 (1990)
5. J. Shamma, M. Athans, Guaranteed properties of gain scheduled control of linear parameter-varying plants. Automatica **27**(3), 559–564 (1991)
6. J.S. Shamma, M. Athans, Gain scheduling: potential hazards and possible remedies. IEEE Contr. Syst. Mag. **12**(3), 101–107 (1992)
7. W.J. Rugh, J.S. Shamma, Research on gain scheduling. Automatica **36**(10), 1401–1425 (2000)
8. J. Mohammadpour, C.W. Scherer (eds.), *Control of Linear Parameter Varying Systems with Applications* (Springer, New York, 2012)
9. J. Daafouz, J. Bernussou, J.C. Geromel, On inexact LPV control design of continuous-time polytopic systems. IEEE Trans. Autom. Control **53**(7), 1674–1978 (2008)
10. W.P.M.H. Heemels, J. Daafouz, G. Millerioux, Observer-based control of discrete-time LPV systems with uncertain parameters. IEEE Trans. Autom. Control **55**(9), 2130–2135 (2010)

11. M. Sato, D. Peaucelle, Gain-scheduled output-feedback controllers using inexact scheduling parameters for continuous-ime LPV systems. Automatica **49**, 1019–1025 (2013)
12. V.L. Syrmos, C.T. Abdallah, P. Dorato, K. Grigoriadis, Static output feedback—a survey. Automatica **33**(2), 125–137 (1997)
13. V. Blondel, J.N. Tsitsiklis, NP-hardness of some linear control problems. SIAM J. Control Optim. **35**(6), 2118–2127 (1997)
14. M. Fu, Pole placement via static output feedback is NP-hard. IEEE Trans. Autom. Control **49**(5), 855–857 (2004)
15. L. El-Ghaoui, F. Oustry, M. Ait Rami, A cone complementary linearization algorithm for static output-feedback and related problems. IEEE Trans. Autom. Control **42**, 1171–1176 (1997)
16. E. Prempain, I. Postlethwaite, Static $\mathcal{H}_\infty$ loop shaping control of a fly-by-wire helicopter. Automatica **41**, 1517–1528 (2005)
17. D. Henrion, J.B. Lasserre, Convergent relaxations of polynomial matrix inequalities and static output feedback. IEEE Trans. Autom. Control **51**(2), 192–202 (2006)
18. P. Apkarian, D. Noll, Nonsmooth H_∞ synthesis. IEEE Trans. Autom. Control **51**(1), 71–86 (2006)
19. A. Packard, Gain scheduling via linear fractional transformations. Syst. Control Lett. **22**, 79–92 (1994)
20. P. Apkarian, P. Gahinet, A convex characterization of gain-scheduled $\mathcal{H}_\infty$ controllers. IEEE Trans. Autom. Control **5**, 853–864 (1995)
21. C.W. Scherer, LPV control and full-block multipliers. Automatica **37**, 361–375 (2001)
22. C.W. Scherer, I.E. Köse, Gain-scheduled control synthesis using dynamic D-scales. IEEE Trans. Autom. Control **57**(9), 2219–2234 (2012)
23. C.W. Scherer, P. Gahinet, M. Chilali, Multiobjective output-feedback control via LMI optimization. IEEE Trans. Autom. Control **42**(7), 896–911 (1997)
24. T. Iwasaki, R.E. Skelton, The X-Y centering algorithm for the dual-LMI problem. Int. J. Control **62**, 1252–1257 (1995)
25. P. Gahinet, Explicit controller formulas for LMI-based H_∞ synthesis. Automatica **32**(7), 1007–1014 (1996)
26. P. Apkarian, R.J. Adams, Advanced gain-scheduling techniques for uncertain systems. IEEE Trans. Autom. Control **6**, 21–32 (1998)
27. R.E. Skelton, T. Iwasaki, K.M. Grigoriadis, *A Unified Algebraic Approach to Linear Control Design* (Taylor & Francis, London, 1997)
28. I. Masubuchi, I. Kurata, Gain-scheduled control via filtered scheduling parameters. Automatica **47**, 1821–1826 (2011)
29. R. Zope, J. Mohammadpour, K. Grigoriadis, M. Franchek, Delay-Dependent Output Feedback Control of Time-Delay LPV Systems, in *Control of Linear Parameter Varying Systems with Applications*, ed. by J. Mohammadpour, C.W. Scherer (Springer, US, 2012), pp. 181–216. M. sato and d. peaucelle
30. P. Apkarian, P. Gahinet, G. Becker, Self-scheduled control of linear parameter varying systems: a design example. Automatica **31**(9), 1251–1261 (1995)
31. P. Apkarian, H.D. Tuan, Parameterized LMIs in control theory. SIAM J. Control Optim. **38**(4), 1241–1264 (2000)
32. R.C.L.F. Oliveira, P.L.D. Peres, LMI conditions for robust stability analysis based on polynomially parameter-dependent Lyapunov functions. Syst. Control Lett. **55**, 52–61 (2006)
33. R.C.L.F. Oliveira, P.L.D. Peres, Parameter-dependent LMIs in robust analysis: Characterization of homogeneous polynomially parameter-dependent solutions via LMI relaxations. IEEE Trans. Autom. Control **57**(7), 1334–1340 (2002)
34. G. Chesi, LMI techniques for optimization over polynomials in control: a survey. IEEE Trans. Autom. Control **55**(11), 2500–2510 (2010)
35. C.W. Scherer, Robust mixed control and linear parameter-varying control with full block scalings, in *Advances in linear matrix inequalities methods in control*, ed. by L. El-Ghaoui, S.-I. Niculescu (SIAM, Philadelphia, 1999), pp. 187–207

Part II
Time-Delay Systems

Chapter 4
Introduction to Time-Delay Systems

Nous pouvons rêver d'équations fonctionelles plus compliquées que les équations classiques parce qu'elles renfermeront en outre des intégrales prises entre le temps passé très éloigné et le temps actuel, qui apporteront la part de l'hérédité. (We may dream about more complicated functional equations than classical equations since they shall in addition contain integrals taken between the distant past time and the current time, which shall bring the share of heredity.)

Emile Picard

Abstract The goal of this chapter is to introduce the main manners for representing time-delay systems. As for parameters in LPV systems, delays can also be classified in different categories depending on the their nature and the way they act on the system. Several real world examples are given to motivate the usefulness and relevance of time-delay systems in science and engineering. We notably discuss about the harmful and beneficial effects of the delays on the stability properties of dynamical systems. Controllers and observers that are specific to time-delay systems are finally briefly presented.

4.1 Representation of Time-Delay Systems

Three main distinct frameworks for representing time-delay systems are presented in this section: functional differential equations, differential equations with coefficients in a ring of operator and abstract representation on infinite-dimensional linear space.

4.1.1 Functional Differential Equations

The most common way for representing time-delay systems is by means of functional differential equations; see e.g. [1–5]. Perhaps the most simple, yet general, linear time-delay system is given in this framework by

C. Briat, *Linear Parameter-Varying and Time-Delay Systems*,
Advances in Delays and Dynamics 3, DOI 10.1007/978-3-662-44050-6_4

$$\begin{aligned} \dot{x}(t) &= Ax(t) + A_h x(t-h) \\ x(s) &= \varphi(s), \ s \in [-h, 0] \end{aligned} \tag{4.1}$$

where x is the state of the system, $h > 0$ is the constant delay and $\varphi \in C([-h, 0], \mathbb{R}^n)$ is the functional initial condition. Unlike, LPV systems, or finite-dimensional systems, discussed in the first part of this monograph, the solution of a time-delay system is not uniquely defined by the sole knowledge of the pointwise initial condition x_0 at $t = 0$ but by a functional initial condition $\varphi(\cdot)$ defined over the interval $[-h, 0]$ [3]. This critical difference tells us that time-delay systems are not finite-dimensional systems, but infinite-dimensional ones and that their state, i.e. the minimal information needed to properly define the notion of solutions, is not a single point $x(t)$ in $\mathbb{R}^n$, but a *function* x_t defined as $x_t(s) = x(t+s)$, $s \in [-h, 0]$.

4.1.2 Differential Equation with Coefficients in a Ring of Operators

This framework for time-delay systems has been developed quite early and has led to several important algebraic results on the analysis [6, 7], decoupling [8, 9], controllability [10], observability [11], control [9, 12] and observation [11] of time-delay systems.

The counterpart of (4.1) in this framework is given by the following differential equation with coefficients in a ring

$$\dot{x}(t) = \bar{A}(\nabla)x(t)$$

where $\bar{A}(\nabla) = A + A_h\nabla$ and ∇ is the shift operator defined as

$$(\nabla x)(t) = x(t-h). \tag{4.2}$$

It is important to mention here that the fact that the operator ∇ belongs to a ring is primordial from an engineering perspective. The inverse of ∇, denoted by ∇^{-1}, is the advance operator

$$(\nabla^{-1}x)(t) = x(t+h) \tag{4.3}$$

which violates causality since it is not possible to predict the future (in principle). If, however, the future were predictable, ∇ would have been considered as invertible and the coefficients of $\bar{A}$ to lie in a field.

4.1.3 Abstract Representation Over an Infinite Dimensional Linear Space

The framework is typically the general framework of infinite-dimensional systems [13–16] applied to the special case of time-delay systems, see e.g. [17–20]. In this framework, the state-space is defined as $\mathbb{R}^n \times L_2([-h, 0], \mathbb{R}^n)$ and the state writes

$$\tilde{x} = \begin{bmatrix} x(t) \\ x_t(s) \end{bmatrix}$$

where $s \in [-h, 0]$ and $x_t(s) = x(t+s)$. System (4.1) admits the following abstract representation

$$\frac{d}{dt}\begin{bmatrix} x(t) \\ x_t(\cdot) \end{bmatrix} = \mathcal{A}\begin{bmatrix} x(t) \\ x_t(\cdot) \end{bmatrix}$$

where the operator $\mathcal{A}$ is defined as

$$\mathcal{A}\begin{bmatrix} x(t) \\ x_t(\cdot) \end{bmatrix} = \begin{bmatrix} Ax(t) + A_h x_t(-h) \\ \dfrac{dx_t(\theta)}{d\theta} \end{bmatrix}.$$

The operator $\mathcal{A}$ is the infinite dimensional counterpart of the finite dimensional operator A describing LTI dynamical systems of the form $\dot{x} = Ax$. Many tools have been developed to deal with such abstract systems, lots of them relying on functional analysis and operator theory.

4.2 Bestiary of Time-Delay Systems and Delays

In this section, we shall focus on the functional differential equations framework which is the one we consider throughout this monograph. When considering this formulation, different types of delay systems and different types of delays can be distinguished. They are exposed below.

4.2.1 Types of Time-Delay Systems

Following the commonly accepted denomination partly introduced by Kamenskii [21], four types of time-delay systems may be defined, namely systems with discrete delay, with distributed delay, neutral delay and scale delay.

4.2.1.1 Systems with Discrete Delays

Systems with discrete delays (see e.g. the monographs [2–5, 22–24]) or pointwise delays, are systems where delayed signals are shifted pointwisely in time:

$$\begin{aligned}\dot{x}(t) &= Ax(t) + A_h x(t - h_x) + Bu(t) + B_h u(t - h_u)\\ y(t) &= C_h x(t - h_y)\\ x(s) &= \varphi(s),\ \ s \in [-h, 0]\end{aligned} \tag{4.4}$$

where x, u and y are respectively the state, the input and the output of the system, respectively. To every signal, a different delay has been assigned. To distinguish between systems where the delays act on different signals, the denominations of *state-delay systems*, *input-delay systems* or *output-delay systems* are very often used. Systems with discrete delays arise in many processes and engineering problems such as networked control systems [25, 26], communication networks [27–30], epidemiology [31, 32], systems biology [33, 34], water flow control [35], etc.

Systems with discrete delays exhibit a more complicated behavior than finite-dimensional linear systems. A striking difference is that they possess an infinite (countable) number of characteristic roots (or eigenvalues), as opposed to a finite number for LTI systems without delays. To illustrate this, let us consider the system

$$\dot{x}(t) = \alpha x(t - h) \tag{4.5}$$

where $\alpha \in \mathbb{R}$ and $h > 0$. Two different eigenvalue profiles have been computed using the method described in [36], the results are depicted in Figs. 4.1 and 4.2. We can see that the system is unstable when $\alpha = -1$ and $h = 2$, and stable when $\alpha = -1$ and $h = 1$. It is, furthermore, possible to prove that when $\alpha < 0$, the system is stable for any constant delay belonging to $[0, \bar{h})$ where

$$\bar{h} = \frac{\pi}{2|\alpha|}.$$

When $\alpha > 0$, the system is unstable regardless of the delay value. An important fact about eigenvalues is that the number of unstable ones can only be in finite number since *the number of characteristic roots located to the right of any vertical line in the complex plane is finite*. Based on the above formula, stability regions in the (α, h)-plane can be easily derived; see Fig. 4.3. It is very important to mention that, for retarded delay systems, stability is a continuous property when seen as a function of the system parameters [24]. This has deep practical and theoretical consequences since this means that retarded delay systems are intrinsically robust with respect to small changes in the delay.

4.2.1.2 Systems with Distributed Delays

Systems with distributed delays (see e.g. the monographs [2–5, 22–24]) are systems involving continuous-delays of the form

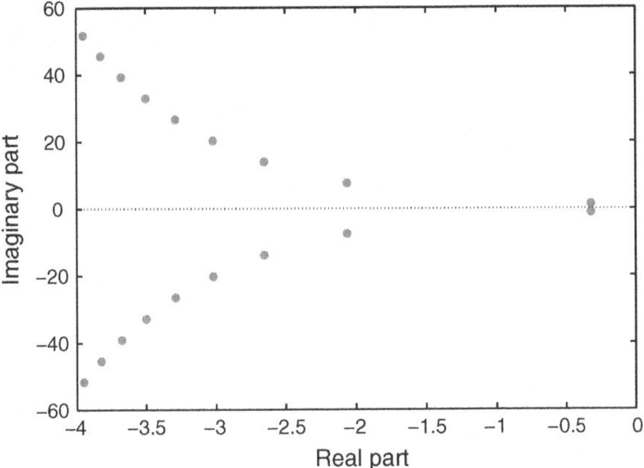

Fig. 4.1 Eigenvalues of system (4.5) with $\alpha = -1$ and $h = 1$. The system is asymptotically stable since all the characteristic roots have negative real part

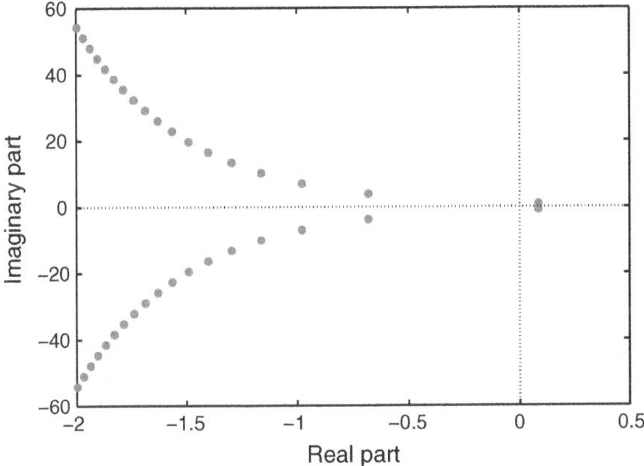

Fig. 4.2 Eigenvalues of system (4.5) with $\alpha = -1$ and $h = 2$. The system is unstable since there is a pair of complex characteristic roots in the right-half plane

$$\dot{x}(t) = Ax(t) + \int_{-h_x}^{0} A_h(\theta)x(t+\theta)d\theta + Bu(t) + \int_{-h_u}^{0} B_h(\theta)u(t+\theta)d\theta \qquad (4.6)$$

where x and u are the state and the input of the system, respectively. In the system above, we can clearly see that past values of the state and the input influence the evolution of the system through a continuous weighted sum (an integral). Similarly to as discrete-delay systems, distributed-delay systems have an infinite number of characteristic roots.

When the matrix functions $A_h(\cdot)$ or $B_h(\cdot)$ admit a rational Laplace transform, the system can be reformulated as a linear time-delay system with discrete-delays

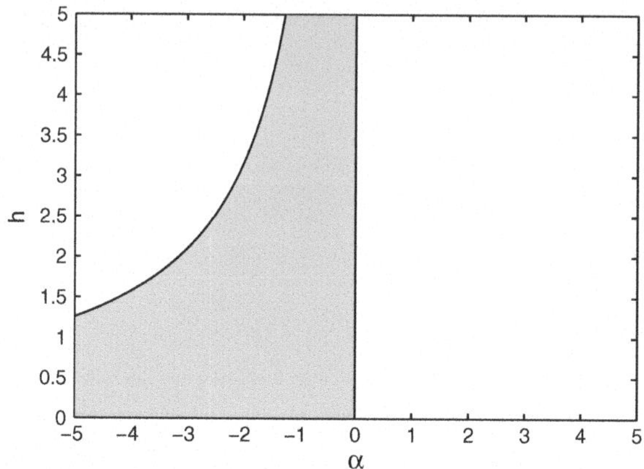

Fig. 4.3 Stability regions of system (4.5). Unstable region in *white*; stable region in *grey*

by suitably augmenting the number of state variables [37]. When, however, kernel functions are polynomial or rational, numerical methods can be used to analyze them [38, 39].

Distributed delay systems arise in combustion systems in rockets [40, 41], epidemiology [42], traffic model and control [43], biology [44, 45], etc.

4.2.1.3 Neutral Systems

Neutral systems (see e.g. [2–4, 46, 47]) are systems where the delay acts on the higher order derivative. A simple example is given by

$$\dot{x}(t) - F\dot{x}(t-h) = Ax(t). \tag{4.7}$$

This type of systems typically exhibit a more complicated behavior than the other time-delay systems. While systems with distributed and discrete delays may only have a finite number of unstable characteristic roots, neutral systems may have them in infinite number. Additionally, stability is not a continuous property in terms of the delay. There indeed exists systems which can be destabilized with arbitrarily small changes in the delay value [48]. The concept of strong stability [3] has been therefore introduced to overcome this difficulty and obtain robust stability results. In the example above, strong stability refers to as exponential stability of the difference equation $x(t) = Fx(t-h)$, which is equivalent to the condition $\varrho(F) < 1$.

Neutral delay systems arise, for instance, in the modeling of lossless transmission lines [49], combustion systems [50], partial element equivalent circuits [51], implementation schemes of predictive controllers [52], control systems with derivative feedback and delay [53], boundary controlled partial differential equations with delays [54], ecological systems [55–57], population dynamics [56, 58] and epidemiological models [59, 60].

4.2.1.4 Scale-Delay Systems

Scale-delay systems (see e.g. [61–67]) are less known than the former ones and much less attention has been devoted to them, at least in the systems and control community. An example of such systems is given by the *pantograph equation*:

$$\dot{x}(t) = Ax(t) + A_h x(\alpha t) \tag{4.8}$$

where $\alpha \in (0, 1]$. We can clearly see that the scaled argument αt takes smaller values than t, and therefore can be understood as a delayed argument. Letting $h(t) = (1 - \alpha)t$, (4.8) can be rewritten as

$$\dot{x}(t) = Ax(t) + A_h x(t - h(t)) \tag{4.9}$$

and therefore system (4.8) can be viewed as a time-delay system with discrete ramp-shaped time-varying delay. Note that the delay inexorably grows unbounded for this class of systems.

Historically, the term "pantograph" dates back to the seminal paper [68], where such equations emerged in a mathematical model for the dynamics of an overhead current collection system on an electric locomotive. The pantograph equation has also been previously obtained in [69, 70] to describe the absorption of light by the interstellar matter. This equation is also found in [71] on a certain partition problem in number theory and also in [72] on a special ruin problem. We find again this equation in quantum theory [73, 74] and cell-growth biology [75].

4.2.2 Families of Delays

According to their behavior and dependence, time-delays can be assigned to different families, and specific approaches are usually necessary to analyze the system they are involved within.

4.2.2.1 Constant Delays

Constant time-delays are the first class of time-delays that have been considered. The reasons for that were easier analysis, yet difficult enough, and no real motivation for considering a more general class of time-delay. Linear systems with constant delays benefit of a very rich and complete theory based on many different tools, both in the frequency and time domains. Even more importantly, linear time-delay systems with constant delays are part of the few systems for which there exist constructive necessary and sufficient conditions for characterizing their stability.

Some examples of systems with constant time-delays are given in Sect. 4.3. Some more examples can also be found in the textbooks [4, 5].

4.2.2.2 Time-Varying Delays

The advent of communication networks and networked control systems where delays are time-varying partly motivated their analysis. Time-varying delays are more harmful to stability than time-invariant ones. It is indeed possible to find systems that are stable for constant-delays but become unstable when the delay starts to be time-varying: this is very often referred to as the *quenching phenomenon* [76, 77]. Notably, when a linear time-invariant system with constant delay is asymptotically stable up to a delay $\bar{h}$, it is most of the time asymptotically stable up to a delay smaller than $\bar{h}$ when the delay becomes time-varying. The rate of variation of the delay indeed plays an important role is reducing the maximal admissible delay value: the faster the delay, the most harmful the delay is.

It is also important to distinguish between smooth (or at least absolutely continuous) delays and delays that can be possibly discontinuous. Whereas many works consider delays having derivative smaller than 1, which actually preserves invertibility of the map $t \mapsto t - h(t)$ where $h(t)$ is the delay, several other works relax this constraint to allow for larger bounds on the delay derivative, or even no bound at all. Having fast varying delays may be responsible of undesirable behavior of the system such as loss of uniqueness of solutions, loss of causality, existence of small solutions (i.e. non-complete solutions), etc.; see e.g. [78–80] are references therein.

Time-varying delays arise, among others, in networked control systems [25], sampled-data systems and control [81–83].

4.2.2.3 State-Dependent Delays

State-dependent delays [84] is the last class of time-delays and is certainly the least well understood. The state-dependence makes the overall system strongly nonlinear, even when the system expression is affine in the state-variables. State-dependent delays arise, for instance, in communication networks modeling and analysis [29, 30], networked control systems [26], water flow control [35], soft landing [85, 86].

There is, at this time, no general framework for dealing with such systems and every instance of them should be considered as a particular case. A lot of works have been, however, devoted to these systems, e.g. on well-posedness [87], stability analysis [85, 86, 88–91], linearization [92], numerical analysis and integration [93–95], and control [96–98].

4.3 Examples

Different examples from various fields are discussed here in order to present the main types of delay systems introduced in the previous section. Constant, time-varying and state-dependent delays are covered as well.

The first example is a biological example taken from [33] and is presented in Sect. 4.3.1. In this example, a delayed protein degradation process is studied both in the deterministic and stochastic frameworks, leading to models with a constant discrete delay.

The second example, treated in Sect. 4.3.2 pertains on the representation of aperiodic sampled-data systems as systems with sawtooth-shaped discrete delay. This approach has been first considered in [81, 82].

In Sect. 4.3.3, the epidemiological SIR-model of [42] is discussed. This model involves a constant distributed delay modeling the recovery time after infection.

A neutral-delay with constant delay model describing forests evolution is discussed in Sect. 4.3.4.

A recent communication network model is discussed in Sect. 4.3.5. This model is taken from [29, 30, 99] and consists of a nonlinear model with constant and state-dependent discrete delays.

4.3.1 Delays in Biological Reaction Networks

Reaction networks are very powerful modeling tools aiming at describing complex interactions between different agents (or species). Examples of applications range from chemistry, biology, population dynamics, epidemiology, to communication, opinion and social networks; see e.g. [100].

We consider here a very simple biological network [33] where a protein P is produced at constant rate k and degraded at rate γ. A second degradation process also takes place at rate γ_d after a delay of τ seconds. This setup corresponds to the following schematic representation:

$$
\begin{aligned}
\phi &\xrightarrow{k} P \\
P &\xrightarrow{\gamma} \phi \\
P &\xRightarrow{\gamma_d} \phi.
\end{aligned}
$$

The first reaction represents the production reaction which occurs at rate k while the second one is the degradation reaction which occurs at rate γ. The last one represents the delayed degradation which occurs at rate γ_d but is effective only after τ seconds.

In what follows, both deterministic and stochastic models are considered.

4.3.1.1 Deterministic Approach

The deterministic approach is based on *reaction network theory* that has been developed in the 70s, see e.g. [101–103]. In this framework, the quantity of species interacting with each other are described in terms of their concentration, a continuous quantity. Hence the state takes real nonnegative real values. The evolution of these concentrations is governed by differential equations referred to as *reaction rate equations*.

Let $x(t) \in \mathbb{R}_{\geq 0}$ be the concentration of the protein P at time t. We then have the following deterministic representation for the network

$$\dot{x}(t) = -\gamma x(t) - \gamma_d x(t-\tau) + k \tag{4.10}$$

which is a linear functional differential equation with constant discrete delay. The unique equilibrium point of the system is given by

$$x^* = \frac{k}{\gamma + \gamma_d}. \tag{4.11}$$

Global stability of the above equilibrium point can be inferred using frequency domain analysis:

Proposition 4.3.1 *The system* (4.10) *is globally asymptotically stable if and only if one of the following statements holds:*

- *either* $\gamma > \gamma_d$; *or*
- $\gamma < \gamma_d$ *and* $\tau \in [0, \tau^*)$ *where*

$$\tau^* = \frac{1}{\sqrt{\gamma_d^2 - \gamma^2}}\left[\pi - \arctan\left(\frac{\sqrt{\gamma_d^2 - \gamma^2}}{\gamma}\right)\right]. \tag{4.12}$$

Proof The characteristic polynomial of the system (4.10) is given by $P(s) + e^{-\tau s}Q(s)$ where $P(s) := s + \gamma$ and $Q(s) := \gamma_d$. The idea is to find the critical values for the constant delay such that we have characteristic roots on the imaginary axis. Therefore, we look for pairs $(\omega, \tau) \in \mathbb{R}_{>0} \times \mathbb{R}_{>0}$ such that $P(j\omega) + Q(j\omega)e^{-j\tau\omega} = 0$ holds. It is immediate to see that there exists a $\tau > 0$ such that this equation holds if and only if there is a positive solution to the polynomial equation $|P(j\omega)|^2 - |Q(j\omega)|^2 = 0$. We have that

$$|P(j\omega)|^2 - |Q(j\omega)|^2 = \omega^2 + \gamma^2 - \gamma_d^2. \tag{4.13}$$

Case 1 When $\gamma > \gamma_d$, then there is no positive solution ω and thus the system is stable for all $\tau \geq 0$.

Case 2 When $\gamma < \gamma_d$, then the unique positive solution is given by $\omega^* = \left(\gamma_d^2 - \gamma^2\right)^{1/2}$ and the critical delay value τ^* solves the equation

$$\frac{-Q(j\omega^*)}{P(j\omega^*)} = e^{j\omega^*\tau^*} \tag{4.14}$$

which yields the result. ■

4.3.1.2 Stochastic Approach

In the stochastic approach, the exact count of molecules is tracked by the model, that is, the states takes nonnegative integer values. Under several assumptions, it is possible to show that stochastic chemical reaction networks without delay can be represented as Markov processes [104]. The presence of a delay, however, destroys the Markov property of the problem, but similar ideas can still be applied [33]. Under some assumptions on the delay value and various approximations, it is possible to prove that the first and second order moments obey the expressions

$$\begin{aligned}\frac{dm_1(t)}{dt} &= -\gamma m_1(t) - \gamma_d m_1(t-\tau) + k\\ \frac{dm_2(t)}{dt} &= -2\gamma m_2(t) + 2km_1(t) - 2\gamma_d m_1(t-\tau)(1 - m_1(t)).\end{aligned} \tag{4.15}$$

Clearly, the above model is a nonlinear delay system with constant discrete delay. Note also that the dynamics of the first order moment is identical to the dynamics in the deterministic case. In this respect, the deterministic model also describes the evolution of the average number of protein molecules over time. Note that this is not a general fact, i.e. the deterministic model may not describe the average number of molecules of the stochastic model.

The equilibrium point of this model is given by

$$\begin{aligned}m_1^* &= \frac{k}{\gamma + \gamma_d}\\ m_2^* &= m_1^*\frac{k + \gamma_d(m_1^* - 1)}{\gamma}.\end{aligned} \tag{4.16}$$

We have the following stability result:

Proposition 4.3.2 *The dynamical model* (4.15) *is globally asymptotically stable if and only if the conditions of Proposition 4.3.1 hold.*

Proof Global stability of this equilibrium point is easily inferred from the stability result in the deterministic setting. Indeed, with the same conditions as in the deterministic case, the equilibrium point of the first order moment is globally asymptotically stable. Noting then that the dynamics of the second-order moment is of the form $\dot{m}_2(t) = -2\gamma m_2(t) + f(m_1(t), m_1(t-\tau))$ and is simply a low-pass filter with input depending on $m_1(t)$ and $m_1(t-\tau)$. The input $f(m_1(t), m_1(t-\tau))$ is bounded since the first-order moment is bounded when the conditions of Proposition 4.3.1 are met, and therefore the second-order moment globally converges to its equilibrium point. ∎

4.3.2 Aperiodic Sampled-Data Systems

In sampled-data control, continuous-time systems are controlled by discrete-time controllers implemented in digital devices. The output of the controller is a discrete-time sequence of numbers that is converted to a continuous signal by a hold function in order to be driven to the input of the continuous-time system. When the hold-function is a zero-order hold function and the controller is a sampled-data state-feedback controller, the overall closed-loop system writes

$$\begin{aligned} \dot{x}(t) &= Ax(t) + Bu(t) \\ u(t) &= Kx(t_k), \; t \in [t_k, t_{k+1}) \end{aligned} \tag{4.17}$$

where $x \in \mathbb{R}^n$, $u \in \mathbb{R}^m$ are the state of the system and the control input, respectively. The gain K is the stabilizing state-feedback gain to be determined. Each control input value $Kx(t_k)$ is maintained over $[t_k, t_{k+1})$ where the sequence $\{t_k\}$ of *sampling instants* is assumed to be strictly increasing and to grow without bound (no accumulation point). When periodic sampling is considered, the points of the sequence are equidistant, e.g. $t_{k+1} = t_k + T$, for some constant $T > 0$ called the *sampling period.* However, in recent applications such as networked control systems [25, 105, 106] or event-triggered control [107–109], the sampling-period is not constant anymore and must be considered as time-varying. Such systems are referred to as *aperiodic sampled-data systems*.

Several approaches have been developed to deal with aperiodic sampled-data systems: discrete-time approaches [110–113], input-delay approaches [81–83, 114–116], robust analysis techniques [117–119], impulsive systems formulation [120–126], LPV techniques [127], looped-functionals [124–126, 128], or clock-dependent Lyapunov functions [129].

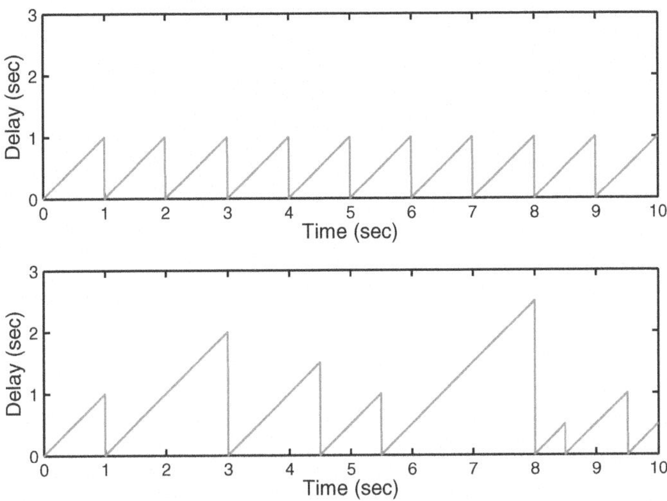

Fig. 4.4 Example of delays describing the zero-order hold function: periodic hold-function (*top*); aperiodic hold-function (*bottom*)

The input-delay representation, which is the one in which we are interested in, is simply based on the identity

$$x(t_k) = x(t - h(t)), \ t \in [t_k, t_{k+1}) \tag{4.18}$$

where $h(t) = t_k - t, t \in [t_k, t_{k+1})$. Therefore, the sampled-state $x(t_k)$ can be viewed as a delayed version of $x(t)$ with sawtooth delay $h(t) = t - t_k, t \in [t_k, t_{k+1})$, which is periodic whenever the sampling is periodic, aperiodic otherwise; see Fig. 4.4.

Based on this fact, the sampled-data system (4.19) can be rewritten as

$$\begin{aligned} \dot{x}(t) &= Ax(t) + BKx(t - h(t)) \\ h(t) &= t - t_k, \ t \in [t_k, t_{k+1}) \end{aligned} \tag{4.19}$$

which is clearly a linear time-delay system with time-varying delay acting on the state of the system. Techniques for analyzing time-delay systems can therefore be applied to aperiodic sampled-data systems with a particular attention on the fact that $\dot{h}(t) = 1$ almost everywhere, in contrast to the condition $\dot{h}(t) < 1$ of Sect. 4.2.2. Due to this peculiarity, specific methods have to be considered in order to accurately characterize the stability of sampled-data systems in the time-delay systems framework.

For illustration, we consider the following scalar sampled-data system

$$\dot{x}(t) = x(t) - Kx(t_k), \ t \in [t_k, t_{k+1}) \tag{4.20}$$

where K is the gain of the controller. We then have the following result:

Proposition 4.3.3 *The aperiodic sampled-data system* (4.20) *is asymptotically stable if and only if* $K > 1$ *and* $T_k \in (0, \bar{T})$ *where*

$$\bar{T} = \log\left(\frac{1+K}{K-1}\right). \tag{4.21}$$

Proof The proof is based on the computation of the equivalent discrete-time system. ∎

We also have the following analogous result:

Proposition 4.3.4 *The time-delay system with constant delay*

$$\dot{x}(t) = x(t) - Kx(t-h). \tag{4.22}$$

is asymptotically stable if and only if $K > 1$ *and* $h \in [0, \bar{h})$ *where*

$$\bar{h} = \frac{1}{\sqrt{K^2-1}}\arctan(\sqrt{K^2-1}). \tag{4.23}$$

Using the results above, we get the bounds depicted in Fig. 4.5 where we can see that the system with constant delay has a smaller stability region than the sampled-data system, i.e. $\bar{h} < \bar{T}$. This essentially means that representing a sampled-data system by a time-delay system and only considering the delay upper-bound is not a good strategy since it may be conservative. This demonstrates that it is crucial to consider the time-varying nature (sawtooth shape) of the delay in order to develop accurate stability results. A similar remark pertaining on the norm of certain integral operators involved in the analysis of time-delay and sampled-data systems is made in [117].

4.3.3 Delay-SIR Model

SIR models (see e.g. [59, 130–136]) and the like, are ubiquitous in epidemiology. They are instances of a broader family of systems referred to as *compartmental systems* [137] whose main paradigm is to

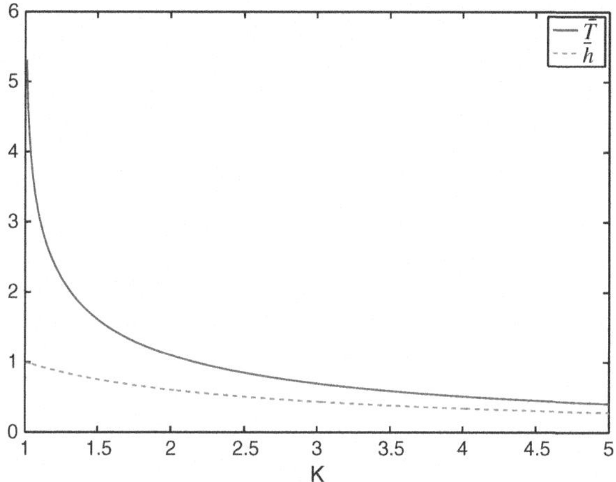

Fig. 4.5 Stability regions of the systems (4.20) and (4.22)

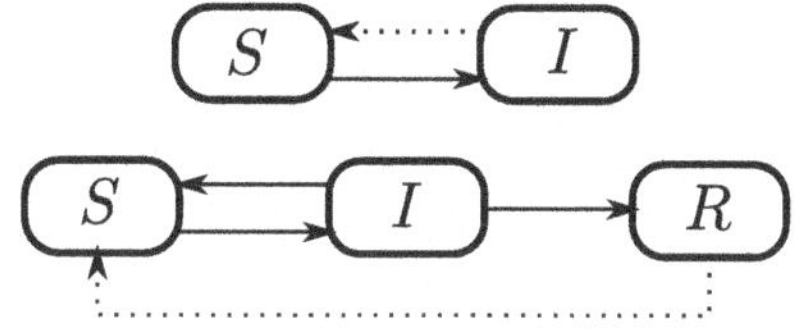

Fig. 4.6 Block-diagrams corresponding to the SIS and SIR model

- represent networks of interactions within a certain population by gathering individuals having common properties together into *compartments*, and to
- describe their compartment-to-compartment interactions and migrations.

Compartmental models have applications in population dynamics, epidemiology, biology, chemistry, ecology, etc.

In epidemiological models, such as SIR-, SIS- or SEIR-models, the compartments S, E, I and R refer to compartments of *susceptible*, *exposed*, *infectious* and *recovered people* (Fig. 4.6). Susceptible are healthy people that can be contaminated when exposed to the disease by contact with infectious people. Unlike infectious people, exposed people are contaminated by the disease but cannot transmit it; this is a latent state such as an incubation state. Finally, infected people may go to a recovered state, a healthy state, where they are, temporarily or definitively, immune to the disease.

In SIS models, the immunity after recovery is very short and the recovered state can hence be neglected. This is the case, for instance, of common cold. SIR models with no feedback from the R-state to the S-state are more adapted to disease like measles, mononucleosis, mumps or rubella where immunity after infection is permanent.

Historically, these models were considered as deterministic and delay-free such as the widely known SIR-model [136] given by

$$\begin{aligned}\dot{S}(t) &= -\beta S(t)I(t)\\ \dot{I}(t) &= \beta S(t)I(t) - \alpha I(t)\\ \dot{R}(t) &= \alpha I(t)\end{aligned} \tag{4.24}$$

where $S(t)$, $I(t)$ and $R(t)$ are the number of individuals being susceptible, infectious and recovered at time t. The rate constants β and α characterize the rate of infection and the rate of recovery. The rationale for modeling contamination by the product $S(t)I(t)$ follows from mass-action law models [138–141], which are reasonable whenever the population is assumed to be well-mixed. The recovering process is assumed to be exponential with rate α.

More recently, the importance of considering delays have grown in importance [59] since delays can model more accurately incubation times, recovery times, etc. A distributed delay can be introduced in the models in order to incorporate some latent behavior, for instance the time for an infectious person to recover [59, 60]:

$$\begin{aligned}\dot{S}(t) &= -\beta S(t)I(t)\\ \dot{I}(t) &= \beta S(t)I(t) - \beta\int_h^\infty \gamma(\tau)S(t-\tau)I(t-\tau)d\tau\\ \dot{R}(t) &= \beta\int_h^\infty \gamma(\tau)S(t-\tau)I(t-\tau)d\tau.\end{aligned}$$

To account for the fact that the delay may be different from one person to another, the function $\gamma(\tau)$ somehow serves the role of probability distribution. Under the assumption that the function γ has a rational, stable and strictly proper Laplace transform, the system can be reformulated as a system with discrete-delay [37] as

$$\dot{S}(t) = -\beta S(t)I(t) \tag{4.25}$$

$$\dot{I}(t) = \beta S(t)I(t) - cq(t) \tag{4.26}$$

$$\dot{q}(t) = Aq(t) + bS(t-h)I(t-h). \tag{4.27}$$

where $\widehat{\gamma}(s) := c(sI - A)^{-1}b$ is the Laplace transform of γ. For instance, choosing $\gamma(\theta) = \xi(1+\delta\theta)e^{-\lambda\theta}$ with $\xi = \frac{\lambda^2}{\delta+\lambda+\delta\lambda h}e^{\lambda h}$ gives the model

$$\begin{aligned}\dot{S}(t) &= -\beta S(t)I(t)\\ \dot{I}(t) &= \beta S(t)I(t) - \beta\mathcal{N}(q_1(t) + \delta q_2(t)\\ \dot{R}(t) &= \beta\xi(q_1(t) + \delta q_2(t))\\ \dot{q}_1(t) &= -\lambda q_1(t) + e^{-\lambda h}S(t-h)I(t-h)\\ \dot{q}_2(t) &= q_1(t) - \lambda q_2(t) + he^{-\lambda h}S(t-h)I(t-h).\end{aligned} \tag{4.28}$$

Table 4.1 Evolution of the number of infected boys during the outbreak

Day	Number infected	Day	Number infected
1	3	8	237
2	6	9	191
3	25	10	125
4	73	11	69
5	222	12	27
6	294	13	11
7	258	14	4

For model validation, we use the influenza epidemic scenario reported in the British Medical Journal of the 4th of March 1978 [142]. This epidemic occurred in a boy's boarding school in the north of England where 763 boys between the ages of 10 and 18 were at risk. Using the data reported in [143, 144], see Table 4.1, and assuming that at day 0, only one individual is infected, i.e. $I(0) = 1$, $R(0) = 0$ and $S(0) = 762$, we find the parameters

$$h = 0.69, \ \beta = 0.00177, \ \ \delta = 0.3 \ \text{ and } \lambda = 0.75. \tag{4.29}$$

The trajectories of the identified model are depicted in Fig. 4.7 where it is possible to see that the delay-SIR model can be used to describe the evolution of the influenza among the population. An even better matching between the trajectories and the real data should be possible to obtain by refining the parameter values.

4.3.4 Neutral Pearl-Verhulst Equation and Ecology

An example of dynamical system governed by a neutral delay equation is the evolution of forests. The model is based on an extension of the Pearl-Verhulst equation [55, 136, 145, 146] where effects such as soil depletion and erosion have been introduced in the model to give

$$\dot{x}(t) = rx(t)\left[1 - \frac{x(t-\tau) + c\dot{x}(t-\tau)}{K}\right] \tag{4.30}$$

where x is the tree population, r is the intrinsic growth rate, K is the environmental carrying capacity and c is a constant parameter that weights how new growth exhausts resources. This model has been considered for analysis in [56], and for control in [57].

A simple analysis shows that the system exhibits two distinct equilibrium points:

$$x^* = 0 \quad \text{and} \quad x^* = K. \tag{4.31}$$

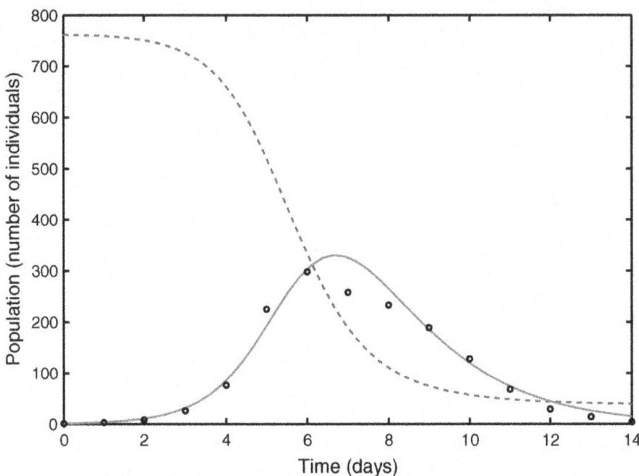

Fig. 4.7 Trajectories of the identified delay-SIR model: Susceptible (*plain*), Infected (*dashed*) and real data (*circles*)

It seems therefore interesting to study the stability of these equilibrium points. This leads us to the following result:

Theorem 4.3.5 ([147]) *The system* (4.30) *with* $r, K, c > 0$ *has the following properties:*

- *The equilibrium point* $x^* = 0$ *is unstable.*
- *The equilibrium point* $x^* = K$ *is locally asymptotically stable if*
  1. $rc < 1$; *and*
  2. $\tau \in [0, \bar{\tau})$ *where*

$$\bar{\tau} := \sqrt{\frac{1 - r^2c^2}{r^2}} \left[\frac{\pi}{2} + \arctan\left(\sqrt{\frac{c^2r^2}{1 - c^2}} \right) \right]. \tag{4.32}$$

- *If* $rc > 1$ *or* $\tau > \bar{\tau}$, *the equilibrium point* $x^* = K$ *is unstable.*

For illustration, we choose $K = 10$ and the initial population is set to 1. Setting different values for c, r and τ, we obtain the Figs. 4.8, 4.9 and 4.10 where we can observe several different types of behavior for the system (4.30).

4.3.5 Networks and Congestion Control Modeling

In communication networks like Internet, congestion is an important efficiency-limiting phenomenon responsible of large communication delays and data loss. To

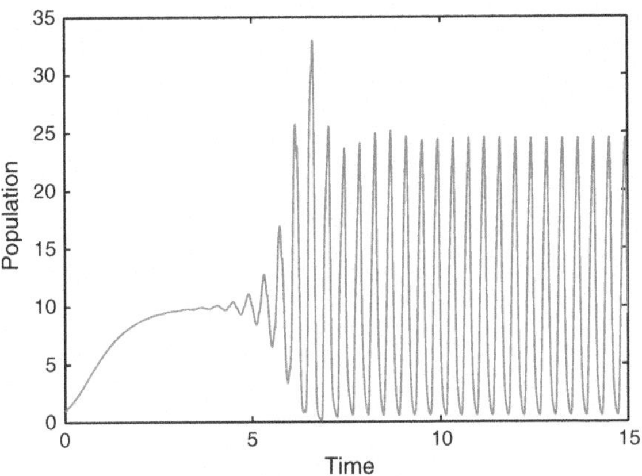

Fig. 4.8 Trajectories of the ecological model (4.30) for $c = 1/2$, $r = 3$ and $\tau = 0.2$

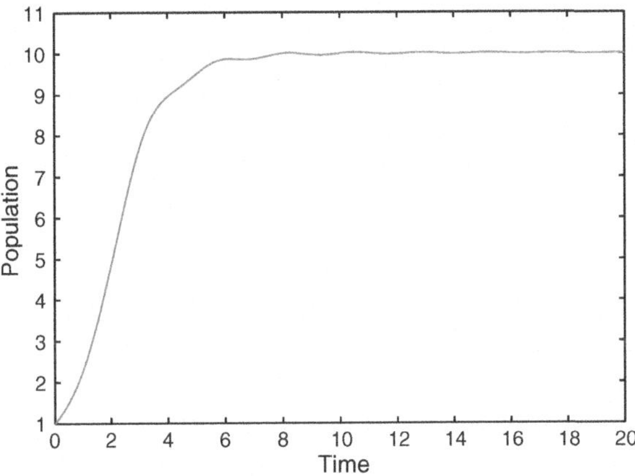

Fig. 4.9 Trajectories of the ecological model (4.30) for $c = 3/4$, $r = 1$ and $\tau = 0.1$ ($\bar{\tau} = 2.6389$)

try to reduce the effects of congestion (ideally to control it), congestion control algorithms have been implemented in transmission protocols such as TCP [148].

Congestion control [148, 149] is truly a control problem, in the control theory sense; see Table 4.2. The network is the system, congestion is the controlled output, protocols are controllers and the user sending rates, i.e. the rate at which users are sending information, are the control inputs. The sending rates are computed from

1. congestion windows which correspond to the number of packets the users would like to maintain in the network,[1] and from
2. a congestion measure representing the level of congestion of the network.

[1] The definition of the congestion window size may differ from one protocol to another.

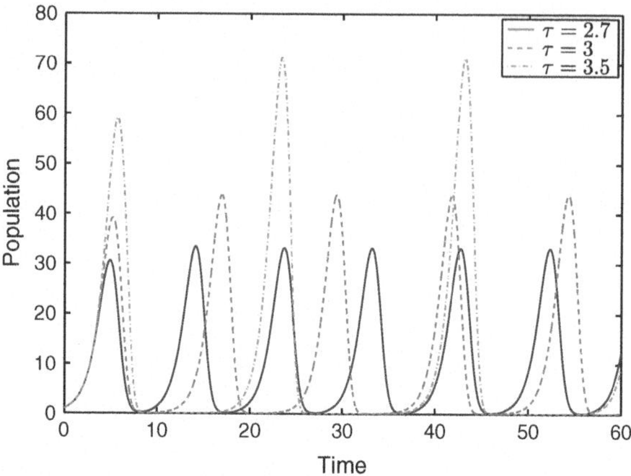

Fig. 4.10 Trajectories of the ecological model (4.30) for $c = 3/4$, $r = 1$ and several values for τ ($\bar{\tau} = 2.6389$)

Table 4.2 Correspondence between control theory and communication networks terminologies

Control theory	Communication networks
System	Network
Controlled output	Congestion (through flight size)
Reference	Congestion window size
Controller	Protocol
Control input	User sending rate
Measured output	Congestion measure (delays, data loss)

Usual congestion measures are data loss (as in TCP) or delays (as in FAST-TCP, [150]): we refer to them as *loss-based* and *delay-based protocols*, respectively.

Modeling congestion is therefore of great importance for multiple reasons. First of all, models can be used for theoretical analysis of congestion and may allow for a better understanding of the underlying mechanisms. Model-based protocol design is another interesting benefit since protocols could then be analytically designed to guarantee certain objectives, FAST-TCP is one such example. The availability of accurate mathematical models for networks may also give rise to a new generation of *model-based simulators*, opposed for instance to NS-2,[2] a discrete-event simulator.

4.3.5.1 A Congestion Model Based on the Conservation of Information

Many different classes of models have been proposed: stochastic continuous-time models [151], discrete-time models [152, 153], continuous-time models [27, 28,

[2] More information on NS-2 available here http://nsnam.isi.edu/nsnam/index.php/Main_Page.

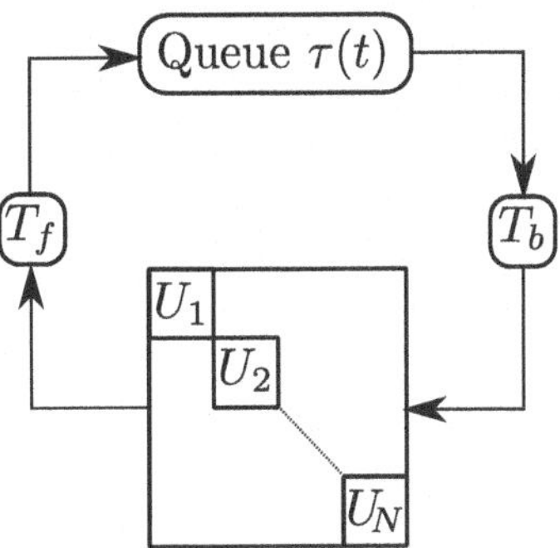

Fig. 4.11 Simple multiple-users/single buffer topology

154–159], hybrid models [160] and, finally, conservation-law-based models [29, 30, 99, 161] which take the form of hybrid systems with state-dependent delays.

In this example, we shall focus on continuous-time systems also referred to as *fluid-flow models* in the literature [162, 163]. In this framework, transmitted information through the network is viewed as a fluid flowing through pipes. This liquid has however few things in common with water or gas, since it somehow shares properties with both liquids and solids at the same time.

The model described below is a simplification of the one from [30] where the hybrid nature of the model has been neglected for simplicity of exposure. Let us consider the topology depicted in Fig. 4.11 where N users, denoted by U_i, communicate via a single FIFO queue.[3] Users' flows experience a constant propagation delay T_f, called *forward propagation delay*, before reaching the queue. Symmetrically, flows leaving the queue experience a *backward propagation delay* T_b when they return to the users. The total propagation delay is simply defined as the sum $T := T_f + T_b$.

Queue model. When the queue is assumed to be non-empty over time, an acceptable model [28] is given by

$$\dot{\tau}(t) = \frac{1}{c}\left(\sum_{i=1}^{N} \phi_i(t - T_f) - c\right) \tag{4.33}$$

where τ is the queuing delay, c is the capacity of the downstream link, i.e. the maximal rate at which the server can process stored information, and ϕ_i is the sending rate of user i (Fig. 4.12). Note the presence of the forward propagation delay T_f. The flows leaving the queue, denoted by ϕ_i^o, have been proved to obey the formula [29, 30, 159, 164]

$$\phi_i^o(t) = \frac{\phi_i(t - d(t))c}{\sum_{j=1}^{N} \phi_j(t - d(t))} \tag{4.34}$$

[3] FIFO stands for "First-In-First-Out" and means that the packets leave the queue in the same order they entered. This is an order preserving queue.

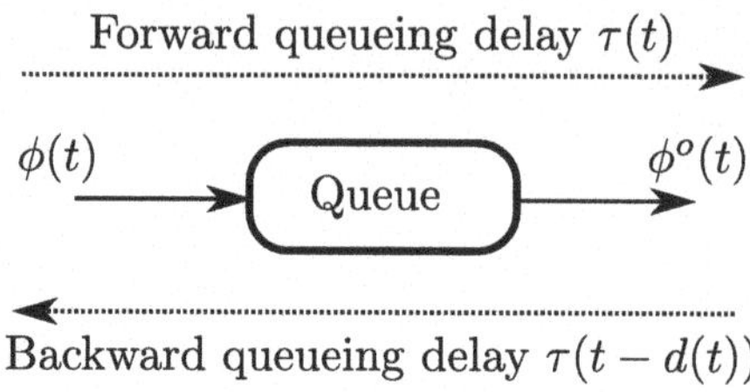

Fig. 4.12 Block diagram of a FIFO queue

where the delay $d(t)$ solves the implicit functional equation

$$d(t) = \tau(t - d(t)),\ t \geq 0. \tag{4.35}$$

For illustration, let us consider two square input flows

$$\begin{aligned} \phi_1(t) &= (1+\beta)c(1+\text{Sq}(\omega t))/2 \\ \phi_2(t) &= (1+\beta)c(1-\text{Sq}(\omega t))/2 \end{aligned} \tag{4.36}$$

where $\beta > 0$ is a given parameter and $\text{Sq}(\omega t) := \text{sgn}(\sin(\omega t))$ is a square function of period $T = 2\pi/\omega$ ($\text{sgn}(\cdot)$ denotes the signum function). Note that $\phi_1(t)+\phi_2(t) = (1+\beta)c$ for all $t \geq 0$. Choosing, for instance, $\omega = 2\pi$, $\beta = 1$, $c = 100$ Mb/s and $\tau(0) = 0$, we get the results depicted in Fig. 4.13 where we can clearly see that the model predicts the output flows quite accurately since the number of packets leaving the queue coincides with the number of packets predicted by NS-2.

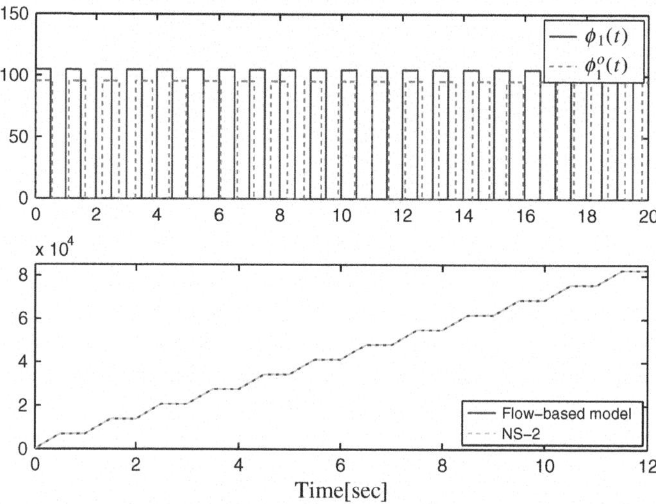

Fig. 4.13 Trajectories of the input and output flows $\phi_1(t)$ and $\phi_1^o(t)$ (*top*) and the number of transmitted packets by the queue for the model (4.33)–(4.34) and the event-based simulator NS-2 (*bottom*)

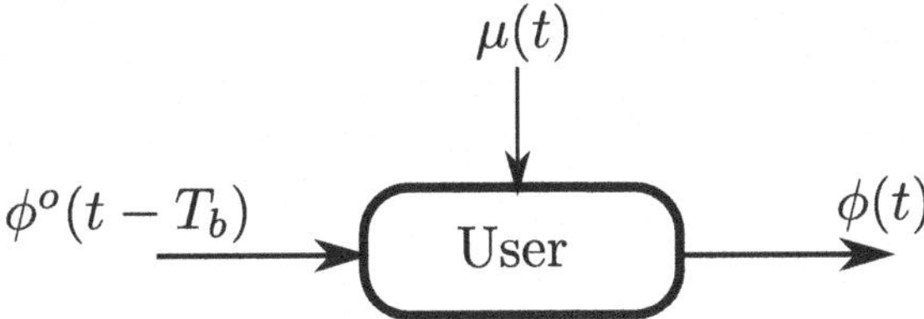

Fig. 4.14 Block diagram of a user

Protocol/user model. Under some technical simplifying, quite restrictive, assumptions it is possible to show that the users sending rate can be computed from the derivative of the congestion window size as

$$\begin{aligned} \dot{w}_i(t) &= f(w_i(t), \mu(t)) \\ \phi_i(t) &= \dot{w}_i(t) + \phi_i^o(t - T_b) \end{aligned} \tag{4.37}$$

where w_i is the congestion window size of user i and μ is the congestion measure (Fig. 4.14). The congestion window size is defined here as the number of packets to maintain throughout the network at any time (in the single queue case).

An example of protocol is FAST-TCP whose behavior can be described as

$$\dot{w}_i(t) = \gamma \left(-w_i(t) + \frac{T}{T + \mu(t)} w_i(t) + \alpha \right) \tag{4.38}$$

where $\gamma, \alpha > 0$ are some tuning parameters of the protocol. This protocol uses the *perceived queuing delay*, i.e. a delayed version of the queueing delay, as congestion measure. The explicit form of the congestion measure is given by

$$\mu(t) = \tau(t - T_b - d(t - T_b)). \tag{4.39}$$

The term γ can be tuned in order to shape reactivity and robustness with respect to delays, whereas α is the desired number of packets (or bytes) to maintain in the queue at equilibrium.

Network. In the simple case described above, the queuing delay can be explicitly solved to get

$$\tau(t) = \frac{1}{c} \sum_{i=1}^{N} w_i(t - T_f) - T \tag{4.40}$$

under the assumption that the right-hand side is always nonnegative. After substitution in the protocol model, we obtain the network model

$$\dot{w}_i(t) = \gamma \left(-w_i(t) + \frac{cT}{\sum_{i=1}^{N} w_i(t - T - d(t - T_b))} w_i(t) + \alpha \right), \; i = 1, \dots, N \tag{4.41}$$

where $d(t)$ is now given by

$$d(t) = \frac{1}{c} \sum_{i=1}^{N} w_i(t - T_f - d(t)) - T. \tag{4.42}$$

The overall network model is therefore a N-dimensional nonlinear delay system with constant and *state-dependent* delays due to the relation (4.42). Global analysis of such systems is mainly an open problem. For completeness, we will prove local stability of the unique equilibrium point

$$w_i^* = w^* := \alpha + \frac{cT}{N} \text{ and } \tau^* = \frac{N\alpha}{c}. \tag{4.43}$$

Theorem 4.3.6 *The network model* (4.41) *is locally exponentially stable if one of the following statements hold:*

- $T < \tau^*$,
- $T > \tau^*$ *and* $\tau^* + T < \tau_c$ *where*

$$\tau_c = \frac{1}{\gamma}\sqrt{\frac{N\alpha + cT}{cT - N\alpha}}\left[\pi - \arctan\left(\sqrt{\left(\frac{cT}{N\alpha}\right)^2 - 1}\right)\right]. \tag{4.44}$$

Proof Local approximation of the system about the equilibrium point obeys [92]

$$\dot{\tilde{w}}(t) = \gamma\left(-\frac{N\alpha}{N\alpha + cT}\tilde{w}(t) - \frac{cT}{N(N\alpha + cT)}\mathbb{1}_N\mathbb{1}_N^{\mathrm{T}}\tilde{w}(t - \tau^* - T)\right) \tag{4.45}$$

where $\mathbb{1}_N$ is the N-dimensional vector of ones. Since the matrix acting on the delayed term is of rank one, then stability can be inferred from a small-gain argument on the transfer function

$$\begin{aligned} G(s) &= \frac{cT\gamma}{N(N\alpha + cT)}\mathbb{1}_N^{\mathrm{T}}\left(sI - \frac{\gamma N\alpha}{N\alpha + cT}\right)^{-1}\mathbb{1}_N \\ &= \frac{cT\gamma}{(N\alpha + cT)\left(s + \frac{\gamma N\alpha}{N\alpha + cT}\right)}. \end{aligned} \tag{4.46}$$

It is easy to see that the system (4.45) is stable for any delay if and only if $||G||_{H_\infty} < 1$ or, equivalently, if and only if $T/\tau^* < 1$. Otherwise, stability depends on the delay value. To prove the delay upper-bound τ_c, let us consider the change of variables

$$y_1(t) = \sum_{k=1}^{N} \tilde{w}_k(t)$$
$$y_{i+1}(t) = \tilde{w}_{i+1}(t) - \tilde{w}_i(t) \tag{4.47}$$

for $i = 1, \ldots, N-1$. The derivatives of the y_i's are given by

$$\dot{y}_1(t) = -\gamma \frac{N\alpha}{N\alpha + cT} y_1(t) - \gamma \frac{cT}{N\alpha + cT} y_1(t-h)$$
$$\dot{y}_{i+1} = -\frac{N\alpha\gamma}{N\alpha + cT} y_{i+1}(t) \tag{4.48}$$

for $i = 1, \ldots, N-1$ where we can see that the state components are now totally decoupled. The states $y_2, \ldots, y_N$ are easily seen to be asymptotically stable. It is therefore only necessary to analyze the stability of the state y_1. Using similar results as in [4], it is possible to show that the system (4.48) is delay-dependent stable provided that $h < \tau_c$ where τ_c is given in (4.44). ■

Example 4.3.7 We consider in this example the interconnection depicted in Fig. 4.15 where two users communicate through a single FIFO queue. The implemented protocol is a simple resend protocol which sends a new packet each time an acknowledgment packet is received. Therefore, the number of packets a user is maintaining in the communication path is constant unless the congestion window size changes.

For simulation purposes, the bottleneck has capacity $c = 100$ Mb/s and the packet size including headers is 1590 bytes. The congestion windows sizes are initially $w_1^0 = 50$ and $w_2^0 = 550$ packets, respectively. At $t = 3$ s, w_1 is increased to 150 packets. The propagation delays are $T_1 = 3.2$ and $T_2 = 117$ ms for users 1 and 2, respectively. Note that since the propagation delays are different, the theoretical analysis carried out above is not valid anymore. Simulations yield the queue trajectories depicted in Fig. 4.16 where we also compare with several other models that have been proposed in the literature[a] and the results obtained with NS-2. It is easily seen that the proposed model predicts the same trajectories as NS-2, which is known to be very accurate for simple topologies. For completeness, the trajectories of the input and output flows of the queue are depicted in Figs. 4.17 and 4.18.

[a] See [155] for the static-link model [27, 28] for the ratio-link model and [157, 165, 166] for the joint-link model.

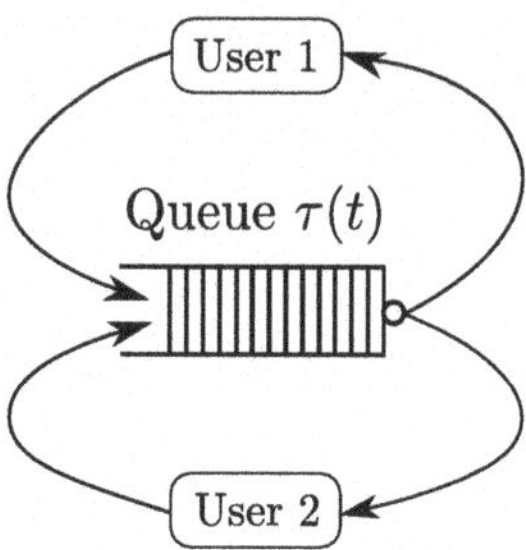

Fig. 4.15 Single-buffer/multiple-user topology considered in Example 4.3.7

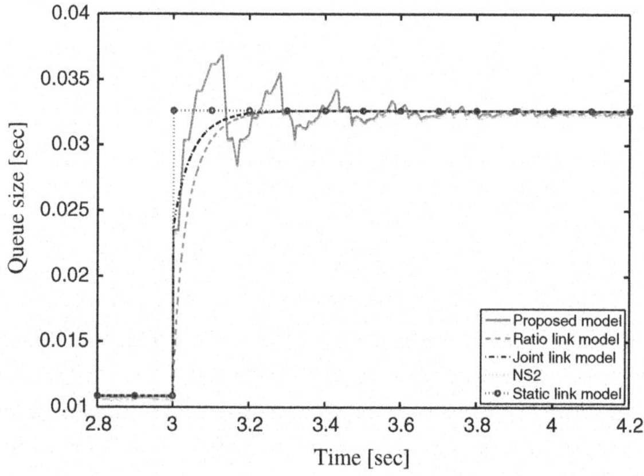

Fig. 4.16 Evolution of the queue-size in the scenario of Example 4.3.7

4.4 Control, Observation and Filtering of Time-Delay Systems

The goal of this section is to briefly expose the types of filters, observers and controllers that can be designed for time-delay systems with discrete-delays. The extensions to neutral systems and to systems with distributed delay follow from the same ideas, i.e. the incorporation of delayed terms of the same types as the ones in the system in the controller or observer expression. The main reason for not entering into deep details and not explicitly providing design criteria lies in the fact that the results developed for LPV time-delay systems in Chaps. 7 and 8 can be directly applied to the particular case of time-delay systems without parameters. So, to avoid redundancy and save paper, design problems will only be addressed in the aforementioned chapters.

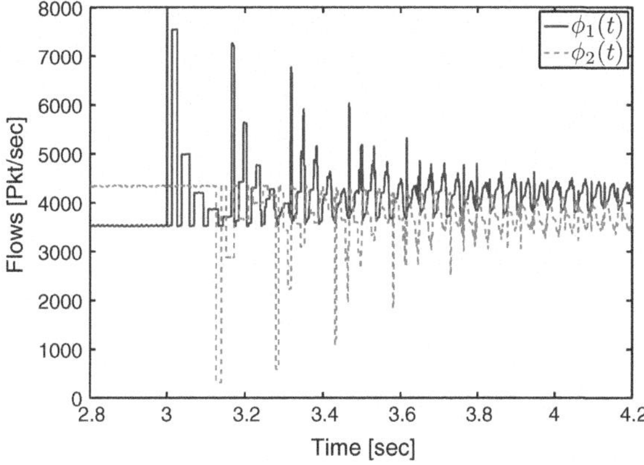

Fig. 4.17 Evolution of the input flows of the queue in the scenario of Example 4.3.7

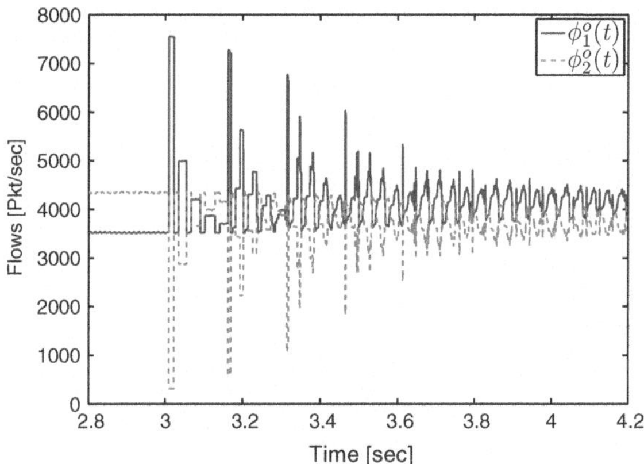

Fig. 4.18 Evolution of the output flows of the queue in the scenario of Example 4.3.7

4.4.1 Observation and Filtering

The main difference between filters and observers lies in the fact that observers are usually designed such that the observation error is asymptotically stable, i.e. the observer tracks the state of the system, whereas filters are aimed to find the best estimate of any signal in a certain sense, e.g. in the L_2-norm sense. In this respect, observers can be designed for unstable systems, while filters can not. The similarity of their function makes not very surprising that their structure be quite similar.

In what follows, we shall consider the generic linear time-delay system

$$
\begin{aligned}
\dot{x}(t) &= Ax(t) + A_h x(t - h(t)) + Bu(t) + Ew(t) \\
z(t) &= Cx(t) + C_h x(t - h(t)) + Fw(t) + Du(t) \\
y(t) &= C_y x(t) + C_{yh} x(t - h(t)) + F_y w(t)
\end{aligned} \tag{4.49}
$$

where $x \in \mathbb{R}^n$, $u \in \mathbb{R}^m$, $w \in \mathbb{R}^p$, $z \in \mathbb{R}^q$ and $y \in \mathbb{R}^r$ are the state of the system, the control input, the exogenous input, the output to be estimated (observer or filtered) and the measured output, respectively. The delay h is not assumed to satisfy any condition.

4.4.1.1 Filters for Time-Delay Systems

Filters with memory involve a delayed component in their dynamical model as seen below

$$
\begin{aligned}
\dot{x}_F(t) &= A_F x_F(t) + A_{Fh} x_F(t - h(t)) + B_{Fy} y(t) + B_{Fy} u(t) \\
z_F(t) &= C_F x_F(t) + C_{Fh} x_F(t - h(t)) + D_{Fy} y(t) + D_{Fu} u(t)
\end{aligned} \tag{4.50}
$$

where $x_F \in \mathbb{R}^n$ is the state of the filter. Note that the dimension of the state of the filter is the same as the one of the system. The goal of the filter is to make the gain of the transfer $w \to z - z_F$, e.g. the L_2-gain, as small as possible. Since the dimension of $x_F(t)$ is n, the above filter is then a *full-order filter*. When the dimension of the state of the observer is smaller than the one of the system, the filter is said to be a *reduced-order filter*.

It is also important to stress that the filter implements the same delay as the one in the system, which may be unrealistic from a practical point of view due to the inherent difficulty of measuring or estimating delays; see e.g. [167–170]. Implementing an approximate delay, say $d(t)$, seems to be more relevant. When this is the case, the filter is said to be a *filter with approximate memory*. When the filter does not implement any delayed component, it is said to be *memoryless*. An example of memoryless filter is given below

$$
\begin{aligned}
\dot{x}_F(t) &= A_F x_F(t) + B_{Fy} y(t) + B_{Fu} u(t) \\
z_F(t) &= C_F x_F(t) + D_{Fy} y(t) + D_{Fu} u(t)
\end{aligned} \tag{4.51}
$$

where $x_F \in \mathbb{R}^n$ is the state of the filter.

Design approaches for such filters have been, for instance, proposed in [171–175].

4.4.1.2 Observers for Time-Delay Systems

Before talking about observation, it seems important to mention a one thing about observability. Since delay-systems are infinite-dimensional systems, several notions of observability exist depending on the considered framework, i.e. functional differential equations, ordinary differential equations with coefficients in a ring of operators or abstract dynamical systems. Since we are not going to enter into details about all of this, the interested readers should, for instance, look at these references [176–180], and references therein.

In a similarly way as for filters, observers with exact memory for the system (4.49) with $y = C_y x$ and $z = Tx$ take the form

$$\begin{aligned}\dot{\xi}(t) &= M_0\xi(t) + M_h\xi(t - h(t)) + N_0 y(t) + N_h y(t - h(t)) + Su(t)\\ \hat{z}(t) &= \xi(t) + Hy(t)\end{aligned} \tag{4.52}$$

where $\xi \in \mathbb{R}^q$ is the state of the observer. The goal of the observer is to make the estimation error $e := z - \hat{z}$ asymptotically stable and such that the gain of the transfer $w \to z - \hat{z}$ as small as desired. The terminology of filters applies to observers as well. Note, moreover, that the above observer structure is a quite general one. Some more restrictive structures, sometimes easier to design and implement, may be considered. For instance, the observer

$$\begin{aligned}\dot{\xi}(t) &= A\xi(t) + A_h\xi(t - h(t)) + Bu(t) + L(y(t) - C_y\xi(t) + C_{yh}\xi(t - h(t)))\\ \hat{z}(t) &= \xi(t)\end{aligned} \tag{4.53}$$

where $\xi \in \mathbb{R}^n$ is a possible one for the system (4.49).

Memoryless observers can also be considered, they take the form

$$\begin{aligned}\dot{\xi}(t) &= M_0\xi(t) + N_0 y(t) + Su(t)\\ \hat{z}(t) &= \xi(t) + Hy(t).\end{aligned} \tag{4.54}$$

where $\xi \in \mathbb{R}^q$ is the state of the observer.

Design methods for time-delay systems observers have been proposed in the literature; see e.g. [181] in the Riccati framework, [182–185] in the algebraic framework, [186–191] in the LMI framework, and also [192] in a more functional analytic setting.

4.4.2 Control

As for finite-dimensional LTI systems, two different classes of controllers can be designed, namely static and dynamic controllers. On the top of that, the time-delay nature of the dynamics of the system allows us to consider controllers that may or may not implement a delayed part. As for observability, several controllability concepts can be defined. See e.g. [10, 20, 177, 179, 193–198].

In what follows, the following generic linear time-delay system is considered

$$\begin{aligned}\dot{x}(t) &= Ax(t) + A_h x(t-h(t)) + Bu(t) + Ew(t)\\ z(t) &= Cx(t) + C_h x(t-h(t)) + Du(t) + Fw(t)\\ y(t) &= C_y x(t) + C_{yh} x(t-h(t)) + F_y w(t)\end{aligned} \tag{4.55}$$

where $x \in \mathbb{R}^n$, $u \in \mathbb{R}^m$, $w \in \mathbb{R}^p$, $z \in \mathbb{R}^q$ and $y \in \mathbb{R}^r$ are the state of the system, the control input, the exogenous input, the output to be controlled and the measured output, respectively.

4.4.2.1 Static Controllers with and without Memory

State feedback or static output feedback are part of the family of static controllers:

- Memoryless and exact-memory static output-feedback controllers are given by

$$u(t) = Ky(t) \quad \text{and} \quad u(t) = Ky(t) + K_h y(t-h(t)),$$

  respectively.
- Static output-feedback controllers with approximate memory are given by

$$u(t) = Ky(t) + K_d y(t-d(t)) \tag{4.56}$$

  where the delay d belongs to the set

$$\mathscr{D}_\delta := \left\{d : \mathbb{R}_{\geq 0} \to \mathbb{R}_{\geq 0} : \ |d(t) - h(t)| \leq \delta\right\}$$

  for some $\delta > 0$.
- Memoryless and exact-memory state-feedback controllers are described by

$$u(t) = Kx(t) \quad \text{and} \quad u(t) = Kx(t) + K_h x(t-h(t)),$$

  respectively.
- State-feedback controllers with approximate memory are given by

$$u(t) = Kx(t) + K_d x(t-d(t)) \tag{4.57}$$

where the delay $d \in \mathscr{D}_\delta$ for some $\delta > 0$.

Static output-feedback controllers have been considered in [199–203]. The design of such controllers is known to be NP-hard in certain conditions, see e.g. [204–206]. State-feedback results have been for instance obtained [207–209] in the memoryless and exact memory case, and in [210, 211] for the approximate memory case.

4.4.2.2 Dynamic Controllers

Dynamic controllers may be classified in two main categories: observer-based output-feedback controllers and dynamic output-feedback controllers.

Observer-Based Controllers
As the name indicates, these controllers are made of an observer part which estimates the state of the system and state-feedback part that computes the control input based on the estimated state value. Observer-based output controllers take one of the following forms

$$\left|\begin{aligned} \dot{\xi}(t) &= M_0\xi(t) + N_0 y(t) + Su(t) \\ \hat{x}(t) &= \xi(t) + Hy(t) \\ u(t) &= K_0\hat{x}(t) \end{aligned}\right.$$

$$\left|\begin{aligned} \dot{\xi}(t) &= M_0\xi(t) + M_h\xi(t-h(t)) + N_0 y(t) + N_h y(t-h(t)) + Su(t) \\ \hat{x}(t) &= \xi(t) + Hy(t) \\ u(t) &= K_0\hat{x}(t) + K_h\hat{x}(t-h(t)) \end{aligned}\right.$$

$$\left|\begin{aligned} \dot{\xi}(t) &= M_0\xi(t) + M_d\xi(t-d(t)) + N_0 y(t) + N_d y(t-d(t)) + Su(t), \ d \in \mathscr{D}_\delta \\ \hat{x}(t) &= \xi(t) + Hy(t) \\ u(t) &= K_0\hat{x}(t) + K_d\hat{x}(t-d(t)) \end{aligned}\right.$$

depending on whether the controller is memoryless, with exact memory or with approximate memory. Some of these control law structures have been considered in [181, 212–216].

Dynamic Output-Feedback Controllers
Dynamic output feedback controllers have a similar structure to observer-based ones with the difference that the state is not aimed to be estimated. The controller is a one-block structure whose goal is only the computation of a suitable control input from the measured output. These controllers take the form:

$$\left|\begin{aligned} \dot{x}_c(t) &= A_c x_c(t) + B_c y(t) \\ u(t) &= C_c x_c(t) + D_c y(t) \end{aligned}\right.$$

$$\left|\begin{aligned} \dot{x}_c(t) &= A_c x_c(t) + A_{hc} x_c(t-h(t)) + B_c y(t) \\ u(t) &= C_c x_c(t) + C_{hc} x_c(t-h(t)) + D_c y(t) \end{aligned}\right.$$

$$\left|\begin{aligned} \dot{x}_c(t) &= A_c x_c(t) + A_{hd} x_c(t-d(t)) + B_c y(t), \ d \in \mathscr{D}_\delta \\ u(t) &= C_c x_c(t) + C_{hd} x_c(t-d(t)) + D_c y(t) \end{aligned}\right.$$

depending on whether the controller is without memory, with exact memory or with approximate memory. Such controllers have been considered, for instance, in [207, 217, 218].

4.4.2.3 Delay-Scheduled Controllers

Delay-scheduled controllers are a class of controllers apart from the previous ones since their structure depends on the delay value but does not implement a memory term. These controllers are actually gain-scheduled controllers with the delay as scheduling parameter, whence the name *delay-scheduled controllers*. Two main classes of controllers can be distinguished: *smoothly scheduled controllers* that depend continuously on the delay value, and *switched controllers* that switch among a finite collection of controllers.

Smoothly Scheduled Controllers
These controllers have been relatively few studied in the literature. Most of the results have been reported in [219–222] in the continuous-time framework. Delay-scheduled controllers in the discrete-time setting can be obtained by representing first the discrete-time system with time-varying into a switched system and by then designing a mode-dependent static output-feedback controller.[4] In continuous-time, delay-scheduled controllers take the forms

$$u(t) = K(h(t))x(t)$$

and

$$\begin{aligned}\dot{x}_c(t) &= A_c(h(t))x_c(t) + B_c(h(t))y(t)\\ u(t) &= C_c(h(t))x_c(t) + D_c(h(t))y(t)\end{aligned}$$

where the matrices are assumed to be continuous functions of $h(t)$. The main idea behind this type of control law is to consider the delay-information in a way that does not need memory. These controllers kind of therefore lie between memoryless controllers and controllers with exact memory. Note moreover, that when the delay is not well-known, it is easier to consider uncertainties than in the case of controllers implementing a memory. This feature is quite interesting since measuring or estimating the delay in real-time is not an easy task; see e.g. [167–170].

Switched controllers
The other class of delay-scheduled controllers is the class of switched controllers where the switching signal sequence depends on the value of the delay. Assume that $h(t) \in (0, \bar{h}]$ and let $0 = h_0 < h_1 < h_2 < \ldots < h_N = \bar{h}$. Then, define the signal

$$\sigma(t) = i \quad \text{if} \quad h(t) \in (h_i, h_{i+1}]. \tag{4.58}$$

For all these delay intervals, we design time-invariant controllers that are finally scheduled according to the value of the delay. That is, we finally obtain controllers of the form

[4] See e.g. [223–225] for discrete-time switched systems and time-delay systems. These references do not consider delay-scheduled controllers.

$$u(t) = K_{\sigma(t)}x(t)$$

or

$$\dot{x}_c(t) = A_{c,\sigma(t)}x_c(t) + B_{c,\sigma(t)}y(t)$$
$$u(t) = C_{c,\sigma(t)}x_c(t) + D_{c,\sigma(t)}y(t)$$

that adapt to changes in the delay value. Such controllers have been considered for instance in [226–228].

References

1. R.E. Bellman, K.L. Cooke, *Differential Difference Equations* (Academic Press, New York, 1963)
2. V.B. Kolmanovskii, A.D. Myshkis, *Applied Theory of Functional Differential Equations* (Kluwer, Dordrecht, 1992)
3. J.K. Hale, S.M. Verduyn Lunel, *Introduction to Functional Differential Equations* (Springer, New York, 1991)
4. S.I. Niculescu, *Delay Effects on Stability: A Robust Control Approach* (Springer, Heidelbeg, 2001)
5. K. Gu, V.L. Kharitonov, J. Chen, *Stability of Time-Delay Systems* (Birkhäuser, Boston, 2003)
6. A.S. Morse, Ring models for delay differential systems. Automatica **12**, 529–531 (1976)
7. E. W. Kamen, Lectures on algebraic system theory: linear systems over rings. Contractor report 3016, NASA, 1978
8. G. Conte, A.M. Perdon, The decoupling problem for systems over a ring, in *34th IEEE Conference on Decision and Control*, (New Orleans, USA, 1995), pp. 2041–2045
9. G. Conte, A. M. Perdon, A. Lombardo, The decoupling problem withweak ouptput controllability for systems over a ring. in *36th IEEE Conferenceon Decision and Control*, (San Diego, USA, 1997), pp. 313–317
10. O. Sename, J.F. Lafay, R. Rabah, Controllability indices of linear systems with delays. Kybernetika **6**, 559–580 (1995)
11. P. Picard, J.F. Lafay, Weak observability and observers for linear systems with delays, in *MTNS 96*, (Saint Louis, USA, 1996)
12. G. Conte, A.M. Perdon, Noninteracting control problems for delay-diffrerential systems via systems over rings, in *Colloque Analyse et Commande des systèmes avec retards* (Nantes, France, 1996), pp. 101–114
13. R. Curtain, H. Logemann, S. Townley, H. Zwart, Well-posedness, stabilizability and admissibility for pritchard-slamon systems. J. Math. Syst. Estimation Control **4**(4), 1–38 (1994)
14. G. Meinsma, H. Zwart, On $\mathcal{H}_\infty$ control for dead-time systems. IEEE Trans. Autom. Control **45**(2), 272–285 (2000)
15. O.V. Iftime, H.J. Zwart, R.F. Curtain, representation of all solutions of the control algebraic ricatti equations for infinite-dimensional systems. Int. J. Control **78**(7), 505–520 (2005)
16. A. Bensoussan, G. Da Prato, M.C. Delfour, S. K. Mitter, Representation and Control of Infinite Dimensional Systems, 2nd Edn. (Birkhäuser, Boston, USA, 2007)
17. J.G. Borisovic, A.S. Turbabin, On the Cauchy problem for linear non-homogeneous differential equations with retarded argument. Soviet Math. Dokl. **10**, 401–405 (1969)
18. M.C. Delfour, S.K. Mitter, Controllability, observability and optimal feedback control of affine hereditary differential systems. SIAM J. Control Optim. **10**, 298–328 (1972)

19. C. Bernier, A. Manitius, On semigroups in $\mathbb{R}^n \times L^p$ corresponding to differential equations with delays. Canadian J. Math. **5**, 897–914 (1978)
20. A. Manitius, R. Triggiani, Function space controllability of linear retarded systems: a derivation from abstract operator conditions. SIAM J. Control Optim. **16**(4), 599–645 (1978)
21. V. Kolmanovskii, A. Myshkis, *Introduction to the Theory and Applications of Functional Differential Equations* (Kluwer Academic Publishers, Dordrecht, 1999)
22. N.N. Krasovskiĭ, *Stability of Motion. Applications of Lyapunov's Second Method to Differential Systems and Equations with Delay (Translated from Russian)* (Stanford University Press, Stanford, 1963)
23. O. Diekmann, S.A. van Gils, S.M. Verduyn Lunel, H.O. Walther, *Delay Equations: Functional-, Complex-, and Nonlinear Analysis* (Springer, New York, 1995)
24. W. Michiels, S.I. Niculescu, *Stability and Stabilization of Time-Delay Systems. An Eigenvalue based Approach* (SIAM Publication, Philadelphia, 2007)
25. J.P. Hespanha, P. Naghshtabrizi, Y. Xu, A survey of recent results in networked control systems. Proc. IEEE **95**(1), 138–162 (2007)
26. E. Witrant, C. Canudas-de-Wit, D. Gerorges, M. Alamir, Remote stabilization via communication networks with a distributed control law. IEEE Trans. Autom. Control **52**(8), 1480–1485 (2007)
27. G. Vinnicombe, On the stability of networks operating TCP-like congestion control, in *15th IFAC World Congress* (Barcelona, Spain, 2002) pp. 217–222
28. F. Pagaganini, J.C. Doyle, S. Low, A control theoretical look at internet congestion control, in *Multidisciplinary Research in Control*, Lecture Notes in Control and Information Sciences, ed. by L. Giarré, B. Bamieh, vol. 289 (Springer, Berlin Heidelberg, 2003), pp. 17–31
29. C. Briat, H. Hjalmarsson, K.H. Johansson, G. Karlsson, U.T.Jönsson, H. Sandberg, Nonlinear state-dependent delay modeling and stability analysis of internet congestion control. In *49th IEEE Conference on Decision and Control* (Atlanta, USA, 2010), pp. 1484–1491
30. C. Briat, E.A. Yavuz, G. Karlsson, A conservation-law-based modular fluid-flow model for network congestion modeling, in *31st IEEE International Conference on Computer Communications (INFOCOM)* (Orlando, Florida, USA, 2012), pp. 2050–2058
31. H.W. Hethcote, M.A. Lewis, P. van den Driessche, An epidemiological model with a delay and a nonlinear incidence rate. J. Math. Biol. **27**, 49–64 (1998)
32. H.W. Hethcote, P. van den Driessche, An SIS epidemic model with variable population size and a delay. J. Math. Biol. **34**, 177–194 (1995)
33. D. Bratsun, D. Volfson, L.S. Tsimring, J. Hasty, Delay-induced stochastic regulation oscillations in gene regulation. Proc. Nat. Acad. Sci. U.S.A. **102**(42), 14593–14598 (2005)
34. M.E. Ahsen, H. Özbay, S.-I. Niculescu, On the analysis of a dynamical model representing gene regulatory networks under negative feedback. Int. J. Robust Nonlinear Control, **24**, 1609–1627 (2014)
35. G. Besançon, D. Georges, Z. Benayache, Towards nonlinear delay-based control for convection-like distributed systems: the example of water flow control in open channel systems. Networks Heterogen. Media **4**(2), 177–194 (2009)
36. Z. Wu, W. Michiels, Reliably computing all characteristic roots of delay differential equations in a given right half plane using a spectral method. J. Comput.Appl. Math. **236**, 2499–2514 (2011)
37. E.I. Verriest, Linear systems with rational distributed delay: reduction and stability, in *European Control Conference, ECC'99* (Karlsruhe, Germany, 1999)
38. U. Münz, J.M. Rieber, F. Allgöwer, Robust stabilization and H_∞ control of uncertain distributed delay systems, in *Topics in Time-Delay Systems*, vol. 388 of Lecture Notes in Control and Information Sciences (Springer, Berlin Heidelberg, 2009), pp. 221–231
39. F. Gouaisbaut, Y. Ariba, Delay range stability of a class of distributed time delay systems. Syst. Control Lett. **60**, 211–217 (2011)
40. L. Crocco, Aspects of combustion stability in liquid propellant rocket motors, Part I. Fundamentals—Low frequency instability with monopropellants. J. Am. Rocket Soc. **21**, 163–178 (1951)

41. Y.A. Fiagbedzi, A.E. Pearson, A multistage reduction technique for feedback stabilizing distributed time-lag systems. Automatica **23**, 311–326 (1987)
42. C. Briat, E.I. Verriest, A new delay-SIR model for pulse vaccination. Biomed. Signal Process. Control **4**(4), 272–277 (2009)
43. W. Michiels, C.-I. Morărescu, S.-I. Niculescu, Consensus problems with distributed delays, with application to traffic flow models. SIAM J. Control Optim. **48**(1), 77–101 (2009)
44. S.-I. Niculescu, C.-I. Morărescu, W. Michiels, K. Gu, Geometric ideas in the stability analysis of delay models in biosciences, in *Biology and Control Theory: Current Challenges*, vol. 357, Lecture Notes in Control and Information Sciences, ed. by I. Queinnec, S. Tarbouriech, G. Garcia, S.-I. Niculescu (Springer, Berlin, 2007), pp. 217–259
45. H. Özbay, C. Bonnet, H. Benjelloun, J. Clairambault, Stability analysis of cell dynamics in leukemia. Math. Model. Nat. Phenom. **7**(1), 203–234 (2012)
46. R.K. Brayton, Bifurcation of periodic solutions in a nonlinear difference-differential equation of neutral type. Quart. Appl. Math **24**, 215–224 (1966)
47. P. Picard, O. Sename, J.F. Lafay, Weak controllability and controllability indices for linear neutral systems. Math. Comput. Simul. **45**, 223–233 (1998)
48. W. Michiels, K. Engelborghs, D. Roose, D. Dochain, Sensitivity to infinitesimal delays in neutral equations. SIAM J. Control Optim. **40**(4), 1134–1158 (2002)
49. V.B. Kolmanovskii, V.R. Nosov, *Stability of Functional Differential Equations* (Academic Press, London, 1986)
50. R.M. Murray, C.A. Jacobsson, R. Casas, A.I. Khibnik, C.R. Johnson Jr., R. Bitmead, A.A. Peracchio, W.M. Proscia, System identification for limit cycling systems: a case study for combustion instabilities, in *American Control Conference* (1998), pp. 2004–2008
51. A. Bellen, N. Guglielmi, Methods for linear systems of circuit delay differential equations of neutral type. IEEE Trans. Circuits Syst. I **76**(1), 212–215 (1999)
52. K. Engelborghs, M. Dambrine, D. Roose, Limitations of a class of stabilizing methods for delay equations. IEEE Trans. Autom. Control **46**(2), 336–339 (2001)
53. T. Vyhlídal, W. Michiels, P. Zítek, P. McGrahan, Stability impact of small delays in proportional-derivative state feedback. Control Eng. Pract. **17**, 382–393 (2009)
54. W. Michiels, K. Engelborghs, D. Roose, D. Dochain, Sensitivity to infinitesimal delays in neutral equations. SIAM J. Control Optim. **40**(4), 1134–1158 (2001)
55. E.C. Pielou, *Mathematical ecology* (Wiley Interscience, New York, 1977)
56. K. Gopalsamy, B.G. Zhang, On a neutral delay-logistic equation. Dyn. Stab. Syst. Int. J. **2**(3–4), 183–195 (1988)
57. E.I. Verriest, P. Pepe, Time optimal and optimal impulsive control for coupled differential difference point delay systems with an application in forestry, in *Topics in Time Delay Systems*, vol. 388, Lecture Notes in Control and Information Sciences, ed. by J.J. Loiseau, W. Michiels, S.I. Niculescu, R. Sipahi (Springer, Berlin, 2009), pp. 255–265
58. Y. Kuang, *Delay Differential Equations with Applications in Population Dynamics* (Academic, San Diego, 1993)
59. H.J. Wearing, P. Rohani, M.J. Keeling, Appropriate models for the management of infectious diseases. PLoS Med **2**(7), e174 (2005)
60. C. Briat, E.I. Verriest, A new delay-SIR model for pulse vaccination, in *17th IFAC World Congress* (Seoul, South Korea, 2008), pp. 10295–10300
61. T. Kato, J.B. McLeod, The functional-differential equation $y(x) = ay(\lambda x) + by(x)$. Bull. Am. Math. Soc. **77**(6), 891–937 (1971)
62. L. Fox, D.F. Mayers, J.R. Ockendon, A.B. Tayler, On a functional differential equation. J. Inst. Maths. Applics. **8**, 271–307 (1971)
63. A. Iserles, On the generalized pantograph functional-differential equation. Eur. J. Appl. Math. **4**(1), 1–38 (1993)
64. A. Iserles, On nonlinear delay differential equations. Trans. Am. Math. Soc. **344**(1), 441–477 (1994)
65. A.P. Zhabko, A.A. Laktionov, V.I. Zubov, Robust stability of differential-difference systems with linear time-delay, in *IFAC Symposium on Robust Control Design* (Budapest, Hungary, 1997), pp. 97–101

66. A.A. Laktionov, A.P. Zhabko, Method of difference transformations for differential systems with linear time-delay, in *IFAC Workshop on Linear Time Delay Systems* (Grenoble, France, 1998), pp. 201–205
67. E.I. Verriest, Robust stability, adjoints, and lq control of scale-delay systems, in *38th IEEE Conference on decision and control* (Phoenix, Arizona, USA, 1999), pp. 209–214
68. J.R. Ockendon, A.B. Tayler, The dynamics of a current collection system for an electric locomotive. Proc. R. Soc. London A **322**, 447–468 (1971)
69. V.A. Ambartsumian, On the theory of brightness fluctuations in the milky way (in russian). *Doklady Akad. Nauk SSSR*, 44:244–247, 1944 (Trans. Compt. Rend. (Doklady) Acad. Sci. URSS 44 (1944), 223–226)
70. V.A. Ambartsumian, *Theoretical astrophysics* (Pergamon Press, NY, 1958)
71. K. Mahler, On a special nonlinear functional equation. Proc. R. Soc. London. Ser. A, Math. Phys. Sci. **378**(1773), 155–178 (1981)
72. D.P. Gaver Jr, An absorption probability problem. J. Math. Anal. Appl. **9**, 384–393 (1964)
73. G.A. Derfel, Kato problem for functional-differential equations and difference schrödinger operator, in *Operator Theory*, ed. by P. Exner, P. Seba (Birhhäuser Verlag, Basel, 1990), pp. 319–321
74. V. Spiridonov, Universal superpositions of coherent states and self-similar potentials. Phys. Rev. A **52**, 1909–1935 (1995)
75. G.C. Wake, S. Cooper, H.K. Kim, B. van Brunt, Functional differential equations for cell-growth models with dispersion. Commun. Appl. Anal. **4**, 561–573 (2000)
76. J. Louisell, New examples of quenching in delay differential equations having time-varying delay, in *4th European Control Conference* (1999)
77. A. Papachristodoulou, M.M. Peet, S.I. Niculescu, Stability analysis of linear systems with time-varying delays: Delay uncertainty and quenching, in *46th Conference on Decision and Control* (LA, USA, New Orleans, 2007)
78. E.I. Verriest, Well-posedness of problems involving time-varying delays. in *19th International Symposium on Mathematical Theory and Networks and Systems* (Budapest, Hungary, 2010), pp. 1203–1210
79. W. Michiels, E.I. Verriest, *A look at fast varying and state dependent delays from a system theory point of view* (K. U. Leuven, Internal report, 2011)
80. E.I. Verriest, State space realization for systems with state dependent delay, in *11th IFAC Workshop on Time-Delay Systems* (Grenoble, France, 2013), pp. 446–451
81. A.R. Teel, D. Nešić, P.V. Kokotović, A note on input-to-state stability of sampled-data nonlinear systems, in *37th IEEE Conference on Decision and Control* (1998), pp. 2473–2478
82. E. Fridman, A. Seuret, J.P. Richard, Robust sampled-data stabilization of linear systems: an input delay approach. Automatica **40**, 1441–1446 (2004)
83. A. Seuret, Stability analysis for sampled-data systems with a time-varying period, in *48th Conference on Decision and Control* (Shanghai, China, 2009), pp. 8130–8135
84. F. Hartung, T. Krisztin, H.O. Walther, J. Wu, Functional differential equations with state-dependent delays: Theory and applications, in A. Ca nana, P. Drábek, and A. Fonda, editors, Handbook of differential equations—Ordinary differential equations, Vol. 3, (Elsevier, 2006), pp. 435–546.
85. H.O. Walther, On a model for soft landing with state-dependent delay. J. Dyn. Differ. Equ. **19**(3), 593–622 (2007)
86. H.O. Walther, On a model of soft landing with state-dependent delay. Int. J. Non-Linear Mech. **43**(2), 140–149 (2008)
87. H.O. Walther, Differentiable Semiflows for Differential Equations with State-Dependent Delays. Technical report, Universitatis Lagellonicae Acta Mathematica, 2003
88. M. Bartha, Convergence of solutions for an equation with state-dependent delay. J. Math. Anal. Appl. **254**, 410–432 (2001)
89. E.I. Verriest, Stability of systems with state-dependent and random delays. IMA J. Math. Control Inf. **19**(1–2), 103–114 (2002)

90. I. Györi, F. Hartung, Exponential stability of a state-dependent delay system. Discrete Continuous Dyn. Syst. **18**(4), 773–791 (2007)
91. M. Louihi, M.L. Hbid, Exponential stability for a class of state-dependent delay equations via the crandall-liggett approach. J. Math. Anal. Appl. **329**, 1045–1063 (2007)
92. K.L. Cooke, W. Huang, On the problem of linearization for state-dependent delay differential equations. Proc. Am. Math. Soc. **124**(5), 1417–1426 (1996)
93. T. Luzianina, K. Engelborghs, D. Roose, Diiferential equations with state-dependent delay—a numerical study, in *16th IMACS World Congress* (2000)
94. T. Luzianina, K. Engelborghs, D. Roose, Numerical bifurcation analysis of differential equations with state-dependent delay. Int. J. Bifurcation Chaos **11**(3), 737–753 (2001)
95. A. Feldstein, K.W. Neves, S. Thompson, Sharpness results for state dependent delay differential equations: an overview. Appl. Numer. Anal. **56**, 472–487 (2005)
96. N. Bekiaris-Liberis, M. Krstic, Compensation of time-varying input and state delays for nonlinear systems. J. Dyn. Syst. Measur. Control **134**, 011009 (2012)
97. N. Bekiaris-Liberis, M. Jankovich, M. Krstic, Compensation of state-dependent state delay nonlinear systems. Syst. Control Lett. **61**, 849–856 (2012)
98. N. Bekiaris-Liberis, M. Krstic, Compensation of state-dependent input delay for nonlinear systems. IEEE Trans. Autom. Control **58**(2), 275–289 (2013)
99. C. Briat, H. Hjalmarsson, K.H. Johansson, G. Karlsson, U.T. Jönsson, H. Sandberg, E.A. Yavuz, An axiomatic fluid-flow model for congestion control analysis, in *50th IEEE Conference on Decision and Control* (Orlando, Florida, USA, 2011), pp. 3122–3129
100. J. Goutsias, G. Jenkinson, Markovian Dynamics on Complex Reaction Networks. Physical reports, **529**, 199–264 (2013)
101. F. Horn, R. Jackson, General mass action kinetics. Arch. Ration. Mech. Anal. **47**(2), 81–116 (1972)
102. M. Feinberg, Complex balancing in general kinetic systems. Arch. Ration. Mech. Anal. **49**(3), 187–194 (1972)
103. P. Érdi, J. Tóth, *Mathematical Models of Chemical Reactions: Theory and Applications of Deterministic and Stochastic Models* (Princeton University Press, Princeton, 1989)
104. D. Anderson, T.G. Kurtz, Continuous time Markov chain models for chemical reaction networks. *Design and analysis of biomolecular circuits—Engineering Approaches to Systems and Synthetic Biology*, ed. by H. Koeppl, D. Densmore, G. Setti, M. di Bernardo (Springer Science+Business Media, 2011), pp. 3–42
105. W. Zhang, M.S. Branicky, S.M. Phillips, Stability of Networked Control Systems. IEEE Control Syst. Mag. **21**(1), 84–99 (2001)
106. S. Zampieri, A survey of recent results in Networked Control Systems, in *17th IFAC World Congress* (Seoul, South Korea, 2008), pp. 2886–2894
107. P. Tabuada, Event-triggered real-time scheduling of stabilizing control tasks. IEEE Trans. Autom. Control **52**(9), 1680–1685 (2007)
108. A. Anta, P. Tabuada, To sample or not to sample: Self-triggered control for nonlinear systems. IEEE Trans. Autom. Control **55**(9), 2030–2042 (2010)
109. D. Dimarogonas, E. Frazzoli, K.H. Johansson, Distributed event-triggered control for multi-agent systems. IEEE Trans. Autom. Control **57**(12), 1291–1297 (2012)
110. Y.S. Suh, Stability and stabilization of nonuniform sampling systems. Automatica **44**, 3222–3226 (2008)
111. H. Fujioka, A discrete-time approach to stability analysis of systems with aperiodic sample-and-hold devices. IEEE Trans. Autom. Control **54**(10), 2440–2445 (2009)
112. Y. Oishi, H. Fujioka, Stability and stabilization of aperiodic sampled-data control systems: an approach using robust linear matrix inequalities, in *48th Conference on Decision and Control* (Shanghai, China, 2009), pp. 8142–8147
113. L. Hetel, J. Daafouz, C. Iung, Stabilization of arbitrary switched linear systems with unknown time-varying delays. IEEE Trans. Autom. Control **51**(10), 1668–1674 (2006)
114. E. Fridman, A refined input delay approach to sampled-data control. Automatica **46**(2), 421–427 (2010)

115. K. Liu, V. Suplin, E. Fridman, Stability of linear systems with general sawtooth delay. IMA J. of Math. Control Inf. **27**(4), 419–436 (2010)
116. K. Liu, E. Fridman, Wirtinger's inequality and Lyapunov-based sampled-data stabilization. Automatica **48**(1), 102–108 (2012)
117. L. Mirkin, Some remarks on the use of time-varying delay to model sample-and-hold circuits. IEEE Trans. Autom. Control **52**(6), 1109–1112 (2007)
118. H. Fujioka, Stability analysis of systems with aperiodic sample-and-hold devices. Automatica **45**, 771–775 (2009)
119. C.-Y. Kao, H. Fujioka, On stability of systems with aperiodic sampling devices. IEEE Trans. Autom. Control **58**(3), 2085–2090 (2013)
120. W. Sun, K.M. Nagpal, P.P. Khargonekar, H_∞ control and filtering with sampled measurements, in *American Control Conference, 1991* (1991), pp. 1652–1657
121. N. Sivashankar, P.P. Khargonekar, Characterization of the $\mathcal{L}_2$-induced norm for linear systems with jumps with applications to sampled-data systems. SIAM J. Control Optim. **32**(4), 1128–1150 (1994)
122. G.E. Dullerud, S. Lall, Asynchronous hybrid systems with jumps—analysis and synthesis methods. Syst. Control Lett. **37**(2), 61–69 (1999)
123. P. Naghshtabrizi, J.P. Hespanha, A.R. Teel, Exponential stability of impulsive systems with application to uncertain sampled-data systems. Syst. Control Lett. **57**, 378–385 (2008)
124. C. Briat, A. Seuret, A looped-functional approach for robust stability analysis of linear impulsive systems. Syst. Control Lett. **61**, 980–988 (2012)
125. C. Briat, A. Seuret, Robust stability of impulsive systems: a functional-based approach, in *4th IFAC Conference on analysis and design of hybrid systems* (Eindovhen, the Netherlands, 2012), pp. 412–417
126. C. Briat, A. Seuret, Convex dwell-time characterizations for uncertain linear impulsive systems. IEEE Trans. Autom. Control **57**(12), 3241–3246 (2012)
127. D. Robert, O. Sename, D. Simon, An $\mathcal{H}_\infty$ LPV design for sampling varying controllers: experimentation with a T-inverted pendulum. IEEE Trans. Control Syst. Technol. **18**(3), 741–749 (2010)
128. A. Seuret, A novel stability analysis of linear systems under asynchronous samplings. Automatica **48**(1), 177–182 (2012)
129. C. Briat, Convex conditions for robust stability analysis and stabilization of linear aperiodic impulsive and sampled-data systems under dwell-time constraints. Automatica **49**, 3449–3457 (2013)
130. K. Wickwire, Mathematical models for the control of pests and infectious diseases: a survey. Theor. Popul. Biol. **11**(2), 182–238 (1977)
131. R.M. Anderson, R.M. May, Directly transmitted infectious diseases: control by vaccination. Science **215**, 1053–1060 (1982)
132. J. Arino, P. van den Driessche, Time delays in epidemic models: modeling and numerical considerations, in *Delay Differential Equations and Applications*, ed. by O. Arino, M. L. Hbid and E. Ait Dads (Springer, Netherlands, 2006), pp. 539–578
133. O. Diekmann, J.A.P. Heesterbeek, *Mathematical Epidemiology of Infectious Diseases* (Wiley, Chichester, 2000)
134. H.W. Hethcote, The mathematics of infectious diseases. SIAM Rev. **42**(4), 599–653 (2002)
135. R.M. Anderson, R.M. May, *Infectious Diseases of Humans: Dynamics and Control* (Oxford University Press, New York, 2002)
136. J.D. Murray, *Mathematical Biology Part I. An Introduction*, 3rd edn. (Springer, Berlin, 2002)
137. W.M. Haddad, V. Chellaboina, Q. Hui, *Nonnegative and Compartmental Dynamical Systems* (Princeton University Press, New Jersey, 2010)
138. C.M. Guldberg, P. Waage, Studies concerning affinity. Forhandlinger: Videnskabs-Selskabet i Christiana, 35, 1864
139. P. Waage, Experiments for determining the affinity law. Forhandlinger: Videnskabs-Selskabet i Christiana, 92, 1864

140. C.M. Guldberg, Concerning the laws of chemical affinity. Forhandlinger: Videnskabs-Selskabet i Christiana, 111, 1864
141. L. Trevor, *Affinity and Matter—Elements of Chemical Philosophy 1800–1865* (Gordon and Breach Science Publishers, San Fracisco, 1971)
142. Unknown author, Influenza in a boarding school. Br. Med. J., March **4**, 587 (1978)
143. Unknown author, Tashkent influenza ('red flu') January 1978. Ampleforth J., **4**, 587 (1978)
144. G. de Vries, J. Müller, T. Hillen, B. Schönfisch, M. Lewis, *A Course in Mathematical Biology: Quantitative Modeling with Mathematical and Computational Methods* (SIAM, Philadelphia, 2006)
145. R. Pearl, *The biology of Population Growth* (Knopf, New York, 1930)
146. P.F. Verhulst, Notice sur la loi que la population suit dans son accroissement (french). Correspondance mathématique et physique **10**, 113–121 (1938)
147. H.I. Freedman, Y. Kuang, Stability switches in linear scalar neutral delay equations. Funkcialaj Ekvacioj **34**, 187–209 (1991)
148. V. Jacobson, Congestion avoidance and control. SIGCOMM Comput. Commun. Rev. **18**(4), 314–329 (1988)
149. R. Srikant, *The Mathematics of Internet Congestion Control* (Birkhäuser, Boston, 2004)
150. D.X. Wei, C. Jin, S.H. Low, S. Hedge, FAST TCP: Motivation, architecture, algorithms, performance. IEEE/ACM Trans. Networking **14**(6), 1246–1259 (2006)
151. V. Misra, W.B. Gong, D. Towsley, Fluid-based analysis of a network of AQM routers supporting TCP flows with an application to RED, in *ACM SIGCOMM'00* (Stockholm, Sweden, 2000)
152. R. Johari, D. Tan, End-to-End Congestion Control for the Internet: Delay and Stability. Technical report, (Statistical Laboratory, University of Cambridge, 2000)
153. R. Johari, D.K. Tan, End-to-end congestion control for the Internet—Delays and Stability. IEEE/ACM Trans. Networking **19**(6), 818–832 (2001)
154. F. Paganini, Z. Wang, J.C. Doyle, S.H. Low, Congestion control for high performance, stability, and fairness in general networks. IEEE/ACM Trans. Networking **13**(1), 43–56 (2005)
155. J. Wang, D.X. Wei, S. H. Low, Modelling and stability of FAST TCP, in *28th IEEE Conference on Computer Communications (INFOCOM)* (2005), pp. 938–948
156. K. Jacobsson, L.L.H. Andrew, A.K. Tang, K.H. Johansson, H. Hjalmarsson, S.H. Low, ACK-clock dynamics: Modeling the interaction between ACK-clock and network, in *27th IEEE Conference on Computer Communications (INFOCOM)* (Phoenix, Arizona, USA, 2008), pp. 181–185
157. K. Jacobsson, Dynamic Modeling of Internet Congestion Control. PhD thesis, KTH School of Electrical Engineering, 2008
158. N. Möller, Window-based congestion control. Doctoral thesis, KTH, Stockholm, Sweden, 2008
159. Y. Liu, F.L. Presti, V. Misra, D.F. Towsley, Scalable fluid models and simulations for large-scale ip networks. ACM Trans. Model. Comput. Simul. **14**(3), 305–324 (2004)
160. J.P. Hespanha, S. Bohacek, K. Obraczka, J. Lee, Hybrid modeling of TCP congestion control, in *Hybrid Systems: Computation and Control*, Lecture Notes in Computer Science, ed. by M. Di Benedetto, A. Sangiovanni-Vincentelli (Springer, Berlin, 2001), pp. 291–304
161. C. Briat, E.A. Yavuz, H. Hjalmarsson, K.-H. Johansson, U.T. Jönsson, G. Karlsson, H. Sandberg, The conservation of information, towards an axiomatized modular modeling approach to congestion control, in *IEEE Transactions on Networking*, 23(3) (2015). doi:10.1109/TNET.2014.2308272
162. D. Mitra, Stochastic theory of a fluid model of producers and consumers coupled by a buffer. Adv. Appl. Probab. **20**(3), 646–676 (1988)
163. D.V. Lindley, The theory of queues with a single server. Math. Proc. Cambridge Philos. Soc. **48**, 277–289 (1952)
164. C. Ohta, F. Ishizaki, Output processes of shaper and switch with self-similar traffic in ATM networks. IEICE Trans. Commun. **E81–B**(10), 1936–1940 (1998)

165. A. Tang, K. Jacobsson, L.L.H. Andrew, S.H. Low, An accurate link model and its application to stability analysis of FAST TCP, in *26th IEEE Conference on Computer Communications (INFOCOM)* (Anchorage, Alaska, USA, 2007), pp. 161–169
166. K. Jacobsson, H. Hjalmarsson, N. Möller, ACK-clock dynamics in network congestion control—An inner feedback loop with implications on inelastic flow impact, in *45th IEEE Conference on Decision and Control* (San Diego, California, USA, 2006), pp. 1882–1887
167. Y. Orlov, L. Belkoura, J.-P. Richard, M. Dambrine, Adaptive identification of linear time-delay systems. Int. J. Robust Nonlinear Control **13**(9), 857–872 (2003)
168. S. Drakunov, S. Perruquetti, J.P. Richard, L. Belkoura, Delay identification in time-delay systems using variable structure control. Annual Rev. Control **30**(2), 143–158 (2006)
169. L. Belkoura, J.P. Richard, M. Fliess, Real time identification of time-delay systems, in *IFAC Workshop on Time-Delay Systems* (Nantes, France, 2007)
170. L. Belkoura, J.P. Richard, M. Fliess, A convolution approach for delay systems identification, in *IFAC World Congress* (Seoul, South Korea, 2008)
171. A.W. Pila, U. Shaked, C.E. de Souza, $\langle_\infty$ filtering for continuous-time linear systems with delay. IEEE Trans. Autom. Control **44**(7), 1412–1417 (1999)
172. E. Fridman, U. Shaked, A new H_∞ filter design for linear time delay systems. IEEE Trans. on Signal Process. **49**(11), 2839–2843 (2001)
173. C.E. de Souza, R.M. Palhares, P.L.D. Peres, Robust $\mathcal{H}_\infty$ filters design for uncertain linear systems with multiple time-varying state delays. IEEE Trans. Signal Process. **49**(3), 569–576 (2001)
174. H. Gao, C. Wang, Delay-dependent robust H_∞ and $L_2 - L_\infty$ filtering for a class of uncertain nonlinear time-delay systems. IEEE Trans. Autom. Control **48**(9), 1661–1666 (2003)
175. E. Fridman, U. Shaked, L. Xie, Robust H_∞ filtering of linear systems with time-varying delays. IEEE Trans. Autom. Control **48**(1), 159–165 (2003)
176. K.P.M. Bhat, H.N. Koivo, An observer theory for time-delay systems. IEEE Trans. Autom. Control **216**(2), 166–169 (1976)
177. A.W. Olbrot, Stabilizability, detectability, and spectrum assignment for linear autonomous systems with general time delays. IEEE Trans. Autom. Control **23**(5), 887–890 (1978)
178. E.B. Lee, A.W. Olbrot, Observability and related structural results for linear hereditary systems. Int. J. Control **34**, 1061–1078 (1981)
179. D. Salamon, On controllability and observability of time delay systems. IEEE Trans. Autom. Control **29**(5), 432–439 (1984)
180. O. Sename, New trends in design of observers for time-delay systems. Kybernetika **37**(4), 427–458 (2001)
181. A. Fattouh, O. Sename, J.M. Dion, Robust observer design for time-delay systems: A riccati equation approach. Kybernetika **35**(6), 753–764 (1999)
182. O. Sename, Unknown input robust observers for time-delay systems, in *36th IEEE Conference on Decision and Control* (San Diego, California, USA, 1997), pp. 1629–1630
183. A. Fattouh, O. Sename, J. M. Dion, An unknown input observer design for linear time-delay systems, in *Proceedings of 38th IEEE Conference on Decision & Control* (Phoenix, Arizona, USA, 1999), pp. 4222–4227
184. A. Fattouh, O. Sename, J. M. Dion, $\mathcal{H}_\infty$ observer design for time-delay systems, in *Proceedings of 37th IEEE Conference on Decision & Control* (Tampa, Florida, USA, 1998), pp. 4545–4546
185. O. Sename, A. Fattouh, J-M. Dion, Further results on unknown input observers design for time-delay systems, in *40th IEEE Conference on Decision and Control* (Orlando, Florida, USA, 2001)
186. W. Aggoune, M. Boutayeb, M. Darouach, Observers design for a class of non linear systems with time-varying delay, in *Proceedings of 38th IEEE Confererence on Decision & Control* (Phoenix, Arizona, USA, 1999), pp. 2912–2913
187. A. Fattouh, O. Sename, J.M. Dion, An LMI approach to robust observer design for linear time-delay systems, in *Proceedings 39th IEEE Conference on Decision & Control* (Sydney, Australia, December, 2000), pp. 12–15

188. M. Darouach, Linear functional observers for systems with delays in state variables. IEEE Trans. Autom. Control **46**(3), 491–496 (2001)
189. D. Koenig, B. Marx, O. Sename, Unknown inputs proportional integral observers for descriptors systems with multiple delays and unknown inputs, in *American Control Conference* (Boston, Massachusetts, USA, 2004)
190. M. Darouach, Reduced-order observers for linear neutral delay systems. IEEE Trans. Autom. Control **50**(9), 1407–1413 (2005)
191. D. Koenig, D. Jacquet, S. Mammar, Delay dependent $\mathcal{H}_\infty$ observer for linear delay descriptor systems, in *American Control Conference 2006* (Minneapolis, USA, 2006)
192. Y.A. Fiagbedzi, A.E. Pearson, A state observer for systems described by functional differential equations. Automatica **26**(2), 321–331 (1990)
193. A.W. Olbrot, On controllability of linear systems with time delay in control. IEEE Trans. Autom. Control **17**(5), 664–666 (1972)
194. M.W. Spong, T.J. Tarn, On the spectral controllability of delay-differential equations. IEEE Trans. Autom. Control **26**(2), 527–528 (1981)
195. O. Sename, J.F. Lafay, A sufficient condition for static decoupling without prediction of linear time-invariant systems with delays, in *ECC 93, European Control Conference* (Groningen, The Netherlands, 1993), pp. 673–678
196. O. Sename, *On controllability and decoupling of linear systems with delays (in French)*. PhD thesis, Ecole Centrale Nantes, France, 1994
197. O. Sename, J.F. Lafay, Decoupling of linear systems with delays, in *33th IEEE Conference on Decision and Control* (Orlando, Floride, USA, 1994)
198. O. Sename, J.F. Lafay, R. Rabah, Decoupling without prediction of linear systems with delays: a structural approach. Syst. Control Lett. **25**, 387–395 (1995)
199. X. Li, C.E. de Souza, A. Trofino, Delay dependent robust stabilization of uncertain linear state-delayed systems via static output feedback, in *IFAC Workshop on Linear Time-Delay Systems* (Grenoble, France, 1998), pp. 1–6
200. S.I. Niculescu, C.T. Abdallah, Delay effects on static output feedback stabilization, in *39th IEEE Conference on Decision and Control* (Sydney, Australia, 2000), pp. 2811–2816
201. W. Michiels, S.I. Niculescu, L. Moreau, Using delays and time-varying gains to improve the static output feedback of linear systems: a comparison. IMA J. Math. Control Inf. **21**(4), 393–418 (2004)
202. V.L. Kharitonov, S.-I. Niculescu, J. Moreno, W. Michiels, Static output feedback stabilization: necessary conditions for multiple delay controllers. IEEE Trans. Autom. Control **50**(1), 82–86 (2005)
203. A. Seuret, C. Edwards, S.K. Spurgeon, E. Fridman, Static output feedback sliding mode control design via an artificial stabilizing delay. IEEE Trans. Autom. Control **54**(2), 256–265 (2009)
204. V.L. Syrmos, C.T. Abdallah, P. Dorato, K. Grigoriadis, Static output feedback—a survey. Automatica **33**(2), 125–137 (1997)
205. V. Blondel, J.N. Tsitsiklis, NP-hardness of some linear control problems. SIAM J. Control Optim. **35**(6), 2118–2127 (1997)
206. M. Fu, Pole placement via static output feedback is NP-hard. IEEE Trans. Autom. Control **49**(5), 855–857 (2004)
207. F. Wu, Robust quadratic performance for time-delayed uncertain linear systems. Int. J. Robust Nonlinear Control **13**, 153–172 (2003)
208. C. Briat, O. Sename, J.-F. Lafay, A full-block $\mathcal{S}$-procedure application to delay-dependent $\mathcal{H}_\infty$ state-feedback control of uncertain time-delay systems, in *17th IFAC World Congress* (Seoul, South Korea, 2008), pp. 12342–12347
209. M.M. Peet, Full-state feedback of delayed system using SOS: A new theory of duality, in *11th IFAC Workshop on Time-Delay Systems* (Grenoble, France, 2013), pp. 24–29
210. C. Briat, O. Sename, J.-F. Lafay, Memory resilient gain-scheduled state-feedback control of time-delay systems with time-varying delays, in *6th IFAC Symposium on Robust Control Design* (Haifa, Israel, 2009), pp. 202–207

211. C. Briat, O. Sename, J.-F. Lafay, Memory resilient gain-scheduled state-feedback control of uncertain LTI/LPV systems with time-varying delays. Syst. Control Lett. **59**, 451–459 (2010)
212. J.-H. Ge, P.M. Frank, C.-F. Lin, H_∞ control via output feedback for state delayed system. Int. J. Control **64**(1), 1–7 (1996)
213. A. Fattouh, O. Sename, J. M. Dion, $\mathcal{H}_\infty$ controller and observer design for linear systems with point and distributed time-delays: An LMI approach, in *2nd IFAC Workshop on Linear Time Delay Systems* (Ancône, Italy, 2000)
214. O. Sename, C. Briat, Observer-based $\mathcal{H}_\infty$ control for time-delay systems: a new LMI solution, in *6th IFAC Workshop on Time Delay Systems* (L'Aquila, Italy, 2006), pp. 114–119
215. O. Sename, C. Briat, Observer-based $\mathcal{H}_\infty$ control for time-delay systems: a new LMI solution, in *9th European Control Conference* (Kos, Greece, 2007), pp. 5123–5130
216. O. Sename, Is a mixed design of observer-controllers for time-delay systems interesting ? Asian J. Control **9**(2), 180–189 (2007)
217. E.T. Jeung, J.H. Kim, H.B. Park, H^∞-output feedback controller design for linear systems with time-varying delayed state. IEEE Trans. Autom. Control **43**(7), 971–974 (1998)
218. J.M. Gomes da Silva, I. Ghiggi, S. Tarbouriech, Non-rational dynamic ouput feedback for time-delay systems with saturating inputs. Int. J. Control **81**(4), 557–570 (2008)
219. C. Briat, O. Sename, J.-F. Lafay, A LFT/$\mathcal{H}_\infty$ state-feedback design for linear parameter varying time delay systems, in *9th European Control Conference* (Kos, Greece, 2007), pp. 4882–4888
220. C. Briat, O. Sename, J.-F. Lafay, Delay-scheduled state-feedback design for time-delay systems with time-varying delays, in *17th IFAC World Congress* (Seoul, South Korea, 2008), pp. 1267–1272
221. C. Briat, O. Sename, J.-F. Lafay, $\mathcal{H}_\infty$ delay-scheduled control of linear systems with time-varying delays. IEEE Trans. Autom. Control **42**(8), 2255–2260 (2009)
222. C. Briat, O. Sename, J.-F. Lafay, Delay-scheduled state-feedback design for time-delay systems with time-varying delays—a LPV approach. Syst. Control Lett. **58**(9), 664–671 (2009)
223. J. Daafouz, J. Bernussou, Parameter dependent Lyapunov functions for discrete time systems with time varying parametric uncertainties. Syst. Control Lett. **43**, 355–359 (2001)
224. J. Daafouz, P. Riedinger, C. Iung, Stability analysis and control synthesis for switched systems: a switched Lyapunov function approach. IEEE Trans. Autom. Control **47**(11), 1883–1887 (2002)
225. L. Hetel, J. Daafouz, C. Iung, Equivalence between the Lyapunov-Krasovskii functionals approach for discrete delay systems and that of the stability conditions for switched systems. Nonlinear Anal. Hybrid Syst. **2**(3), 697–705 (2008)
226. W. Jiang, E. Fridman, A. Kruszewski, J.-P. Richard, Switching controller for stabilization of linear systems with switched time-varying delays, in *IEEE Conference on Decision and Control* (2009), pp. 7923–7928
227. B. Demirel, C. Briat, M. Johansson, Supervisory control design for networked systems with time-varying communication delays, in *4th IFAC Conference on analysis and design of hybrid systems* (Eindovhen, the Netherlands, 2012), pp. 133–140
228. B. Demirel, C. Briat, M. Johansson, Supervisory control design for networked systems with time-varying communication delays. Nonlinear Anal. Hybrid Syst. **10**, 94–110 (2013)

Chapter 5
Stability Analysis of Time-Delay Systems

Time is what prevents everything from happening at once.
John Archibald Wheeler

Abstract This chapter presents the main stability and instability results for general time-delay systems. These results are further adapted to the analysis of linear time-delay system using the extensions of Lyapunov theory, namely the Lyapunov-Krasovskii and Lyapunov-Razumikhin theorems, and input-output analysis techniques such small-gain techniques, integral quadratic constraints and quadratic separation. Theoretical results regarding the conservatism of model-transformations and bounding techniques are also derived. The different approaches are compared with each other based on their corresponding stability criteria. Some discussions about complexity reduction are also provided. As for LPV systems, all the obtained stability criteria take the form of matrix inequalities.

5.1 Chapter Outline

The analysis of time-delay systems is a well-developed field gathering a lot of different techniques. These methods can be categorized to either belong to frequency-domain or time-domain techniques.

Frequency-domain approaches are mostly devoted to linear time-invariant systems, yet under some circumstances, it is possible to adapt them to address the case of varying delays using, for instance, model transformations. Time-domain approaches can, however, be applied to any type of systems: linear or nonlinear, with constant or time-varying delays, etc. In the following, only time-domain analysis techniques will be discussed since they are more suitable for dealing with LPV time-delay systems. Many excellent monographs are, moreover, already devoted to frequency-domain techniques; see e.g. [1–3].

C. Briat, *Linear Parameter-Varying and Time-Delay Systems*,
Advances in Delays and Dynamics 3, DOI 10.1007/978-3-662-44050-6_5

In the following, general notions of stability for time-delay systems as well as stability/instability results are introduced in Sect. 5.2. Notably, the Lyapunov-Krasovskii and Lyapunov-Razumikhin Theorems, and an instability result due to Haddock are presented. *Complete Lyapunov-Krasovskii functionals* are also introduced in this section. Section 5.3 is devoted to the definition of *delay-independent stability* and *delay-dependent stability*, and also introduces the *quenching phenomenon*. The concepts of *model-transformations* and *additional dynamics* are defined and discussed in Sect. 5.4. The analysis of linear time-delay systems using the Lyapunov-Razumikhin Theorem is addressed in Sect. 5.5 whereas the Lyapunov-Krasovskii Theorem is considered in Sect. 5.6 using various model-transformations and bounding techniques. The analysis of the conservatism of certain bounding techniques is also carefully made. Sections 5.7 and 5.8 are devoted to the application of robust analysis techniques to the analysis of time-delay systems. Notably, Sect. 5.7 focuses on small-gain results in the L_2-norm whereas Sect. 5.8 considers the L_∞-norm. Still in a robust analysis perspective, Sects. 5.9 and 5.10 consider Integral Quadratic Constraints and quadratic separation, respectively.

5.2 General Notions of Stability for Time-Delay Systems

In this section, general statements about time-delay systems are provided. We, therefore, consider the following time-delay system

$$\begin{aligned} \dot{x}(t) &= f(t, x_t), \quad t \geq t_0 \\ x(t_0 + s) &= \varphi(s), \quad s \in [-h, 0] \end{aligned} \tag{5.1}$$

where $h > 0$ is the delay and $\varphi \in C([-h, 0], \mathbb{R}^n)$ is the functional initial condition. The state of the system, denoted by $x_t \in C([-h, 0], \mathbb{R}^n)$, is defined as

$$x_t(\theta) = x(t + \theta).$$

We assume that the system (5.1) has a unique solution, see e.g. [4–6]. In the following, $x_t(t_0, \varphi)$ denotes the state-value at time t with initial condition $x_{t_0} = \varphi$. We finally assume, without loss of generality, that (5.1) admits the solution $x(t) = 0$, i.e. $f(t, 0) = 0$, generally referred to as the *trivial solution*.

5.2.1 Definitions

Definition 5.2.1 (Uniform norm) Let $\phi \in C([a, b], \mathbb{R}^n)$, then the **uniform norm** of ϕ is defined as

$$||\phi||_c = \max_{s \in [a,b]} ||\phi(s)|| \tag{5.2}$$

where $||\cdot||$ is any vector-norm, e.g. the vector 2-norm.

Definition 5.2.2 Consider the time-delay system (5.1). The trivial solution is said to be

- **stable** if for any $t_0 \geq 0$ and any $\varepsilon > 0$, there exists $\delta = \delta(t_0, \varepsilon) > 0$ such that
$$||x_{t_0}||_c \leq \delta \Rightarrow ||x(t)|| \leq \varepsilon \tag{5.3}$$
for all $t \geq t_0$.
- **attractive** if for any $t_0 \geq 0$ and any $\varepsilon > 0$, there exists $\delta_a = \delta_a(t_0, \varepsilon) > 0$ with the property that
$$||x_{t_0}||_c \leq \delta_a \Rightarrow \lim_{t\to\infty} ||x(t)|| = 0. \tag{5.4}$$
- **asymptotically stable** (in the sense of Lyapunov) if it is both stable and attractive.
- **uniformly stable** if it is stable and $\delta(t_0, \varepsilon)$ can be chosen independently of t_0.
- **exponentially stable** if there exist $\delta, \alpha > 0$ and $\beta \geq 1$ such that
$$||x_{t_0}||_c \leq \delta \Rightarrow ||x(t)|| \leq \beta e^{-\alpha t}||x_0|| \tag{5.5}$$
for all $t \geq 0$.
- **unstable** if it is not stable in the sense of Lyapunov.

5.2.2 *General Stability and Instability Results*

As for finite-dimensional systems, stability of time-delay systems can be also characterized using Lyapunov theory. Two important extensions have been obtained in this respect: the Lyapunov-Razumikhin theorem, and the Lyapunov-Krasovskii theorem. The first result involves Lyapunov-Razumikhin function as main ingredients, whereas Krasovskii's theorem relies on the use of functionals. The main difficulty in characterizing stability of time-delay systems lies in the fact that the system is infinite-dimensional and, therefore, the sole knowledge of $x(t)$ is not a sufficient statistic for establishing the stability of the system. The infinite-dimensional state x_t however contains all the information we need, and must be thus considered instead. Both aforementioned approaches do capture this subtlety in two very different ways.

5.2.2.1 Lyapunov-Krasovskii Theorem

Krasovskii's idea was to extend Lyapunov's result to account for the infinite-dimensionality of the state through the use of functionals [7] and obtained the following result, restated from [2]:

Theorem 5.2.3 (Lyapunov-Krasovskii Stability Theorem) *Suppose that the function*

$$f : \mathbb{R}_{\geq t_0} \times C([-h, 0], \mathbb{R}^n) \rightarrow \mathbb{R}^n$$

in (5.1) *maps* $\mathbb{R}_{\geq t_0} \times$ *(bounded sets of* $C([-h, 0], \mathbb{R}^n)$*) into bounded sets of* $\mathbb{R}^n$*, and* u*,* v*,* $w : \mathbb{R}_{\geq 0} \rightarrow \mathbb{R}_{\geq 0}$ *are continuous nondecreasing functions,* $u(s)$ *and* $v(s)$ *are positive for* $s > 0$*, and* $u(0) = v(0) = 0$.

Assume further that there exists a continuous differentiable functional

$$V : \mathbb{R} \times C([-h, 0], \mathbb{R}^n) \rightarrow \mathbb{R}$$

such that

$$u(||\phi(0)||) \leq V(t, \phi) \leq v(||\phi||_c) \tag{5.6}$$

and

$$\begin{aligned} \dot{V}(t, \phi) &:= \limsup_{\epsilon \rightarrow 0^+} \frac{1}{\epsilon}[V(t + \epsilon, x_{t+\epsilon}(t, \phi)) - V(t, \phi)] \\ &\leq -w(||\phi(0)||). \end{aligned} \tag{5.7}$$

Then, the trivial solution of (5.1) *is uniformly stable. Moreover, if* $w(s) > 0$ *for* $s > 0$*, then it is uniformly asymptotically stable. If, in addition,* $\lim_{s \rightarrow +\infty} u(s) = +\infty$*, then it is globally uniformly asymptotically stable.*

Proof The proof of this result can be found, for instance, in [2, 7]. ■

When the functional V is chosen such that it represents the size of the state x_t at any time t, i.e. it satisfies condition (5.6), a nonpositive $\dot{V}$ indicates that x_t does not "grow" with t, which in turn means that the considered system is stable. To clear up readers' mind, a simple illustrative example is given below. More general and advanced results are discussed in Sect. 5.6.

Example 5.2.4 Let us consider the scalar system

$$\dot{x}(t) = -ax(t) + bx(t - h) \tag{5.8}$$

where $a > 0$. The Lyapunov-Krasovskii functional

$$V(x_t) = x(t)^2 + q \int_{t-h}^{t} x(s)^2 ds, \quad q > 0 \tag{5.9}$$

trivially satisfies the condition (5.6) with $u(s) = s^2$ and $v(s) = (1 + qh)s^2$. The derivative of V along the trajectories of the system (5.8) is given by

$$\dot{V}(x_t) = \begin{bmatrix} x(t) \\ x(t-h) \end{bmatrix}^{\mathrm{T}} \begin{bmatrix} -2a + q & b \\ b & -q \end{bmatrix} \begin{bmatrix} x(t) \\ x(t-h) \end{bmatrix}. \tag{5.10}$$

The system is therefore globally asymptotically stable if the matrix

$$\begin{bmatrix} -2a + q & b \\ b & -q \end{bmatrix}$$

is negative definite or, equivalently, if the inequality $-2a + q + q^{-1}b^2 < 0$ holds. In this case, the function w can be chosen as $w(s) = \varepsilon s^2$ for some sufficiently small $\varepsilon > 0$.

We can then see that each $q > 0$ defines a stability region in the (a, b)-plane which is given by

$$\frac{q + q^{-1}b^2}{2} < a. \tag{5.11}$$

To have the maximal stability region, it seems natural to try to minimize the left-hand side with respect to q. Simple calculations show that the minimum is attained for $q = b$, which yields the "maximal" stability region given by $a > 0$ and $a > b$.

It can actually be verified that this stability region is exact whenever the delay is allowed to take any value in $[0, \infty)$.

5.2.2.2 Lyapunov-Razumikhin Theorem

Whereas Krasovskii's theory relies on the use of functionals in order to consider the infinite-dimensionality of the state, Razumikhin's theory rather involves functions V that are representative of the size of the state $x(t)$. For such functions V, the functional

$$\bar{V}(x_t) = \max_{\theta \in [-h, 0]} V(x(t + \theta))$$

serves as a measure of the size of x_t. The key idea behind Lyapunov-Razumikhin theorem is to note that whenever $V(x(t)) < \bar{V}(x_t)$, then the function $\bar{V}(x_t)$ does not grow when we have $\dot{V}(x(t)) > 0$. For $\bar{V}(x_t)$ to grow, it is necessary and sufficient to have $\dot{V}(x(t)) > 0$ when $\bar{V}(x_t) = V(x(t))$. This is formalized in the result below:

Theorem 5.2.5 (Lyapunov-Razumikhin Stability Theorem) *Suppose that the function*

$$f : \mathbb{R} \times C([-h, 0], \mathbb{R}^n) \to \mathbb{R}^n$$

in (5.1) *takes* $\mathbb{R}_{\geq t_0} \times$ (*bounded sets of* $C([-h, 0], \mathbb{R}^n)$) *into bounded sets of* $\mathbb{R}^n$, *and* $u, v, w : \mathbb{R}_{\geq 0} \to \mathbb{R}_{\geq 0}$ *are continuous nondecreasing functions,* $u(s)$ *and* $v(s)$ *are positive for* $s > 0$, *and* $u(0) = v(0) = 0$, v *strictly increasing. Assume further that there exists a continuously differentiable function*

$$V : \mathbb{R} \times \mathbb{R}^n \to \mathbb{R}$$

verifying

$$u(||x||) \leq V(t, x) \leq v(||x||), \quad \text{for } t \geq t_0 \text{ and } x \in \mathbb{R}^n \tag{5.12}$$

and such that the derivative of V *along the solution of* (5.1) *satisfies*

$$\dot{V}(t, x(t)) \leq -w(||x(t)||) \text{ whenever } V(t+\theta, x(t+\theta)) \leq V(t, x(t)) \tag{5.13}$$

for all $\theta \in [-h, 0]$. *Then, the system* (5.1) *is uniformly stable.*

- *If, moreover,* $w(s) > 0$ *for* $s > 0$ *and there exists a continuous nondecreasing function* $p(s) > s$ *for* $s > 0$ *such that condition* (5.13) *is strengthened to*

  $$\dot{V}(t, x(t)) \leq -w(||x(t)||) \text{ if } V(t+\theta, x(t+\theta)) \leq p(V(t, x(t)))$$

  for all $\theta \in [-h, 0]$, *then the system* (5.1) *is uniformly asymptotically stable.*
- *If, in addition,* $\lim_{s\to+\infty} u(s) = +\infty$, *then the system* (5.1) *is globally uniformly asymptotically stable.*

Proof The proof of this result can be found, for instance, in [2, 8]. ∎

Theoretical applications of the above theorem are given in Sect. 5.5.

Example 5.2.6 Taking back the scalar system of Example 5.2.4. The Lyapunov-Razumikhin function $V(x) = x^2/2$ satisfies the condition (5.12) with $u(s) = v(s) = s^2/2$. Its time-derivative along the solutions of the system is given by

$$\dot{V}(t) = -ax(t)^2 + bx(t)x(t-h). \tag{5.14}$$

According to Lyapunov-Razumikhin Theorem, we just demand that the derivative is negative define whenever $V(x(t-h)) < p^2 V(x(t))$ for some $p > 1$. This is equivalent to the condition $|x(t-h)| < p|x(t)|$. Thus we obtain the condition

$$\begin{aligned} \dot{V}(t) &\le -ax(t)^2 + |b||x(t)||x(t-h)| \\ &\le (-a + |b|p)x(t)^2 < 0 \end{aligned} \tag{5.15}$$

where the last inequality has been obtained by upper-bounding $|x(t-h)|$ by $p|x(t)|$. Noting that if $-a + |b|p < 0$ for $p = 1$, then it also holds for a sufficiently small $p > 1$. We finally obtain the stability region defined by $a > 0$ and $|b| < a$, which turns out to be identical to the one obtained with the Lyapunov-Razumikhin theorem in Example 5.2.4.

5.2.2.3 An Instability Theorem

Instability results can also be found in the literature, even though they are scarcer due to the difficulty of checking the resulting conditions. For completeness, one such result is provided here together with the references [9–12]. The following result is taken from [11] and relies on Lyapunov-Razumikhin-type conditions:

Theorem 5.2.7 *Let G be an open subset of $C([-h, 0], \mathbb{R}^n)$ containing 0 and suppose that there exists a function $V : \mathbb{R}^n \to \mathbb{R}_{\ge 0}$ such that $V(0) = 0$ and $V(x) > 0$ when $x \neq 0$. Let us further define the functional*

$$\bar{V}'(\phi) := \limsup_{h \to 0^+} \frac{V(\phi(0) + hf(\phi)) - V(\phi(0))}{h}$$

and assume that one of the following statements hold

1. $\bar{V}'(\phi) > 0$ for all $\phi \in G$ with

$$V(\phi(0)) = \max_{s \in [-h,0]} V(\phi(s)) > 0, \tag{5.16}$$

2. $\bar{V}'(\phi) > 0$ *for all* $\phi \in G$ *with*

$$V(\phi(0)) = \min_{s\in[-h,0]} V(\phi(s)) > 0, \tag{5.17}$$

then the solution $x = 0$ *of the system* (5.1) *is unstable.*

Checking the conditions of the above result is not an easy task since there is, as usual, no constructive way for finding a "good" function V together with a 'good' set G. Moreover, assuming that we have suitable V and G, accurately checking the condition $\bar{V}'(\phi) > 0$ for all $\phi \in G$ may also be quite tricky.

Example 5.2.8 Let us consider the scalar nonlinear delay system

$$\begin{aligned} \dot{x}(t) &= af(x(t)) + bf(x(t-h)) \\ x(s) &= \varphi(s), \; s \in [-h, 0] \end{aligned} \tag{5.18}$$

where the function $f : \mathbb{R} \to \mathbb{R}$ is continuous, strictly increasing with $f(0) = 0$. We propose to show that, when $a + b > 0$, the zero solution is unstable. Let $V(x) := |x|$, then we have two different cases: either $b \leq 0$ or $b > 0$.

1. Suppose $b \leq 0$ and assume that $t \geq 0$ is such that

$$V(x(t)) = \max_{\theta\in[-h,0]} V(x(t+\theta)),$$

whence $|x(t)| = \max_{\theta\in[-h,0]} |x(t+\theta)|$. Then, we have the following cases:

- If $x(t) > 0$, then we have $x(t) \geq x(t-h)$ and

$$\begin{aligned} V'(x(t)) &= af(x(t)) + bf(x(t-h)) \\ &\geq (a+b)f(x(t)) > 0. \end{aligned} \tag{5.19}$$

- If $x(t) < 0$, then we have $x(t) \leq x(t-h)$ and

$$\begin{aligned} V'(x(t)) &= -af(x(t)) - bf(x(t-h)) \\ &\geq -(a+b)f(x(t)) > 0. \end{aligned} \tag{5.20}$$

Therefore, statement 1 of Theorem 5.2.7 holds.

2. Suppose $b > 0$ and suppose $t \geq 0$ is such that

$$V(x(t)) = \min_{\theta\in[-h,0]} V(x(t+\theta)),$$

whence $|x(t)| = \min_{\theta \in [-h,0]} |x(t+\theta)|$. Then, we have the following cases:

- If $x(t) > 0$, then $x(t) \leq x(t-h)$ and

$$\begin{aligned} V'(x(t)) &= af(x(t)) + bf(x(t-h)) \\ &\geq (a+b)f(x(t)) > 0. \end{aligned} \tag{5.21}$$

- If $x(t) < 0$, then $x(t) \geq x(t-h)$ and

$$\begin{aligned} V'(x(t)) &= -af(x(t)) - bf(x(t-h)) \\ &\geq -(a+b)f(x(t)) > 0. \end{aligned} \tag{5.22}$$

Therefore, statement 2 of Theorem 5.2.7 holds.

In conclusion, the zero solution is unstable when $a + b > 0$ by virtue of Theorem 5.2.7.

5.2.3 LTI System Case

When LTI systems are considered, explicit Lyapunov-Krasovskii functionals referred to as *complete Lyapunov-Krasovskii functionals* have been proven to yield necessary and sufficient stability conditions for asymptotic stability. Several different complete functionals have been proposed over the past years, see e.g. [13–17].

Before stating the main result, let us first consider the LTI time-delay system:

$$\begin{aligned} \dot{x}(t) &= Ax(t) + A_h x(t-h) \\ x(\theta) &= \varphi(\theta), \quad \theta \in [-h, 0] \end{aligned} \tag{5.23}$$

where $x(t) \in \mathbb{R}^n$, $\varphi \in C([-h,0],\mathbb{R}^n)$ and $h \in \mathbb{R}_{\geq 0}$ are the system state, the functional initial condition and the constant time-delay, respectively. We have the following result:

Theorem 5.2.9 ([2, 16]) *The system* (5.23) *with constant delay* $h > 0$ *is asymptotically stable if and only if there exist a constant matrix* $P \in \mathbb{S}^n$, *a scalar* $\varepsilon > 0$ *and continuously differentiable matrix functions*

$$\begin{aligned} &Q : [-h, 0] \to \mathbb{R}^{n\times n} \\ &R : [-h, 0]^2 \to \mathbb{R}^{n\times n}, \quad R(\xi,\eta) = R(\eta,\xi)^T, \\ &S : [-h, 0] \to \mathbb{S}^n \end{aligned}$$

such that

$$
\begin{aligned}
V(x_t) = x(t)^T P x(t) + 2x(t)^T \int_{-h}^{0} Q(\xi)x(t+\xi)d\xi \\
+ \int_{-h}^{0} \left[\int_{-h}^{0} x(t+\xi)^T R(\xi,\eta)x(t+\eta)d\eta \right] d\xi \\
+ \int_{-h}^{0} x(t+\xi)^T S(\xi)x(t+\xi)d\xi \geq \varepsilon ||x(t)||^2
\end{aligned}
$$

is a Lyapunov-Krasovskii functional and the derivative of V evaluated along the trajectories of (5.23) *verifies*

$$
\begin{aligned}
\dot{V}(x_t) = & x(t)^T [PA + A^T P + Q(0) + Q^T(0) + S(0)]x(t) \\
& - x(t-h)^T S(-h)x(t-h) + 2x(t)^T [PA_h - Q(-h)]x(t-h) \\
& - \int_{-h}^{0} x(t+\xi)^T \dot{S}(\xi)x(t+\xi)d\xi \\
& - \int_{-h}^{0}\int_{-h}^{0} x(t+\xi)^T \left[\frac{\partial}{\partial\xi} R(\xi,\eta) + \frac{\partial}{\partial\eta} R(\xi,\eta) \right] x(t+\eta)d\eta d\xi \\
& + 2x(t)^T \int_{-h}^{0} [A^T Q(\xi) - \dot{Q}(\xi) + R(0,\xi)]x(t+\xi)d\xi \\
& + 2x(t-h)^T \int_{-h}^{0} [A_h^T Q(\xi) - R(-h,\xi)]x(t+\xi)d\xi \leq -\varepsilon ||x(t)||^2.
\end{aligned}
$$

Finding a suitable matrix P and suitable matrix functions Q, R and S is not an easy task since the decision variables are functions. To overcome this difficulty, discretization schemes involving piecewise affine matrix functions have been considered in [2, 18, 19] whereas sum-of-squares techniques, that consider polynomial matrix functions, have been applied in [20–25]. A different approach based on a parametrization of functionals using polynomials is also proposed in [26].

5.3 Delay-Related Notions of Stability

The stability conditions obtained in the Examples 5.2.4 and 5.2.6 do not depend on the delay value, whereas Theorem 5.2.9 basically characterizes the stability of system (5.23) for some given delay $h > 0$. This tells us that two types of stability results may be distinguished based on whether they depend on the delay value. Suppose further that we would also be interested in assessing stability of system (5.23) for a range of delay values or even obtain stability results for another relevant family of delays. This leads us to the concepts of delay-independent, delay-dependent and delay-range stability.

5.3.1 Delay-Independent Stability

Delay-independent stability, as coined in [27], is defined as follows:

Definition 5.3.1 (*Delay-Independent Stability*) A time-delay system is **stable independently of the delay** or **delay-independent stable** if stability does not depend on the delay value, that is, if the system is stable for any delay value in $[0, \infty)$.

The above definition immediately extends to systems with multiple delays and time-varying delays. This concept of stability is quite strong since delays must have no impact on stability. This imposes, in return, strong constraints on the structure of the system. It is therefore expected that time-delay systems are, most likely, not delay-independent stable. However, it is important to point out that these results are still of importance since, from a pragmatic point of view, delay-independent stabilization or observation results (relying on delay-independent stability results) are of great interest in order to make the controlled system or the observer robust with respect to any delay-perturbation.

Example 5.3.2 Let us consider the linear time-delay system with constant delay

$$\dot{x}(t) = \begin{bmatrix} -5 & 1 \\ 0 & -5 \end{bmatrix} x(t) + \begin{bmatrix} -1 & 0 \\ 1 & -2 \end{bmatrix} x(t-h). \tag{5.24}$$

Necessary and sufficient conditions for delay-independent stability [2] are given by (a) A Hurwitz; (b) $\varrho(A^{-1}A_h) < 1$; and (c) $\varrho[(j\omega - A)^{-1}A_h] < 1$ for all $\omega > 0$. These conditions are fulfilled here since

$$\begin{aligned} \lambda(A) &= \{-5, -5\}, \\ \varrho(A^{-1}A_h) &= \frac{\sqrt{2}}{5}, \\ \varrho[(j\omega - A)^{-1}A_h] &< 0.31, \quad \text{for all } \omega > 0. \end{aligned} \tag{5.25}$$

The system is therefore delay-independent stable.

Several methods can be used to prove delay-independent stability of a time-delay systems. To cite a few: 2-D stability tests, pseudo-delay methods, frequency direct methods, frequency sweeping tests, constant matrix tests, matrix pencil tests, Lyapunov methods, algebraic geometry methods, etc. Interested readers should refer, for instance, to [1–3, 28].

5.3.2 Delay-Dependent and Delay-Range Stability

Unlike delay-independent stability, delay-dependent stability is a concept of stability that is actually sensitive to change in the delay values. This is certainly the most realistic notion of stability since delays are, most of the time, influential on the stability of real world systems.

Definition 5.3.3 (*Delay-Dependent Stability*) A time-delay system is *delay-dependent stable* if there exists a (bounded) interval $\mathbb{I} \subset \mathbb{R}_{\geq 0}$ for which the system is stable for any delay in $\mathbb{I}$, and unstable otherwise.

In the delay systems literature, the most common interval $\mathbb{I}$ is given by $[0, \bar{h})$ as illustrated below:

Example 5.3.4 (*Delay-Dependent Stability*) Let us consider the time-delay system

$$\dot{x}(t) = \begin{bmatrix} -2 & 0 \\ 0 & -0.9 \end{bmatrix} x(t) + \begin{bmatrix} -1 & 0 \\ -1 & -1 \end{bmatrix} x(t-h). \tag{5.26}$$

As stated in Example 5.3.2, the condition $\varrho(A^{-1}A_h) < 1$ is necessary for delay-independent stability. For the above system we have that $\varrho(A^{-1}A_h) = 10/9 > 1$, and thus the system is not stable independently of the delay. A frequency domain analysis allows us to prove that the system is delay-dependent stable for all $h \in \mathbb{I} := [0, \bar{h})$ where

$$\bar{h} = \frac{1}{\sqrt{0.19}}\left[\pi - \arctan\left(\frac{\sqrt{0.19}}{0.9}\right)\right] \simeq 6.1726. \tag{5.27}$$

When $h = \bar{h}$, the characteristic equation of the system has zeros on the imaginary axis and the system is not asymptotically stable.

When the system admits a delay stability interval of the form $\mathbb{I} = [h_1, h_2]$ for some $0 < h_1 < h_2$, the term *delay-range stability* is very often employed.

Example 5.3.5 (Delay-range stability [29]) Let us consider the unstable second order system with negative damping

$$\ddot{y}(t) = 0.1\dot{y}(t) - 2y(t) + u(t) \tag{5.28}$$

where $u(t)$ is the control input. Assume that, we would like to stabilize this system with a proportional feedback $u(t) = ky(t)$. This is clearly not possible since a proportional feedback cannot compensate the negative damping responsible for instability. Let us explore now the case of a *delayed proportional feedback* $u(t) = ky(t-h)$. By choosing $k = 1$, we obtain the system

$$\dot{x}(t) = \begin{bmatrix} 0 & 1 \\ -2 & 0.1 \end{bmatrix} x(t) + \begin{bmatrix} 0 & 0 \\ 1 & 0 \end{bmatrix} x(t-h)$$

which we analyze using time-delay systems tools to conclude on the effect of h on the stability of the control system.

First of all, note that this system is not stable for $h = 0$ since the matrix $A + A_h$ is not Hurwitz, which is consistent with the discussion on the delay-free proportional feedback. A frequency-dependent criterion similar to the one used in the previous examples allows us to conclude that the system is stable for all $h \in \mathbb{I} = (h_1, h_2)$ with

$$\begin{aligned} h_1 &:= \frac{1}{\omega_1} \arctan\left(\frac{0.1\omega_1}{2-\omega_1^2}\right) \simeq 0.10018 \\ h_2 &:= \frac{1}{\omega_2}\left[\pi - \arctan\left(\frac{0.1\omega_2}{\omega_2^2 - 2}\right)\right] \simeq 1.7178 \end{aligned} \tag{5.29}$$

where $\omega_1 := \sqrt{\dfrac{3.99 - \sqrt{\delta}}{2}}$ and $\omega_2 := \sqrt{\dfrac{3.99 + \sqrt{\delta}}{2}}$, $\delta = 3.9201$.

This example then shows that delays can also have a stabilizing effect for some systems. In the case treated here, this can be understood through the approximation

$$y(t-h) \simeq y(t) - h\dot{y}(t)$$

emphasizing that delays can be used to mimic/approximate a derivative action, see e.g. [29, 30], which is actually needed here in order to counteract the negative damping. This can be seen from the closeness of the delay lower bound $h_1 \simeq 0.10018$ to the negative damping value 0.1. Note, however, that when the delay is too large, the value $y(t-h)$ becomes independent of $\dot{y}(t)$, and the derivative action is lost.

When several delays are involved, the stability maps can be much more complicated, consisting of several disjoint compact sets, see e.g. [28, 31–33].

5.3.3 Time-Varying Delays and the Quenching Phenomenon

When delays turn out to be time-varying, the rate of variation has a non-negligible impact on stability. In most of the practical situations, the delay upper-bound is a nonincreasing function of the maximal rate of variation. The quenching phenomenon [21, 34] refers to as the property of a system with constant delay to be stable over a certain range of delay values, but unstable over the same range when the delay is time-varying; or vice-versa. Since mathematical tools for analytically and exactly analyzing time-delay systems with time-varying delays are unavailable, we will illustrate this phenomenon through an example taken from [34].

Example 5.3.6 ([34]) The time-delay system

$$\dot{x}(t) = ax(t) + bx(t-h) \tag{5.30}$$

where $a = -1$, $b = -1.5$ is exponentially stable for any constant delay $h \in [0, \bar{h})$ where

$$\bar{h} = \frac{2}{\sqrt{5}} \arccos\left(-\frac{2}{3}\right) \simeq 2.05765.$$

Assume now that the delay is time-varying and given by $h(t) = t - \alpha k$, $k\alpha \leq t < (k+1)\alpha$, $\alpha > 0$, $k \in \mathbb{N}$. This sawtooth delay corresponds to a zero-order hold function with period α; see Sect. 4.3.2. Therefore, with this time-varying delay, the delay-system becomes the discrete-time system

$$x_{k+1} = a_d x_k$$

with $x_k \equiv x(k\alpha)$ and $a_d = e^{a\alpha}(1+a^{-1}b) - a^{-1}b$. Stability of the discrete-time system is ensured if and only if $|a_d| < 1$. Using the numerical values $a = -1$, $b = -1.5$, we get the stability condition $\alpha < \log(5) \simeq 1.6094 < \bar{h}$. This shows that, even though the time-varying delay takes values inside the stability interval of constant time-delays, the system with time-varying delay can be unstable. This illustrates the quenching phenomenon and the harmful impact of delay variations. It is important to stress than the presence of discontinuities in the delay function is not responsible of the quenching phenomenon since it is possible to construct a continuous delay-function for which the same phenomenon occurs [34].

5.4 Model Transformations, Comparison Systems and Additional Dynamics

Model transformation is a very common procedure introduced quite early in the analysis of time-delay systems, but not restricted to. The rationale behind model transformations is to turn a time-delay system into another system, referred to as a *comparison system* or *comparison model*, which may or may not be a time-delay system. Analysis tools are then applied on the comparison system in order to draw conclusions on the stability of the original time-delay system. Model transformations lie at the core of many efficient analysis techniques such as robust analysis techniques based on LFT, IQCs or well-posedness, or even Lyapunov-Razumikhin and Lyapunov-Krasovskii approaches. Comparison models may take various forms: uncertain finite-dimensional linear systems [2, 35–39], time-delay systems [2, 40], or even uncertain LPV systems [41, 42].

The goal of model transformations is to simplify the analysis of time-delay systems. The compensation for this is that the comparison system may exhibit *additional dynamics* leading to a possible loss of equivalence, in terms of stability, between the original and the comparison system. Additional dynamics consist of supplementary zeros in the characteristic equation of the comparison model. When at least one of these additional zeros is unstable, the comparison model is unstable and the stability of the original system cannot be inferred from the comparison model. Additional dynamics have been studied in [2, 43–46]. Some additional details can also be found in [47, 48].

Since many different model transformation procedures have been proposed in the literature, it is difficult to give a complete picture here. We will, however, focus on three important model transformations to illustrate the notion of additional dynamics. The first one, the *Newton-Leibniz model transformation*, see. e.g. [49–53], will be shown to add additional zeros whose location only depends on the system. The second one, the *parametrized Newton-Leibniz model transformation*, will be proved to be a refinement of the former where the additional zeros do not directly depend on the system; see [54, 55]. The last one, the *descriptor model transformation* proposed more recently in [40, 56], will be shown to yield a comparison model having no additional dynamics, thus equivalent to the original one.

5.4.1 Newton-Leibniz Model Transformation

The Newton-Leibniz model transformation based on the identity

$$x(t-h) = x(t) - \int_{t-h}^{t} \dot{x}(\theta)d\theta$$

is certainly the first model-transformation to have been introduced for the analysis of time-delay systems [46, 57]. This model transformation allows us to substitute the delayed term $x(t-h)$ in the system (5.23) by the right-hand side of the above expression to yield the comparison system

$$\dot{x}(t) = (A + A_h)x(t) - A_h \int_{t-h}^{t} [Ax(s) + A_h x(s-h)]ds. \tag{5.31}$$

Note that, unlike system (5.23), this comparison model requires an initial condition in $C([-2h, 0], \mathbb{R}^n)$ due to the delayed term in the integral. The comparison model can now be analyzed using Lyapunov-Krasovskii functionals, Lyapunov-Razumikhin functions or some robust analysis techniques. This will addressed later in this chapter.

5.4.1.1 Additional Dynamics

Let us now analyze the exactness of the Newton-Leibniz model transformation in terms of stability characterization. The characteristic equation of system (5.31) given by

$$\Delta_n(s) := \det\left[s^2 I - (A + A_h)s + A_h A(1 - e^{-sh}) + A_h^2 e^{-sh}(1 - e^{-sh})\right]$$

admits the factorization $\Delta_n(s) = \Delta_a(s)\Delta_o(s)$ where

$$\Delta_o(s) := \det(sI - A - A_h e^{-sh}) \text{ and } \Delta_a(s) := \det\left(I - \frac{1 - e^{-sh}}{s} A_h\right). \tag{5.32}$$

The set of characteristic roots of $\Delta_c(s)$ is then simply given by the union of the sets of characteristic roots of $\Delta_a(s)$ and $\Delta_o(s)$, the latter being nothing else but the characteristic equation of the original system (5.23). Therefore, assuming that $\Delta_o(s)$ is stable, equivalence in terms of stability holds if and only if $\Delta_a(s)$ does not have any zeros in the closed right half-plane. It is immediate to see that we have

$$\Delta_a(s) = \prod_{i=1}^{n} \left(1 - \lambda_i \frac{1 - e^{-sh}}{s}\right)$$

where λ_i is the ith eigenvalue of matrix A_h. Let $\mathcal{S}_i$ be defined as

$$\mathcal{S}_i := \left\{s \in \mathbb{C} : 1 - \lambda_i \frac{1 - e^{-sh}}{s} = 0\right\}.$$

We then have the following proposition:

Proposition 5.4.1 ([2, 46]) *For any given A_h, all the additional zeros $s \in \mathcal{S}_i$, $i = 1, \ldots, n$, of $\Delta_a(s)$ satisfy*

$$\lim_{h \downarrow 0} \Re(s) = -\infty.$$

The above result means that all the additional zeros are stable provided that the delay is sufficiently small. Therefore, the model transformation preserves stability for sufficiently small delays. The next question is the preservation of stability for arbitrarily large finite delays. This question is answered by the following theorem:

Theorem 5.4.2 ([46]) *Let λ_i be an eigenvalue of A_h. Then, the following statements hold:*

1. *there is an additional zero of $\Delta_a(s)$ corresponding to the an eigenvalue λ_i, $\Im(\lambda_i) \neq 0$, on the imaginary axis if and only if the time-delay satisfies*

$$h = h_{i,k} = \frac{k\pi + \arg(\lambda_i)}{\Im(\lambda_i)} > 0, \quad k = 0, \pm 1, \pm 2, \ldots$$

2. *there is an additional zero of $\Delta_a(s)$ corresponding to the an eigenvalue λ_i, $\Im(\lambda_i) = 0$, $\Re[\lambda_i] > 0$, on the imaginary axis if and only if the time-delay satisfies*

$$h = \frac{1}{\lambda_i}.$$

3. *no additional zero of $\Delta_a(s)$ corresponding to a real negative eigenvalue λ_i will reach the imaginary axis for any finite delay.*

To summarize, when the spectrum of A_h lies on the negative real line, then the model-transformation is always stability preserving. Otherwise, the comparison model may be unstable while the original system is not.

5.4.2 Parametrized Newton-Leibniz Model Transformation

This model transformation [54, 55] generalizes the Newton-Leibniz model transformation by introducing a free parameter $C \in \mathbb{R}^{n \times n}$ as

$$Cx(t-h) = Cx(t) - C\int_{t-h}^{t} \dot{x}(\theta)d\theta.$$

The corresponding comparison model is given in this case by the following time-delay system with discrete and distributed delays:

$$\dot{x}(t) = (A+C)x(t) + (A_h - C)x(t-h) - C\int_{t-h}^{t} [Ax(s) + A_h x(s-h)]ds. \quad (5.33)$$

When $C = 0$, the original system (5.23) is recovered whereas letting $C = A_h$ yields the comparison model (5.31) obtained from the Newton-Leibniz model transformation. This model transformation therefore defines a continuous family of comparison systems comprising, among others, the systems (5.23) and (5.31).

5.4.2.1 Additional Dynamics

Additional dynamics can be studied exactly in the same way as for the Newton-Leibniz model transformation. In this case, the characteristic equation can be factorized as $\Delta_p(s) = \Delta_a(s)\Delta_o(s)$ where

$$\begin{aligned}\Delta_o(s) &:= \det(sI - A - A_h e^{-sh}),\\ \Delta_a(s) &:= \det\left(I - C\frac{1-e^{-sh}}{s}\right).\end{aligned} \quad (5.34)$$

It is immediate to see that, with a judicious choice for the matrix C, unstable additional dynamics may be avoided. This feature makes this model transformation more interesting than the usual Newton-Leibniz model transformation. When, moreover, C can be embedded as a decision variable in an LMI stability test, the solver will automatically find a suitable value for C. Note, however, that the quite intricate structure of the comparison system may require the use of complex analysis tools, e.g. complicated Lyapunov-Krasovskii functionals.

5.4.3 Descriptor Model Transformation

The *descriptor model transformation*, introduced more recently in [40, 56], yields the comparison model

$$\begin{bmatrix} I & 0 \\ 0 & 0 \end{bmatrix}\begin{bmatrix} \dot{x}(t) \\ \dot{y}(t) \end{bmatrix} = \begin{bmatrix} 0 & I \\ A + A_h & -I \end{bmatrix}\begin{bmatrix} x(t) \\ y(t) \end{bmatrix} + \int_{t-h}^{t} \begin{bmatrix} 0 & 0 \\ 0 & -A_h \end{bmatrix}\begin{bmatrix} x(s) \\ y(s) \end{bmatrix} ds \quad (5.35)$$

where $y(t) = \dot{x}(t)$. Simple calculations show that the system (5.23) is retrieved after eliminating the variable y in the above comparison model.

5.4.3.1 Additional Dynamics

The characteristic equation of system (5.35) is given by

$$\begin{aligned}
\Delta_d(s) &:= \det\left(\begin{bmatrix} sI & -I \\ -(A + A_h) & I + A_h\dfrac{1 - e^{-sh}}{s} \end{bmatrix}\right) \\
&= \det(sI_n)\det\left(I + A_h\frac{1 - e^{-sh}}{s} - \frac{1}{s}(A + A_h)\right) \\
&= \det\left(sI + A_h(1 - e^{-sh}) - (A + A_h)\right) \\
&= \det\left(sI - A - A_h e^{-sh}\right)
\end{aligned}$$

where we have used the Schur determinant formula; see Appendix A.1. It is immediate to see that the characteristic equation of the transformed model (5.35) is identical to the one of the original system (5.23). Therefore, no additional dynamics, stable or unstable, are introduced by this model transformation. Note, however, that the system is changed into a singular system with distributed delay which may require the use of slightly more complex analysis tools than by considering the initial time-delay system.

5.4.4 Other Model Transformations

Many other types of model transformations have been introduced in the literature. For instance, Padé approximants are considered in [37–39], model transformations by means of operators are reported in [58–64], whereas implicit model and second order transformations are discussed in [1, 2]. Some of them will be discussed in the sections on input/output stability analysis techniques.

5.5 Lyapunov-Razumikhin Stability Results

The Lyapunov-Razumikhin theorem, stated in Sect. 5.2.2, can be used to derive both delay-independent and delay-dependent stability results for systems with time-invariant and time-varying delays. Interestingly, the obtained stability conditions, taking the form of matrix inequalities, do not depend on the rate of variation of the delay. This has made the use of Lyapunov-Razumikhin functions quite appealing in fields such as networked control systems where it may be difficult to a priori define an upper-bound on the delay derivative.

The use of Lyapunov-Razumikhin functions is very often considered as leading to conservative stability conditions, and is always confronted to the Lyapunov-Krasovskii approach. Even if this is generally the case when the goal is to derive generic analysis tools aiming at proving stability for a wide class of systems, e.g. the class of linear systems, it is very important to keep in mind that these two approaches do not consider the same stability measure, and should not be directly compared.[1]

However, when a particular system is considered, e.g. a specific type of nonlinear system as in [67] where a nonlinear congestion control mechanism is analyzed, it is sometimes much easier to find a Lyapunov-Razumikhin function than a Lyapunov-Krasovskii functional since the latter has, in general, a more complicated structure.

5.5.1 Delay-Independent Stability

A simple test on delay-independent stability can be obtained from a direct application of the Lyapunov-Razumikhin Theorem and the S-procedure as shown below:

Theorem 5.5.1 *The time-delay system* (5.23) *is stable independently of the delay if there exist a matrix* $P \in \mathbb{S}^n_{\succ 0}$ *and a scalar* $\tau > 0$ *such that the matrix inequality*

$$\begin{bmatrix} A^T P + PA + \tau P & PA_h \\ A_h^T P & -\tau P \end{bmatrix} \prec 0 \tag{5.36}$$

holds.

Proof Let us consider the Lyapunov-Razumikhin function $V(x) = x^{\mathrm{T}} Px$. The time derivative of V along the trajectories of system (5.23) is given by

$$\dot{V}(x_t) = \begin{bmatrix} x(t) \\ x(t-h) \end{bmatrix}^{\mathrm{T}} \begin{bmatrix} A^{\mathrm{T}} P + PA & PA_h \\ A_h^{\mathrm{T}} P & 0 \end{bmatrix} \begin{bmatrix} x(t) \\ x(t-h) \end{bmatrix}.$$

According to the Lyapunov-Razumikhin Theorem, i.e. Theorem 5.2.5, the derivative $\dot{V}(x(t))$ must be negative only whenever $V(x(t+\theta)) < pV(x(t))$ for some $p > 1$ and for all $\theta \in [-h, 0]$. Since, $\dot{V}$ only depends on $x(t-h)$, we only have to consider

[1] It will be shown later that, in the case of linear systems, Lyapunov-Krasovskii functionals can be connected to robust stability analysis in the L_2-norm, and Lyapunov-Razumikhin functions to robust stability analysis in the L_∞-norm: see also [65, 66].

the inequality $V(x(t-h)) < pV(x(t))$. Using the S-procedure (see [68] or Appendix C.8), which is lossless in this case, we get the matrix inequality

$$\begin{bmatrix} A^{\mathrm{T}}P + PA + \tau pP & PA_h \\ A_h^{\mathrm{T}}P & -\tau P \end{bmatrix} \prec 0 \tag{5.37}$$

for some $p > 1$ and $\tau > 0$. Noting finally that if the above matrix inequality holds for $p = 1$, then it also holds for any sufficiently small $p > 1$ completes the proof. ■

Note that condition (5.36) of Theorem 5.5.1 is not an LMI due to the bilinear term τP. However, when τ is fixed, the condition becomes an LMI and can be checked efficiently. Rewriting first condition (5.36) as

$$A^{\mathrm{T}}P + PA + \tau P + \tau^{-1}PA_hP^{-1}A_h^{\mathrm{T}}P \prec 0,$$

it is immediate to see that a too large or too small τ will yield an infeasible problem. An upper bound on τ can be determined by solving the generalized eigenvalue problem [68]

$$\begin{aligned} \bar{\tau} := & \max_{P \in \mathbb{S}^n_{\succ 0}, \tau > 0} \quad \tau \\ & \text{s.t.} \quad A^{\mathrm{T}}P + PA + \tau P \prec 0. \end{aligned}$$

The lower bound is, on the other hand, determined by solving the problem

$$\begin{aligned} \underline{\tau} := & \min_{P \in \mathbb{S}^n_{\succ 0}, \tau > 0} \quad \tau \\ & \text{s.t.} \quad \begin{bmatrix} A^{\mathrm{T}}P + PA & PA_h \\ \star & -\tau P \end{bmatrix} \prec 0. \end{aligned}$$

With these bounds in mind, it is enough to search for $\tau \in (\underline{\tau}, \bar{\tau})$ when solving condition (5.36). Note, however, that the search procedure may be computationally expensive.

5.5.2 Delay-Dependent Stability

The Lyapunov-Razumikhin Theorem can also be used to derive several delay-dependent results. A simple one is presented below:

Theorem 5.5.2 *The system (5.23) is stable for all $h \in [0, \bar{h}]$ if there exist a matrix $P \in \mathbb{S}^n_{\succ 0}$ and scalars $\epsilon_1, \epsilon_2 > 0$ such that the matrix inequality*

$$\begin{bmatrix} \mathrm{He}[P(A+A_h)]+\bar{h}(\epsilon_1+\epsilon_2)P & -\bar{h}PA_hA & -\bar{h}PA_h^2 \\ \star & -\bar{h}\epsilon_1 P & 0 \\ \star & \star & -\bar{h}\epsilon_2 P \end{bmatrix} \prec 0 \tag{5.38}$$

holds.

Proof This result is based on the Newton-Leibniz model transformation and thus considers the comparison model

$$\dot{x}(t) = (A + A_h)x(t) - A_h \int_{t-h}^{t} [Ax(s) + A_h x(s-h)]\,ds.$$

The time-derivative of the Lyapunov-Razumikhin function $V(x) = x^{\mathrm{T}} Px$ along the trajectories of the above comparison model is given by

$$\dot{V} = x(t)^{\mathrm{T}}\mathrm{He}[P(A+A_h)]x(t) - 2\int_{t-h}^{t} x(t)^{\mathrm{T}} PA_h\,[Ax(s) + A_h x(s-h)]\,ds.$$

Completing the squares[2] yields the bounds

$$\begin{aligned} -2x(t)^{\mathrm{T}}PA_hAx(s) &\le \epsilon_1^{-1}x(t)^{\mathrm{T}}PA_hAP^{-1}A^{\mathrm{T}}A_h^{\mathrm{T}}Px(t) + \epsilon_1 x(s)^{\mathrm{T}}Px(s) \\ -2x(t)^{\mathrm{T}}PA_h^2x(s-h) &\le \epsilon_2^{-1}x(t)^{\mathrm{T}}PA_h^2P^{-1}A_h^{2\mathrm{T}}Px(t) + \epsilon_2 x(s)^{\mathrm{T}}Px(s) \end{aligned}$$

where the parameters $\epsilon_1, \epsilon_2 > 0$ are arbitrary. Substituting the above bounds in the Lyapunov-Razumikhin function derivative yields

$$\begin{aligned} \dot{V} \le\; & x(t)^{\mathrm{T}}\left[\mathrm{He}[P(A+A_h)] + h\epsilon_1^{-1}PA_hAP^{-1}A^{\mathrm{T}}A_h^{\mathrm{T}}P\right]x(t) \\ & + h\epsilon_2^{-1}x(t)^{\mathrm{T}}PA_h^2P^{-1}A_h^{2\mathrm{T}}Px(t) + \epsilon_1 \int_{t-h}^{t} x(s)^{\mathrm{T}}Px(s)ds \\ & + \epsilon_2 \int_{t-h}^{t} x(s-h)^{\mathrm{T}}Px(s-h)ds \end{aligned} \tag{5.39}$$

We need to prove negativity of the derivative whenever $V(x(t-h)) \le pV(x(t))$, $p > 1$, therefore only when

$$\begin{aligned} \int_{t-h}^{t} x(s)^{\mathrm{T}}Px(s)ds &\le phx(t)^{\mathrm{T}}Px(t) \\ &\le p\bar{h}x(t)^{\mathrm{T}}Px(t) \end{aligned}$$

[2] Let X be symmetric positive definite with positive definite square root $X^{1/2}$, then $(X^{1/2}x + X^{-1/2}y)^{\mathrm{T}}(X^{1/2}x + X^{-1/2}y) \ge 0$ implies that $-2x^{\mathrm{T}}y \le x^{\mathrm{T}}Xx + y^{\mathrm{T}}X^{-1}y$ holds. A more thorough discussion on this type of bounds is provided in Sects. 5.6.2 and 5.6.4.

and

$$\begin{aligned}\int_{t-h}^{t} x(s-h)^{\mathrm{T}} P x(s-h) ds &\leq p \int_{t-h}^{t} x(s)^{\mathrm{T}} P x(s)^{\mathrm{T}} ds \\ &\leq h p^2 x(t)^{\mathrm{T}} P x(t) \\ &\leq \bar{h} p^2 x(t)^{\mathrm{T}} P x(t).\end{aligned}$$

Using these bounds in (5.39) and performing two Schur complements yield

$$\begin{bmatrix} \mathrm{He}[P(A+A_h)] + (\bar{h} p \epsilon_1 + \bar{h} p^2 \epsilon_2) P & -\bar{h} P A_h A & -\bar{h} P A_h^2 \\ \star & -\bar{h} \epsilon_1 P & 0 \\ \star & \star & -\bar{h} \epsilon_2 P \end{bmatrix} \prec 0.$$

Noting finally that when the above matrix inequality holds for $p = 1$ it also holds for a sufficiently small $p > 1$ completes the proof. ■

As in the delay-independent case, tuning the scaling terms ϵ_1 and ϵ_2 may be very tricky. A procedure for assigning suitable values to them is discussed in [2].

5.6 Lyapunov-Krasovskii Stability Results

Methods based on Lyapunov-Krasovskii functionals (LKFs) are certainly the most popular for analyzing and controlling time-delay systems in the time-domain framework. Dozens of different functionals have been proposed in the literature and it is clearly neither possible nor even interesting to detail them all here. A recurrent problem in LKF-based results stems from the fact that the benefit from using a given term in an LKF is theoretically unclear. This is mainly due to the facts that few works have actually been focused on comparisons of LKFs, and that most of the comparisons are made on simple numerical examples. Revealing the connections between LKFs (or their corresponding criteria) in terms of dominance [69], equivalence or disjointness, is a critical open question whose answer would allow us to clearly characterize and categorize Lyapunov-Krasovskii functionals, and yield a neat theory of Lyapunov-Krasovskii functionals. Preliminary comparison results have been obtained in [70] where several LMI conditions derived from different functionals are shown to be equivalent. In [35], the equivalence between Lyapunov-Krasovskii and robust approaches is emphasized. Equivalence and convergence results for certain types of bounds have also been obtained in [71].

The goal of this section is to introduce and give a clear picture of certain key LKF-based results in a chronological order or evolutionary way. We will indeed see how model transformations, see Sect. 5.4, can be used together with LKFs and various bounding techniques [72, 73] to obtain tractable delay-dependent stability conditions. Finally, methods avoiding model transformations and making use of more accurate bounding techniques will be introduced and shown to potentially yield more interesting results, together with a reduced computational complexity.

In what follows, many different stablity results will be provided together with a detailed and fully analyzed proof. The various sources of conservatism, such as the use of model transformations or bounding techniques, will be clearly pointed out and some ways to avoid/overcome the conservatism increase will be proposed. It is very crucial to understand here that every Lyapunov-Krasovskii functional possesses an inherent conservatism (which is not very well-understood, except for complete LKFs) and when deriving stability conditions by differentiation, it is, most of the time, necessary to process the functional derivative, by using bounding or approximation techniques, in order to obtain tractable conditions. Every single bound or single approximation technique used in the proof is likely to introduce some extra conservatism. It is therefore very important to carefully choose the functional, the model transformation (can be actually avoided) and the bounding techniques in order to preserve accuracy or, in more adapted terms, limit the increase of conservatism.

5.6.1 Delay-Independent Stability

Several delay-independent stability results can be obtained using Lyapunov-Krasovskii functionals. The goal of this section is to present two of them. The first one is obtained from a simple 2-term Lyapunov-Krasovskii functional and has been proposed in [74, 75]. The second one is an LMI-based necessary and sufficient condition for strong delay-independent stability of delay systems that has been derived in [76].

5.6.1.1 A Simple Delay-Independent Stability Test

The following result is probably the simplest delay-independent stability test that can be obtained using the Lyapunov-Krasovskii Theorem. Despite its simplicity, the characterization of all pairs (A, A_h) satisfying the following theorem is still an open problem to date; see [77].

Theorem 5.6.1 ([7, 74, 75]) *The system* (5.23) *is stable independently of the delay if there exist matrices* $P, Q \in \mathbb{S}^n_{\succ 0}$ *such that the LMI*

$$\begin{bmatrix} A^T P + PA + Q & PA_h \\ \star & -Q \end{bmatrix} \prec 0 \tag{5.40}$$

holds.

Proof Let us consider the LKF

$$V(x_t) = x(t)^{\mathrm{T}} P x(t) + \int_{t-h}^{t} x(\theta)^{\mathrm{T}} Q x(\theta) d\theta \tag{5.41}$$

defined for $P, Q \in \mathbb{S}^n_{\succ 0}$. The derivative of V along the trajectories of the system (5.23) is given by

$$\dot{V}(x_t) = \begin{bmatrix} x(t) \\ x(t-h) \end{bmatrix}^{\mathrm{T}} \begin{bmatrix} A^{\mathrm{T}} P + PA + Q & PA_h \\ \star & -Q \end{bmatrix} \begin{bmatrix} x(t) \\ x(t-h) \end{bmatrix}.$$

Negative definiteness of the central matrix therefore ensures delay-independent asymptotic stability of the system by virtue of the Lyapunov-Krasovskii Theorem. ■

It is worth pointing out the similarities between the matrix inequalities obtained from the Lyapunov-Razumikhin and Lyapunov-Krasovskii theorems, namely the conditions (5.36) and (5.40), respectively. The Lyapunov-Krasovskii result is actually more general since the matrix Q is decoupled from P, unlike in the Lyapunov-Razumikhin result and, therefore, condition (5.40) is an LMI condition, unlike (5.36). Note, however, that the latter remains valid for a time-varying delay while the Lyapunov-Krasovskii condition needs to be adapted to cope with time-varying delays:

Theorem 5.6.2 *The system* (5.23) *with time-varying delay* $h(t)$ *satisfying* $\dot{h}(t) \leq \mu < 1$ *is stable independently of the delay if there exist matrices* $P, Q \in \mathbb{S}^n_{\succ 0}$ *such that the LMI*

$$\begin{bmatrix} A^T P + PA + Q & PA_h \\ \star & -(1-\mu)Q \end{bmatrix} \prec 0 \tag{5.42}$$

holds.

Proof The proof relies on the use of the functional

$$V(x_t) = x(t)^{\mathrm{T}} P x(t) + \int_{t-h(t)}^{t} x(\theta)^{\mathrm{T}} Q x(\theta) d\theta$$

defined for $P, Q \in \mathbb{S}^n_{\succ 0}$. The rest of the proof follows the same lines as the one for constant delay. ■

We can clearly see that the rate of variation of the delay is harmful to stability. A Schur complement indeed yields

$$A^{\mathrm{T}}P + PA + Q + (1-\mu)^{-1}PA_hQA_h^{\mathrm{T}}P \prec 0$$

and shows that when μ increases from 0 to 1, the μ-dependent term to the right grows without bound.

5.6.1.2 Strong Delay-Independent Stability

The notion of strong delay-independent stability [78] is defined as follows:

Definition 5.6.3 The time-delay system (5.23) with constant delay is strongly delay-independent stable if

$$\det(sI_n - A - zA_h) \neq 0$$

for all $(s,z) \in \{(s,z) \in \mathbb{C}^2 : \Re[s] \geq 0,\ |z| \leq 1\}$.

Note the main difference with the usual delay-independent stability (see e.g. [79–83]) that would simply require that

$$\det(sI_n - A - e^{-sh}A_h) \neq 0$$

holds for all $s \in \mathbb{C}$, $\Re[s] \geq 0$ and all $h \geq 0$.

The following result, proved in [76], provides a necessary and sufficient condition for strong delay-independent stability:

Theorem 5.6.4 *The following statements are equivalent:*

1. *The system (5.23) with constant delay is strongly delay-independent stable.*
2. *There exists $k \in \mathbb{N}_{>0}$ and matrices $P_k, Q_k \in \mathbb{S}^{kn}_{>0}$ such that the LMI*

$$\Psi_k^T \begin{bmatrix} P_k(I_k \otimes A) + (I_k \otimes A)^T P_k + Q_k & P_k(I_k \otimes A_h) \\ \star & -Q_k \end{bmatrix} \Psi_k \prec 0 \quad (5.43)$$

holds with

$$\Psi_k := \begin{bmatrix} I_{kn} & 0_{kn\times n} \\ 0_{kn\times n} & I_{kn} \end{bmatrix}.$$

The above theorem can be viewed as a generalization of Theorem 5.6.1 where the state involved in the LKF (5.41) has been changed from $x(t)$ to

$$\begin{bmatrix} x(t) \\ \vdots \\ x(t-(k-1)h) \end{bmatrix}$$

and the matrices P, Q to P_k, Q_k. When $k = 1$, the result of Theorem 5.6.1 is retrieved.

The price to pay for exactness is a polynomial increase of the computational complexity. The number of variables indeed evolves according to $O(k^2n^2)$ whereas the size of the LMI constraint is equal to $(k+1)n$.

5.6.2 *Delay-Dependent Stability—Newton-Leibniz Model Transformation*

In this section, the Lyapunov-Krasovskii functional

$$V = x(t)^{\mathrm{T}}Px(t) + \int_{t-h}^{t} x(\theta)^{\mathrm{T}}Qx(\theta)d\theta + \int_{-h}^{0}\int_{t+\theta}^{t} \dot{x}(\eta)^{\mathrm{T}}R\dot{x}(\eta)d\eta d\theta \tag{5.44}$$

where $P, Q, R \in \mathbb{S}^n_{\succ 0}$ is considered. We can easily recognize, in the two first terms, the functional used for delay-independent stability analysis in Sect. 5.6.1. The last term, as we shall see later, will make the stability condition delay-dependent. Note that many of other terms can be used for this purpose, see e.g. [1, 2, 59, 84].

Using the above Lyapunov-Krasovskii functional and the Newton-Leibniz model transformation of Sect. 5.4.1, the following result is obtained:

Theorem 5.6.5 *The system (5.23) is stable for all* $h \in [0, \bar{h}]$ *if there exist* $P, Q, R \in \mathbb{S}^n_{\succ 0}$ *such that the LMI*

$$\begin{bmatrix} (A+A_h)^T P + P(A+A_h) + Q + \bar{h}A^T RA & \bar{h}A^T RA_h & \bar{h}PA_h \\ \star & -Q + \bar{h}A_h^T RA_h & 0 \\ \star & \star & -\bar{h}R \end{bmatrix} \prec 0 \tag{5.45}$$

holds.

Proof Computing the derivative of the LKF (5.44) along the trajectories of system (5.31) yields

$$\begin{aligned}\dot{V} &= x(t)^{\mathrm{T}}[(A+A_h)^{\mathrm{T}}P + P(A+A_h) + Q]x(t) - 2x(t)^{\mathrm{T}}PA_h \int_{t-h}^{t} \dot{x}(\theta)d\theta \\ &\quad - x(t-h)^{\mathrm{T}}Qx(t-h) + h\dot{x}(t)^{\mathrm{T}}R\dot{x}(t) - \int_{t-h}^{t} \dot{x}(\theta)^{\mathrm{T}}R\dot{x}(\theta)d\theta.\end{aligned}$$

The cross-term

$$-2x(t)^{\mathrm{T}}PA_h\int_{t-h}^{t}\dot{x}(\theta)d\theta \tag{5.46}$$

introduced by the model transformation cannot be simply substituted by

$$-2x(t)^{\mathrm{T}}PA_h(x(t)-x(t-h))$$

since we would recover the original system and the effect of the model transformation would be lost. This term must therefore be judiciously incorporated in the conditions. A way to do so consists of replacing the cross-term by an upper-bound of it. Completing the squares allows us to obtain the following (coarse) inequality

$$-2x(t)^{\mathrm{T}}PA_h\dot{x}(\theta)\le x(t)^{\mathrm{T}}PA_hR^{-1}A_h^{\mathrm{T}}Px(t)+\dot{x}(\theta)^{\mathrm{T}}R\dot{x}(\theta) \tag{5.47}$$

which is valid for any $R\in\mathbb{S}^n_{\succ 0}$. We therefore get the inequality

$$-2x(t)^{\mathrm{T}}PA_h\int_{t-h}^{t}\dot{x}(\theta)d\theta\le hx(t)^{\mathrm{T}}PA_hR^{-1}A_h^{\mathrm{T}}Px(t)+\int_{t-h}^{t}\dot{x}(\theta)^{\mathrm{T}}R\dot{x}(\theta)d\theta.$$

Substituting this bound in $\dot{V}$ gives

$$\begin{aligned}\dot{V}\le{}& x(t)^{\mathrm{T}}[(A+A_h)^{\mathrm{T}}P+P(A+A_h)+Q+\bar{h}PA_hR^{-1}A_h^{\mathrm{T}}P]x(t)\\&-x(t-h)^{\mathrm{T}}Qx(t-h)+\bar{h}\dot{x}(t)^{\mathrm{T}}R\dot{x}(t).\end{aligned}$$

Expanding, finally, the term $\dot{x}(t)^{\mathrm{T}}R\dot{x}(t)$ and performing a Schur complement yield the result. ■

Two important facts deserve to be pointed out in the proof above. The first one concerns the use of the model transformation subsequently imposing the use of a cross-term bounding technique. The considered bound, based on square completion, is easily seen to be conservative since the right-hand side of (5.47) is always positive, while the left-hand side may be negative.

The second fact is about the cancellation of the integral term

$$-\int_{t-h}^{t}\dot{x}(\theta)^{\mathrm{T}}R\dot{x}(\theta)d\theta \tag{5.48}$$

in the functional derivative after bounding the cross-term. At first sight, this seems to be a rather interesting aftereffect since this integral term looks difficult to deal with. However, by compensating this term, we lose important information on the stability of the system since this integral term is negative definite (and therefore contributes to the negative definiteness of the derivative of the functional).

We can therefore conclude that the use of the Newton-Leibniz model transformation may yield some conservative stability conditions due to the loss of important information on the stability of the system and the potential introduction of additional dynamics. Ways to improve this certainly relies on the use of less restrictive model-transformations (or even no model-transformation at all) and/or more accurate bounding techniques. The next section considers the use of a less conservative model transformation whereas Sect. 5.6.4 introduces a better cross-terms bounding technique.

5.6.3 Delay-Dependent Stability—Parametrized Newton-Leibniz Model Transformation

The advantage of using the parametrized Newton-Leibniz formula model transformation over the standard one lies in the control of additional dynamics. It is shown here how the free parameter C of the parametrized Newton-Leibniz model transformation can be automatically determined by convex programming. To this aim, let us consider the following Lyapunov-Krasovskii functional

$$\begin{aligned} V(x_t) = x(t)^{\mathrm{T}}Px(t) + \int_{t-h}^{t} x(s)^{\mathrm{T}}Qx(s)ds + \int_{-h}^{0}\int_{t+s}^{t} x(\theta)^{\mathrm{T}}R_1x(\theta)d\theta ds \\ + \int_{-2h}^{-h}\int_{t+s}^{t} x(\theta)^{\mathrm{T}}R_2x(\theta)d\theta ds. \end{aligned} \tag{5.49}$$

Using the above LKF together with the parametrized Newton-Leibniz model transformation, we obtain the following result [85]:

Theorem 5.6.6 *The system (5.23) is stable for all* $h \in [0, \bar{h}]$ *if there exist* $P, Q, R_1, R_2 \in \mathbb{S}^n_{\succ 0}$ *and* $W \in \mathbb{R}^{n\times n}$ *such that the LMI*

$$\begin{bmatrix} He[PA+W]+\bar{h}(R_1+R_2)+Q & PA_h - W & \bar{h}WA & \bar{h}WA_h \\ \star & -Q & 0 & 0 \\ \star & \star & -\bar{h}R_1 & 0 \\ \star & \star & \star & -\bar{h}R_2 \end{bmatrix} \prec 0 \tag{5.50}$$

holds. Moreover, a suitable matrix C *is given by the formula* $C = WP^{-1}$.

Proof The proof is only sketched for simplicity since the procedure is similar to the one of the Newton-Leibniz model transformation. Computing the derivative of the LKF (5.49) along the solutions of system (5.33) yields the expression

$$\dot{V}(x_t) = \begin{bmatrix} x(t) \\ x(t-h) \end{bmatrix}^{\mathrm{T}} \begin{bmatrix} \mathrm{He}[P(A+C)] + \bar{h}R + Q & P(A_h - C) \\ \star & -Q \end{bmatrix} \begin{bmatrix} x(t) \\ x(t-h) \end{bmatrix}$$
$$+ 2x(t)^{\mathrm{T}} PCA \int_{t-h}^{t} x(s)ds + 2x(t)^{\mathrm{T}} PCA_h \int_{t-2h}^{t-h} x(s)ds$$
$$- \int_{t-h}^{t} x(s)^{\mathrm{T}} R_1 x(s)ds - \int_{t-2h}^{t-h} x(s)^{\mathrm{T}} R_2 x(s)ds$$

where $R = R_1 + R_2$. Using the bounds

$$2x(t)^{\mathrm{T}} PCA \int_{t-h}^{t} x(s)ds \le \bar{h}x(t)^{\mathrm{T}} PCAR_1^{-1} A^{\mathrm{T}} C^{\mathrm{T}} Px(t) + \int_{t-h}^{t} x(s)^{\mathrm{T}} R_1 x(s)ds$$

$$2x(t)^{\mathrm{T}} PCA_h \int_{t-2h}^{t-h} x(s)ds \le \bar{h}x(t)^{\mathrm{T}} PCA_h R_2^{-1} A_h^{\mathrm{T}} C^{\mathrm{T}} Px(t) + \int_{t-2h}^{t-h} x(s)^{\mathrm{T}} R_2 x(s)ds$$

and performing successive Schur complements yield the result with $W := PC$. ■

The above results (partly) overcome the problem of additional dynamics by letting the optimization solver finding the 'best' matrix C. The approach, however, still relies on a bounding technique that is very conservative. A more accurate bounding technique is discussed in the next section.

5.6.4 Delay-Dependent Stability—Park's Inequality

The key idea behind Park's inequality on cross-terms [72, 73] is to limit the increase of conservatism when bounding cross-terms such as (5.46) by proposing a more accurate bound than the one used in Sects. 5.6.2 and 5.6.3. This bound is stated in the following result:

Lemma 5.6.7 ([73]) *Assume that $a, b : \Omega \to \mathbb{R}^n$ are given vector functions. Then, for any matrices $X \in \mathbb{S}^n_{\succ 0}$ and $M \in \mathbb{R}^{n\times n}$, the following inequality*

$$-2\int_{\Omega} b(s)^T a(s)ds \le \int_{\Omega} \begin{bmatrix} a(s) \\ b(s) \end{bmatrix}^T \Psi \begin{bmatrix} a(s) \\ b(s) \end{bmatrix} ds \tag{5.51}$$

holds with

$$\Psi = \begin{bmatrix} X & XM \\ M^T X & (M^T X + I)X^{-1}(XM + I) \end{bmatrix}.$$

Using this bound, the following result is obtained in [73]:

Theorem 5.6.8 *The system (5.23) is asymptotically delay-dependent stable for all $h \in [0, \bar{h}]$ if there exist $P, Q, R, V \in \mathbb{S}^n_{\succ 0}$ and $W \in \mathbb{R}^{n\times n}$ such that the LMI*

$$\begin{bmatrix} M_{11} & -W^T A_h & A^T A_h^T V & \bar{h}(W^T + P) \\ \star & -Q & A_h^T A_h^T V & 0 \\ \star & \star & -V & 0 \\ \star & \star & \star & -V \end{bmatrix} \prec 0 \tag{5.52}$$

holds with $M_{11} = (A + A_h)^T P + P(A + A_h) + W^T A_h + A_h^T W + Q$.

Proof The proof is based on the use of the LKF

$$V(x_t, \dot{x}_t) = x(t)^{\mathrm{T}} P x(t) + \int_{t-h}^{t} x(\theta)^{\mathrm{T}} Q x(\theta) d\theta + \int_{-h}^{0}\int_{t+\theta}^{t} \dot{x}(\eta)^{\mathrm{T}} A_h^{\mathrm{T}} X A_h \dot{x}(\eta) d\eta d\theta \tag{5.53}$$

whose derivative along the trajectories of the comparison system (5.31) is given by

$$\begin{aligned}\dot{V}(x_t) &\le 2x(t)^{\mathrm{T}} P \dot{x}(t) + x(t)^{\mathrm{T}} Q x(t) - x(t-h)^{\mathrm{T}} Q x(t-h) + \bar{h}\dot{x}(t)^{\mathrm{T}} R \dot{x}(t) \\ &\quad - \int_{t-h}^{t} \dot{x}(s)^{\mathrm{T}} A_h^{\mathrm{T}} X A_h \dot{x}(s) ds \\ &= 2x(t)^{\mathrm{T}} P(A + A_h)x(t) - 2x(t)^{\mathrm{T}} P A_h \int_{t-h}^{t} \dot{x}(s) ds + x(t)^{\mathrm{T}} Q x(t)\end{aligned}$$

$$- x(t-h)^{\mathrm{T}} Qx(t-h) + \bar{h}\dot{x}(t)^{\mathrm{T}} R\dot{x}(t) - \int_{t-h}^{t} \dot{x}(s)^{\mathrm{T}} A_h^{\mathrm{T}} X A_h \dot{x}(s) ds.$$

Letting $a(s) = A_h\dot{x}(s)$ and $b(s) = Px(t)$ in Lemma 5.6.7, we obtain the inequality

$$\begin{aligned} -2x(t)^{\mathrm{T}} P A_h \int_{t-h}^{t} \dot{x}(s) ds \leq & \int_{t-h}^{t} \dot{x}(s)^{\mathrm{T}} A_h^{\mathrm{T}} X A_h \dot{x}(s) ds \\ & +2x(t)^{\mathrm{T}} P M^T X A_h \int_{t-h}^{t} \dot{x}(s) ds \\ & +\bar{h}x(t)^{\mathrm{T}} P(XM + I)^{\mathrm{T}} X^{-1}(XM + I)Px(t). \end{aligned}$$

which, substituted in the expression of $\dot{V}$, yields the result after the changes of variables $W = XMP$ and $V = \bar{h}X$, and Schur complements. ■

Although this technique allows us to reduce the conservatism by coping with cross-terms more accurately, it is still limited by the use of the Newton-Leibniz model-transformation and the necessity of using bounding techniques. Note, however, that this bounding technique can also be used with the parameterized Newton-Leibniz model transformation, or basically any other one. As a final remark, it seems also important to mention that the computational complexity of the above result is not optimal since the matrix W can be eliminated using the projection lemma; see [86] or Appendix C.12. To avoid redundancy in exposure, complexity reduction is not performed here but will be described in Sects. 5.6.5 and 5.6.6.

5.6.4.1 Theoretical Comparison of Bounds on Cross-Terms

It seems important to carry out a theoretical analysis of the bounding techniques discussed above. It is shown below that some of them are irremediably conservative and that they can be compared with each other in this respect. As a concluding remark, a non-conservative cross-term upper-bound is proposed.

Before moving on to this right away, it seems important to mention the following generalization of Lemma 5.6.7 initially proposed in [87]:

Lemma 5.6.9 *Assume that $a : \Omega \to \mathbb{R}^{n_a}$ and $b : \Omega \to \mathbb{R}^{n_b}$ are given matrix functions and $N \in \mathbb{R}^{n_a \times n_b}$ is a given matrix. Then, for any matrices $X \in \mathbb{S}^{n_a}$, $Y \in \mathbb{R}^{n_a \times n_b}$ and $Z \in \mathbb{S}^{n_b}$ verifying*

$$\begin{bmatrix} X & Y \\ \star & Z \end{bmatrix} \succeq 0 \qquad (5.54)$$

the following inequality

$$-2\int_{\Omega} b(s)^T N^T a(s) ds \leq \int_{\Omega} \begin{bmatrix} a(s) \\ b(s) \end{bmatrix}^T \begin{bmatrix} X & Y-N \\ \star & Z \end{bmatrix} \begin{bmatrix} a(s) \\ b(s) \end{bmatrix} ds \tag{5.55}$$

holds.

It is easily seen that when $n_a = n_b$, setting $N = I$, $Y = I$ and $Z = X^{-1}$ allows us to recover the trivial bound on cross-term $-2x^{\mathrm{T}}y \leq x^{\mathrm{T}}Xx + y^{\mathrm{T}}X^{-1}y$ used in Sects. 5.6.2 and 5.6.3. When, on the other hand, $N = I$, $Y = I + XM$ and $Z = (XM + I)^{\mathrm{T}}X^{-1}(XM + I)$, we recover the bound of Lemma 5.6.7. The bound of Lemma 5.6.9 is therefore a generalization of the previous ones.

The question now is: how to measure the conservatism of a bound? To answer this, let us assume that the following inequality

$$-2x^{\mathrm{T}}y \leq \begin{bmatrix} x \\ y \end{bmatrix}^{\mathrm{T}} \mathcal{M} \begin{bmatrix} x \\ y \end{bmatrix} \tag{5.56}$$

holds for all $x, y \in \mathbb{R}^n$ and for some matrix $\mathcal{M} \in \mathbb{S}^n$. Then, is clear that the conservatism can be measured as the maximal eigenvalue of the matrix

$$\mathcal{M}_s := \mathcal{M} + \begin{bmatrix} 0 & I \\ I & 0 \end{bmatrix} \succeq 0. \tag{5.57}$$

If the maximal eigenvalue is 0, then the bound is exact in the sense that the right-hand side of (6.56) is equal to the left-hand side for all $x, y \in \mathbb{R}^n$. If, however, the maximal eigenvalue is positive, then there exists a pair $(\bar{x}, \bar{y})$ for which we have

$$\begin{bmatrix} \bar{x} \\ \bar{y} \end{bmatrix}^{\mathrm{T}} \mathcal{M}_s \begin{bmatrix} \bar{x} \\ \bar{y} \end{bmatrix} > 0 \tag{5.58}$$

and thus

$$-2\bar{x}^{\mathrm{T}}\bar{y} < \begin{bmatrix} \bar{x} \\ \bar{y} \end{bmatrix}^{\mathrm{T}} \mathcal{M} \begin{bmatrix} \bar{x} \\ \bar{y} \end{bmatrix}, \tag{5.59}$$

i.e. the bound is conservative.

From now on, the maximal eigenvalue of $\mathcal{M}_s$, $\lambda_{max}(\mathcal{M}_s)$, will be referred to as the *bounding gap*.

Proposition 5.6.10 *The bounding gap of the bound*

$$-2x^T y \le x^T X x + y^T X^{-1} y \tag{5.60}$$

is given by $\max_{\zeta \in \lambda(X)} \{\zeta + \zeta^{-1}\}$. *Moreover, the gap is minimal when* X *is chosen as* $X = I$ *and the minimum value is 2.*

Proof In this case, we simply have

$$\mathcal{M}_s = \begin{bmatrix} X & I_n \\ I_n & X^{-1} \end{bmatrix}. \tag{5.61}$$

The characteristic polynomial of this matrix is given by

$$\begin{aligned} \det(\xi I - \mathcal{M}_s) &= \det(\xi I_n - X)\det(\xi I_n - X^{-1} - (\xi I_n - X)^{-1}) \\ &= \det(\xi^2 I_n + \xi(X + X^{-1})) \\ &= \xi^n \det(\xi I_n + (X + X^{-1})) \\ &= \xi^n \prod_{\zeta \in \lambda(X)} (\xi - \zeta - \zeta^{-1}) \end{aligned} \tag{5.62}$$

where the last row has been obtained using the fact that the eigenvalues ζ of X and ζ^{-1} of X^{-1} have the same eigenvectors. Finally, it is immediate to see that the minimum of the function $\zeta + \zeta^{-1}$ is given by 2 and is attained for $\zeta = 1$. The proof is complete. ■

Proposition 5.6.11 *The bounding gap of the bound of Lemma 5.6.7 is given by* $\lambda_{max}(X)$. *Moreover, this gap can be made as small as desired.*

Proof In this case, we simply have

$$\mathcal{M}_s = \begin{bmatrix} X & XM + I_n \\ M^{\mathrm{T}} X + I_n & (M^{\mathrm{T}} X + I_n) X^{-1} (XM + I_n) \end{bmatrix}. \tag{5.63}$$

By factorizing this matrix as

$$\mathcal{M}_s = \begin{bmatrix} I_n & 0 \\ 0 & I_n + XM \end{bmatrix}^{\mathrm{T}} \begin{bmatrix} X & I_n \\ I_n & X^{-1} \end{bmatrix} \begin{bmatrix} I_n & 0 \\ 0 & I_n + XM \end{bmatrix} \tag{5.64}$$

we can clearly see that the central matrix is identical to the matrix of the bounding technique of Proposition 5.6.10. The bound of Lemma 5.6.7 can hence be understood as a scaled version of (5.60) and the scaling is optimum when $M = -X^{-1}$. In such a case, $\mathcal{M}_s$ reduces to

$$\begin{bmatrix} X & 0 \\ 0 & 0 \end{bmatrix} \tag{5.65}$$

and the result follows. Since $X \succ 0$ is arbitrary, then X can be made arbitrarily small. ■

Proposition 5.6.12 *The bounding gap of the bound of Lemma 5.6.9 is given by*

$$\lambda_{max}\left(\begin{bmatrix} X & Y \\ \star & Z \end{bmatrix}\right) \tag{5.66}$$

and can be made as small as desired.

Proof The proof is simply based on the fact the matrix $\mathcal{M}_s$ fully consists of free variables that can arbitrarily set to any value. The value 0 for the bounding gap can only be attained by letting $X = Y = Z = 0$. ■

Note, however, that setting $X = Y = Z = 0$ does not help when deriving stability conditions since no bounding is actually performed in this case.

All the bounding techniques described above have a nonzero bounding gap, even though some of them can approach arbitrarily closely this value. Note that the latter statement has only a theoretical value since numerical solvers work in finite precision. The following bound, which does not seem to have been proposed anywhere before, is shown to have zero bounding gap:

Proposition 5.6.13 *For any matrices $N_1, N_2 \in \mathbb{R}^{n \times n}$ and $X \in \mathbb{S}^n_{\succ 0}$, the following inequality*

$$-2x^T y \le \begin{bmatrix} x \\ y \end{bmatrix}^T \begin{bmatrix} X + XN_1 + N_1^T X & N_1^T + XN_2 \\ \star & X^{-1} + N_2^T + N_2 \end{bmatrix} \begin{bmatrix} x \\ y \end{bmatrix} \tag{5.67}$$

holds for all $x, y \in \mathbb{R}^n$. Moreover, the bounding gap is equal to 0.

Proof In this case, the matrix $\mathcal{M}_s$ is given by

$$\mathcal{M}_s := \begin{bmatrix} X + XN_1 + N_1^{\mathrm{T}}X & N_1^{\mathrm{T}} + XN_2 + I_n \\ \star & X^{-1} + N_2^{\mathrm{T}} + N_2 \end{bmatrix} \tag{5.68}$$

and can be decomposed as

$$\mathcal{M}_s = \begin{bmatrix} X & I_n \\ I_n & X^{-1} \end{bmatrix} + \mathrm{He}\left(\begin{bmatrix} X \\ I_n \end{bmatrix} \begin{bmatrix} N_1 & N_2 \end{bmatrix} \right). \tag{5.69}$$

So, applying the elimination lemma (or Finsler Lemma, see [88] or Appendix C.11) on the matrix inequality $\mathcal{M}_s \succeq 0$ yields the equivalent condition

$$\begin{bmatrix} I_n \\ -X \end{bmatrix}^{\mathrm{T}} \begin{bmatrix} X & I_n \\ I_n & X^{-1} \end{bmatrix} \begin{bmatrix} I_n \\ -X \end{bmatrix} \succeq 0. \tag{5.70}$$

Evaluating the left-hand side yields 0, which means that there always exist $N_1, N_2 \in \mathbb{R}^{n\times n}$ such that $\mathcal{M}_s = 0$ and, therefore, that the bound is exact. It is, moreover, easy to see that the 0 bounding gap is attained with the values $N_1 = -I_n/2$ and $N_2 = -X^{-1}/2$. ■

5.6.5 Delay-Dependent Stability—Descriptor Model Transformation

Unlike the model transformations described in the previous sections, the descriptor model transformation yields a comparison system which is identical to the original one, from the stability viewpoint. On the strength of this fact, it may be possible to obtain more efficient stability conditions than by using inaccurate model transformations. To address the stability analysis of the comparison system (5.35), the following LKF in proposed in [40, 56]:

$$V(x_t, y_t) = \begin{bmatrix} x(t) \\ y(t) \end{bmatrix}^{\mathrm{T}} E^{\mathrm{T}} P \begin{bmatrix} x(t) \\ y(t) \end{bmatrix} + \int_{-h}^{0} \int_{t+\theta}^{t} y(s)^{\mathrm{T}} R y(s) ds d\theta \tag{5.71}$$

where

$$E = \begin{bmatrix} 1 & 0 \\ 0 & 0 \end{bmatrix}, \quad P = \begin{bmatrix} P_1 & 0 \\ P_2 & P_3 \end{bmatrix}$$

and $P_1, R \in \mathbb{S}^n_{\succ 0}$, $P_2, P_3 \in \mathbb{R}^{n\times n}$. The following result can then be obtained:

Theorem 5.6.14 *The System (5.23) is delay-dependent stable for all $h \in [0, \bar{h}]$ if there exist matrices $P_1, R \in \mathbb{S}^n_{\succ 0}$ and $P_2, P_3 \in \mathbb{R}^{n\times n}$ such that the LMI*

$$\begin{bmatrix} \mathrm{He}[P_2^T(A+A_h)] & P_1 - P_2^T + (A+A_h)^T P_3 & \bar{h}P_2^T A_h \\ \star & -P_3 - P_3^T + \bar{h}R & \bar{h}P_3^T A_h \\ \star & \star & -\bar{h}R \end{bmatrix} \prec 0 \qquad (5.72)$$

holds.

Proof The derivative of the LKF (5.71) along the trajectories of the system (5.35) is given by

$$\begin{aligned} \dot{V}(t) \le & \begin{bmatrix} x(t) \\ y(t) \end{bmatrix}^{\mathrm{T}} \begin{bmatrix} \mathrm{He}[P_2^{\mathrm{T}}(A+A_h)] & P_1 - P_2^{\mathrm{T}} + (A+A_h)^{\mathrm{T}} P_3 \\ \star & -P_3 - P_3^{\mathrm{T}} + \bar{h}R \end{bmatrix} \begin{bmatrix} x(t) \\ y(t) \end{bmatrix} \\ & - \int_{t-h}^{t} y(s)^{\mathrm{T}} R y(s) ds - 2 \int_{t-h} \begin{bmatrix} x(t) \\ y(t) \end{bmatrix}^{\mathrm{T}} \begin{bmatrix} P_2^{\mathrm{T}} A_h \\ P_3^{\mathrm{T}} A_h \end{bmatrix} y(s) ds. \end{aligned} \qquad (5.73)$$

The cross-term is simply bounded as

$$\begin{aligned} -2 \int_{t-h} \begin{bmatrix} x(t) \\ y(t) \end{bmatrix}^{\mathrm{T}} \begin{bmatrix} P_2^{\mathrm{T}} A_h \\ P_3^{\mathrm{T}} A_h \end{bmatrix} y(s) ds \le \bar{h} & \begin{bmatrix} x(t) \\ y(t) \end{bmatrix}^{\mathrm{T}} \begin{bmatrix} P_2^{\mathrm{T}} A_h \\ P_3^{\mathrm{T}} A_h \end{bmatrix} R^{-1} \begin{bmatrix} P_2^{\mathrm{T}} A_h \\ P_3^{\mathrm{T}} A_h \end{bmatrix} \begin{bmatrix} x(t) \\ y(t) \end{bmatrix} \\ & + \int_{t-h}^{t} y(s)^{\mathrm{T}} R y(s) \end{aligned}$$

which leads to the result after performing a Schur complement. ■

As seen in the proof, this result relies on the naive cross-term bounding technique based on squares completion. The result in [89], however, improves this by considering the bound of Lemma 5.6.7. Although this method relies on a nonconservative model transformation and leads to interesting results, it still has to cope with cross-terms which are inexorably leading to an increase of conservatism. The computational complexity of the above result is also suboptimal since several matrices can be eliminated.

5.6.5.1 Complexity Reduction

From the structure of the LMI condition (5.72), it is easily seen that the matrices P_2 and P_3 can be fully eliminated from the result. By elimination, it is not meant here that these matrices are set to 0, but that they are set to their optimal value. This can be implicitly done using the projection lemma; see [86] or Appendix C.12. By doing so, we obtain the following result:

Theorem 5.6.15 *The following statements are equivalent:*

1. *There exist a scalar* $\bar{h} > 0$, *matrices* $P_1, R \in \mathbb{S}^n_{\succ 0}$, $P_2, P_3 \in \mathbb{R}^{n\times n}$ *such that the LMI* (5.72) *of Theorem 5.6.14 is feasible.*
2. *The LMI*

$$\begin{bmatrix} M_{11} & \bar{h}P_1A_h + \bar{h}^2(A+A_h)^T RA_h \\ \star & \bar{h}^3 A_h^T RA_h - \bar{h}R \end{bmatrix} \prec 0 \tag{5.74}$$

where $M_{11} = \mathrm{He}\,[P_1(A+A_h)] + \bar{h}(A+A_h)^T R(A+A_h)$ *is feasible with the same matrices* $P_1, R \in \mathbb{S}^n_{\succ 0}$ *and same scalar* $\bar{h} > 0$.

Proof Let us first rewrite condition (5.72) as

$$\begin{bmatrix} 0 & P_1 & 0 \\ \star & \bar{h}R & 0 \\ \star & \star & -\bar{h}R \end{bmatrix} + \mathrm{He}\left(\begin{bmatrix} I & 0 \\ 0 & I \\ 0 & 0 \end{bmatrix} \begin{bmatrix} P_2^{\mathrm{T}} \\ P_3^{\mathrm{T}} \end{bmatrix} \begin{bmatrix} A+A_h & -I & \bar{h}A_h \end{bmatrix} \right) \prec 0. \tag{5.75}$$

By virtue of the projection lemma (see Appendix C.12), this condition is equivalent to $-\bar{h}R \prec 0$ and

$$\begin{bmatrix} I & 0 \\ A+A_h & \bar{h}B \\ 0 & I \end{bmatrix}^{\mathrm{T}} \begin{bmatrix} 0 & P_1 & 0 \\ \star & \bar{h}R & 0 \\ \star & \star & -\bar{h}R \end{bmatrix} \begin{bmatrix} I & 0 \\ A+A_h & \bar{h}B \\ 0 & I \end{bmatrix} \prec 0. \tag{5.76}$$

Noting, finally, that the above LMI is equivalent to (5.74) completes the proof. ■

The LMI (5.74) in the above result is very instructive, independently of the complexity reduction. The negative definiteness of the (2, 2) block of (5.74) is indeed equivalent to the condition $\varrho(A_h) < 1/\bar{h}$. This condition on the delay is, therefore, *necessary for the LMI* (5.74) to hold. Consequently, this LMI test cannot predict accurate delay upper-bounds whenever the actual upper bound is greater than $\varrho(A_h)$.

This result will be shown in Sect. 5.7.3 to be fully interpreted as a result obtained from the application of the scaled small-gain theorem.

5.6.6 Delay-Dependent Stability—Method of Free-Weighting Matrices

Finsler's lemma and the projection lemma have been used in previous sections to eliminate superfluous matrices from the stability conditions. The present approach, introduced in [90], makes use of this result to incorporate algebraic constraints relating different signals. Instead of eliminating matrices, the procedure is inverse here and introduces additional matrices, justifying then the denomination of *free weighting matrices method*. Note, however, that these additional matrices are more generally referred to as *slack-variables* or *lifting variables* in the "LMI community".

The main difference with the previous approaches lies in the (partial) avoidance of any model transformation. The considered LKF [90] is given by

$$V(x_t, \dot{x}_t) = x(t)^{\mathrm{T}} P x(t) + \int_{t-h}^{t} x(\theta)^{\mathrm{T}} Q x(\theta) d\theta + \int_{-h}^{0} \int_{t+\theta}^{t} \dot{x}(\eta)^{\mathrm{T}} R \dot{x}(\eta) d\eta d\theta \tag{5.77}$$

where $P, Q, R \in \mathbb{S}^n_{\succ 0}$. We then have the following result:

Theorem 5.6.16 *The system (5.23) is delay-dependent stable for all $h \in [0, \bar{h}]$ if there exist matrices $P, Q, R \in \mathbb{S}^n_{\succ 0}$ and $N \in \mathbb{R}^{3n \times 2n}$ such that the LMI*

$$\Psi + U^T N V + V^T N^T U \prec 0 \tag{5.78}$$

holds with

$$\Psi = \begin{bmatrix} Q & 0 & P & 0 \\ \star & -Q & 0 & 0 \\ \star & \star & \bar{h}R & 0 \\ \star & \star & \star & -\bar{h}R \end{bmatrix}, \; U^T = \begin{bmatrix} I & 0 & 0 \\ 0 & I & 0 \\ 0 & 0 & I \\ 0 & 0 & 0 \end{bmatrix} \; \textit{and} \; V^T = \begin{bmatrix} -A^T & I \\ -A_h^T & -I \\ I & 0 \\ 0 & \bar{h}I \end{bmatrix}.$$

Proof The derivative of the LKF (5.77) along the system (5.23) is given by

$$\dot{V} \le \zeta(t)^{\mathrm{T}} \begin{bmatrix} Q & P & 0 \\ \star & -Q & 0 \\ \star & \star & \bar{h}R \end{bmatrix} \zeta(t) - \int_{t-h}^{t} \dot{x}(s)^{\mathrm{T}} R \dot{x}(s) ds$$

where $\zeta(t) = \mathrm{col}(x(t), x(t-h), \dot{x}(t))$. Note now that the following identities

$$2\zeta(t)^{\mathrm{T}}N_1\left[x(t)-x(t-h)-\int\limits_{t-h}^{t}\dot{x}(s)ds\right]=0$$
$$2\zeta(t)^{\mathrm{T}}N_2\left[\dot{x}(t)-Ax(t)-A_hx(t-h)\right]=0$$
$$\bar{h}\zeta(t)^{\mathrm{T}}X\zeta(t)-\int\limits_{t-h}^{t}\zeta(t)^{\mathrm{T}}X\zeta(t)d\theta\geq 0$$

hold for any matrices $N_1, N_2 \in \mathbb{R}^{3n\times n}$ and $X \in \mathbb{S}^{3n}_{\succeq 0}$. The first constraint specifies the Newton-Leibniz formula; the second one, the system's dynamics and the last one incorporates information on the maximal delay value. Adding these constraints to the derivative yields

$$\dot{V}(t)\leq\zeta(t)^{\mathrm{T}}\left(\Xi+\bar{h}X\right)\zeta(t)-\int_{t-h}^{t}\begin{bmatrix}\zeta(t)\\ \dot{x}(s)\end{bmatrix}^{\mathrm{T}}\begin{bmatrix}X & N_1\\ \star & R\end{bmatrix}\begin{bmatrix}\zeta(t)\\ \dot{x}(s)\end{bmatrix}ds$$

where Ξ is the $3n\times 3n$ principal submatrix of the matrix (5.78). Note that since the matrix X is positive semidefinite then $\Xi+\bar{h}X\succeq\Xi$, then the ideal value for X would be the smallest one. On the other hand, the integral term should be positive semidefinite as well and this can be imposed by choosing X such that $X\succeq N_1^{\mathrm{T}}R^{-1}N_1$. Since we are looking for a 'small' X, the optimal choice is then $X=N_1^{\mathrm{T}}R^{-1}N_1$. Substitution of this value in the sum $\Xi+\bar{h}X$ gives, after a Schur complement, LMI (5.78) with $N=\begin{bmatrix}N_2 & N_1\end{bmatrix}$. The proof is complete. ■

5.6.6.1 Complexity Reduction

The presence of slack variables in stability conditions is actually beneficial when robustness analysis is of interest, e.g. with respect to polytopic types uncertainties, since the overall LMI can be expressed as a linear combination of matrices when the matrices P, Q and R are also parameter dependent; see e.g. Theorem 2.5.6. The presence of additional variables, however, increases the computational complexity of the method and makes the derivation of design results, such as control design, rather difficult. As in Sect. 5.6.5, we perform now an elimination procedure on the matrices N_1 and N_2 using the projection lemma; see Appendix C.12. This yields the following result:

Lemma 5.6.17 *The following statements are equivalent:*

1. *There exist a scalar $\bar{h}>0$ and matrices $P, Q, R\in\mathbb{S}^n_{\succ 0}$ and $N\in\mathbb{R}^{3n\times 2n}$ such that the LMI* (5.78) *holds.*

2. *The LMI*

$$\begin{bmatrix} A^TP + PA + Q - \bar{h}^{-1}R & PA_h + \bar{h}^{-1}R \\ \star & -Q - \bar{h}^{-1}R \end{bmatrix} + \bar{h} \begin{bmatrix} A^T \\ A_h^T \end{bmatrix} R \begin{bmatrix} A^T \\ A_h^T \end{bmatrix}^T \prec 0 \tag{5.79}$$

is feasible with the same $\bar{h} > 0$, $P, Q, R \in \mathbb{S}^n_{\succ 0}$.

Proof Note that LMI (5.78) is already in 'projection lemma form' and, therefore, the projection lemma can be applied directly. Bases for the null-spaces of U and V are given by

$$\mathcal{N}_U = \begin{bmatrix} 0 \\ 0 \\ 0 \\ I \end{bmatrix} \text{ and } \mathcal{N}_V = \begin{bmatrix} I & 0 \\ 0 & I \\ A & A_h \\ -\bar{h}^{-1}I & \bar{h}^{-1}I \end{bmatrix},$$

respectively. Applying the projection lemma, we get the two equivalent LMIs $-\bar{h}R \prec 0$ and (5.79). Since the first one is trivially satisfied, the feasibilities of (5.78) and (5.79) are thus equivalent. The proof is complete. ■

The main interest of the result of this section lies in the fact that the approach is free of model transformation and hence no conservatism is added in this respect. The main drawback is the computational complexity, but has been resolved by eliminating the extra variables, leaving only 3 decision matrices, namely P, Q and R. The procedure to obtain the 'optimal' result of Lemma 5.6.17 is quite cumbersome since we have to use Finsler's lemma to incorporate constraints, then eliminate them using a similar result; this way of proceeding does not really seem to be appropriate. Would not it be possible instead to obtain Lemma 5.6.17 directly? This question is answered in the following section.

5.6.7 Delay-Dependent Stability—Jensen's Inequality

The use of Jensen's inequality in time-delay systems can be traced back to [91], and has been proved to be very useful since then. In what follows, the inequality is formally stated first and used then to derive a delay-dependent stability result. The source of conservatism of Jensen's inequality is finally discussed.

5.6.7.1 Jensen's Inequality

This inequality, proved in [92], was first inspired from the inequality of arithmetic and geometric means introduced by Cauchy in [93]. Jensen's result has found many applications in statistical physics, information theory, statistics and probability theory and certainly many other fields. In systems and control theory, it can be used to evaluate norms of certain integral operators involved in time-delay systems analysis [2] or to provide bounds on certain integral terms present in functionals for time-delay systems [59, 94], sampled-data systems [95, 96] and impulsive systems [97].

Lemma 5.6.18 (Jensen's inequality) *Let ϕ be a convex integrable function and $z : [a, b] \to \mathbb{R}$, $a < b$, be integrable over its domain of definition. Then, the following inequality*

$$\phi\left(\int_a^b z(s)ds\right) \le (b-a)\int_a^b \phi(z(s))ds$$

holds.

Proof The proof can be found in [92, 98]. ■

Whenever linear time-delay systems are considered, quadratic functions ϕ are of interest:

Corollary 5.6.19 *Let $Z \in \mathbb{S}^n_{\succ 0}$ and $z : [a, b] \to \mathbb{R}^n$ be an integrable function on its domain. Then, the following inequality*

$$\left(\int_a^b z(\theta)d\theta\right)^T Z\left(\int_a^b z(\theta)d\theta\right) \le (b-a)\int_a^b z(\theta)^T Z z(\theta)d\theta \tag{5.80}$$

holds.

Proof This is a simple application of Jensen's inequality with $\phi(z) = z^{\mathrm{T}} Z z$. ■

5.6.7.2 Delay-Dependent Stability Condition Using Jensen's Inequality

By considering the LKF (5.77) and Corollary 5.6.19, we are in position to derive the following result:

Theorem 5.6.20 ([59]) *The system (5.23) is delay-dependent stable for all $h \in [0, \bar{h}]$ if there exist matrices $P, Q, R \in \mathbb{S}^n_{\succ 0}$ such that the LMI*

$$\begin{bmatrix} A^T P + PA + Q - \bar{h}^{-1} R & PA_h + \bar{h}^{-1} R \\ \star & -Q - \bar{h}^{-1} R \end{bmatrix} + \bar{h} \begin{bmatrix} A^T \\ A_h^T \end{bmatrix} R \begin{bmatrix} A^T \\ A_h^T \end{bmatrix}^T \prec 0 \quad (5.81)$$

holds.

Proof The derivative of the LKF (5.77) along the system (5.23) gives

$$\begin{aligned} \dot{V}(t) = & \begin{bmatrix} x(t) \\ x(t-h) \end{bmatrix}^{\mathrm{T}} \begin{bmatrix} A^{\mathrm{T}} P + PA + Q & PA_h \\ A_h^{\mathrm{T}} P & -Q \end{bmatrix} \begin{bmatrix} x(t) \\ x(t-h) \end{bmatrix} \\ & - \int_{t-h}^{t} \dot{x}(\theta)^{\mathrm{T}} R \dot{x}(\theta) d\theta. \end{aligned}$$

Using Jensen's inequality, i.e. Corollary 5.6.19, on the integral term yields

$$\begin{aligned} - \int_{t-h}^{t} \dot{x}(\theta)^{\mathrm{T}} R \dot{x}(\theta) d\theta & \leq -\frac{1}{\bar{h}} \left(\int_{t-h}^{t} \dot{x}(s) ds \right)^{\mathrm{T}} R \left(\int_{t-h}^{t} \dot{x}(s) ds \right) \\ & = -\frac{1}{\bar{h}} (x(t) - x(t-h))^{\mathrm{T}} R (x(t) - x(t-h)). \end{aligned} \quad (5.82)$$

Substituting the bound in the LKF derivative yields the result. ■

This result deserves several important remarks. First, no model transformation is used, vanishing then the conservatism of the approach on this level. Second, the negative integral term that was systematically compensated in the results of Sects. 5.6.2–5.6.5 is here preserved and partly taken into account through Jensen's inequality. Finally, condition (5.81) is identical to condition (5.79), which shows that the method of free-weighting matrices of Sect. 5.6.6 is essentially equivalent to using Jensen's inequality. The benefit of the current approach lies in the simplicity and directness of the proof.

5.6.7.3 Conservatism of Jensen's Inequality

With any bounding technique, conservatism is associated. It is therefore important to address this point for Jensen's inequality. This can be performed using Grüss inequality. A simplified, but sufficient for our problem, version of the Grüss inequality [99] is defined as follows:

Lemma 5.6.21 (Grüss Inequality) *Assume that there exist constant scalars* $f^-, f^+, g^-, g^+ \in \mathbb{R}$ *such that* $f^- \leq f^+$, $g^- \leq g^+$ *and functions* $f, g : [a, b] \to \mathbb{R}$ *satisfying* $f^- \leq f \leq f^+$ *and* $g^- \leq g \leq g^+$ *almost everywhere on* $[a, b]$. *Then, the following inequality*

$$\left| \frac{1}{b-a} \int_a^b f(s)g(s)ds - \frac{1}{(b-a)^2} \int_a^b f(s)ds \int_a^b g(s)ds \right| \leq \frac{1}{4}\delta_f \delta_g \quad (5.83)$$

holds where $\delta_f = f^+ - f^-$, $\delta_g = g^+ - g^-$. *Moreover the constant term* $1/4$ *in the right-hand side is sharp and is obtained for the functions* $f(s) = g(s) = sgn\left(s - \frac{a+b}{2}\right)$ *where* $sgn(\cdot)$ *is the signum function.*

Similarly to as for bounds on cross-terms, we define the Jensen's gap to be the maximal distance between the bounded expression and Jensen's bound. Specialized to the quadratic case, the following result on the Jensen's inequality gap is obtained using Grüss inequality:

Theorem 5.6.22 [71] *Given a function* $z : [a, b] \to \mathbb{R}^n$, *then the Jensen's gap is given by*

$$\delta_J := \frac{(b-a)^2}{4}(z^+ - z^-)^T Z(z^+ - z^-) \quad (5.84)$$

where $z^- \leq z(s) \leq z^+$ *almost everywhere on* $s \in [a, b]$. *Therefore, the inequality*

$$0 \leq (b-a) \int_a^b z(s)^T Zz(s)ds - \left(\int_a^b z(s)ds \right)^T Z \left(\int_a^b z(s)ds \right) \leq \delta_J \quad (5.85)$$

holds. Moreover, the constant term $1/4$ *in* δ_J *is sharp and the gap* δ_J *is attained for the functions* $z_i(s) = sgn\left(s - \frac{a+b}{2}\right)$, $i = 1, \ldots, n$.

Proof The proof is an immediate consequence of Corollary 5.6.19 and the Grüss inequality of Lemma 5.6.21. ■

It can be easily verified that Jensen's inequality is nonconservative for the class of constant functions. As stated in Theorem 5.6.22, the Jensen's gap is attained for a class of discontinuous functions and therefore, when using Jensen's bound for deriving stability conditions, we implicitly consider the entire class of signals tackled by

Jensen's inequality, i.e. the class of integrable functions which includes discontinuous ones. This bound thus includes a much wider class of functions than necessary since linear time-delay systems with constant time-delays have differentiable trajectories and will never exhibit discontinuous solutions. Jensen's inequality therefore introduces conservatism that can be characterized in terms of *the class of functions* it considers. Possible refinements of the approach rely on delay-fragmentation, see e.g. [59, 71, 100], or the consideration of a bound that is more adapted to the continuous trajectories of time-delay systems.

5.6.7.4 Affine Version of Jensen's Inequality

We show here that a bound very often used in the literature is actually equivalent to Jensen's inequality. It is also emphasized that this bound has interesting conservatism reduction properties over Jensen's inequality since the obtained LMI conditions remain well-posed when the measure of the interval of integration, i.e. the quantity $b - a$, tends to 0. This property is very convenient when the size of the interval of integration is a time-varying or uncertain data of the problem, which turns out to be the case when analyzing time-delay, sampled-data and impulsive systems [95, 97, 101].

The following result is useful for deriving the main one:

Lemma 5.6.23 (Relaxation lemma [71]) *Let the matrices $M_{22} \in \mathbb{S}^m_{\succ 0}$, $M_{11} \in \mathbb{S}^n$ and $M_{12} \in \mathbb{R}^{n \times m}$ be given. Then, the following statements are equivalent:*

1. *The matrix inequality*
$$M_{11} - M_{12}M_{22}^{-1}M_{12}^T \prec 0 \tag{5.86}$$
holds.
2. *There exists a matrix $N \in \mathbb{R}^{m \times n}$ such that the matrix inequality*
$$M_{11} + N^T M_{12}^T + M_{12}N + N^T M_{22} N \prec 0 \tag{5.87}$$
holds.

A generalization of this result to the case where M_{22} is indefinite is provided in Appendix C.4.

Proof A proof is given in [102] and is quite involved. We provide here an alternative one (another proof relies on the use of the projection lemma). To see the equivalence between the statements, it is enough to show that we have

$$\min_{N\in\mathbb{R}^{m\times n}}\left\{N^{\mathrm{T}}M_{12}^{\mathrm{T}}+M_{12}N+N^{\mathrm{T}}M_{22}N\right\}=-M_{12}M_{22}^{-1}M_{12}^{\mathrm{T}}$$

where the minimum is understood in the sense that

$$-M_{12}M_{22}^{-1}M_{12}^{\mathrm{T}}\preceq N^{\mathrm{T}}M_{12}^{\mathrm{T}}+M_{12}N+N^{\mathrm{T}}M_{22}N \text{ for all } N\in\mathbb{R}^{m\times n}$$

and that there exists a N such that equality holds. It is easy to see that (5.87) is convex in N since $M_{22}\succ 0$. Completing then the squares, we find that the minimum $-M_{12}M_{22}^{-1}M_{12}^{\mathrm{T}}$ is attained for $N=-M_{22}^{-1}M_{12}^{\mathrm{T}}$. The proof is complete. ■

The interest of the above result is twofold. First, it can be used to transform complex nonlinear matrix inequalities in a more convenient form [103–105]. Second, it can be used to prove the equivalence between several results. The following theorem addresses the latter problem:

Lemma 5.6.24 (Affine Jensen's inequality [71]) *Let us consider a function $z:\mathbb{R}_{\ge 0}\to\mathbb{R}^n$ integrable over $[a,b]$, $0\le a<b$, a matrix $R\in\mathbb{S}^n_{\succ 0}$, and a function $w:\mathbb{R}_{\ge 0}\to\mathbb{R}^{n+m}$ verifying $\int_a^b z(s)ds=Mw(a,b)$ for some known matrix $M\in\mathbb{R}^{n\times(n+m)}$. Then, the following statements are equivalent:*

1. *The inequality*

$$-\int_a^b z(s)^T Rz(s)ds\le -\frac{1}{b-a}w(a,b)^T M^T RMw(a,b)$$

holds.

2. *The inequality*

$$-\int_a^b z(s)^T Rz(s)ds\le w(a,b)^T Q(N)w(a,b)$$

holds for all $N\in\mathbb{R}^{n\times(n+m)}$ with

$$Q(N)=N^T M+M^T N+(b-a)N^T R^{-1}N.$$

Proof The first inequality is a consequence of Jensen's inequality. The equivalence of the bounds is a consequence of Lemma 5.6.23. ■

A first integral inequality. Let us consider a differentiable function $x:\mathbb{R}_{\ge 0}\to\mathbb{R}^n$ verifying

$$\int_{t_k}^{t}\dot{x}(s)ds=Mw(t_k,t)$$

where $M = \begin{bmatrix} I & -I \end{bmatrix}$ and $w(t_k, t) = \text{col}(x(t), x(t_k))$. In [95], the following bound is used:

$$-\int_{t_k}^{t} \dot{x}(s)^{\mathrm{T}} R\dot{x}(s)ds \leq w(t_k, t)^{\mathrm{T}} \mathcal{R} w(t_k, t), \quad R \in \mathbb{S}^n_{\succ 0}, \ t > t_k \tag{5.88}$$

where $\mathcal{R} = N^{\mathrm{T}}M + M^{\mathrm{T}}N + (t - t_k)N^{\mathrm{T}}R^{-1}N$ and $N \in \mathbb{R}^{n\times 2n}$ is an additional matrix to be determined. Then, according to Lemma 5.6.24, we can conclude on the equivalence with Jensen's inequality:

$$-\int_{t_k}^{t} \dot{x}(s)^{\mathrm{T}} R\dot{x}(s)ds \leq \frac{-1}{t - t_k} w(t_k, t)^{\mathrm{T}} M^{\mathrm{T}} R M w(t_k, t). \tag{5.89}$$

A second integral inequality. In [106], the following bound is considered:

$$-\int_{t-\tau}^{t} \dot{x}(s)^{\mathrm{T}} R\dot{x}(s)ds \leq w(t)^{\mathrm{T}}(\mathcal{M} + \tau N^{\mathrm{T}}R^{-1}N)w(t) \tag{5.90}$$

where $w(t) = \text{col}(x(t), \bullet, x(t-\tau), \bullet)$ is a vector of signals involved in the system, $N = \begin{bmatrix} N_1 & N_2 & N_3 & N_4 \end{bmatrix}$ and

$$\mathcal{M} = \begin{bmatrix} N_1 + N_1^{\mathrm{T}} & N_2 & -N_1^{\mathrm{T}} + N_3 & N_4 \\ \star & 0 & -N_2^{\mathrm{T}} & 0 \\ \star & \star & -N_3 - N_3^{\mathrm{T}} & -N_4 \\ \star & \star & \star & 0 \end{bmatrix}.$$

Note that we also have that

$$\int_{t-\tau}^{t} \dot{x}(s)ds = Mw(t)$$

where $M = \begin{bmatrix} I & 0 & -I & 0 \end{bmatrix}$. We show now that this bound is equivalent to Jensen's inequality. Noting that $\mathcal{M} = N^{\mathrm{T}}M + M^{\mathrm{T}}N$, the right-hand side of (5.90) then writes

$$w(t)^{\mathrm{T}}(N^{\mathrm{T}}M + M^{\mathrm{T}}N + \tau N^{\mathrm{T}}R^{-1}N)w(t)$$

and, by virtue of Lemma 5.6.24, we have equivalence with Jensen's inequality

$$
\begin{aligned}
-\int_{t-\tau}^{t} \dot{x}(s)^{\mathrm{T}} R\dot{x}(s)ds &\leq -\frac{1}{\tau} w(t)^{\mathrm{T}} M^{\mathrm{T}} R M w(t) \\
&= -\frac{1}{\tau}(x(t) - x(t-\tau))^{\mathrm{T}} R(x(t) - x(t-\tau)).
\end{aligned} \tag{5.91}
$$

Explanation of the advantage of the affine formulation over the rational one. Even though the bounds are theoretically equivalent, they actually exhibit differences from a computational perspective. Well-posedness of the affine inequality makes it indeed more adequate for deriving less conservative results. The presence of the rational term in Jensen's inequality, e.g. $-1/\tau$ in (5.91), forces us to upper-bound the rational term in order to derive tractable LMI-based results, e.g. $-1/\tau \leq -1/\bar{\tau}$ where $\bar{\tau}$ is the maximal value for τ. By doing so, we actually neglect all the possible intermediary values of this rational term and only consider the worst-case value. This conservative bounding procedure is not needed when the affine bound is considered.

To illustrate this, let us consider the sampled-data system

$$
\dot{x}(t) = \begin{bmatrix} 0 & 1 \\ 0 & -0.1 \end{bmatrix} x(t) + \begin{bmatrix} 0 & 0 \\ 0.375 & -1.15 \end{bmatrix} x(t_k), \ t \in [t_k, t_{k+1}) \tag{5.92}
$$

and let us informally compare the result obtained with Theorem 1 of [95] which uses the affine bound, to the result that would have been obtained with Jensen's inequality. The former result shows that the system remains stable provided that $t_{k+1} - t_k \leq 1.6894$ whereas the latter yields the bound 0.8691. Clearly, the criterion based on the affine formulation is less conservative than the one based on Jensen's inequality.

As a concluding statement, we can say that *even though the bounds are theoretically equivalent, the manipulations performed in order to obtain tractable (LMI) conditions introduce some additional conservatism.*

5.6.8 Fragmented Lyapunov-Krasovskii Functional

The functionals considered in the previous section is rather simple and may yield to conservative results. In order to improve the results, an idea is to generalize the functional (5.77) to

$$
\begin{aligned}
V(x_t) = x(t)^{\mathrm{T}} P x(t) &+ \int_{-h(t)}^{0} x(t+\theta)^{\mathrm{T}} Q(\theta) x(t+\theta) d\theta \\
&+ \bar{h} \int_{-\bar{h}}^{0} \int_{\theta}^{0} \dot{x}(t+\eta)^{\mathrm{T}} R(\theta) \dot{x}(t+\eta) d\eta d\theta
\end{aligned} \tag{5.93}
$$

where the decision variables are now the matrix $P \in \mathbb{S}^n_{\succ 0}$ and the matrix functions

$$\begin{aligned} Q &: [-\bar{h}, 0] \to \mathbb{S}^n_{\succ 0} \\ R &: [-\bar{h}, 0] \to \mathbb{S}^n_{\succ 0}. \end{aligned} \tag{5.94}$$

Whenever, the functions Q and R are chosen to be piecewise constant, i.e. as

$$\begin{aligned} Q(\theta) &= Q_i,\ \theta \in \left[-(i+1)\frac{h(t)}{f}, -i\frac{h(t)}{f}\right],\ i = 0, \dots, f-1 \\ R(\theta) &= R_i,\ \theta \in \left[-(i+1)\frac{\bar{h}}{f}, -i\frac{\bar{h}}{f}\right],\ i = 0, \dots, f-1 \end{aligned} \tag{5.95}$$

where $f > 0$ is a positive integer, we obtain a so-called *fragmented Lyapunov-Krasovskii functional*[3] [100]:

$$\begin{aligned} V(x_t, \rho(t)) = x(t)^{\mathrm{T}} P x(t) + \sum_{i=0}^{f-1} \int_{t-(i+1)h_f(t)}^{t-ih_f(t)} x(\theta)^{\mathrm{T}} Q_i x(\theta) d\theta \\ + \bar{h}_f \sum_{i=0}^{f-1} \int_{-(i+1)\bar{h}_f}^{-i\bar{h}_f} \int_{t+\theta}^{t} \dot{x}(\eta)^{\mathrm{T}} R_i \dot{x}(\eta) d\eta d\theta \end{aligned} \tag{5.96}$$

where $h_f(t) := \dfrac{h(t)}{f}$ and $\bar{h}_f := \dfrac{\bar{h}}{f}$ and $f \in \mathbb{N}$ is the fragmentation order.

This functional gives better results than the functional (5.77) due to the presence of additional decision variables, which makes the functional more flexible, and the fact that the conservatism introduced by Jensen's inequality is effectively reduced. This last point is a consequence of the fact that fragmentation reduces the length of integration intervals. Indeed, according to Theorem 5.6.22, the gap is proportional to the measure of the interval and the variability of the integrated function over that interval. Thus, when intervals get smaller, Jensen's bounds get tighter; see [71].

5.6.9 Delay-Dependent Stability—Wirtinger's Inequality

Wirtinger's inequality has been introduced very recently in the stability analysis of sampled-data systems using functionals [107], and subsequently in the stability of time-delay systems [108] and linear impulsive systems. We briefly recall this inequality below and then provide a stability result relying upon it.

[3] Sometimes called *discretized Lyapunov-Krasovskii functional*.

5.6.9.1 Wirtinger's Inequality

There actually exist several Wirtinger's inequalities but only the following one, taken from [109], is relevant here:

Lemma 5.6.25 *Let $z : [a, b] \to \mathbb{R}^n$ be a differentiable function over (a, b) having square integrable first-order derivative and such that $z(a) = z(b) = 0$. Then, for any $Z \in \mathbb{S}^n_{\succ 0}$, the inequality*

$$\int_a^b \dot{z}^T(s) Z \dot{z}(s) ds \geq \frac{\pi^2}{(b-a)^2} \int_a^b z^T(s) Z z(s) ds \tag{5.97}$$

holds.

We can easily recognize an integral form which is similar to the one involved in the derivative of Lyapunov-Krasovskii functionals in the previous sections. The above inequality, however, cannot be directly applied to the integral terms involved in LKF derivatives due to the constraint $z(a) = z(b) = 0$ which is, in general, not satisfied for time-delay systems. To circumvent this difficulty, the following result based on Wirtinger's inequality has been proposed in [108]:

Lemma 5.6.26 *Let $\omega : [a, b] \to \mathbb{R}^n$ be a differentiable function over (a, b) having square integrable first order derivative. Then, for any $Z \in \mathbb{S}^n_{\succ 0}$, we have*

$$\int_a^b \dot{\omega}(u) Z \dot{\omega}(u) du \geq \frac{1}{b-a}(\omega(b) - \omega(a))^T Z(\omega(b) - \omega(a)) + \frac{\pi^2}{b-a} \psi^T Z \psi \tag{5.98}$$

where

$$\psi = \frac{\omega(b) + \omega(a)}{2} - \frac{1}{b-a} \int_a^b \omega(u) du.$$

Proof Let $\omega : [a, b] \to \mathbb{R}$ be a differentiable function over (a, b) and let $z(u) = (b-u)\omega(a) + (u-a)\omega(b) - (b-a)\omega(u)$ where $u \in [a, b]$. Since the function z satisfies the boundary condition $z(a) = z(b) = 0$, Lemma 5.6.25 therefore applies. The derivative of $z(u)$ with respect to u is given by:

$$\dot{z}(u) = (\omega(b) - \omega(a)) - (b-a)\dot{\omega}(u),$$

and hence the right-hand side of (5.97) writes

$$\begin{aligned}\int_a^b \dot{z}^{\mathrm{T}}(u) Z \dot{z}(u) \mathrm{d}u &= (b-a)(\omega(b)-\omega(a))^{\mathrm{T}} Z(\omega(b)-\omega(a)) \\ &\quad -(b-a)^2 \int_a^b \dot{\omega}^{\mathrm{T}}(u) Z \dot{\omega}(u) \mathrm{d}u.\end{aligned} \tag{5.99}$$

Note that the nonnegativity of the above expression immediately yields the well-known bound obtained from Jensen's inequality. From Lemma 5.6.25, we obtain

$$\begin{aligned}\int_a^b \dot{z}^{\mathrm{T}}(u) Z \dot{z}(u) \mathrm{d}u &\geq \frac{\pi^2}{(b-a)^2} \int_a^b z^{\mathrm{T}}(u) Z z(u) \mathrm{d}u \\ &\geq \frac{\pi^2}{(b-a)^3} \left(\int_a^b z(u) \mathrm{d}u \right)^{\mathrm{T}} Z \left(\int_a^b z(u) \mathrm{d}u \right)\end{aligned} \tag{5.100}$$

where the last inequality has been obtained using Jensen's inequality. By combining then (5.99) and (5.100) we get

$$\begin{aligned}\int_a^b \dot{\omega}^{\mathrm{T}}(u) Z \dot{\omega}(u) \mathrm{d}u &\geq \frac{1}{b-a}(\omega(b)-\omega(a))^{\mathrm{T}} Z(\omega(b)-\omega(a)) \\ &\quad + \frac{\pi^2}{(b-a)^5} \left(\int_a^b z(u) \mathrm{d}u \right)^{\mathrm{T}} Z \left(\int_a^b z(u) \mathrm{d}u \right).\end{aligned} \tag{5.101}$$

Noting finally that

$$\int_a^b z(u) \mathrm{d}u = (b-a)^2 \left(\frac{\omega(b)+\omega(a)}{2} - \nu_\omega(a,b) \right) \tag{5.102}$$

where $\nu_\omega(a,b) = \dfrac{1}{b-a} \displaystyle\int_a^b \omega(u) \mathrm{d}u$ yields the result. ■

It is immediate to see in the above result that the right-hand side is made of two terms: the first one is actually the bound obtained by using Jensen's inequality whereas the second term is specific to Wirtinger's inequality. Note that the integral term is negative and thus makes the upper bound tighter, improving then over Jensen's bound. The reason for this improvement is that C^1 functions are considered in Wirtinger's inequality (actually functions in the Sobolev space $W^{1,2}$) where as Jensen's inequality considers integrable functions (in our case, L_2 functions). By considering a more adapted set of functions, i.e. differentiable functions, we can see that Wirtinger's inequality allows us to obtain a more accurate bound on the integral quadratic term.

5.6.9.2 A Stability Result Using Wirtinger's Inequality

The following LKF is considered in [108]:

$$\begin{aligned}V(x_t) &= x(t)^{\mathrm{T}}Px(t) + 2x(t)^{\mathrm{T}}Q\int_{t-h}^{t} x(s)ds + h\int_{t-h}^{t} x(\theta)^{\mathrm{T}}Sx(\theta)d\theta \\ &+ \left(\int_{t-h}^{t} x(s)ds\right)^{\mathrm{T}} Z\left(\int_{t-h}^{t} x(s)ds\right) + h\int_{-h}^{0}\int_{t+\theta}^{t} \dot{x}(\eta)^{\mathrm{T}}R\dot{x}(\eta)d\eta d\theta\end{aligned} \tag{5.103}$$

where $P, S, R, Z \in \mathbb{S}^n$ and $Q \in \mathbb{R}^{n\times n}$. Using this functional, the following result can be obtained

Theorem 5.6.27 *The system (5.23) is stable for all $h \in [\bar{h}_1, \bar{h}_2]$ if there exist matrices $P, S, R, Z \in \mathbb{S}^n$ and $Q \in \mathbb{R}^{n\times n}$ such that the LMIs*

$$\begin{bmatrix} P & Q \\ \star & Z+S \end{bmatrix} \succ 0 \tag{5.104}$$

$$\begin{bmatrix} M_{11} & PA_h - Q + R & h(A^TQ+Z) \\ \star & -hS & h(A_h^TQ - Z) \\ \star & \star & 0 \end{bmatrix} + h^2\begin{bmatrix} A^T \\ A_h^T \\ 0 \end{bmatrix} R \begin{bmatrix} A^T \\ A_h^T \\ 0 \end{bmatrix}^T + \frac{1}{\bar{h}}\Psi \prec 0 \tag{5.105}$$

hold for all $h \in \{\bar{h}_1, \bar{h}_2\}$ where $M_{11} = \mathrm{He}[PA + Q] + hS - R$ and

$$\Psi = \begin{bmatrix} R & -R & 0 \\ \star & R & 0 \\ \star & \star & 0 \end{bmatrix} - \frac{\pi^2}{4}\begin{bmatrix} R & R & -2R \\ \star & R & -2R \\ \star & \star & 4R \end{bmatrix}. \tag{5.106}$$

Proof The proof relies on simple manipulations almost identical to the result based on Jensen's inequality with the difference that the inequality of Lemma 5.6.26 is used instead. A detailed proof can be found in [108]. ■

5.7 L_2 Scaled Small-Gain Theorem Based Results

Input/output approaches heavily rely on the use model transformation procedures and, as we shall see later, these model transformations are generally formulated in terms of interconnections of operators. These operator-based model transformations can be much more advanced and accurate than the ones generally used in Lyapunov approaches (even though they may be adapted to). Using such model transformations, the original time-delay system is rewritten as an uncertain system in LFT-form (or the like) which may or may not involve time-delays, and where operators are considered as uncertainties. The resulting interconnection is then analyzed using input/output analysis techniques, such as small-gain results, IQCs or topological separation arguments; see Sect. 2.6.

According to the type and combination of operators, several different stability criteria can be obtained in this framework. Moreover, it turns out that constructing operators for deriving comparison models is very often more intuitive than building an LKF. This makes the overall framework quite neat since operators can be analyzed and understood independently of the others.

It is shown in this section that several results obtained in Sect. 5.6 can be retrieved using simple operators and the scaled small-gain theorem; see Sect. 2.6.3.

5.7.1 *Delay Operators*

Several different "delay-operators" can be found in the literature. For instance, operators based on Padé approximants are considered in [37, 39], delay-difference operators in [60] and integral operators in [2, 63, 64, 110–112].

In this section, we will focus on the following simple operators

$$\begin{aligned} \mathscr{D}_h &: w(t) \to w(t-h(t)) \\ \mathscr{S}_h &: w(t) \to \int\limits_{t-h(t)}^{t} w(s)ds \end{aligned} \tag{5.107}$$

where $w \in L_2$. More advanced operators will be considered in Sects. 5.9 and 5.10.

5.7.1.1 Characterization of the Operator $\mathscr{D}_h$

This operator is the "pure-delay operator" and is certainly the simplest operator that can be considered for studying time-delay systems. We have the following result:

Proposition 5.7.1 *Let us consider the operator $\mathscr{D}_h$ defined in (5.107). Then,*

1. *when the delay is time-invariant, i.e. $\dot{h} \equiv 0$, we have*
$$||\mathscr{D}_h||_{L_2-L_2} = 1. \tag{5.108}$$

2. *when the delay is time-varying and such that $\dot{h} \leq \mu < 1$, we have*
$$||\mathscr{D}_h||_{L_2-L_2} = \frac{1}{\sqrt{1-\mu}}. \tag{5.109}$$

Proof In the constant-delay case, the operator can be described by the transfer function $\widehat{\mathscr{D}}_h(s) = e^{-hs}$ which defines a bounded-input bounded-output stable system. In this case, the L_2-gain coincides with the H_∞-norm of the transfer function and is therefore given by 1.

When the delay is time-varying, we have to evaluate the (scaled) L_2-norm of the output of the operator, that is, we look at the quantity $||X\mathscr{D}_h(w)||_{L_2}$, defined for any nonsingular matrix X. This gives

$$\begin{aligned} ||X\mathscr{D}_h(w)||_{L_2}^2 &= \int_0^\infty w(\theta - h(\theta))^{\mathrm{T}} X^{\mathrm{T}} X w(\theta - h(\theta)) d\theta \\ &= \int_{-h}^\infty w(s)^{\mathrm{T}} X^{\mathrm{T}} X w(s) \frac{ds}{1 - \dot{h}(q(s))} \end{aligned} \tag{5.110}$$

where we have used the change of variables $s = \theta - h(\theta)$ and used the fact that the function $p(t)$ defined as $s = p(t) := t - h(t)$ is invertible with inverse $q := p^{-1}$. Invertibility of p indeed follows from the fact that $p'(t) > 0$ for all $t \geq 0$ when assumed that $\dot{h} \leq \mu < 1$. Assuming further that $w(s) = 0$, $s \leq 0$, we get that

$$||X\mathscr{D}_h(w)||_{L_2}^2 \leq \frac{1}{1-\mu} ||Xw||_{L_2}^2. \tag{5.111}$$

To show that the upper-bound on the gain is tight, it is enough to pick $h(t) = \mu t + h_0$, where $h_0 > 0$, and check that, for this delay value, we indeed have

$$||\mathscr{D}_h(w)||_{L_2}^2 = \frac{1}{1-\mu} ||w||_{L_2}^2 \tag{5.112}$$

for any $w \in L_2$, $w(s) = 0$, $s \leq 0$. The proof is complete. ∎

It is easily seen that the operator $\mathscr{D}_h$ does not embed any information on the size of the delay since the norm only depends on the rate of variation of the delay. This operator alone may be therefore suitable for characterizing delay-independent stability, but not delay-dependent stability.

5.7.1.2 Characterization of the Operator $\mathscr{S}_h$

Unlike the previously studied operator, the operator $\mathscr{S}_h$ turns out to embed some information on the delay amplitude. It is therefore suitable for performing delay-dependent stability analysis.

Proposition 5.7.2 *Assume that the delay is either time-invariant or time-varying, and such that $\dot{h}(t) \leq \mu < 1$. Then, the L_2-gain of the operator $\mathscr{S}_h$ is given by*

$$||\mathscr{S}_h||_{L_2-L_2} = \bar{h} \tag{5.113}$$

where $\bar{h}$ is the value of the constant delay or the maximal value of the time-varying delay.

Proof In the constant-delay case, the operator can be described by the transfer function

$$\widehat{\mathscr{S}_h}(s) = \frac{1-e^{-\bar{h}s}}{s}.$$

Thanks to the pole/zero cancellation at 0, this transfer function has no pole in the closed right-half plane (actually no pole at all) and thus defines a bounded-input bounded-output stable system. Therefore, the L_2-gain coincides with the H_∞-norm of the transfer function which is given by $\bar{h}$.

When the delay is time-varying, the scaled L_2-norm of the signal $\mathscr{S}_h(w)$ must be evaluated and we get

$$\begin{aligned}
||X\mathscr{S}_h(w)||_{L_2}^2 &= \int_0^\infty \left(\int_{t-h(t)}^t w(s)ds \right)^\mathrm{T} X^\mathrm{T} X \left(\int_{t-h(t)}^t w(s)ds \right) dt \\
&\leq \int_0^\infty h(t) \left(\int_{t-h(t)}^t w(s)^\mathrm{T} X^\mathrm{T} X w(s) ds \right) dt \\
&\leq \bar{h} \int_0^\infty \left(\int_{t-h(t)}^t w(s)^\mathrm{T} X^\mathrm{T} X w(s) ds \right) dt
\end{aligned} \tag{5.114}$$

where the first inequality has been obtained using Jensen's inequality; see Lemma 5.6.18 and Corollary 5.6.19. The idea now is to exchange the order of integration and to do so, we define $p(t)$ as $s = p(t) := t - h(t)$. Since $\dot{h} \leq \mu < 1$, then we have that $p'(t) > 0$ for all $t \geq 0$ and thus the function p is invertible. Let $q := p^{-1}$, then we get

$$\begin{aligned}||X\mathscr{S}_h(w)||_{L_2}^2 &\leq \bar{h}\int_0^\infty\left(\int_s^{q(s)} w(s)^{\mathrm{T}}X^{\mathrm{T}}Xw(s)dt\right)ds\\ &= \bar{h}\int_0^\infty (q(s)-s)w(s)^{\mathrm{T}}X^{\mathrm{T}}Xw(s)ds\\ &\leq \bar{h}^2||Xw||_{L_2}^2\end{aligned} \tag{5.115}$$

where we have used the fact that $q(s) - s = h(q(s))$ for all $s \geq 0$. To show that the bound is tight, it is enough to consider the delay $h(t) = \bar{h}$. The proof is complete. ■

Unlike the operator $\mathscr{D}_h$, the operator $\mathscr{S}_h$ embeds information on the delay magnitude. It does not, however, explicitly capture information on the delay derivative, although *some conditions on the delay-derivative are necessary to obtain the above result*! When the delay violates the condition $\dot{h}(t) \leq \mu < 1$, the following result should be considered instead:

Lemma 5.7.3 ([110]) *Let us consider the operator*

$$\mathscr{S}_h' : w(t) \to \int_{-\bar{h}-\eta(t)}^{-\bar{h}} w(s)ds \tag{5.116}$$

where $\dot{\bar{h}} = 0$, $|\eta(t)| \leq \bar{h}$ and $\dot{\eta}(t) \leq \mu$. Then, we have that

$$||\mathscr{S}_h||_{L_2-L_2} \leq \bar{h}\sqrt{F(\mu)} \tag{5.117}$$

where

$$F(\mu) = \begin{cases} 1 & \text{if } \mu \in [0, 1)\\ 2 - \dfrac{1}{\mu} & \text{if } \mu \in [1, 2)\\ 2 - \dfrac{\mu}{4(\mu-1)} & \text{if } \mu \in [2, \infty).\end{cases}$$

The above result shows that the norm of the operator is influenced by the time-varying nature of the delay, and makes this relationship explicit. It is, however, unclear whether these bounds are tight or not. As a concluding remark, it is interesting to note that $F(\mu) \to 7/4$ as $\mu \to \infty$. This bound must therefore be used when no upper-bound on the delay-derivative can be considered.

5.7.2 Delay-Independent Stability

Delay-independent stability can be analyzed by means of the operator $\mathscr{D}_h$. The first step towards the derivation of this result lies in the reformulation of the system (5.23) as an uncertain system in LFT-form involving the operator $\mathscr{D}_h$ as uncertainty. It is immediate to see that this reformulation is given by

$$\begin{aligned} \dot{x}(t) &= Ax(t) + A_h w(t) \\ z(t) &= x(t) \\ w(t) &= \mathscr{D}_h(z)(t). \end{aligned} \tag{5.118}$$

The original system is easily retrieved by eliminating w in the above equations. By noting further that the operator $\mathscr{D}_h$ is a norm-bounded operator, D-scalings can be used to analyze stability of the system (5.118); see the list of scalings in Sect. 2.6.5. Invoking the scaled small-gain theorem yields the following results:

Theorem 5.7.4 *The system* (5.23) *with constant delay is delay-independent stable if there exist matrices* $P, L \in \mathbb{S}^n_{\succ 0}$ *such that the LMI*

$$\begin{bmatrix} A^TP + PA + L & PA_h \\ \star & -L \end{bmatrix} \prec 0 \tag{5.119}$$

holds.

Theorem 5.7.5 *The system* (5.23) *with time-varying delay satisfying* $\dot{h}(t) \leq \mu < 1$ *is delay-independent stable if there exist matrices* $P, L \in \mathbb{S}^n_{\succ 0}$ *such that the LMI*

$$\begin{bmatrix} A^TP + PA + L & PA_h \\ \star & -(1-\mu)L \end{bmatrix} \prec 0 \tag{5.120}$$

holds.

Proof The proofs are straightforward applications of the scaled small-gain theorem of Sect. 2.6.3. ∎

It is immediate to recognize the LMIs obtained using the LKF (5.41) in Sect. 5.6.1, emphasizing the link between Lyapunov-Krasovskii functionals and the use of the L_2-scaled small-gain. Note, moreover, that the necessary and sufficient condition for strong delay-independent stability of Theorem 5.6.4, taken from [76], can also be retrieved by using the augmented operator $\widetilde{\mathscr{D}}_h = I_{nk} \otimes \mathscr{D}_h$, where $\otimes$ is the Kronecker product, and symmetric matrices P, L of appropriate dimensions.

5.7.3 Delay-Dependent Stability

Delay-dependent stability analysis can be performed using the operator $\mathscr{S}_h$. With this operator, the system (5.23) reformulates as

$$\begin{aligned} \dot{x}(t) &= (A + A_h)x(t) - A_h w(t) \\ z(t) &= \bar{h}(A + A_h)x(t) - \bar{h} A_h w(t) \\ w(t) &= \frac{1}{\bar{h}} \mathscr{S}_h(z)(t) \end{aligned} \tag{5.121}$$

where the factor $1/\bar{h}$ normalizes the norm of the uncertainty. The application of the scaled small-gain theorem leads to the following result:

Theorem 5.7.6 *The system (5.23) is delay-dependent stable for all $h \in [0, \bar{h}]$ if there exist matrices $P, L \in \mathbb{S}^n_{\succ 0}$ such that the LMI*

$$\begin{bmatrix} (A + A_h)^T P + P(A + A_h) & -PA_h & \bar{h}(A + A_h)^T L \\ \star & -L & -\bar{h} A_h^T L \\ \star & \star & -L \end{bmatrix} \prec 0 \tag{5.122}$$

holds.

Proof The proof is a straightforward application of the scaled small-gain theorem. ■

Similarly to as in the delay-independent case, the above result can be connected to LKFs. Stated as it is, the LMI (5.122) is equivalent to the LMI obtained using the descriptor model transformation; see Theorem 5.6.15. By using indeed the comparison model

$$\begin{aligned} \begin{bmatrix} \dot{x}(t) \\ z(t) \end{bmatrix} &= \begin{bmatrix} A + A_h & \bar{h} A_h \\ A + A_h & \bar{h} A_h \end{bmatrix} \begin{bmatrix} x(t) \\ w(t) \end{bmatrix} \\ w(t) &= \frac{-1}{\bar{h}} \mathscr{S}_h(z)(t) \end{aligned} \tag{5.123}$$

together with the scaling $L = \bar{h} R$, $R \in \mathbb{S}^n_{\succ 0}$, we obtain the LMI (5.74). Note also that by using the LKF

$$V(x_t) = x(t)^{\mathrm{T}} P x(t) + \int_{-h}^{0} \int_{t+s}^{t} \dot{x}(\theta)^{\mathrm{T}} \bar{h} R \dot{x}(\theta) d\theta ds \tag{5.124}$$

and bounding the negative integral term in the derivative by 0, the same LMI would have been obtained.

5.7.4 *Final Remarks*

Even though connections with Lyapunov-Krasovskii functionals have been only emphasized through several examples, it is a well-known fact that many other results can be connected to scaled-small gain results, see e.g. [35]. Connections have been also pointed out in [113] in a slightly different, yet relevant, framework. It will be shown later in Sects. 5.9 and 5.10 that more general results can be related to small-gain results and, consequently, to Lyapunov-Krasovskii functionals.

5.8 QL_∞ Scaled Small-Gain Theorem Based Results

In the section above, results based on the scaled-small gain defined in terms of the L_2-norm have been obtained and connected to Lyapunov-Krasovskii functionals. In this section, we are interested in scaled small-gain results based on the L_∞-norm. The L_∞-norm that we are considering here is, however, not the usual L_∞-norm but a slightly different one which will be referred to as the QL_∞-norm in order to avoid confusion. It will be shown that the obtained results based on the QL_∞-norm may be somehow connected to Lyapunov-Razumikhin results. The use of the L_∞-norm has been proven to be very useful for the analysis of nonlinear systems through the notion of input-to-state stability (ISS); see e.g. [114–116]. More specifically, it is proved in [65] that the Lyapunov-Razumikhin Theorem can be understood as a nonlinear small-gain result in the ISS-framework. We are interested now in the linear version of this statement using the QL_∞-norm as main tool. This norm is defined as[4]

$$||w||_{QL_\infty} := \sup_{t \geq 0} ||w(t)||_2 \tag{5.125}$$

and should be contrasted with

$$||w||_{L_\infty} := \sup_{t \geq 0} ||w(t)||_\infty. \tag{5.126}$$

---

[4] Note that usually they are defined using the "essential supremum" operator. We assume here that the considered signals are bounded on all intervals of measure zero.

The QL_∞-norm defines the same space of bounded functions than the L_∞-norm since $w \in L_\infty \Leftrightarrow w \in QL_\infty$. The induced-topology is therefore identical since

$$||w||_{L_\infty} \le ||w||_{QL_\infty} \le \sqrt{n}||w||_{L_\infty}. \tag{5.127}$$

The QL_∞-norm has been first introduced in [117] in order to provide tractable conditions for peak-to-peak gain minimization. It is indeed well-known that the design of controllers that minimize the L_1-norm is a difficult problem [118, 119], except in some very particular cases; see e.g. [120, 121]. The associated induced-norm, referred here to as $*$-norm, induces the same topology as the L_∞-induced norm (the so-called L_1-norm) on the space of asymptotically stable linear systems.

Proposition 5.8.1 *For any given bounded operator H (finite L_1- and $*$-norms) mapping p inputs to q outputs, we have*

$$p^{-1/2}||H||_{L_\infty - L_\infty} \le ||H||_* \le q^{1/2}||H||_{L_\infty - L_\infty}. \tag{5.128}$$

Proof The proof follows from inequality (5.127). ∎

When the system is single-input single-output, the two norms obviously coincide. Moreover, when the dimension of the output is one, then the $*$-norm is always smaller or equal to the L_1-norm. In such circumstances, when considering the stability of an interconnection of a system H with an uncertain term Δ verifying $||\Delta||_* = ||\Delta||_{L_1}$, the use of the $*$-norm may be beneficial since $||H||_* \le ||H||_{L_1}$, authorizing then a larger set of uncertainties. It is however difficult to conclude on anything in the general multiple-input multiple-output case.

5.8.1 *-Bounded Real Lemma and QL_∞ Scaled Small-Gain Theorem

Let us consider here an LTI system H with state-space representation given by

$$\begin{aligned} \dot{x}(t) &= Ax(t) + Ew(t) \\ z(t) &= Cx(t) + Fw(t) \\ x(0) &= x_0 \end{aligned} \tag{5.129}$$

where $x, x_0 \in \mathbb{R}^n$, $w \in \mathbb{R}^p$ and $z \in \mathbb{R}^q$ are respectively the system state, the initial condition, the exogenous input and the controlled output.

A Riccati inequality approach has been proposed in [117] to compute an upper-bound on the $*$-norm. Later, a quasi-LMI solution has been proposed in [122]. We continue in the same vein and consider the matrix inequality framework.

Lemma 5.8.2 ($*$-Bounded Real Lemma, [122]) *The LTI system H with state-space representation* (5.129) *is asymptotically stable if there exist a matrix $P \in \mathbb{S}^n_{\succ 0}$ and scalars $\xi, \eta, \delta > 0$ such that the matrix inequalities*

$$\begin{bmatrix} A^T P + PA + \xi P & PE \\ \star & -\delta I \end{bmatrix} \prec 0 \tag{5.130}$$

$$\begin{bmatrix} \xi P & 0 & C^T \\ \star & (\eta - \delta) I & F^T \\ \star & \star & \eta I \end{bmatrix} \succeq 0 \tag{5.131}$$

hold. Moreover, in such a case, we have $||H||_* \le \eta$.

Proof The proof is taken from [122] and detailed below for completeness. Assume that inequalities (5.130) and (5.131) hold. Then, defining $V(t) := x(t)^{\mathrm{T}} P x(t)$, the inequality (5.130) implies that

$$\dot{V}(t) + \xi V(t) - \delta w(t)^{\mathrm{T}} w(t) \le 0$$

holds for all $x(t) \in \mathbb{R}^n$, $w(t) \in \mathbb{R}^p$. Assuming further that $x(0) = 0$ and $||w||_{QL_\infty} \le 1$, we get that

$$\begin{aligned} V(t) &\le e^{-\xi t} V(0) + \delta \int_0^t e^{-\xi(t-s)} w(s)^{\mathrm{T}} w(s) ds \\ &= \delta \int_0^t e^{-\xi(t-s)} w(s)^{\mathrm{T}} w(s) ds \\ &\le \delta ||w||_{QL_\infty} \int_0^t e^{-\xi(t-s)} ds \\ &= \delta \xi^{-1} (1 - e^{-\xi t}) ||w||_{QL_\infty} \\ &\le \delta \xi^{-1}. \end{aligned} \tag{5.132}$$

From inequality (5.131), we get that

$$\eta^{-1} \begin{bmatrix} C^{\mathrm{T}} \\ F^{\mathrm{T}} \end{bmatrix} \begin{bmatrix} C^{\mathrm{T}} \\ F^{\mathrm{T}} \end{bmatrix}^{\mathrm{T}} - \begin{bmatrix} \xi P & 0 \\ 0 & (\eta - \delta) I \end{bmatrix} \succeq 0 \tag{5.133}$$

and thus

$$\begin{aligned} z(t)^{\mathrm{T}}z(t) &\leq \eta\xi x(t)^{\mathrm{T}}Px(t) + \eta w(t)^{\mathrm{T}}(\eta - \delta)w(t) \\ &\leq \eta\delta + \eta w(t)^{\mathrm{T}}(\eta - \delta)w(t) \\ &\leq \eta^2 \end{aligned} \tag{5.134}$$

and hence $||z||_{QL_\infty} \leq \eta||w||_{QL_\infty}$. The proof is complete. ■

The $*$-Bounded Real Lemma can be viewed as a starting point for deriving the scaled small-gain theorem in the QL_∞-norm setting. To this aim, let us consider the uncertain LTI system in LFT-form:

$$\begin{aligned} \dot{x}(t) &= Ax(t) + Bw(t) \\ z(t) &= Cx(t) + Fw(t) \\ w(t) &= \Delta(z)(t) \\ x(0) &= x_0 \end{aligned} \tag{5.135}$$

where $x, x_0 \in \mathbb{R}^n$, $w \in \mathbb{R}^{n_0}$ and $z \in \mathbb{R}^{n_0}$ are the system state, the initial condition, the robustness-channel input and output, respectively. The uncertain operator Δ is assumed to be bounded, i.e. $||\Delta||_* \leq \eta^{-1}$, $\eta > 0$. Similarly as in [123], see also Sect. 2.6.5, the set of constant D-scalings associated with Δ is defined as

$$\mathcal{D}(\Delta) := \left\{ U \in \mathbb{S}^{n_0}_{\succ 0} : U^{1/2}\Delta = \Delta U^{1/2} \right\} \tag{5.136}$$

where $U^{1/2}$ is the unique positive square root of $U = U^T \succ 0$. Mixing D-scalings and the $*$-Bounded Real Lemma above yields the following result:

Theorem 5.8.3 *(QL_∞ Scaled Small-Gain Theorem) The uncertain system (5.135) is asymptotically stable if there exist symmetric matrices $P \succ 0$, $S \in \mathcal{D}(\Delta)$ and scalars $\varepsilon, \xi, \delta > 0$ such that the matrix inequalities*

$$\begin{bmatrix} A^TP + PA + \xi P & PE \\ \star & -\delta S \end{bmatrix} \prec 0 \tag{5.137}$$

$$\begin{bmatrix} \xi P & 0 & C^TS \\ \star & (\zeta - \delta)S & F^TS \\ \star & \star & \zeta S \end{bmatrix} \succeq 0 \tag{5.138}$$

hold.

Proof Following [123], and as also done in Sect. 2.6.3, we introduce a nonsingular matrix L such that $\Delta L = L\Delta$, thus $\Delta = L\Delta L^{-1}$. Incorporating the scalings in the system (5.135), we obtain the 'scaled' system

$$\begin{aligned}\dot{\tilde{x}}(t) &= A\tilde{x}(t) + BL^{-1}\tilde{w}(t)\\ \tilde{z}(t) &= LC\tilde{x}(t) + LDL^{-1}\tilde{w}(t)\end{aligned} \tag{5.139}$$

where $\tilde{w}(t) = Lw(t)$ and $\tilde{z}(t) = Lz(t)$. Substituting then the above system inside inequalities (5.130) and (5.131), and performing a congruence transformation with respect to $\mathrm{diag}(I, L)$ and $\mathrm{diag}(I, L, L)$, respectively, yield the result with $S := L^{\mathrm{T}}L \in \mathcal{D}(\Delta)$. ■

5.8.2 *Norms of Delay-Operators*

Before applying our results, it is necessary to provide theoretical results on our delay operators in the QL_∞ framework, similarly to as in the L_2 case.

Proposition 5.8.4 *The operator $\mathscr{D}_h$ defined in* (5.107) *satisfies*

$$||\mathscr{D}_h||_* = 1 \tag{5.140}$$

for any delay $h : \mathbb{R}_{\geq 0} \to \mathbb{R}_{\geq 0}$.

Proof Under the standard assumption of zero initial conditions, it is clear that

$$\sup_{t\geq 0} ||w(t - h(t))||_2^2 = \sup_{s\geq 0} ||w(s)||_2^2$$

for any $h : \mathbb{R}_{\geq 0} \to \mathbb{R}_{\geq 0}$ and any $w \in QL_\infty$. This bound is trivially seen to be attained. ■

Proposition 5.8.5 *The operator $\mathscr{S}_h$ defined in* (5.107) *satisfies*

$$||\mathscr{S}_h||_* = \bar{h} \tag{5.141}$$

for any $h : \mathbb{R}_{\geq 0} \to [0, \bar{h}]$.

Proof Considering again zero initial conditions, we have

$$\left\| \int_{t-h(t)}^{t} w(s)ds \right\|_2^2 \leq h(t) \int_{t-h(t)}^{t} ||w(s)||_2^2 ds \leq h(t)^2 \sup_{s \leq t} ||w(s)||_2^2 \tag{5.142}$$

where the first inequality has been obtained using Jensen's inequality. Hence, we have $||\mathscr{S}_h||_* \leq \bar{h}$. To see that this bound is attained, it is enough to choose the constant input signal $w \equiv 1$ and the constant delay $h \equiv \bar{h}$. ■

The main difference with L_2-norm results lies in the fact that operator gains do not depend on the delay derivative, even in the case of fast varying delays. Stability results obtained via this framework are therefore expected to be applicable to any type of delay trajectories. Note, however, that certain conditions on the delay derivative may be necessary to ensure well-posedness of the considered time-delay systems, see e.g. [124, 125].

5.8.3 Delay-Independent Stability

Using the $\mathscr{D}_h$ operator defined in (5.107), the system (5.23) can be equivalently rewritten as

$$\begin{aligned} \dot{x}(t) &= Ax(t) + Bw(t) \\ z(t) &= x(t) \\ w(t) &= \mathscr{D}_h(z)(t) \end{aligned} \tag{5.143}$$

where the operator $\mathscr{D}_h$ is considered as a norm-bounded uncertainty with $*$-norm equal to 1. By applying then Theorem 5.8.3, we get the following result for delay-independent stability:

Theorem 5.8.6 *The system* (5.23) *is asymptotically stable independently of the delay if there exist a matrix* $P \in \mathbb{S}_{\succ 0}^n$ *and a scalar* $\xi > 0$ *such that the matrix inequality*

$$\begin{bmatrix} A^TP + PA + \xi P & PB \\ \star & -\xi P \end{bmatrix} \prec 0 \tag{5.144}$$

holds.

Proof Substituting the system (5.143) into the matrix inequalities (5.137) and (5.138) with $\eta = 1$ yields

$$\begin{bmatrix} A^{\mathrm{T}}P + PA + \xi P & PB \\ \star & -\delta S \end{bmatrix} \prec 0 \tag{5.145}$$

and

$$\begin{bmatrix} \xi P & 0 & S \\ \star & (1-\delta)S & 0 \\ \star & \star & S \end{bmatrix} \succeq 0. \tag{5.146}$$

The second inequality is equivalent to $\delta \leq 1$ and $\xi P - S \succeq 0$. Choosing the best value for S, i.e. $S = \xi P$, we get that

$$\begin{bmatrix} A^{\mathrm{T}}P + PA + \xi P & PB \\ \star & -\delta \xi P \end{bmatrix} \prec 0. \tag{5.147}$$

The best value for $\delta \leq 1$ is given by $\delta = 1$, which leads to condition (5.144). The proof is complete. ■

We can recognize in the above result the matrix inequality condition for delay-independent stability obtained using the Lyapunov-Razumikhin Theorem in Sect. 5.5.1.

5.8.4 Delay-Dependent Stability

Consider now the comparison system

$$\begin{aligned} \begin{bmatrix} \dot{\tilde{x}}(t) \\ \hline z(t) \end{bmatrix} &= \left[\begin{array}{c|cc} A+B & 0 & -\bar{h}B \\ \hline I & 0 & 0 \\ A & B & 0 \end{array}\right] \begin{bmatrix} \tilde{x}(t) \\ \hline w(t) \end{bmatrix} \\ w(t) &= \operatorname{diag}\left(\mathscr{D}_h, \frac{1}{\bar{h}}\mathscr{S}_h\right)(z)(t). \end{aligned} \tag{5.148}$$

where $h(t) \in [0, \bar{h}]$ for some $\bar{h} > 0$. We obtain the following theorem:

Theorem 5.8.7 *The system (5.23) is asymptotically stable for all $h(t) \in [0, \bar{h}]$, $\bar{h} > 0$ if there exist matrices P, $Q \in \mathbb{S}^n_{\succ 0}$ and scalars $\xi > 0$, $\delta \in (0, 1)$ such that the matrix inequality*

$$\begin{bmatrix} (A+B)^T P + P(A+B) + \xi P & -\bar{h}PBA & -\bar{h}PB^2 \\ \star & -\delta(\xi P - S_1) & 0 \\ \star & \star & -\delta(1-\delta)S_1 \end{bmatrix} \prec 0 \tag{5.149}$$

holds.

Proof The D-scaling corresponding to the uncertainty structure is given by

$$S := \operatorname{diag}(S_1, S_2) \succ 0.$$

Substitution of the comparison system (5.148) into (5.137) yields that

$$\begin{bmatrix} (A+B)^{\mathrm{T}}P + P(A+B) + \xi P & 0 & -\bar{h}PB \\ \star & -\delta S_1 & 0 \\ \star & \star & -\delta S_2 \end{bmatrix} \prec 0 \tag{5.150}$$

or equivalently

$$\begin{bmatrix} (A+B)^{\mathrm{T}}P + P(A+B) + \xi P & -\bar{h}PB \\ \star & -\delta S_2 \end{bmatrix} \prec 0. \tag{5.151}$$

Substituting now the comparison system (5.148) into (5.138) with $\eta = 1$, we get that

$$\begin{bmatrix} \xi P & 0 & 0 & S_1 & A^{\mathrm{T}}S_2 \\ \star & (1-\delta)S_1 & 0 & 0 & B^{\mathrm{T}}S_2 \\ \star & \star & (1-\delta)S_2 & 0 & 0 \\ \star & \star & \star & S_1 & 0 \\ \star & \star & \star & 0 & S_2 \end{bmatrix} \succeq 0. \tag{5.152}$$

This is equivalent to saying that $\delta \leq 1$ and

$$\begin{bmatrix} \xi P - S_1 & 0 & A^{\mathrm{T}} \\ \star & (1-\delta)S_1 & B^{\mathrm{T}} \\ \star & \star & S_2^{-1} \end{bmatrix} \succeq 0 \tag{5.153}$$

which are equivalent in turn to $\delta \leq 1$ and

$$S_2^{-1} \succeq A(\xi P - S_1)^{-1}A^{\mathrm{T}} + (1-\delta)^{-1}BS_1^{-1}B^{\mathrm{T}}. \tag{5.154}$$

Inequality (5.151) is equivalent to

$$(A+B)^{\mathrm{T}}P + P(A+B) + \xi P + \delta^{-1}\bar{h}^2 PBS_2^{-1}B^{\mathrm{T}}P \prec 0 \tag{5.155}$$

and substituting the smallest possible value for S_2^{-1} defined by (5.154) yields

$$\begin{aligned} &P(A+B) + (A+B)^{\mathrm{T}}P + \xi P \\ &\quad + \delta^{-1}\bar{h}^2 PB\left((1-\delta)^{-1}BS_1^{-1}B^{\mathrm{T}} + A(\xi P - S_1)^{-1}A^{\mathrm{T}}\right)B^{\mathrm{T}}P \prec 0. \end{aligned} \tag{5.156}$$

A Schur complement yields the inequalities $\delta < 1$ and

$$\begin{bmatrix} (A+B)^{\mathrm{T}}P + P(A+B) + \xi P & -\bar{h}PBA & -\bar{h}PB^2 \\ \star & -\delta(\xi P - S_1) & 0 \\ \star & \star & -\delta(1-\delta)S_1 \end{bmatrix} \prec 0. \tag{5.157}$$

The proof is complete. ∎

This result is very similar to the one obtained with the Lyapunov-Razumikhin Theorem in Sect. 5.5.2 but does not seem to be equivalent to it.

5.9 Integral Quadratic Constraints

Time-delay systems approaches based on IQCs have been proposed in [60, 126–131]. The reader should refer to Sect. 2.6.7 or [126] for some details about the IQC approach. In this section, we will simply describe two different stability tests that can be obtained in this framework. More advanced ones may be found in the references above.

5.9.1 Delay-Independent Stability

The following proposition provides an IQC for the operator $\mathscr{D}_h$ defined in (5.107):

Proposition 5.9.1 ([60]) *Assume that $\dot{h}(t) \le \mu < 1$. Then, the operator $\mathscr{D}_h$ satisfies the IQC*

$$\int_{-\infty}^{+\infty} \begin{bmatrix} z(t) \\ \mathscr{D}_h(z)(t) \end{bmatrix}^T \begin{bmatrix} X_1 & 0 \\ 0 & -(1-\mu)X_1 \end{bmatrix} \begin{bmatrix} z(t) \\ \mathscr{D}_h(z)(t) \end{bmatrix} dt \ge 0$$

defined for any $X_1 \in \mathbb{S}^n_{\succeq 0}$ and all $z \in L_2$.

Invoking then the Kalman-Yakubovich-Popov Lemma and the IQC of Proposition 5.9.1, a delay-independent stability test can be obtained:

Theorem 5.9.2 *The time-delay system* (5.23) *with time-varying delay satisfying $\dot{h}(t) \le \mu < 1$ is asymptotically stable if there exist $P, X_1 \in \mathbb{S}^n_{\succ 0}$ such that the LMI*

$$\begin{bmatrix} A^T P + PA + X_1 & PA_h \\ \star & -(1-\mu)X_1 \end{bmatrix} \prec 0 \tag{5.158}$$

holds.

Proof Let us rewrite first the system (5.23) as

$$\begin{aligned} \dot{x}(t) &= Ax(t) + A_h w(t) \\ z(t) &= x(t) \\ w(t) &= \mathscr{D}_h(z)(t). \end{aligned} \tag{5.159}$$

From the main IQC theorem, i.e. Theorem 2.6.30, stability of the interconnected system (5.159) is ensured if

$$\begin{bmatrix} G_1(j\omega) \\ I \end{bmatrix}^* \begin{bmatrix} X_1 & 0 \\ 0 & -(1-\mu)X_1 \end{bmatrix} \begin{bmatrix} G_1(j\omega) \\ I \end{bmatrix} \prec 0 \tag{5.160}$$

for all $w \in \mathbb{R}$ and where $G_1(s) = (sI - A)^{-1}A_h$. Invoking then the Kalman-Yakubovich-Popov Lemma we get the LMI condition

$$\begin{bmatrix} A^{\mathrm{T}}P + PA & PA_h \\ \star & 0 \end{bmatrix} + \begin{bmatrix} I & 0 \\ 0 & I \end{bmatrix} \begin{bmatrix} X_1 & 0 \\ 0 & -(1-\mu)X_1 \end{bmatrix} \begin{bmatrix} I & 0 \\ 0 & I \end{bmatrix} \prec 0 \tag{5.161}$$

and the result is obtained. ∎

We can clearly recognize the delay-independent stability test obtained in the Sects. 5.6.1 and 5.7.2.

5.9.2 Delay-Dependent Stability

In order to obtain a delay-dependent stability test, the following operator is considered in [60]:

$$\mathscr{T}_h : u(t) \to u(t) - u(t - h(t)). \tag{5.162}$$

Note the difference with the operator $\mathscr{S}_h$ in (5.107) since no integral is involved. An IQC for this operator is given in the following proposition[5]:

Proposition 5.9.3 ([60]) *Suppose $h(t) \in [0, \bar{h}]$ and $\dot{h}(t) \leq \mu < 1$. Then, the operator $\mathscr{T}_h$ satisfies the IQC defined by*

$$\Pi(j\omega) = \begin{bmatrix} |\psi(j\omega)|^2 X_2 & 0 \\ 0 & -X_2 \end{bmatrix}$$

where $X_2 \in \mathbb{S}^n_{\succeq 0}$ and where $\psi(s)$ is any bounded rational transfer function satisfying

[5] Additional IQCs can be found in [60, 126].

$$\begin{aligned}|\phi(j\omega)| &> 1+\frac{1}{\sqrt{1-\mu}} && \text{if } \bar{h}|\omega| > 1+\frac{1}{\sqrt{1-\mu}}\\ |\phi(j\omega)| &> \bar{h}|\omega| && \text{if } \bar{h}|\omega| \leq 1+\frac{1}{\sqrt{1-\mu}}.\end{aligned} \tag{5.163}$$

A suitable choice for $\psi(s)$ *is given by*

$$\psi(s) = k\frac{\bar{h}^2 s^2 + c\bar{h}s}{\bar{h}^2 s^2 + a\bar{h}s + kc} \tag{5.164}$$

where $k = 1 + 1/\sqrt{1-\mu}$, $a = \sqrt{2kc}$ *and* c *is any positive real number.*

Based on the above IQC, we can formulate the following result:

Theorem 5.9.4 *The system (5.23) with time-varying delay* $h(t) \in [0, \bar{h}]$, $\dot{h}(t) \leq \mu < 1$ *is asymptotically stable if there exist* $P \in \mathbb{S}^{3n}_{\succ 0}$ *and* $X_2 \in \mathbb{S}^{n}_{\succ 0}$ *such that the LMI*

$$\begin{bmatrix} A_\psi^T P + PA_\psi + C_\psi^T X_2 C_\psi & PB_\psi + C_\psi^T X_2 D_\psi \\ \star & D_\psi^T X_2 D_\psi - X_2 \end{bmatrix} \prec 0 \tag{5.165}$$

holds where $C_\psi(sI - A_\psi)^{-1}B_\psi + D_\psi$ *is a minimal representation of the transfer function* $-\psi(s)(sI - A - A_h)^{-1}A_h$.

Proof Using the operator $\mathscr{T}_h$, we can rewrite the system (5.23) as

$$\begin{aligned}\dot{x}(t) &= (A + A_h)x(t) - A_h w(t)\\ z(t) &= x(t)\\ w(t) &= \mathscr{T}_h(z)(t).\end{aligned} \tag{5.166}$$

From the main IQC theorem, i.e. Theorem 2.6.30, stability of the interconnected system (5.166) is ensured if

$$\begin{bmatrix} G_2(j\omega) \\ I \end{bmatrix}^* \begin{bmatrix} |\psi(j\omega)|^2 X_2 & 0 \\ 0 & -X_2 \end{bmatrix} \begin{bmatrix} G_2(j\omega) \\ I \end{bmatrix} \prec 0 \tag{5.167}$$

for all $w \in \mathbb{R}$ and where $G_2(s) = (-sI + A + A_h)^{-1}A_h$. Invoking then the Kalman-Yakubovich-Popov Lemma we get the LMI condition

$$\begin{bmatrix} A_\psi^{\mathrm{T}} P + PA_\psi & PB_\psi \\ \star & 0 \end{bmatrix} + \begin{bmatrix} C_\psi & D_\psi \\ 0 & I \end{bmatrix}^{\mathrm{T}} \begin{bmatrix} X_2 & 0 \\ 0 & -X_2 \end{bmatrix} \begin{bmatrix} C_\psi & D_\psi \\ 0 & I \end{bmatrix} \prec 0. \tag{5.168}$$

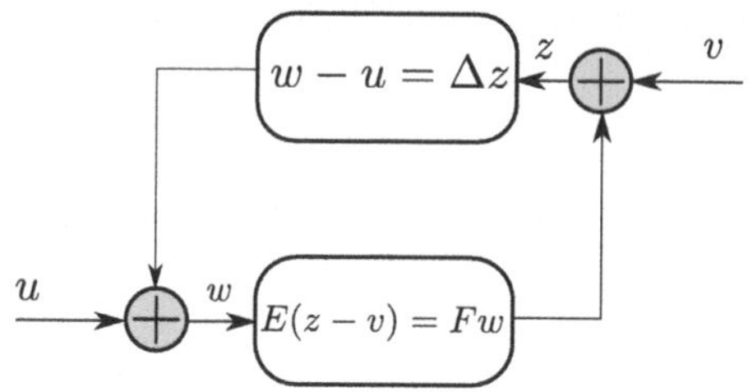

Fig. 5.1 Quadratic separation setup with implicit transformation

The proof is complete. ■

This condition is way different from the other obtained delay-dependent stability conditions due to the presence of the filter $\psi(s)$. An identical delay-dependent stability condition to the condition in Sect. 5.7.3 can be obtained using the operator $\mathscr{S}_h$ defined in (5.107) and the IQC

$$\int_0^{+\infty} \begin{bmatrix} z(t) \\ \mathscr{S}_h(z)(t) \end{bmatrix}^{\mathrm{T}} \begin{bmatrix} \bar{h}^2 X_3 & 0 \\ 0 & -X_3 \end{bmatrix} \begin{bmatrix} z(t) \\ \mathscr{S}_h(z)(t) \end{bmatrix} dt \geq 0$$

defined for any $X_3 \in \mathbb{S}^n_{\succeq 0}$ and all $z \in L_2$.

5.10 Quadratic Separation

This final section on the analysis of time-delay systems pertains on the quadratic separation framework. This framework has been mostly applied to time-delay systems in [58, 61–64, 132]. For more details on the quadratic separation framework, see Sect. 2.6.6 or [132, 133].

5.10.1 Preliminary Results

Let us consider first the system depicted in Fig. 5.1 and described by the following equations

$$\begin{aligned} w - u &= \Delta z \\ E(z - v) &= Fw \end{aligned} \tag{5.169}$$

where w, z are the interconnection signals and u, v are the input signals. The matrices E and F are assumed to be known, constant and complex-valued whereas the constant complex matrix Δ is uncertain and belongs to some set $\boldsymbol{\Delta}$. A generalization to a time-varying matrix Δ of operators can be found in [62]. A striking difference with the usual quadratic separation framework of Sect. 2.6.6 lies in the fact that the second expression of (5.169) is implicit. This will allow us to consider several operators and specify algebraic relations between their input and output signals.

The following result is proved in [132]:

Theorem 5.10.1 *Let us assume for simplicity that E is full-column rank.*[a] *Then, the interconnection* (5.169) *is well-posed if and only if there exists a Hermitian matrix X such that the LMIs*

$$\begin{bmatrix} E & -F \end{bmatrix}_{\perp}^{*} X \begin{bmatrix} E & -F \end{bmatrix}_{\perp} \succ 0 \tag{5.170}$$

and

$$\begin{bmatrix} I \\ \Delta \end{bmatrix}^{*} X \begin{bmatrix} I \\ \Delta \end{bmatrix} \preceq 0 \tag{5.171}$$

hold for all $\Delta \in \boldsymbol{\Delta}$ where $\begin{bmatrix} E & -F \end{bmatrix}_{\perp}$ stands for a basis of the null-space of $\begin{bmatrix} E & -F \end{bmatrix}$.

Additionally, when E and F are real matrices, the equivalence holds for a real matrix $X = X^T$.

[a] For the general case, see [132].

The above result will turn to be useful for developing stability results for time-delay systems.

5.10.2 Delay-Independent Stability

Let us start with a delay-independent stability test:

Theorem 5.10.2 ([134]) *The system* (5.23) *with constant delay is asymptotically stable independently of the delay if there exist $P, Q \in \mathbb{S}^n_{\succ 0}$ such that the LMI*

$$\begin{bmatrix} A^T P + PA + Q & PA_h \\ \star & -Q \end{bmatrix} \prec 0 \tag{5.172}$$

holds.

Proof Let us consider system (5.23) that we rewrite under the form (5.169) with

$$\Delta = \begin{bmatrix} s^{-1} I & 0 \\ 0 & e^{-sh} \end{bmatrix}, \quad E = \begin{bmatrix} I & 0 \\ 0 & I \end{bmatrix}, \quad F = \begin{bmatrix} A & A_h \\ I & 0 \end{bmatrix}. \tag{5.173}$$

The set $\boldsymbol{\Delta}$ is defined here as $\boldsymbol{\Delta} := \{\Delta : s \in \bar{\mathbb{C}}_+\}$.

Applying then Theorem 5.10.1, we get that condition (5.171) considered with

$$X = \left[\begin{array}{cc|cc} 0 & 0 & -P & 0 \\ 0 & -Q & 0 & 0 \\ \hline \star & \star & 0 & 0 \\ \star & \star & 0 & Q \end{array}\right] \tag{5.174}$$

with $P = P^{\mathrm{T}}$ and $Q = Q^{\mathrm{T}}$ is equivalent to the condition

$$\begin{bmatrix} -P(s^{-1} + s^{-*}) & 0 \\ 0 & Q(e^{-2\Re[s]} - 1) \end{bmatrix} \preceq 0. \tag{5.175}$$

The above condition is satisfied for all $s \in \bar{\mathbb{C}}_+$ if $P, Q \in \mathbb{S}^n_{\succ 0}$.

Considering now the condition (5.170), we get that it is equivalent to

$$\begin{bmatrix} -A^{\mathrm{T}}P - PA - Q & -PA_h \\ \star & Q \end{bmatrix} \succ 0. \tag{5.176}$$

The proof is complete. ■

We can clearly recognize, again, the delay-independent stability test obtained in Sects. 5.6.1, 5.7.2 and 5.9.1.

5.10.3 Delay-Dependent Stability

A simple delay-dependent stability analysis criterion identical to the one obtained in Sect. 5.7.3 is given below:

Theorem 5.10.3 ([134]) *The time-delay system* (5.23) *with constant delay h is asymptotically stable if there exist $P, Q, R \in \mathbb{S}^n_{\succ 0}$ such that the LMI*

$$\begin{bmatrix} He[PA] + Q + hA^TRA - h^{-1}R & PA_h + hA^TRA_h + h^{-1}R \\ \star & -Q + hA_h^TRA_h - h^{-1}R \end{bmatrix} \prec 0 \tag{5.177}$$

holds.

Proof Let us consider system (5.23) that we rewrite under the form (5.169) with

$$\Delta = \begin{bmatrix} s^{-1}I & 0 & 0 \\ 0 & e^{-sh}I & 0 \\ 0 & 0 & \dfrac{1-e^{-sh}}{s}I \end{bmatrix}, \quad E = \begin{bmatrix} I & 0 & 0 \\ 0 & I & 0 \\ I & 0 & -I \end{bmatrix}, \quad F = \begin{bmatrix} A & A_h & 0 \\ I & 0 & 0 \\ -I & I & I \end{bmatrix}. \tag{5.178}$$

The set $\boldsymbol{\Delta}$ is defined here as $\boldsymbol{\Delta} := \{\Delta : s \in \bar{\mathbb{C}}_+\}$. The rest of the proof is similar to the one of Theorem 5.10.2 with the difference that the matrix X is chosen as

$$X = \begin{bmatrix} 0 & 0 & 0 & -P & 0 & 0 \\ \star & -Q & 0 & 0 & 0 & 0 \\ \star & \star & -hR & 0 & 0 & 0 \\ \star & \star & \star & 0 & 0 & 0 \\ \star & \star & \star & \star & Q & 0 \\ \star & \star & \star & \star & \star & \frac{1}{h}R \end{bmatrix}. \tag{5.179}$$

■

References

1. S.I. Niculescu, *Delay Effects on Stability: A Robust Control Approach* (Springer, Heidelberg, 2001)
2. K. Gu, V.L. Kharitonov, J. Chen, *Stability of Time-Delay Systems* (Birkhäuser, Boston, 2003)
3. W. Michiels, S.I. Niculescu, *Stability and Stabilization of Time-Delay Systems. An Eigenvalue Based Approach.* (SIAM Publication, Philadelphia, 2007)
4. V.B. Kolmanovskii, A.D. Myshkis, *Applied Theory of Functional Differential Equations* (Kluwer, Dordrecht, 1992)
5. J.K. Hale, S.M. Verduyn Lunel, *Introduction to Functional Differential Equations* (Springer, New York, 1991)
6. O. Diekmann, S.A. van Gils, S.M. Verduyn Lunel, H.O. Walther, *Delay Equations: Functional-, Complex-, and Nonlinear Analysis* (Springer, New York, 1995)
7. N.N. Krasovskiĭ, *Stability of Motion. Applications of Lyapunov's Second Method to Differential Systems and Equations with Delay* (Translated from Russian) (Stanford University Press, Stanford, 1963)
8. B.S. Razumikhin, The application of Lyapunov's method to problems in the stability of systems with delay. Autom. Remote Control (Automatika i Telemekhanika) **21**, 740–748 (1960)
9. J.K. Hale, Suficient conditions for stability and instability of autonomous functional-differential equations. J. Differ. Equ. **1**, 452–482 (1965)
10. D.I. Barnea, A method and new results for stability and instability of autonomous functional differential equations. SIAM J. Appl. Math. **17**(4), 681–697 (1969)
11. J.R. Haddock, Y. Ko, Liapunov-Razumikhin functions and an instability theorem for autonomous functional differential equations with finite delay. Rocky Mt. J. Math. **25**(1), 261–267 (1995)
12. S. Mondié, G. Ochoa, B. Ochoa, Instability conditions for linear time delay systems: a Lyapunov matrix function approach. Int. J. Control. **84**(10), 1601–1611 (2011)
13. Y.M. Repin, Quadratic Lyapunov functionals for systems with delay. Prikl. Mat. Mekh. **29**, 564–566 (1965)

14. R. Datko, An algorithm for computing Liapunov functionals for some differential difference equations, in *Ordinary Differential Equations*, ed. by L. Weiss (Academic Press, New York, 1972), pp. 387–398
15. E.F. Infante, W.V. Castelan, A Lyapunov functional for a matrix difference-differential equation. J. Diff. Equ. **29**, 439–451 (1978)
16. V.L. Kharitonov, A.P. Zhabko, Lyapunov-Krasovskii approach to the robust stability analysis of time-delay systems. Automatica **39**, 15–20 (2003)
17. E. Fridman, Stability of systems with uncertain delays: a new 'complete' Lyapunov-Krasovskii Functional. IEEE Trans. Autom. Control **51**, 885–890 (2006)
18. Q.L. Han, K. Gu, Stability of linear systems with time-varying delay: A generalized discretized Lyapunov functional approach. Asian J. Control **3**, 170–180 (2001)
19. E. Fridman, Descriptor discretized Lyapunov functional method: analysis and design. IEEE Trans. Autom. Control **51**, 890–897 (2006)
20. A. Papachristodoulou, M. Peet, S. Lall, Constructing Lyapunov-Kraasovskii functionals for linear time delay systems. In *IEEE American Control Conference*, 2005
21. A. Papachristodoulou, M.M. Peet, S.I. Niculescu, Stability analysis of linear systems with time-varying delays: delay uncertainty and quenching. in *46th Conference on Decision and Control*, New Orleans, 2007
22. M.M. Peet, A. Papachristodoulou, S. Lall, Positive forms and stability of linear time-delay systems. SIAM J. Control Optim. **47**(6), 3237–3258 (2009)
23. M. M. Peet and A. Papachristodoulou. Inverses of positive linear operators and state feedback design for time-delay systems. In *8th IFAC Workshop on Time-Delay Systems*, Sinaia, Romania, 2009, pp. 278–283
24. M.M. Peet, P.-A. Bliman, On the conservatism of the sum-of-squares method for analysis of time-delayed systems. Automatica **47**, 2406–2411 (2011)
25. M.M. Peet, Full-state feedback of delayed system using SOS: a new theory of duality. In *11th IFAC Workshop on Time-Delay Systems*, Grenoble, 2013, pp. 24–29
26. A. Seuret, Lyapunov-Krasovskii functionals parameterized with polynomials. In *6th IFAC Symposium on Robust Control Conference (RoConD)*, Haifa, 2009, pp. 214–219
27. E.W. Kamen, P.P. Khargonekar, A. Tannenbaum, Stabilization of time-delay systems using finite dimensional compensators. IEEE Trans. Autom. Control **30**(1), 75–78 (1985)
28. R. Sipahi, S.-I. Niculescu, C.T. Abdallah, W. Michiels, K. Gu, Stability and stabilization of systems with time-delay—limitations and opportunities. IEEE Control Syst. Mag. **31**(1), 38–65 (2011)
29. C.T. Abdallah, P. Dorato, J. Benitez-Read, R. Byrne, Delayed-positive feedback can stabilize oscillatory systems. In *38th IEEE American Control Conference*, San Francisco, 1993, pp. 3106–3107
30. S.I. Niculescu, K. Gu, C.T. Abdallah, Some remarks on the delay stabilizing effect in siso systems. In *48th IEEE American Control Conference*, Denver, 2003, pp. 2670–2675
31. N. Olgac, R. Sipahi, An exact method for the stability analysis of time-delayed linear time-invariant (LTI) systems. IEEE Trans. Autom. Control **47**(5), 793–797 (2002)
32. R. Sipahi, N. Olgac, Complete stability robustness of third-order LTI multiple time-delay systems. Automatica **41**, 1413–1422 (2005)
33. R. Sipahi, N. Olgac, A unique methodology for the stability robustness analysis of multiple time delay systems. Syst. Control Lett. **55**, 819–825 (2006)
34. J. Louisell, Delay-differential systems with time-varying delay: new directions for stability theory. Kybernetika **37**, 239–251 (2001)
35. J. Zhang, C.R. Knospe, P. Tsiotras, Stability of time-delay systems: Equivalence between Lyapunov and scaled small-gain conditions. IEEE Trans. Autom. Control **46**, 482–486 (2001)
36. J. Zhang, C.R. Knospe, P. Tsiotras, Stability of linear time-delay systems: a delay-dependent criterion with a tight conservatism bound. In *38th IEEE Conference on Decision and Control*, Phoenix, 1999, pp. 4678–4683
37. M. Roozbehani, C.R. Knospe, Robust stability and $\mathcal{H}_\infty$ performance analysis of interval-dependent time-delay system. In *American Control Conference*, 2005

38. C.R. Knospe, M. Roozbehani, Stability of linear systems with interval time-delay. In *IEEE American Control Conference*, Denver, 2003, pp. 1458–1463
39. C.R. Knospe, M. Roozbehani, Stability of linear systems with interval time delays excluding zero. IEEE Trans. Autom. Control **51**, 1271–1288 (2006)
40. E. Fridman, U. Shaked, New bounded real lemma representations for time-delay systems and their applications. IEEE Trans. Autom. Control **46**(12), 1913–1979 (2001)
41. C. Briat, O. Sename, J.-F. Lafay, A LFT/$\mathcal{H}_\infty$ state-feedback design for linear parameter varying time delay systems. In *9th European Control Conference*, Kos, 2007, pp. 4882–4888
42. C. Briat, O. Sename, J.-F. Lafay, Delay-scheduled state-feedback design for time-delay systems with time-varying delays. In *17th IFAC World Congress*, Seoul, 2008, pp. 1267–1272
43. E.I. Verriest, Linear systems with rational distributed delay: reduction and stability. In *European Control Conference, ECC'99*, Karlsruhe, 1999
44. K. Gu, S.I. Niculescu, Additional dynamics in transformed time-delay systems. In *Conference on Decision and Control*, Phoenix, 1999
45. K. Gu, S.I. Niculescu, Further remarks on additional dynamics in various model transformation of linear delay systems. In *Conference on Decision and Control*, Chicago, Illinois, 2000
46. K. Gu, N.-I. Niculescu, Additional dynamics in tranformed time-delay systems. IEEE Trans. Autom. Control **45**(3), 572–575 (2000)
47. V.L. Kharitonov, D. Melchor-Aguila, Lyapunov-Krasovskii functionals for additional dynamics. Int. J. Robust Nonlinear Control **13**, 793–804 (2003)
48. V.L. Kharitonov, D. Melchor-Aguila, Additional dynamics for general class of time-delay systems. IEEE Trans. Autom. Control **48**(6), 1060–1064 (2003)
49. T.J. Su, C.G. Huang, Robust stability of delay dependence for linear unceratin systems. IEEE Trans. Autom. Control **37**, 1656–1659 (1992)
50. J.H. Su, Further results on the robust stability of linear systems with a single time delay. Syst. Control Lett. **23**, 375–379 (1994)
51. X. Li, C.E. de Souza, Criteria for robust stability and stabilization of uncertain linear systems with state delay. Automatica **33**(9), 1657–1662 (1997)
52. A. Goubet-Batholoméus, M. Dambrine, J.P. Richard, Stability of perturbed systems with time-varying delays. Syst. Control Lett. **31**, 155–163 (1997)
53. V.B. Kolmanovskii, J.P. Richard, Stability of some linear systems with delays. IEEE Trans. Autom. Control **44**, 985–989 (1999)
54. S.I. Niculescu, On some frequency-sweeping tests for delay-dependent stability: A model transformation case study. In *13th IFAC World Congress*, San Francisco, 1999, pp. 137–142
55. S.I. Niculescu, J. Chen, Frequency sweeping tests for asymptotic stability: a model transformation for multiple delays, *38th IEEE Conference on Decision and Control*, Phoenix, 1999, pp. 4678–4683
56. E. Fridman, New Lyapunov-Krasovskii functionals for stability of linear retarded and neutral type systems. Syst. Cont. Lett. **43**(4), 309–319 (2001)
57. S.-I. Niculescu, C.E. de Souza, J.-M. Dion , L. Dugard, Robust stability and stabilization for uncertain linear systems with state delay: single delay case. In *IFAC Workshop on Robust Control Design*, Rio de Janeiro, 1994, pp. 469–474
58. F. Gouaisbaut, D. Peaucelle, Stability of time-delay systems with non-small delay. In *Conference on Decision and Control*, San Diego, California, 2006, pp. 840–845
59. F. Gouaisbaut, D. Peaucelle, Delay dependent robust stability of time delay-systems. In *5th IFAC Symposium on Robust Control Design*, Toulouse, 2006
60. C.Y. Kao, A. Rantzer, Stability analysis of systems with uncertain time-varying delays. Automatica **43**, 959–970 (2007)
61. F. Gouaisbaut, D. Peaucelle, Robust stability of time-delay systems with interval delays, In *46th IEEE Conference on Decision and Control*, New Orleans, 2007
62. Y. Ariba, F. Gouaisbaut, D. Peaucelle, Stability analysis of time-varying delay systems in quadratic separation framework. In *International conference on mathematical problems in engineering, aerospace and sciences (ICNPAA'08)*, Genoa, 2008

63. Y. Ariba, F. Gouaisbaut, Input-output framework for robust stability of time-varying delay systems. In *48th Conference on Decision and Control*, Shanghai, 2009, pp. 274–279
64. Y. Ariba, F. Gouaisbaut, K.H. Johansson, Stability interval for time-varying delay systems. In *49th Conference on Decision and Control*, Atlanta, Georgia, 2010, pp. 1017–1022
65. A.R. Teel, Connections between Razumikhin-type theorems and the ISS nonlinear small-gain theorem. IEEE Trans. Autom. Control **43**(7), 960–964 (1998)
66. C. Briat, Robust stability analysis in the $*$-norm and Lyapunov-Razumikhin functions for the stability analysis and control of time-delay systems. In *50th IEEE Conference on Decision and Control*, Orlando, Florida, 2011, pp. 6319–6324
67. S. Deb, R. Srikant, Global stability of congestion controllers for the Internet. IEEE Trans. Autom. Control **48**(6), 1055–1060 (2003)
68. S. Boyd, L. El-Ghaoui, E. Feron, V. Balakrishnan, *Linear Matrix Inequalities in Systems and Control Theory* (SIAM, Philadelphia, 1994)
69. J.W. Helton Helton, I. Klep, S. McCullough, The matricial relaxation of a linear matrix inequality. Mathematical Programming, pp. 1–45, 2012
70. S. Xu, J. Lam, On equivalence and efficiency of certain stability criteria for time-delay systems. IEEE Trans. Autom. Control **52**(1), 95–101 (2007)
71. C. Briat, Convergence and equivalence results for the Jensen's inequality—application to time-delay and sampled-data systems. IEEE Trans. Autom. Control **56**(7), 1660–1665 (2011)
72. P. Park, Y.M. Moon, W.H. Kwon, A delay-dependent robust stability criterion for uncertain time-delay systems. In *43th IEEE American Control Conference*, Philadelphia, 1998
73. P. Park, A delay-dependent stability criterion for systems with uncertain time-invariant delays. IEEE Trans. Autom. Control **44**(4), 876–877 (1999)
74. E.I. Verriest, A.F. Ivanov, Robust stabilization of systems with delayed feedback. In *Proceedings of the second International Symposium on Implicit and Robust Systems*, Warszawa, 1991
75. E.I. Verriest, A.F. Ivanov, Robust stability of systems with delayed feedback. Circuits Syst. Signal Process. **13**(2–3), 213–222 (1994)
76. P.-A. Bliman, LMI characterization of the strong delay-independent stability of linear delay systems via quadratic Lyapunov-Krasovskii functionals. Syst. Control Lett. **43**, 263–274 (2001)
77. V. Blondel, A. Megretski, Unsolved Problems in Mathematical Systems and Control Theory. Princeton University Press, Princeton, 2004
78. S.-I. Niculescu, J.-M. Dion, L. Dugard, H. Li, Asymptotic stability sets for linear systems with commensurable delays: a matrix pencil approach. In *IEEE/IMACS CESA'96*, Lille, 1996
79. E.W. Kamen, On the relationship between zero criteria for two-variable polynomials and asymptotic stability of delay differential equations. IEEE Trans. Autom. Control **25**(5), 983–984 (1980)
80. E.W. Kamen, Linear systems with commensurate time delays: stability and stabilization independent of delay. IEEE Trans. Autom. Control **27**(2), 367–375 (1982)
81. E.W. Kamen, Correction to "linear systems with commensurate time delays: stability and stabilization independent of delay". IEEE Trans. Autom. Control **28**(2), 248–249 (1982)
82. J.K. Hale, E.F. Infante, F.S.P. Tsen, Stability in linear delay equations. J. Math. Analysis Appl. **115**, 533–555 (1985)
83. D. Hertz, E.I. Jury, E. Zeheb, Stability indepdendent and dependent of delay for delay differential systems. J. Franklin Inst. **318**(3), 143–150 (1984)
84. H. Shao, Improved delay-dependent stability criteria for systems with a delay varying in a range. Automatica **44**, 3215–3218 (2008)
85. S.-I. Niculescu, A model transformation class for delay-dependent stability analysis. In *American Control Conference*, San Diego, 1999, pp. 314–318
86. P. Gahinet, P. Apkarian, A linear matrix inequality approach to $\mathcal{H}_\infty$ control. Int. J. Robust Nonlinear Control **4**, 421–448 (1994)
87. Y.S. Moon, P. Park, W.H. Kwon, Y.S. Lee, Delay-dependent robust stabilization of uncertain state-delayed systems. Int. J. Control **74**, 1447–1455 (2001)

88. R.E. Skelton, T. Iwasaki, K.M. Grigoriadis, *A Unified Algebraic Approach to Linear Control Design* (Taylor and Francis, London, 1997)
89. E. Fridman, U. Shaked, A descriptor approach to $\mathcal{H}_\infty$ control of linear time-delay systems. IEEE Trans. Autom. Control **47**(2), 253–270 (2002)
90. Y. He, M. Wu, J-H. She adn G-P. Liu, Parameter-dependent Lyapunov functional for stability of time-delay systems with polytopic type uncertainties. IEEE Trans. Autom. Control **49**, 828–832 (2004)
91. K. Gu, An integral inequality in the stability problem of linear time-delay systems. In *Proceedings of the Conference on Decision Control*, Sydney, 2000
92. J.L.W.V. Jensen, Sur les fonctions convexes et les inégalités entre les valeurs moyennes. Acta Math. **30**, 175–193 (1905)
93. A-L. Cauchy, Analyse Algébrique (French) Algebraic Analysis (Edition Jacques Gabay, 1821)
94. Q.L. Han, Absolute stability of time-delay systems with sector-bounded nonlinearity. Automatica **41**, 2171–2176 (2005)
95. A. Seuret, Stability analysis for sampled-data systems with a time-varying period. In *48th Conference on Decision and Control*, Shanghai, 2009, pp. 8130–8135
96. A. Seuret, A novel stability analysis of linear systems under asynchronous samplings. Automatica **48**(1), 177–182 (2012)
97. C. Briat, A. Seuret, A looped-functional approach for robust stability analysis of linear impulsive systems. Syst. Control Lett. **61**, 980–988 (2012)
98. D. Chandler, *Introduction to Modern Statistical Mechanics* (Oxford University Press, New York, 1987)
99. S.S. Dragomir, A generalization of Grüss inequality in inner product spaces and applications. J. Math. Analy. Appl. **237**, 74–82 (1999)
100. Q.L. Han, A delay decomposition approach to stability of linear neutral systems. In *17th IFAC World Congress*, Seoul, 2008, pp. 2607–2612
101. P. Naghshtabrizi, J.P. Hespanha, A.R. Teel, Exponential stability of impulsive systems with application to uncertain sampled-data systems. Syst. Control Lett. **57**, 378–385 (2008)
102. C. Briat. Robust Control and Observation of LPV Time-Delay Systems. PhD thesis, Grenoble Institute of Technology, 2008. http://www.briat.info/thesis/PhDThesis.pdf
103. C. Briat, O. Sename, J.-F. Lafay, Parameter dependent state-feedback control of LPV time delay systems with time varying delays using a projection approach. In *17th IFAC World Congress*, Seoul, 2008, pp. 4946–4951
104. C. Briat, J.J. Martinez, Design of $\mathcal{H}_\infty$ bounded non-fragile controllers for discrete-time systems. In *48th IEEE Conference on Decision and Control*, 2009, pp. 2192–2197
105. C. Briat, J.J. Martinez, $\mathcal{H}_\infty$ bounded resilient state-feedback design for linear continuous-time systems - a robust control approach. In *18th IFAC World Congress*, Milano, 2011, pp. 9307–9312
106. X.-M. Zhang, Q.-L. Han, Robust $\mathcal{H}_\infty$ filtering for a class of uncertain linear systems with time-varying delay. Automatica **44**, 157–166 (2008)
107. K. Liu, E. Fridman, Wirtinger's inequality and Lyapunov-based sampled-data stabilization. Automatica **48**(1), 102–108 (2012)
108. A. Seuret, F. Gouaisbaut, On the use of the Wirtinger inequalities for time-delay systems. In *10th IFAC Worhshop on Time Delay Systems*, 2012, pp. 260–265
109. W.D. Kammler, *A First Course in Fourier Analysis* (Cambridge University Press, Cambridge, 2007)
110. E. Shustin, E. Fridman, On delay-derivative-dependent stability of systems with fast-varying delays. Automatica **43**, 1649–1655 (2007)
111. C. Briat, O. Sename, J.-F. Lafay, Delay-scheduled state-feedback design for time-delay systems with time-varying delays - a LPV approach. Syst. Control Lett. **58**(9), 664–671 (2009)
112. C. Briat, O. Sename, J.-F. Lafay, Memory resilient gain-scheduled state-feedback control of uncertain LTI/LPV systems with time-varying delays. Syst. Control Lett. **59**, 451–459 (2010)

113. F. Gouaisbaut, D. Peaucelle, Delay-dependent stability analysis of linear time delay systems. In *6th IFAC Workshop on Time Delay Systems*, L'Aquila, 2006, pp. 54–59
114. E.D. Sontag, The ISS philosophy as a unifying framework for stability-like behavior, in *Nonlinear Control in the Year 2000*, vol. 2, ed. by A. Isidori, F. Lamnabhi-Lagarrigue, W. Respondek, volume 259 of Lecture Notes in Control and Information Sciences (Springer, London, 2001), pp. 443–467
115. E.D. Sontag, Input to State Stability: Basic Concepts and Results, in *Nonlinear and Optimal Control Theory*, Lecture Notes in Mathematics, ed. by P. Nistri, G. Stefani (Springer, Berlin, 2008), pp. 163–220
116. S.N. Dashkovskiy, D.V. Efimov, E.D. Sontag, Input to state stability and allied system properties. Autom. Remote Control **72**(8), 1579–1614 (2011)
117. K. Nagpal, J. Abedor, K. Poolla, An LMI approach to peak-to-peak gain minimization: filtering and control. In *American Control Conference*, Baltimore, Maryland, 1994, pp. 742–746
118. M.A. Dahleh, I.J. Diaz Bobillo, *Control of Uncertain Systems - A Linear Programming Approach* (Prentice-Hall, Englewood Cliffs, 1995)
119. M.A. Dahleh, J.B. Pearson, Jr., L^1-Optimal Compensators for Continuous-Time Systems. IEEE Trans. Automat. Control **32**(10), 889–895 (1987)
120. C. Briat, Robust stability analysis of uncertain linear positive systems via integral linear constraints - L_1- and L_∞-gains characterizations. In *50th IEEE Conference on Decision and Control*, Orlando, Florida, 2011, pp. 3122–3129
121. C. Briat, Robust stability and stabilization of uncertain linear positive systems via integral linear constraints - L_1- and L_∞-gains characterizations. Int. J. Robust Nonlinear Control **23**(17), 1932–1954 (2013)
122. C.W. Scherer, P. Gahinet, M. Chilali, Multiobjective output-feedback control via LMI optimization. IEEE Trans. Autom. Control **42**(7), 896–911 (1997)
123. P. Apkarian, P. Gahinet, A convex characterization of gain-scheduled $\mathcal{H}_\infty$ controllers. IEEE Trans. Autom. Control **5**, 853–864 (1995)
124. E.I. Verriest, Well-posedness of problems involving time-varying delays. In *19th International Symposium on Mathematical Theory and Networks and Systems*, Budapest, 2010, pp. 1203–1210
125. W. Michiels, E.I. Verriest, *A Look at Fast Varying and State Dependent Delays from a System Theory Point of View* (K. U. Leuven, Internal report, 2011)
126. A. Megretski, A. Rantzer, System analysis via integral quadratic constraints. IEEE Trans. Autom. Control **42**(6), 819–830 (1997)
127. M. Fu, H. Li, S.I. Niculescu, In *Robust Stability and stabilization of time-delay systems via intergal quadractic constraint approach*, chap. 4, ed. by Dugard, Verriest, 1998, pp. 101–116
128. M. Jun, M.G. Safonov, IQC robustness analysis for time-delay systems. Int. J. Robust Nonlinear Control **11**, 1455–1468 (2001)
129. M. Jun, M.G. Safonov, Rational multiplier IQCs for uncertain time-delays and LMI stability conditions. IEEE Trans. Autom. Control **47**(11), 1871–1875 (2002)
130. C.Y. Kao, On robustness of discrete-time LTI systems with varying time delays. In *17th IFAC World Congress*, Seoul, 2008, pp. 12336–12341
131. C.-Y. Kao. Stability analysis of discrete-time systems with time-varying delays via integral quadratic constraints. In *19th International Symposium on Mathematical Theory and Networks and Systems*, Budapest, 2010, pp. 2309–2313
132. D. Peaucelle, D. Arzelier, D. Henrion, F. Gouaisbaut, Quadratic separation for feedback connection of an uncertain matrix and an implicit linear transformation. Automatica **43**, 795–804 (2007)
133. T. Iwasaki, S. Hara, Well-posedness of feedback systems: insight into exact robustness analysis and approximate computations. IEEE Trans.Autom. Control **43**, 619–630 (1998)
134. F. Gouaisbaut, D. Peaucelle, A note on stability of time-delay systems. In *5th IFAC symposium on robust control design*, 2006, pp. 555–560

Part III
Linear Parameter-Varying Time-Delay Systems

Chapter 6
Introduction to LPV Time-Delay Systems

You may delay, but time will not.
Benjamin Franklin

Abstract This chapter introduces parameter-dependent functional differential equations as a convenient way for representing linear parameter-varying time-delay systems. Particularities of these systems arising from the coupling between delays and parameters are notably pointed out. Some examples are then given to show the practical importance of linear parameter-varying time-delay systems in the modeling and control of real-world problems. Several stability results involving parameter-dependent linear matrix inequalities are finally provided. These results constitute the groundwork for developing design criteria for filters, observers and controllers for linear parameter-varying time-delay systems.

6.1 Representation of LPV Time-Delay Systems

LPV time-delay systems can be represented in many different ways. Basically, almost every combination of the representations for LPV and time-delay systems may be considered. It is clearly unnecessary to enumerate all the possibilities here since we will only be interested in parameter-dependent functional differential equations.

Such functional differential equations can be reformulated in different ways, as for LPV systems, that is, in terms of generic LPV delay-differential equations, polytopic LPV delay-differential equations and LPV delay-differential equations in LFT-form. We will be mostly interested in generic LPV delay-differential equations since they are the most flexible. Results for polytopic LPV time-delay systems will be omitted since they can be easily obtained by using the same ideas as for generic LPV time-delay systems and polytopic LPV systems; see Chaps. 2 and 3.

LPV delay-differential equations in LFT-form, however, are much more difficult to deal with for technical reasons. This can be easily explained from the facts that several

C. Briat, *Linear Parameter-Varying and Time-Delay Systems*,
Advances in Delays and Dynamics 3, DOI 10.1007/978-3-662-44050-6_6

key tools developed in the context of LPV systems cannot be applied or fail to yield tractable design criteria, e.g. by resulting in nonlinear nonconvex matrix inequalities. As an example, the dualization lemma (see Appendix C.9) can only be applied whenever a certain rank condition is satisfied. This rank condition is unfortunately not fulfilled in most of the interesting problems arising in the control of LPV time-delay systems in LFT-form. On the other hand, the projection lemma (see Appendix C.12) which successfully yields convex design conditions for the design of controllers in the LFT framework (see Sect. 3.5 or [1, 2]) fails to produce tractable design conditions, and most of time yields strongly nonconvex matrix inequalities. This is the reason why so few results in this framework have been reported in the literature. We will, nevertheless, provide some results on the design of gain-scheduled controllers for LPV time-delay systems in LFT-form together with some computational ways to solve the nonlinear resulting conditions.

In the following, we will focus on generic LPV time-delay systems of the form:

$$\begin{aligned} \dot{x}(t) &= A(\rho(t))x(t) + A_h(\rho(t))x(t-h(t)) \\ x(\theta) &= \phi(\theta), \quad \theta \in [-\bar{h}, 0] \end{aligned} \tag{6.1}$$

where $x \in \mathbb{R}^n$ is the state of the system and $\phi \in C([-\bar{h}, 0], \mathbb{R}^n)$ is the functional initial condition. The time-varying delay h will be assumed to belong to the set

$$\mathscr{H}_{\mu,\bar{h}} := \left\{h : \mathbb{R}_{\geq 0} \to [0, \bar{h}] : \dot{h}(t) \leq \mu < 1, \ t \geq 0\right\} \tag{6.2}$$

and the parameters to the set

$$\mathscr{P}^{\nu} := \left\{\rho : \mathbb{R}_{\geq 0} \to \boldsymbol{\Delta}_{\rho} : \dot{\rho}(t) \in \boldsymbol{\Delta}_{\nu}, \ t \geq 0\right\} \tag{6.3}$$

or the set

$$\mathscr{P}_1^{\nu} := \left\{\rho : \mathbb{R}_{\geq 0} \to [-1, 1]^{N_p} : \dot{\rho}(t) \in \boldsymbol{\Delta}_{\nu}, \ t \geq 0\right\} \tag{6.4}$$

where N_p is the number of parameters. By convention, the set $\mathscr{H}_{0,\bar{h}}$ is used for the constant delay case and $\mathscr{H}_{\mu,\infty}$ for the unbounded delay case. The sets $\mathscr{P}^{\infty}$ and $\mathscr{P}_1^{\infty}$ are used to consider parameters with arbitrarily fast variation rates.

6.1.1 Coupling Between Delays and Parameters

LPV time-delay systems admit slight variations compared to pure LPV systems or pure time-delay systems since delays and parameters may interact to yield delayed parameters $\rho(t-h(t))$ and parameter-dependent delays $h(\rho(t))$. Systems with parameter-dependent delays of the form

$$\begin{aligned} \dot{x}(t) &= A(\rho(t))x(t) + A_h(\rho(t))x(t-h(\rho(t))) \\ x(\theta) &= \phi(\theta), \quad \theta \in [-\bar{h}, 0] \end{aligned} \tag{6.5}$$

have been studied for instance in [3–5] whereas results on systems with delayed-parameters represented as

$$\begin{aligned} \dot{x}(t) &= A(\rho(t), \rho(t-h(t)))x(t) + A_h(\rho(t), \rho(t-h(t)))x(t-h(t)) \\ x(\theta) &= \phi(\theta), \quad \theta \in [-\bar{h}, 0] \end{aligned} \tag{6.6}$$

can be found in [6]. A difficulty when considering delayed-parameters lies in the fact that it is a priori unclear what is their exact domain of values. This will be discussed in the example of Sect. 6.2.2.

6.2 Examples

Four examples of LPV time-delay systems are presented in this section. The first one is the well-known milling process considered in [6] which is an LPV time-delay system in an intrinsic way. The second example concerns the control of an LPV system with input-delay using a gain-scheduled controller. In this setup, delayed-parameters are involved in the closed-loop system expression. The third example is about the approximation of a system with state-dependent delay as an LPV system with parameter-dependent delay. The last example finally discusses of the approximation of a nonlinear system with time-varying delays, describing a marine cooling system, into an LPV time-delay system with two parameters which depend on the state of the system. Simulations show that the devised approximation is fairly accurate.

6.2.1 *Milling Process*

Let us consider the example of a milling taken from [7] and depicted in Fig. 6.1. The corresponding model is given by

$$\begin{aligned} m_1\ddot{x}_1(t) + k_1(x_1(t) - x_2(t)) - k\sin(\phi(t)+\beta)\sin(\phi(t))[x_1(t-h) - x_1(t)] &= 0 \\ m_2\ddot{x}_2(t) + c\dot{x}_2(t) + k_1(x_2(t) - x_1(t)) + k_2x_2(t) &= 0 \end{aligned}$$

where k_1 and k_2 are the stiffness coefficients of the two springs, c is the damping coefficient, m_1 is the mass of the cutter, m_2 is the mass of the 'spindle'. The displacements of the blade and tool are x_1 and x_2, respectively. The angle β depends on the particular material and tool used, and is constant. The term $\phi(t)$ denotes the angular position of the blade at time t and k is the cutting stiffness. The delay between successive passes of the blades is denoted by $h = \pi/\omega$ where $\omega > 0$ is the angular velocity, which is assumed to be constant.

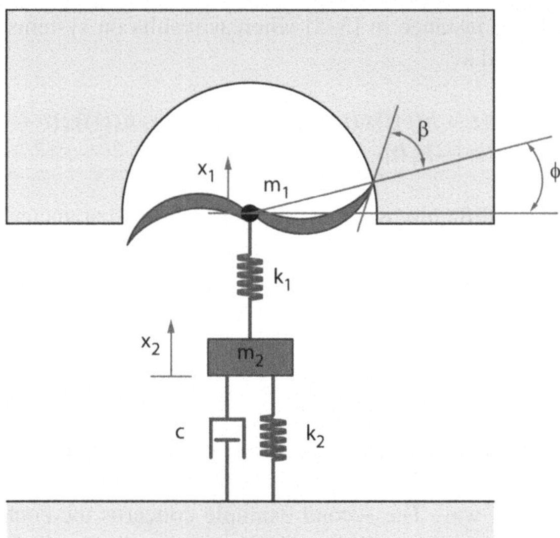

Fig. 6.1 Simplified geometry of a milling process

A suitable LPV representation of the system is therefore given by

$$\dot{x}(t) = (A_0^0 + A_0^1\rho(t))x(t) + (A_h^0 + A_h^1\rho(t))x(t-h) \tag{6.7}$$

where $\rho(t) := \cos(2\phi + \beta) \in [-1, 1]$. This model has been extensively studied in [7] and, notably, a discussion of the relation between the parameter and the delay is provided. It is indeed shown that the following relation

$$\rho(t)^2 + \left(\frac{h\dot{\rho}(t)}{2\pi}\right)^2 = 1 \tag{6.8}$$

holds for all $t \geq 0$. Therefore, as $h \to 0$, we have that $\dot{\rho}(t) \to \infty$, which shows that in order to capture the effects of small delays, we need to allow for arbitrarily large parameter variation rates. The stability of this system should therefore be analyzed in the quadratic stability framework; see Sect. 2.3.1.

6.2.2 LPV Control of a System with Input Delay

Let us consider an LPV system with input delay of the form

$$\dot{x}(t) = A(\rho(t))x(t) + B(\rho(t))u(t-h) \tag{6.9}$$

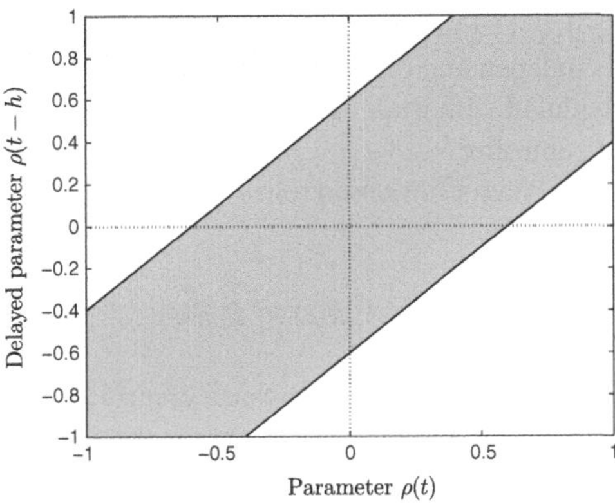

Fig. 6.2 Domain of values of $\rho(t)$ and $\rho(t-h)$ in the case $\nu h = 0.6$

where $x \in \mathbb{R}^n$ and $u \in \mathbb{R}^m$ are the state of the system and the control input, respectively. The input delay h is assumed to be constant and the parameter $\rho(t) \in [-1, 1]$ to be such that $|\dot{\rho}(t)| \leq \nu$, $\nu > 0$. A suitable controller for such a system may be given by

$$u(t) = K(\rho(t))x(t) \tag{6.10}$$

where $K : [-1, 1] \to \mathbb{R}^{m \times n}$ is the parameter-dependent controller matrix gain that has to be designed. The closed-loop system is given in this case by

$$\dot{x}(t) = A(\rho(t))x(t) + B(\rho(t))K(\rho_h(t))x(t-h) \tag{6.11}$$

and consists of an LPV time-delay system with delayed parameter $\rho_h(t) := \rho(t-h)$. It is immediate to see that, since the delay is constant, we have that $|\dot{\rho}_h(t)| \leq \nu$. Since, obviously, $\rho(t)$ and $\rho_h(t)$ are not independent, the real question is: what is the domain of values for $\rho_h(t)$? Using the fact that $|\dot{\rho}(t)| \leq \nu$, and hence that the Lipschitz constant is ν, we can state that

$$|\rho(t) - \rho_h(t)| \leq \nu h \tag{6.12}$$

and therefore that

$$-\nu h + \rho(t) \leq \rho_h(t) \leq \nu h + \rho(t) \tag{6.13}$$

for all $t \geq 0$. Thus, according to the values taken by ν, h and $\rho(t)$, the values taken by $\rho_h(t)$ may not be the complete interval $[-1, 1]$. This is, for instance, the case when $\nu h < 1$ since, in this case, when $\rho(t) = 0$, we have that $|\rho_h(t)| \leq \nu h < 1$. A typical parameter region is depicted in Fig. 6.2.

The main question here is whether it is helpful to use a gain-scheduled controller for controlling LPV systems with input delays. Clearly, when the delayed and non-delayed parameters are highly correlated, i.e. $\nu h \ll 1$, this should definitely help.

However, when $\nu h \geq 1$, $\rho(t)$ and $\rho(t-h)$ become uncorrelated in the sense that the value taken by $\rho(t)$ is independent of the value taken by $\rho(t-h)$. In such a case, the controller will be scheduled with irrelevant information and it seems unnecessary to use a gain-scheduled controller.

A way to overcome the latter limitation relies on the use of the following control-law [8]

$$\dot{u}(t) = A_u u(t) + B_u \tilde{u}(t) \tag{6.14}$$

where A_u is a Hurwitz matrix and $\tilde{u}(t)$ is the "new control input" to be determined. Rewriting then the system in augmented form, we get the following system

$$\dot{z}(t) = \begin{bmatrix} A(\rho(t)) & 0 \\ 0 & A_u \end{bmatrix} z(t) + \begin{bmatrix} 0 & B(\rho(t)) \\ 0 & 0 \end{bmatrix} z(t-h) + \begin{bmatrix} 0 \\ B_u \end{bmatrix} \tilde{u}(t) \tag{6.15}$$

where $z(t) := \text{col}(x(t), u(t))$. Seeking then for a gain-scheduled state-feedback control law of the form

$$\tilde{u}(t) = K_1(\rho(t))x(t) + K_2(\rho(t))u(t) \tag{6.16}$$

we finally get the dynamic state-feedback control-law given by

$$\dot{u}(t) = (A_u + B_u K_2(\rho(t)))u(t) + B_u K_1(\rho(t))x(t). \tag{6.17}$$

Using such a control law, we obtain the closed-loop system

$$\dot{z}(t) = \begin{bmatrix} A(\rho(t)) & 0 \\ B_u K_1(\rho(t)) & A_u + B_u K_2(\rho(t)) \end{bmatrix} z(t) + \begin{bmatrix} 0 & B(\rho(t)) \\ 0 & 0 \end{bmatrix} z(t-h) \tag{6.18}$$

where we can see that the delay does not affect the parameter vector anymore.

6.2.3 Approximation of State-Dependent Delay Systems

Let us consider here the following state-dependent delay system

$$\begin{aligned} \dot{x}(t) &= -\alpha x(t) + x(t - x(t)) \\ x(s) &= \phi(s), \;\; s \in (-\infty, 0] \end{aligned} \tag{6.19}$$

where $x \in \mathbb{R}$ is the state of the system and ϕ is the functional initial condition. Assuming that ϕ is a nonnegative function, then the trajectory solution of the above system will also be nonnegative over time. Note, moreover, that $x^* = 0$ is the unique equilibrium point to this system.

Posing now $\rho(t) = x(t)$ and $h(\rho(t)) = \rho(t)$, then the system becomes

$$\begin{aligned} \dot{x}(t) &= -\alpha x(t) + x(t - h(\rho(t))) \\ x(s) &= \phi(s), \ s \in (-\infty, 0]. \end{aligned} \tag{6.20}$$

Note, however, that the parameter ρ does not, a priori, take bounded values since 1) the system may be unstable or 2) the equilibrium point of the system may be attractive but not stable (hence the transient behavior may be such that the state goes to arbitrarily large values and then comes back at rest). By restricting the value of $\rho(t)$ to lie, for instance, in the interval $[0, \bar{x}]$ for some $\bar{x} > 0$, global asymptotic stability of the system (6.20) for all $\rho \in [0, \bar{x}]$ essentially implies that the original system (6.19) is locally asymptotically stable, provided that $x(t)$ does not goes beyond $\bar{x}$. However, if asymptotic stability can be proved for arbitrarily large $\bar{x} > 0$, the system (6.20) is therefore globally asymptotically stable.

The stability of the state-dependent delay-differential equation (6.19) has been studied, for instance, in [9, 10]. It turns out, however, that the global asymptotic stability of the above system can be easily proved using positive systems theory and the Lyapunov-Razumikhin Theorem. Let us consider indeed the linear copositive Lyapunov-Razumikhin function $V(x) = x$. Differentiating V along the solutions of (6.19) yields $\dot{V}(t) = -\alpha x(t) + x(t - x(t))$ which, according to the Lyapunov-Razumikhin Theorem, must be negative definite whenever $x(t - x(t)) \le px(t)$, $p > 1$. We then obtain that if $\alpha > 1$, then the system (6.19) is globally asymptotically stable. The same result is reported in [9, 10].

6.2.4 *LPV Model of a Marine Cooling System*

We introduce here an LPV model for the cooling system depicted in Fig. 6.3 that has been obtained in [11]. In this system, the SW circuit (sea water circuit) pumps cold sea water in order to reduce the temperature of the coolant in the LT circuit (low temperature circuit). The LT circuit contains n compartments in parallel and the supplied cooling is controlled through the input flow rates $q_{SW}(t)$ and $q_{LT}(t)$.

6.2.4.1 Nonlinear Model of the Cooling System

The dynamics of the temperature of the ith compartment can be described as

$$\dot{T}_i(t) = \frac{\alpha}{V_i}\left[q_i(t)\left[T_{in}(t - h_i(q_{LT}(t))) - T_i(t)\right] + \frac{w_i(t)}{\xi_c \theta_c}\right] \tag{6.21}$$

where q_i is the volumetric flow rate through the compartment i, V_i is the internal volume of the compartment i, T_i is the outlet temperature of the compartment i and T_{in} is the outlet temperature of the heat exchanger. The heat transfer from the compartment i to the coolant is denoted by w_i, whereas ξ_c and θ_c denote the density

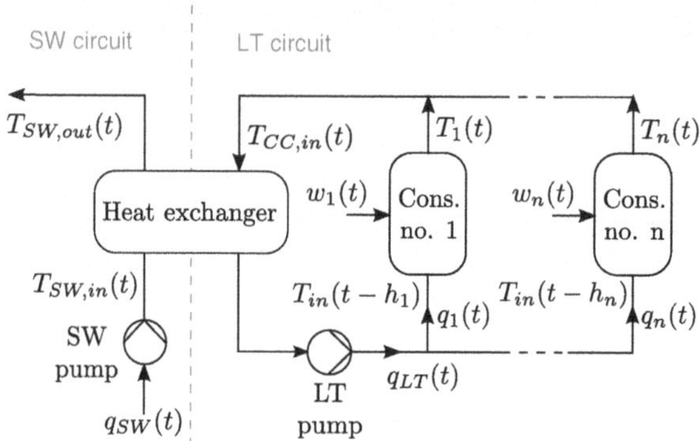

Fig. 6.3 Cooling system of [11]. The sea water circuit is to the *left* whereas the low temperature circuit is to the *right*

of the coolant and the specific heat of the coolant, respectively. The constant α is a scaling factor that depends on the system. The flows $q_1, \ldots, q_n$ are not independent and satisfy the relations $q_i(t) = c_i q_{LT}(t)$ where the positive constants c_i verify $\sum_{i=1}^{n} c_i = 1$.

The dynamical model of the outlet temperature of the heat exchanger T_{in} is given by

$$\dot{T}_{in}(t) = \frac{\alpha}{V_{CC}}\left[q_{LT}(t)\left[T_{CC,in}(t) - T_{in}(t)\right] + q_{SW}(t)\frac{\xi_{sw}\theta_{sw}}{\xi_c\theta_c}\Delta T_{SW}(t)\right] \quad (6.22)$$

where θ_{sw} is the specific heat of sea water, ξ_{sw} is the density of sea water, $T_{CC,in}$ is the temperature of the coolant into the LT side of the heat exchanger and $\Delta T_{SW} := T_{SW,in} - T_{SW,out}$ is the difference between $T_{SW,in}$ and $T_{SW,out}$, which are the temperatures of the sea water in and out of the SW side of the heat exchanger.

The time-varying transport delays can be shown to be described by the expressions

$$h_i(q_{LT}(t)) = \sum_{j=1}^{i}\left(a_{m,j}\sum_{k=j}^{n}\frac{1}{c_k q_{LT}(t)}\right) + \frac{a_{c,i}}{c_i q_{LT}(t)} \quad (6.23)$$

where the $a_{m,j}$'s and $a_{c,i}$'s are positive constants. We thus define the state vector as $x := \text{col}(T_1, \ldots, T_n, T_{in})$, the control input vector as $u := \text{col}(q_{LT}, q_{SW})$ and the disturbance vector as $w := \text{col}(w_1, \ldots, w_n, T_{CC,in})$.

6.2.4.2 LPV Approximation of the Cooling System

The system described by Eqs. (6.21) to (6.23) is clearly nonlinear. In order to obtain an LPV system, we choose $\rho_1(t) = q_{LT}(t) - \delta q_{LT}(t)$ where $\delta q_{LT}(t)$ is the perturbation around some operating point for $q_{LT}(t)$ and $\rho_2(t) = T_{SW,in}(t) - T_{SW,out}(t)$ as scheduling parameters. The model can be therefore reformulated as

$$\begin{aligned}\dot{T}_i(t) &= \frac{\alpha}{V_i}\left[c_i(\rho_1(t)+\delta q_{LT}(t))\left[T_{in}(t-h_i(q_{LT}(t)))-T_i(t)\right]+\frac{w_i(t)}{\xi_c\theta_c}\right]\\ \dot{T}_{in}(t) &= \frac{\alpha}{V_{CC}}\left[(\rho_1(t)+\delta q_{LT}(t))\left[T_{CC,in}(t)-T_{in}(t)\right]+q_{SW}(t)\frac{\xi_{sw}\theta_{sw}}{\xi_c\theta_c}\rho_2(t)\right].\end{aligned}$$

This system is not an LPV system due to the products $\delta q_{LT} T_i$ and $\delta q_{LT} T_{in}$ between the control input and the state. We then make the following approximation

$$\delta q_{LT}(t)\big[T_{in}(t-h_i(q_{LT}(t)))-T_i(t)\big]\approx\delta q_{LT}(t)\Delta_i^*$$

where $\Delta_i^* := T_{in}^* - T_i^*$, T_{in}^* and T_i^* being set-point values for T_{in} and T_i, respectively. This approximation is motivated by the fact that when the system is controlled, the state values are close to the set-point values. We also make the approximation that $q_{LT}(t)\approx\rho_1(t)$ in the dynamical model of T_{in}.

After making these approximations, we finally obtain the following LPV model for the cooling system:

$$\begin{aligned}\dot{\tilde{T}}_i(t) &= \frac{\alpha}{V_i}\left[c_i\rho_1(t)\left[\tilde{T}_{in}(t-\tilde{h}_i(\rho_1(t)))-\tilde{T}_i(t)\right]+c_i\delta q_{LT}(t)\Delta_i^*+\frac{w_i(t)}{\xi_c\theta_c}\right]\\ \dot{\tilde{T}}_{in}(t) &= \frac{\alpha}{V_{CC}}\left[\rho_1(t)\left[T_{CC,in}(t)-\tilde{T}_{in}(t)\right]+q_{SW}(t)\frac{\xi_{sw}\theta_{sw}}{\xi_c\theta_c}\rho_2(t)\right]\end{aligned}\tag{6.24}$$

where the states have been changed to $\tilde{T}_i$ and $\tilde{T}_{in}$ to emphasize that the LPV model is not equivalent to the original nonlinear one. Note that the state of the LPV model is given by $\tilde{x}=\mathrm{col}(\tilde{T}_1,\ldots,\tilde{T}_2,\tilde{T}_{in})$, the control input is given by $\tilde{u}=\mathrm{col}(\delta q_{LT},q_{SW})$ and the disturbance input by $\tilde{w}=\mathrm{col}(w_1,\ldots,w_n,T_{CC,in})$. For the LPV model, the transport delays are given by

$$\tilde{h}_i(\rho_1(t))=\sum_{j=1}^{i}\left(a_{m,j}\sum_{k=j}^{n}\frac{1}{c_k\rho_1(t)}\right)+\frac{a_{c,i}}{c_i\rho_1(t)}\tag{6.25}$$

6.2.4.3 Model Validation

For simulation purposes, we consider the following numerical values: $V_1=20$, $V_2=10$, $V_{cc}=30$, $a_{m,1}=0.06$, $a_{m,2}=0.04$, $a_{c,1}=0.05$, $a_{c,2}=0.008$, $c_1=0.65$, $c_2=0.35$, $\alpha=10000$, $\theta_c=4000$, $\theta_{sw}=3800$, $\xi_c=1000$ and $\xi_{sw}=1200$. The scheduling parameter is obtained by filtering $q_{LT}(t)$ by a first-order low-pass filter with a unitary DC-gain and a time-constant equal to 0.02. We also set $T_1^*=T_1(0)=45$, $T_2^*=T_2(0)=50$ and $T_{in}^*=T_{in}(0)=36$.

The state trajectories of the nonlinear system and LPV systems are depicted in Figs. 6.4, 6.5 and 6.6. We can see that the LPV model approximates very well the

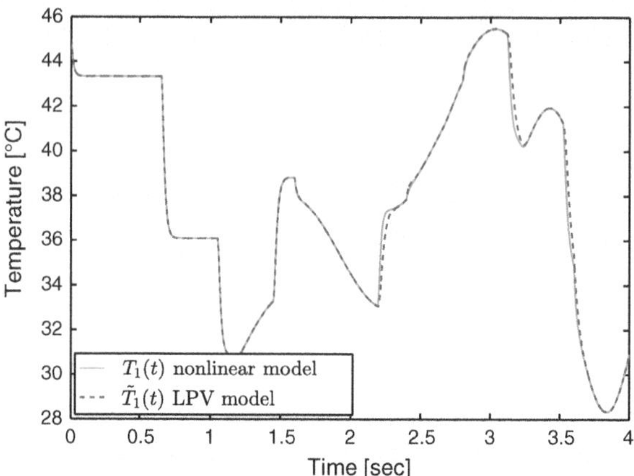

Fig. 6.4 Evolution of $T_1(t)$ of the nonlinear model and $\tilde{T}_1(t)$ of the LPV system

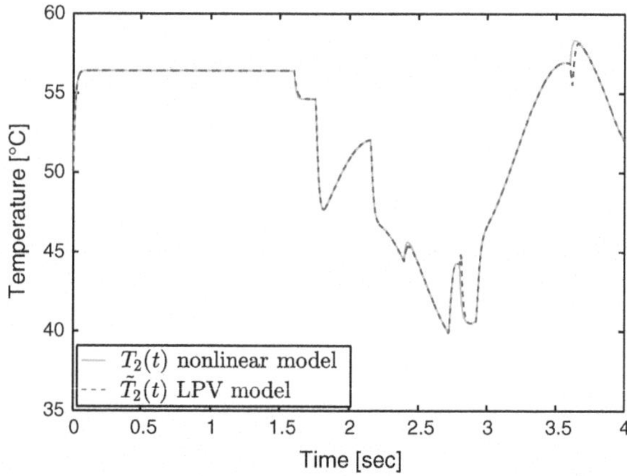

Fig. 6.5 Evolution of $T_2(t)$ of the nonlinear model and $\tilde{T}_2(t)$ of the LPV system

nonlinear system for the scenario corresponding to the signal trajectories depicted in Figs. 6.7, 6.8, 6.9 and 6.10. Other scenarios yield results of similar accuracy.

6.2.5 Other Applications

LPV time-delay can be potentially applied to any type of systems involving time-delays. In spite of this, practical applications of the theory are still quite scarce. In [12–14], LPV time-delay systems are considered for developing a robust fueling strategy for an spark ignition engine. The modeling, identification, fault-detection and control of open-flow canals using LPV time-delay systems have been considered

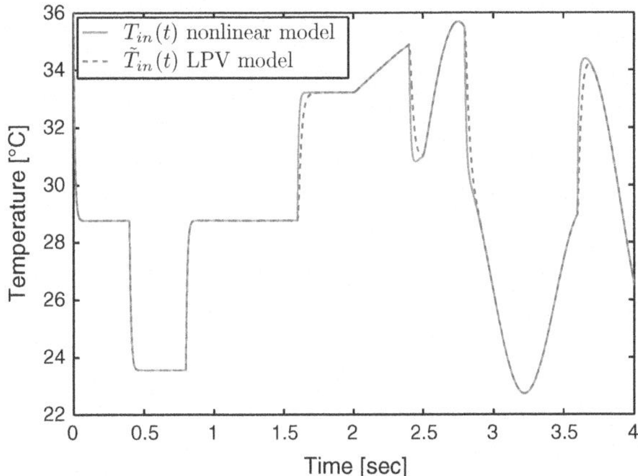

Fig. 6.6 Evolution of $T_{in}(t)$ of the nonlinear model and $\tilde{T}_{in}(t)$ of the LPV system

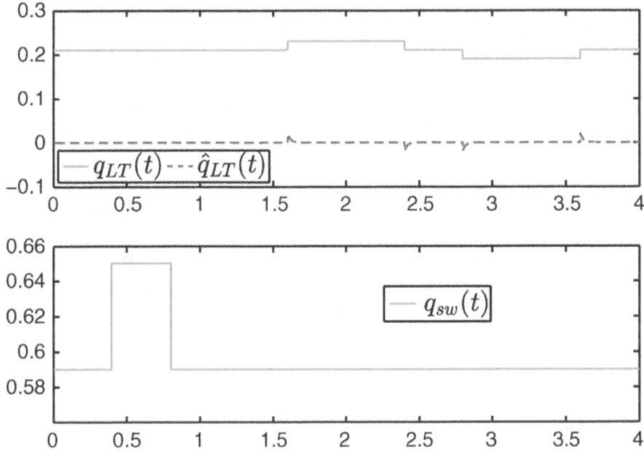

Fig. 6.7 Control inputs trajectories

in [15–21]. The problem of robust synchronization of quadratic chaotic systems with channel time-delay using LPV techniques is solved in [22]. The problems of LPV modeling, analysis and control of TCP/AQM congestion control mechanism for networks have been considered in [23–28].

6.3 Stability Results for LPV Time-Delay Systems

The analysis of LPV time-delay systems mostly relies on the use of parameter-dependent Lyapunov-Krasovskii functionals such as in [7, 8, 29, 30]. In this section, we will focus on a simple Lyapunov-Krasovskii functional from which we will derive results on observation, filtering and control. The goal of this section is not to provide the most accurate stability results for LPV time-delay systems but to expose the main

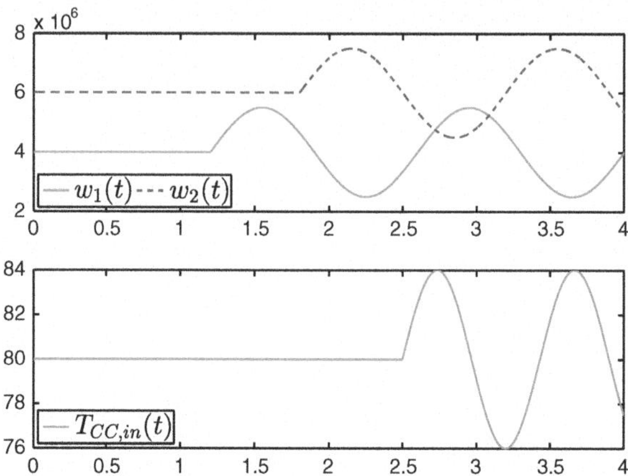

Fig. 6.8 Disturbance inputs trajectories

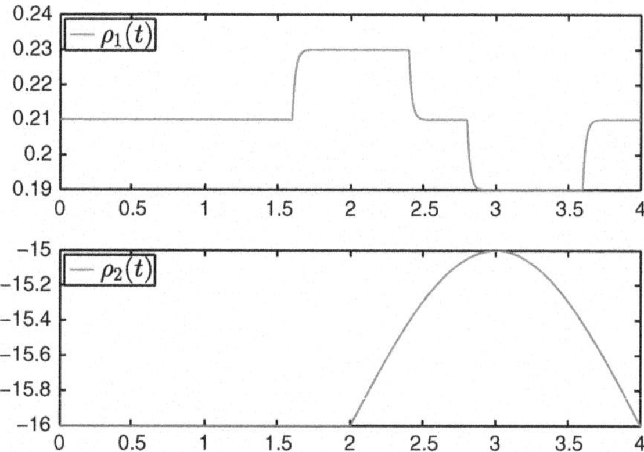

Fig. 6.9 Scheduling parameters trajectories

methodologies. More efficient results can be easily obtained by using more complex Lyapunov-Krasovskii functionals and more accurate bounding techniques such as those based on Wirtinger's inequality; see Sect. 5.6.9.

To this aim, let us then consider the following LPV time-delay system

$$\begin{aligned} \dot{x}(t) &= A(\rho(t))x(t) + A_h(\rho(t))x(t-h(t)) + E(\rho(t))w(t) \\ z(t) &= C(\rho(t))x(t) + C_h(\rho(t))x(t-h(t)) + F(\rho(t))w(t) \\ x(s) &= \phi(s),\ s \in [-\bar{h}, 0] \end{aligned} \tag{6.26}$$

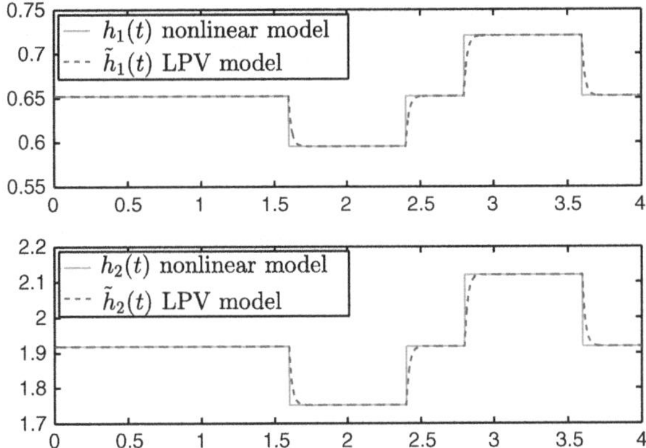

Fig. 6.10 Propagation delays trajectories

where $x \in \mathbb{R}^n$, $w \in \mathbb{R}^p$, $z \in \mathbb{R}^q$ are the state of the system, the exogenous inputs and the controlled/performance outputs, respectively. The time-varying delay and parameter vector satisfy $h \in \mathscr{H}_{\mu,\bar{h}}$ and $\rho \in \mathscr{P}^\nu$.

We have the following stability result:

Theorem 6.3.1 ([31–34]) *Assume that there exist a continuously differentiable matrix function $P : \mathbf{\Delta}_\rho \to \mathbb{S}^n_{\succ 0}$, constant matrices $Q, R \in \mathbb{S}^n_{\succ 0}$ and a scalar $\gamma > 0$ such that the LMI*

$$\begin{bmatrix} \psi_{11} & P(\rho)A_h(\rho) + R & P(\rho)E(\rho) & C(\rho)^{\mathrm{T}} & \bar{h}A(\rho)^{\mathrm{T}}R \\ \star & -(1-\mu)Q - R & 0 & C_h(\rho)^{\mathrm{T}} & \bar{h}A_h(\rho)^{\mathrm{T}}R \\ \star & \star & -\gamma I_p & F(\rho)^{\mathrm{T}} & \bar{h}E(\rho)^{\mathrm{T}}R \\ \star & \star & \star & -\gamma I_q & 0 \\ \star & \star & \star & \star & -R \end{bmatrix} \prec 0 \qquad (6.27)$$

holds for all $(\rho, \nu) \in \mathbf{\Delta}_\rho \times \mathbf{V}_\nu$ where

$$\psi_{11} = \mathrm{He}[P(\rho)A(\rho)] + \sum_i \nu_i \frac{\partial P(\rho)}{\partial \rho_i} + Q - R. \qquad (6.28)$$

Then, the system (6.26) is asymptotically stable for all $(h, \rho) \in \mathscr{H}_{\mu,\bar{h}} \times \mathscr{P}^\nu$ and the L_2-gain of the transfer $w \to z$ is less than γ.

Proof The result is obtained from the use of Lyapunov-Krasovskii functional

$$V(x_t, \rho(t)) = x(t)^{\mathrm{T}} P(\rho(t)) x(t) + \int_{t-h(t)}^{t} x(\theta)^{\mathrm{T}} Q x(\theta) d\theta + \bar{h} \int_{-\bar{h}}^{0} \int_{t+\theta}^{t} \dot{x}(\eta)^{\mathrm{T}} R \dot{x}(\eta) d\eta d\theta \tag{6.29}$$

and Jensen's inequality, as in Sect. 5.6.7. The only difference is that the result above also characterizes the L_2-gain of the transfer $w \to z$. To obtain this characterization, we rely on dissipativity theory (See Appendix C.2), and add the supply-rate $s(w, z) = -\gamma w^{\mathrm{T}} w + \gamma^{-1} z^{\mathrm{T}} z$ to the derivative of the Lyapunov-Krasovskii functional (6.29). Schur complements finally yield the result. □

The result of Theorem 6.3.1 is difficult to consider for design purposes due to the presence of multiple products between the system matrices and the decision variables. This motivates the introduction of the following result:

Theorem 6.3.2 ([31–34]) *Assume that there exist a continuously differentiable matrix function* $P : \mathbf{\Delta}_\rho \to \mathbb{S}^n_{\succ 0}$*, constant matrices* $Q, R \in \mathbb{S}^n_{\succ 0}$*, a matrix function* $X : \mathbf{\Delta}_\rho \to \mathbb{R}^{n \times n}$ *and a scalar* $\gamma > 0$ *such that the LMI*

$$\begin{bmatrix} -\mathrm{He}[X(\rho)] & \psi_{12} & X(\rho)^{\mathrm{T}} A_h(\rho) & X(\rho)^{\mathrm{T}} E(\rho) & 0 & X(\rho)^{\mathrm{T}} & \bar{h} R \\ \star & \psi_{22} & R & 0 & C(\rho)^{\mathrm{T}} & 0 & 0 \\ \star & \star & -Q_\mu - R & 0 & C_h(\rho)^{\mathrm{T}} & 0 & 0 \\ \star & \star & \star & -\gamma I_p & F(\rho)^{\mathrm{T}} & 0 & 0 \\ \star & \star & \star & \star & -\gamma I_q & 0 & 0 \\ \star & \star & \star & \star & \star & -P(\rho) & -\bar{h} R \\ \star & \star & \star & \star & \star & \star & -R \end{bmatrix} \prec 0 \tag{6.30}$$

holds for all $(\rho, \nu) \in \mathbf{\Delta}_\rho \times \mathbf{V}_\nu$ *where* $Q_\mu = (1 - \mu) Q$ *and*

$$\begin{aligned} \psi_{12} &= P(\rho) + X(\rho)^{\mathrm{T}} A(\rho), \\ \psi_{22} &= \sum_i \nu_i \frac{\partial P(\rho)}{\partial \rho_i} - P(\rho) + Q - R. \end{aligned} \tag{6.31}$$

Then, the system (6.26) is asymptotically stable and the L_2*-gain of the transfer* $w \to z$ *is less than* γ *for all* $(h, \rho) \in \mathscr{H}_{\mu, \bar{h}} \times \mathscr{P}^\nu$.

Proof The proof is inspired from [33, 35]. The first step is to decompose the LMI (6.30) as

$$\mathcal{M}(\rho,\nu) + \mathrm{He}\left[\mathcal{P}(\rho)^{\mathrm{T}} X(\rho)\mathcal{Q}\right] \prec 0 \tag{6.32}$$

where

$$\mathcal{M}(\rho,\nu) = \begin{bmatrix} 0 & P(\rho) & 0 & 0 & 0 & 0 & \bar{h}R \\ \star & \psi_{22} & R & 0 & C(\rho)^{\mathrm{T}} & 0 & 0 \\ \star & \star & -Q_{\mu}-R & 0 & C_h(\rho)^{\mathrm{T}} & 0 & 0 \\ \star & \star & \star & -\gamma I_p & F(\rho)^{\mathrm{T}} & 0 & 0 \\ \star & \star & \star & \star & -\gamma I_q & 0 & 0 \\ \star & \star & \star & \star & \star & -P(\rho) & -\bar{h}R \\ \star & \star & \star & \star & \star & \star & -R \end{bmatrix}$$

$$\mathcal{P}(\rho) = \begin{bmatrix} -I_n & A(\rho) & A_h(\rho) & E(\rho) & 0_{n\times q} & I_n & 0_{n\times n} \end{bmatrix}$$

$$\mathcal{Q} = \begin{bmatrix} I_n & 0_{n\times n} & 0_{n\times n} & 0_{n\times p} & 0_{n\times q} & 0_{n\times n} & 0_{n\times n} \end{bmatrix}.$$

Bases of the null-spaces of $\mathcal{P}(\rho)$ and $\mathcal{Q}$ are given by

$$\mathcal{N}_{\mathcal{P}}(\rho) = \begin{bmatrix} A(\rho) & A_h(\rho) & E(\rho) & I_n & 0 & 0 \\ I_n & 0 & 0 & 0 & 0 & 0 \\ 0 & I_n & 0 & 0 & 0 & 0 \\ 0 & 0 & I_p & 0 & 0 & 0 \\ 0 & 0 & 0 & 0 & I_q & 0 \\ 0 & 0 & 0 & I_n & 0 & 0 \\ 0 & 0 & 0 & 0 & 0 & I_n \end{bmatrix}$$

and

$$\mathcal{N}_{\mathcal{Q}} = \begin{bmatrix} 0 & 0 & 0 & 0 & 0 & 0 \\ I_n & 0 & 0 & 0 & 0 & 0 \\ 0 & I_n & 0 & 0 & 0 & 0 \\ 0 & 0 & I_p & 0 & 0 & 0 \\ 0 & 0 & 0 & I_q & 0 & 0 \\ 0 & 0 & 0 & 0 & I_n & 0 \\ 0 & 0 & 0 & 0 & 0 & I_n \end{bmatrix}.$$

Since (6.32) holds, the matrices

$$\mathcal{M}_{\mathcal{P}} := \mathcal{N}_{\mathcal{P}}(\rho)^{\mathrm{T}}\mathcal{M}(\rho,\nu)\mathcal{N}_{\mathcal{P}}(\rho)$$

$$= \begin{bmatrix} \tilde{\psi}_{11} & \tilde{\psi}_{12} & P(\rho)E(\rho) & C(\rho)^{\mathrm{T}} & P(\rho) & \bar{h}A(\rho)^{\mathrm{T}}R \\ \star & \tilde{\psi}_{22} & 0 & C_h(\rho)^{\mathrm{T}} & 0 & \bar{h}A_h(\rho)^{\mathrm{T}}R \\ \star & \star & -\gamma I_p & F(\rho)^{\mathrm{T}} & 0 & \bar{h}E(\rho)^{\mathrm{T}}R \\ \star & \star & \star & -\gamma I_q & 0 & 0 \\ \star & \star & \star & \star & -P(\rho) & 0 \\ \star & \star & \star & \star & \star & -R \end{bmatrix}$$

$$\begin{aligned} \mathcal{M}_{\mathcal{Q}} &:= \mathcal{N}_{\mathcal{Q}}^{\mathrm{T}}\mathcal{M}(\rho,\nu)\mathcal{N}_{\mathcal{Q}} \\ &= \begin{bmatrix} \psi_{22}(\rho,\nu) & R & C(\rho)^{\mathrm{T}} & 0 & 0 & 0 \\ \star & -Q_\mu - R & C_h(\rho)^{\mathrm{T}} & 0 & 0 & 0 \\ \star & \star & -\gamma I_p & F(\rho)^{\mathrm{T}} & 0 & 0 \\ \star & \star & \star & -\gamma I_q & 0 & 0 \\ \star & \star & \star & \star & -P(\rho) & -\bar{h}R \\ \star & \star & \star & \star & \star & -R \end{bmatrix} \end{aligned}$$

where

$$\begin{aligned} \tilde{\psi}_{11} &= \mathrm{He}[P(\rho)A(\rho)] + \sum_i \nu_i \frac{\partial P(\rho)}{\partial \rho_i} + Q - R - P(\rho) \\ \tilde{\psi}_{12} &= P(\rho)A_h(\rho) + R \\ \tilde{\psi}_{22} &= -Q_\mu - R \end{aligned}$$

are therefore both negative definite as well. Noting, finally, that $\mathcal{M}_{\mathcal{P}}$ is equivalent to (6.30) modulo a Schur complement proves that the feasibility of (6.30) implies stability of the system (6.26) and boundedness of the L_2-gain. □

It is important to stress that Theorem 6.3.1 is not equivalent to Theorem 6.3.2 since the condition of Theorem 6.3.2 also implies the feasibility of the auxiliary LMI $\mathcal{M}_{\mathcal{Q}} \prec 0$. This additional LMI restricts the domain of the decision variables and is responsible of the conservatism of the relaxed LMI (6.30) over the initial LMI (6.27). Several other relaxation procedures exist and some of them does not exhibit any additional conservatism; see e.g. [36]. Nevertheless, in many cases, the conditions have to be approximated at some point in order to get tractable synthesis conditions. This is, however, not the case of the one considered here.

6.3.1 Case of Parameter-Dependent Delay

When the delay depends on the parameters, the term $-Q_\mu = -(1-\mu)Q$ in the LMI conditions (6.27) and (6.30) must be replaced by

$$-\left(1-\sum_i \nu_i \frac{\partial h(\rho)}{\partial \rho_i}\right)Q. \tag{6.33}$$

Note that, in this case, the delay-derivative is bounded from above by

$$\dot{h}(t) \leq \sup_{(\rho,\dot{\rho})\in\mathbf{\Delta}_\rho\times\mathbf{\Delta}_\nu} \left\{\sum_i \dot{\rho}_i \frac{\partial h(\rho)}{\partial \rho_i}\right\}.$$

6.3.2 *Case of Delayed Parameters*

For simplicity, we consider here that the rate of variation of the parameters belong to $\mathbf{\Delta}_\nu = [-1, 1]^N$, where N is the number of parameters, and that the delay h is constant. In such a case, it may be preferable to consider a matrix function P that is both a function of $\rho(t)$ and $\rho(t-h)$ and, in a similar way, the matrix Q should also be made parameter-dependent.

Before stating the result, it is convenient to consider the sets

$$\mathscr{P}_h := \left\{(\rho, \rho_h) : \mathbb{R}_{\geq 0} \to \mathbf{\Delta}_\rho \times \mathbf{\Delta}_\rho : \ \rho \in \mathscr{P}^\nu, \ \rho_h(t) = \rho(t-h), \ t \geq 0\right\} \tag{6.34}$$

which contains the trajectories of the parameters and

$$\mathbf{\Delta}_\rho^h := \left\{(\rho, \rho_h) \in \mathbf{\Delta}_\rho \times \mathbf{\Delta}_\rho : \ |\rho_i - \rho_{hi}| \leq \bar{h}, \ i = 1, \ldots, N\right\} \tag{6.35}$$

which contains the parameter and the delayed parameter values. We then have the following result:

Theorem 6.3.3 *Assume that there exist a continuously differentiable matrix function* $P : \mathbf{\Delta}_\rho^h \to \mathbb{S}_{\succ 0}^n$, *a matrix function* $Q : \mathbf{\Delta}_\rho \to \mathbb{S}_{\succ 0}^n$, *a constant matrix* $R \in \mathbb{S}_{\succ 0}^n$ *and a scalar* $\gamma > 0$ *such that the LMI*

$$\begin{bmatrix} \psi_{11} & \psi_{12} & P(\rho,\rho_h)E(\rho,\rho_h) & C(\rho,\rho_h)^{\mathrm{T}} & \bar{h}A(\rho,\rho_h)^{\mathrm{T}}R \\ \star & -Q(\rho_h)-R & 0 & C_h(\rho,\rho_h)^{\mathrm{T}} & \bar{h}A_h(\rho,\rho_h)^{\mathrm{T}}R \\ \star & \star & -\gamma I_p & F(\rho,\rho_h)^{\mathrm{T}} & \bar{h}E(\rho,\rho_h)^{\mathrm{T}}R \\ \star & \star & \star & -\gamma I_q & 0 \\ \star & \star & \star & \star & -R \end{bmatrix} \prec 0$$

holds for all $(\rho, \rho_h) \in \mathbf{\Delta}_\rho^h$ *and all* $(\nu, \nu_h) \in \{-1, 1\}^N \times \{-1, 1\}^N$ *where*

$$\psi_{11} = \text{He}[P(\rho, \rho_h)A(\rho, \rho_h)] + \sum_{i=1}^{N} \left(\nu_i \frac{\partial P(\rho, \rho_h)}{\partial \rho_i} + \nu_{hi} \frac{\partial P(\rho, \rho_h)}{\partial \rho_{hi}} \right)$$
$$+ Q(\rho) - R$$
$$\psi_{12} = P(\rho, \rho_h)A_h(\rho, \rho_h) + R.$$

Then, the system (6.26) is asymptotically stable for all $h \in \mathscr{H}_{0,\bar{h}}$ and $(\rho, \rho_h) \in \mathscr{P}_h$ and the L_2-gain of the transfer $w \to z$ is less than γ.

The relaxed version of this result follows from the same lines as for Theorem 6.3.1. Note also that if the delay is time-varying, terms depending on the time-derivative of the delay will then appear in ψ_{11}.

References

1. P. Apkarian, P. Gahinet, A convex characterization of gain-scheduled $\mathcal{H}_\infty$ controllers. IEEE Trans. Autom. Control **5**, 853–864 (1995)
2. A. Packard, Gain scheduling via Linear fractional transformations. Syst. Control Lett. **22**, 79–92 (1994)
3. F. Wu, Delay dependent induced $_2$-norm analysis and control for LPV systems with state delays. in *ASME International Mechanical Engineering Congress and Exposition*, pp. 1549–1554, 2001
4. K. Tan, K.M. Grigoriadis, F. Wu, $\mathcal{H}_\infty$ and $\mathcal{L}_2$-to-$\mathcal{L}_\infty$ gain control of linear parameter varying systems with parameter varying delays. Control Theory Appl. **150**, 509–517 (2003)
5. J. Mohammadpour and K. M. Grigoriadis, Rate-dependent mixed $\mathcal{H}_2/\mathcal{H}_\infty$ filtering for time varying state delayed LPV systems. in *Conference on Decision and Control*, San Diego, USA, 2006
6. J. Zhang, C. R. Knospe, and P. Tsiotras, Stability of linear time-delay systems: a delay-dependent criterion with a tight conservatism bound. in *38th IEEE Conference on Decision and Control*, pp. 4678–4683, Phoenix, AZ, USA, 1999
7. X. Zhang, P. Tsiotras, C. Knospe, Stability analysis of LPV time-delayed systems. Int. J. Control **75**, 538–558 (2002)
8. F. Wu, K.M. Grigoriadis, LPV systems with parameter-varying time delays: analysis and control. Automatica **37**, 221–229 (2001)
9. I. Györi, F. Hartung, Exponential stability of a state-dependent delay system. Disrete Continuous Dyn. Syst. **18**(4), 773–791 (2007)
10. A.R. Humphries, O.A. DeMasi, F.M.G. Magpantay, F. Upham, Dynamics of a delay differential equation with multiple state-dependent delays. Discrete Continuous Dyn. Syst. **32**(8), 2701–2727 (2012)
11. M. Hansen, J. Stoustrup, J.D. Bendtsen, An LPV model for a marine cooling system with transport delays, in *14th International Conference on Harbor, Maritime & Multimodal Logistics Modelling and Simulation*, Italy, Rome, 2011, pp. 119–124
12. F. Zhang, K. M. Grigoriadis, M. A. Franchek, I. H. Makki, Linear parameter-varying lean burn air-fuel ratio control. in *44th IEEE Conference on Decision and Control*, Seville, Spain, 2005, pp. 2688–2683
13. F. Zhang, K.M. Grigoriadis, M.A. Franchek, I.H. Makki, Linear parameter-varying lean burn air-fuel ratio control for spark ignition engine. IEEE Trans. Circuits Syst. I Regul. Pap. **56**(3), 604–615 (2009)

14. R. Zope, J. Mohammadpour, K. Grigoriadis, M. Franchek, Robust fueling strategy for an SI enfine modeled as a linear parameter varying time-delayed system. in *American Control Conference*, Baltimore, USA, 2010, pp. 4364–4639
15. G. Belforte, F. Dabbene, P. Gay, LPV approximation of distributed parameter systems in an environmental modelling. Environ. Model. Softw. **20**, 1063–1070 (2005)
16. V. Puig, J. Quevedo, T. Escobet, P. Charbonnaud, and E. Duviella, Identification and control of an open-flow canal using LPV models. in *44th IEEE Conference on Decision and Control*, pp. 1893–1898, Seville, Spain, 2005
17. J. Blesa, V. Puig, Y. Bolea, Fault detection using interval LPV models in an open-flow canal. Control Eng. Pract. **18**, 460–470 (2010)
18. Y. Bolea, V. Puig, J. Blesa, Gain-scheduled smith proportional-integral derivative controllers for linear parameter varying first-order plus time-varying delay systems. IET Control Theory Appl. **5**(18), 2142–2155 (2011)
19. Y. Bolea, J. Blesa, and V. Puig. LPV modelling and identification of an open-flow canal for control. in *20th Mediterranean Conference on Control & Automation*, Barcelona, Spain, 2012, pp. 427–432
20. Y. Bolea, N. Chedfor, and A. Grau, MIMO LPV state-space identification of open-flow irrigation canal systems. *Mathematical Problems in Engineering*, Vol. 2012, 2012. Article ID 948936
21. Y. Bolea and V. Puig, Gain-scheduled smith predictor PID-based LPV controller for open-flow canal control. IEEE Trans. Control Syst. Technol. **22**(2), 468–477 (2014)
22. Y. Liang, H.J. Marquez, Robust gain scheduling synchronization method for quadratic chaotic systems with channel time delay. IEEE Trans. Circuits Syst. I Regul. Pap. **56**(3), 604–615 (2009)
23. H. Naitou, T. Azuma, and M. Fujita, Stability analysis for TCP/AQM networks using hybrid system representations. in *41st SICE Annual Conference*, Osaka, Japan, 2002, pp. 1171–1174
24. H. Naito, T. Azuma, M. Fujita, A scheduled controller design of congestion control considering network resource constraints. Electr. Eng. Jpn **149**(1), 1053–1059 (2004)
25. M. Kishi, Y. Funatsu, T. Azuma, and K. Uchida, Gain-scheduling controller design for TCP/AQM considering router queue length and time-delay. in *31th Annual Conference of the IEEE Industrial Electronics Society*, Raleigh, North-Carolina, USA, 2005, pp. 395–399
26. T. Azuma, H. Naitou, M. Fujita, Experimental verification of stabilizing congestion controllers using the network testbed. in *American Control Conference*, Portland, Oregon, USA, 2005, pp. 1841–1846
27. T. Azuma, M. Uchida, A design of gain-scheduled congestion controllers using state predictive observers. Int. J. smart sens. intell. syst. **1**(1), 208–219 (2008)
28. T. Azuma, H. Hirano, and M. Fujita, Stabilizing congestion controller synthesis and development of a computer network testbed. Research report
29. J. Mohammadpour, K.M. Grigoriadis, Less conservative results of delay-depdendent $\mathcal{H}_\infty$ filtering for a class of time-delayed LPV systems. Int. J. Control **80**(2), 281–291 (2007)
30. C. Briat, O. Sename, and J.-F. Lafay, Memory resilient gain-scheduled state-feedback control of time-delay systems with time-varying delays. in *6th IFAC Symposium on Robust Control Design*, Haifa, Israel, 2009, pp. 202–207
31. C. Briat, *Robust Control and Observation of LPV Time-Delay Systems* (Grenoble Institute of Technology, France, 2008). http://www.briat.info/thesis/PhDThesis.pdf
32. C. Briat, O. Sename, and J.-F. Lafay, $\mathcal{H}_\infty$ Filtering of Uncertain LPV systems with time-delays. in *10th European Control Conference* (Budapest, Hungary, 2009)
33. C. Briat, O. Sename, J.-F. Lafay, Memory resilient gain-scheduled state-feedback control of uncertain LTI/LPV systems with time-varying delays. Syst. Control Lett. **59**, 451–459 (2010)
34. C. Briat, O. Sename, and J.-F. Lafay, Design of LPV observers for LPV Time-Delay Systems: An Algebraic Approach. Int. J. Control **84**(9), 1533–1542 (2011)
35. H.D. Tuan, P. Apkarian, T.Q. Nguyen, Robust and reduced order filtering: new LMI-based characterizations and methods. IEEE Trans. Signal Process. **49**, 2875–2984 (2001)

36. R. Zope, J. Mohammadpour, K. Grigoriadis, M. Franchek, Delay-dependent output feedback control of time-delay LPV systems, in *Control of Linear Parameter Varying Systems with Applications*, ed. by J. Mohammadpour, C.W. Scherer (Springer, US, 2012), pp. 279–299

Chapter 7
Observation and Filtering of LPV Time-Delay Systems

No phenomenon is a real phenomenon until it is an observed phenomenon.

John Archibald Wheeler

Abstract This chapter pertains of the observation and filtering of linear parameter-varying time-delay systems in the framework of parameter-dependent delay-differential equations and Lyapunov-Krasovskii functionals. Full-order and reduced order observers are first considered both in the memoryless and with-memory cases. Filters are discussed next. The results of this chapter have both corollaries in the non-delayed LPV systems and parameter-independent time-delay systems settings, and can thus be applied on these types of systems. Several examples with simulations are given for illustration.

7.1 Observation of LPV Time-Delay Systems

In this section, the following class of LPV time-delay systems will be considered

$$
\begin{aligned}
\dot{x}(t) &= A(\rho(t))x(t) + A_h(\rho(t))x(t-h(t)) + B(\rho(t))u(t) + E(\rho(t))w(t) \\
y(t) &= Cx(t) \\
z(t) &= Tx(t) \\
x(\theta) &= \psi(\theta),\ \theta \in \left[-\bar{h}, 0\right]
\end{aligned} \tag{7.1}
$$

where $x \in \mathbb{R}^n$, $w \in \mathbb{R}^m$, $u \in \mathbb{R}^q$, $z \in \mathbb{R}^r$, $y \in \mathbb{R}^s$ and $\psi \in C([-\bar{h}, 0], \mathbb{R}^n)$ are the state of the system, the exogenous input, the known input, the output to estimate, the measured output and the functional initial condition, respectively. The delay and parameter vectors belong to $\mathscr{H}_{\mu,\bar{h}}$ and $\mathscr{P}^\nu$, respectively. The matrices C and T are assumed to be full row rank and parameter-independent. The main reason for this restrictive choice is that, when they are parameter-dependent, exact observation is

C. Briat, *Linear Parameter-Varying and Time-Delay Systems*,
Advances in Delays and Dynamics 3, DOI 10.1007/978-3-662-44050-6_7

only possible if the observer matrices also depend on the parameter derivatives, which are usually unknown. For similar reasons, the measured output is restricted to be a function of $x(t)$ to avoid the presence of the delay-derivative (when $y(t)$ depends on $x(t-h(t))$) or the presence of the derivative of $w(t)$ (when $y(t)$ depends on $w(t)$). Note, however, that by appropriately filtering the measured output of a system, we can always bring back the considered model into the form (7.1).

Two types of observers are considered in this section. The first one is the class of *observers with memory* which implement a delayed term in their dynamical model. Whenever the delay is identical to the one involved in the system, the observer is said to be with *exact memory*. When it is different, but at a certain distance of the one in the model of the system, the observer is said to be with *approximate memory*. In the latter case, the delay implementation error makes the design slightly more involved since the observer has to be robust with respect to its own implementation, this property is called *resilience* or *non-fragility* in the literature; see e.g. [1–4].

The second type of observers considered in this section is the class of *memoryless observers*. This class of observers does not consider any delayed term in its model.

7.1.1 Observer with Exact Memory

Let us consider first the case of observers with exact memory, i.e. the delay in the observer is the same as the one in the system. In this case, a suitable observer structure may be defined as

$$\begin{aligned}\dot{\xi}(t) =& M_0(\rho(t))\xi(t) + M_h(\rho(t))\xi(t-h(t)) + S(\rho(t))u(t)\\ &+ N_0(\rho(t))y(t) + N_h(\rho(t))y(t-h(t))\\ \xi(s) =& \psi_\xi(s),\ s \in \left[-\bar{h}, 0\right]\\ \hat{z}(t) =& \xi(t) + Hy(t)\end{aligned} \tag{7.2}$$

where $\xi \in \mathbb{R}^r$, $\psi_\xi \in C([-\bar{h}, 0], \mathbb{R}^r)$ and $\hat{z} \in \mathbb{R}^r$ are the state of the observer, the functional initial condition and the estimate of the system output z, respectively. Such an observer can be viewed as the generalization of the observer for time-delay considered in [5].

Our objective consists of finding an observer of the form (7.2) for system (7.1) which

1. makes the observation error $e(t) := z(t) - \hat{z}(t)$ asymptotically stable; and
2. ensures that the L_2-gain of the transfer $w \to e$ is less than $\gamma > 0$.

Before being able to provide explicit synthesis conditions for such an observer, several intermediary results have to be stated first. The first one concerns the decoupling of the error dynamics from the other signals involved in the system:

Proposition 7.1.1 ([6]) *Assume that there exist matrix functions* $M_0, M_h : \boldsymbol{\Delta}_\rho \to \mathbb{R}^{r\times r}$, $N_0, N_h : \boldsymbol{\Delta}_\rho \to \mathbb{R}^{r\times s}$ *and a matrix* $H \in \mathbb{R}^{r\times s}$ *such that the nonlinear matrix equalities*

$$TA(\rho) - M_0(\rho)(T - HC) - N_0(\rho)C - HCA(\rho) = 0 \tag{7.3a}$$
$$TA_h(\rho) - M_h(\rho)(T - HC) - N_h(\rho)C - HCA_h(\rho) = 0 \tag{7.3b}$$
$$S(\rho) - (T - HC)B(\rho) = 0 \tag{7.3c}$$

hold for all $\rho \in \boldsymbol{\Delta}_\rho$. *Then, the dynamical model of the observation error writes*

$$\dot{e}(t) = M_0(\rho)e(t) + M_h(\rho)e(t - h(t)) + (T - HC)E(\rho(t))w(t). \tag{7.4}$$

Proof The observation error $e(t) = z(t) - \hat{z}(t)$ defined above is governed by the following differential equation:

$$\begin{aligned}
\dot{e}(t) =& \dot{z}(t) - \dot{\hat{z}}(t) \\
=& M_0(\rho(t))e(t) + M_h(\rho(t))e(t - h(t)) + FE(\rho(t))w(t) \\
&+ (TA(\rho(t)) - M_0(\rho(t))F - N_0(\rho(t))C - HCA(\rho(t)))x(t) \\
&+ (TA_h(\rho(t)) - M_h(\rho(t))F - N_h(\rho(t))C - HCA_h(\rho(t)))x(t - h(t)) \\
&+ (FB(\rho(t)) - S(\rho(t)))u(t)
\end{aligned} \tag{7.5}$$

where $F = T - HC$. Assuming that the matrix equalities (7.3a), (7.3b), (7.3c) are satisfied, we obtain the model (7.4). ■

The algebraic conditions (7.3a), (7.3b), (7.3c) are decoupling conditions of the dynamics of the error from the other signals. If, for instance, the condition (7.3a) is not satisfied, then the observation error will depend on the state of the system which will prevent the observation error from converging to 0 when the inputs to the system are nonzero.

The next result concerns the existence of observer matrices such that the matrix equalities (7.3a), (7.3b), (7.3c) are satisfied.

Lemma 7.1.2 ([6]) *There exist matrix functions* $M_0, M_h : \boldsymbol{\Delta}_\rho \to \mathbb{R}^{r\times r}$, $N_0, N_h : \boldsymbol{\Delta}_\rho \to \mathbb{R}^{r\times s}$ *such that the equalities* (7.3a), (7.3b), (7.3c) *hold if and only if one of the following equivalent statements holds:*

1. *The constant matrix* $H \in \mathbb{R}^{r\times s}$ *is such that the equality*

$$[\varphi(\rho) - H\psi(\rho)]\left[I - \phi^+\phi\right] = 0 \tag{7.6}$$

holds for all $\rho \in \boldsymbol{\Delta}_\rho$ *where*

$$\psi(\rho) = \left[CA(\rho)\ CA_h(\rho)\right],\ \varphi(\rho) = \left[TA(\rho)\ TA_h(\rho)\right] \tag{7.7}$$

and

$$\phi = \begin{bmatrix} T & 0 \\ 0 & T \\ C & 0 \\ 0 & C \end{bmatrix}. \tag{7.8}$$

2. *The constant matrix* $H \in \mathbb{R}^{r\times s}$ *is such that the equality*

$$rank \begin{bmatrix} \phi \\ \varphi(\rho) - H\psi(\rho) \end{bmatrix} = rank\ [\phi] \tag{7.9}$$

holds for all $\rho \in \boldsymbol{\Delta}_\rho$ *with the matrices defined above.*

Proof The condition (7.3c) is trivially satisfied and does not need to be considered further. The conditions (7.3a) and (7.3b) are more involved. The first step consists of rewriting them into the compact form

$$\mathcal{O}(\rho)\phi = \varphi(\rho) - H\psi(\rho) \tag{7.10}$$

where the matrix $\mathcal{O}(\rho)$ is defined as

$$\mathcal{O}(\rho) = \left[M_0(\rho)\ M_h(\rho)\ K_0(\rho)\ K_h(\rho)\right] \tag{7.11}$$

and $K_0(\rho) = N_0(\rho) - M_0(\rho)H$, $K_h(\rho) = N_h(\rho) - M_h(\rho)H$. Note that this bijective change of variables linearizes the equations. According to [7, 8], there exist solutions to such an equation if and only if (7.6) holds for some $H \in \mathbb{R}^{r\times s}$ and for all $\rho \in \boldsymbol{\Delta}_\rho$. The rank condition (7.9) is obtained by equivalence with (7.6). ■

From the above result we can see that when the matrix ϕ is full column rank, then we have $I - \phi^+\phi = 0$ and the equality (7.6) holds for any $H \in \mathbb{R}^{r\times s}$. Since $\phi \in \mathbb{R}^{2(r+s)\times 2n}$, a necessary condition for being full column rank is that $r + s \geq n$, i.e. $\dim(z) + \dim(y) \geq \dim(x)$. As an example, if a full-order observer is sought, then $r = n$, $T = I_n$ and ϕ is automatically full-column rank. When the matrix ϕ is such that $I - \phi^+\phi \neq 0$, then the matrix $H \in \mathbb{R}^{r\times s}$ must be chosen such that one of the statements of Lemma 7.1.2 is satisfied.

One of the benefits of the observer matrices existence results above lies in the fact that they can be used to derive the set of all solutions to the matrix equations (7.3). This is stated below:

Proposition 7.1.3 ([6]) *Assume that the conditions of Lemma 7.1.2 are fulfilled, then, for all $L : \boldsymbol{\Delta}_\rho \to \mathbb{R}^{r\times 2(r+s)}$, the matrices*

$$\begin{aligned} M_0(\rho) &= [\Theta_0(\rho) - H\Theta_H(\rho) - L(\rho)\Xi]\,\Delta_1 \\ M_h(\rho) &= [\Theta_0(\rho) - H\Theta_H(\rho) - L(\rho)\Xi]\,\Delta_2 \\ N_0(\rho) &= [\Theta_0(\rho) - H\Theta_H(\rho) - L(\rho)\Xi]\,\Delta_3 + M_0(\rho)H \\ N_h(\rho) &= [\Theta_0(\rho) - H\Theta_H(\rho) - L(\rho)\Xi]\,\Delta_4 + M_h(\rho)H \\ S(\rho) &= FB(\rho) \end{aligned} \tag{7.12}$$

where $F = T - HC$, $\Xi = (I - \phi\phi^+)$, $\Theta_0(\rho) = T\left[A(\rho)\ A_h(\rho)\right]$, $\Theta_H(\rho) = C\left[A(\rho)\ A_h(\rho)\right]$ and

$$\Delta_1 = \begin{bmatrix} I_r \\ 0 \\ 0 \\ 0 \end{bmatrix},\ \Delta_2 = \begin{bmatrix} 0 \\ I_r \\ 0 \\ 0 \end{bmatrix},\ \Delta_3 = \begin{bmatrix} 0 \\ 0 \\ I_s \\ 0 \end{bmatrix},\ \Delta_4 = \begin{bmatrix} 0 \\ 0 \\ 0 \\ I_s \end{bmatrix}$$

solve the matrix equations (7.3).

Proof Assuming that Eq. (7.6) is satisfied, then the set of all solutions to Eq. (7.10) is parameterized as [8]:

$$\mathcal{O}_s(\rho) = [\varphi(\rho) - H\psi(\rho)]\phi^+ - L(\rho)(I - \phi\phi^+) \tag{7.13}$$

where $L : \boldsymbol{\Delta}_\rho \to \mathbb{R}^{r\times 2(r+s)}$ is arbitrary. With $\mathcal{O}(\rho) = \mathcal{O}_s(\rho)$, the conditions (7.3a) and (7.3b) hold, and the result follows. ■

Now that even if all the observer matrices that satisfy the conditions of Proposition 7.1.1 have been explicitly parametrized in terms of H and $L(\rho)$, it remains to find suitable values for them, that is, values for which the observation error is asymptotically stable. To this aim, first note that the dynamical model of the error writes:

$$\begin{aligned} \dot{e}(t) &= [\Theta_0(\rho(t)) - H\Theta_H(\rho(t)) - L(\rho(t))\Xi]\,\Delta_1 e(t) \\ &\quad + [\Theta_0(\rho(t)) - H\Theta_H(\rho(t)) - L(\rho(t))\Xi]\,\Delta_2 e(t - h(t)) \\ &\quad + FE(\rho(t))w(t) \end{aligned} \tag{7.14}$$

where $L : \boldsymbol{\Delta}_\rho \to \mathbb{R}^{r\times 2(r+s)}$ is arbitrary. Note that, whenever $H \in \mathbb{R}^{r\times s}$ can be chosen such that $(T - HC)E(\rho) = 0$ and (7.14) is asymptotically stable, then the observation error will be decoupled from the exogenous input. In such a case, it is clear that the L_2-gain of the transfer of w to e will be equal to 0.

The following result provides a way for computing suitable matrices $L(\rho)$ and H such that the observation error is asymptotically stable. It assumed below that the matrix H is a free variable that has not been assigned to any specific value in Lemma 7.1.2.

Theorem 7.1.4 ([6]) *Assume that there exist a continuously differentiable matrix function $P : \boldsymbol{\Delta}_\rho \to \mathbb{S}^r_{\succ 0}$, a matrix function $\bar{L} : \boldsymbol{\Delta}_\rho \to \mathbb{R}^{r\times(2r+2s)}$, constant matrices $Q, R \in \mathbb{S}^r_{\succ 0}$, $X \in \mathbb{R}^{r\times r}$, $\bar{H} \in \mathbb{R}^{r\times r}$ and a positive scalar γ such that the LMI:*

$$\begin{bmatrix} -\mathrm{He}[X] & \Sigma_{12} & \Sigma_{13} & \Sigma_{14} & 0 & X^{\mathrm{T}} & \bar{h}R \\ \star & \Sigma_{22} & R & 0 & I_r & 0 & 0 \\ \star & \star & \Sigma_{33} & 0 & 0 & 0 & 0 \\ \star & \star & \star & -\gamma I_m & 0 & 0 & 0 \\ \star & \star & \star & \star & -\gamma I_r & 0 & 0 \\ \star & \star & \star & \star & \star & -P(\rho) & -\bar{h}R \\ \star & \star & \star & \star & \star & \star & -R \end{bmatrix} \prec 0 \tag{7.15}$$

holds for all $(\rho, \nu) \in \boldsymbol{\Delta}_\rho \times \mathbf{V}_\nu$ where

$$\begin{aligned} \Sigma_{12} &= X^{\mathrm{T}}\Theta_0(\rho)\Delta_1 - \bar{H}\Theta_H(\rho)\Delta_1 - \bar{L}(\rho)\Xi\Delta_1 \\ \Sigma_{13} &= X^{\mathrm{T}}\Theta_0(\rho)\Delta_2 - \bar{H}\Theta_H(\rho)\Delta_2 - \bar{L}(\rho)\Xi\Delta_2 \\ \Sigma_{14} &= \left(X^{\mathrm{T}}T - \bar{H}C\right)E(\rho) \\ \Sigma_{22} &= \textstyle\sum_i \dfrac{\partial P(\rho)}{\partial \rho_i}\nu_i - P(\rho) + Q - R \\ \Sigma_{33} &= -(1-\mu)Q - R. \end{aligned} \tag{7.16}$$

Then, there exists an r-order observer of the form (7.2) for system (7.1) such that, for all $(\rho, h) \in \mathscr{P}^\nu \times \mathscr{H}_{\mu,\bar{h}}$, the observation error is asymptotically stable and the L_2-gain of the transfer $w \to e$ is less than γ. Moreover, such an observer is given by the matrices in Proposition 7.1.3 where $L(\rho) = X^{-T}\bar{L}(\rho)$ and $H = X^{-T}\bar{H}$.

Proof The proof is based on the substitution of the observer model (7.14) into the stability condition (6.30). The LMI (7.15) is then obtained by setting X as parameter-independent (since H is parameter-independent) and by considering the linearizing change of variables $\bar{H} = X^{\mathrm{T}}H$, $\bar{L}(\rho) = X^{\mathrm{T}}L(\rho)$. ■

In the case H has been assigned a specific values so that the conditions of Lemma 7.1.2 are satisfied, the above result can be adapted by simply changing $\bar{H}$ into $X^{\mathrm{T}}H$. Note that, in this case, the matrix X can be made parameter-dependent.

7.1.2 Examples

7.1.2.1 Design of Full-Order Observers with Exact Memory

Let us consider the following system adapted from [9]:

$$\begin{aligned}\dot{x}(t) &= A(\rho(t))x(t) + A_h(\rho(t))x(t-h(t)) + E(\rho(t))w(t) + B(\rho(t))u(t)\\ y(t) &= \begin{bmatrix}1 & 0\end{bmatrix}x(t)\\ z(t) &= x(t)\end{aligned} \tag{7.17}$$

where

$$A(\rho) = \begin{bmatrix}0 & 1+0.2\rho\\ -2 & -3+0.1\rho\end{bmatrix}, \qquad E(\rho) = \begin{bmatrix}-0.2\\ -0.2\end{bmatrix}, \tag{7.18}$$

$$A_h(\rho) = \begin{bmatrix}0.2\rho & 0.1\\ -0.2+0.1\rho & -0.3\end{bmatrix}, \qquad B(\rho) = \begin{bmatrix}1+\rho\\ 2+\rho\end{bmatrix}. \tag{7.19}$$

The parameter and delay are assumed to verify $\rho(t) \in [-1, 1]$, $\dot{\rho}(t) \in [-1, 1]$, $h(t) \in [0, \bar{h}]$ and $\dot{h}(t) \leq \mu < 1$ for all $t \geq 0$. In the present case, the matrix ϕ in Lemma 7.1.2 is full-column rank, thus H is arbitrary. Since the system is affine in the parameter, we choose the following polynomial structure for the decision matrix functions

$$\begin{aligned}L(\rho) &= L_2\rho^2 + L_1\rho + L_0,\\ P(\rho) &= P_0 + P_1\rho + P_2\rho^2/2.\end{aligned} \tag{7.20}$$

and, using Theorem 7.1.4 with a constant delay $\bar{h} = 2$, we find that the minimal γ that can be achieved with this result is $\gamma^* = 1.36 \cdot 10^{-4}$ along with the observer matrices

$$M_0(\rho) = \begin{bmatrix}-4.4439 & 0\\ 0 & -0.1\rho - 4\end{bmatrix}, \quad H = \begin{bmatrix}1\\ 1\end{bmatrix}, \quad S(\rho) = \begin{bmatrix}0\\ 1\end{bmatrix},$$

$$M_h(\rho) = \begin{bmatrix}0 & 0\\ 0 & -0.4\end{bmatrix}, \quad N_h(\rho) = \begin{bmatrix}0\\ -0.1\rho - 0.6\end{bmatrix}, \quad N_0(\rho) = \begin{bmatrix}0\\ -0.1\rho - 6\end{bmatrix}.$$

It is interesting to note that we have $(T - HC)E(\rho) = 0$ for all $\rho \in \boldsymbol{\Delta}_\rho$ and hence we have full decoupling between the disturbance w and the observation error e. It is thus expected to have perfect observation for this example.

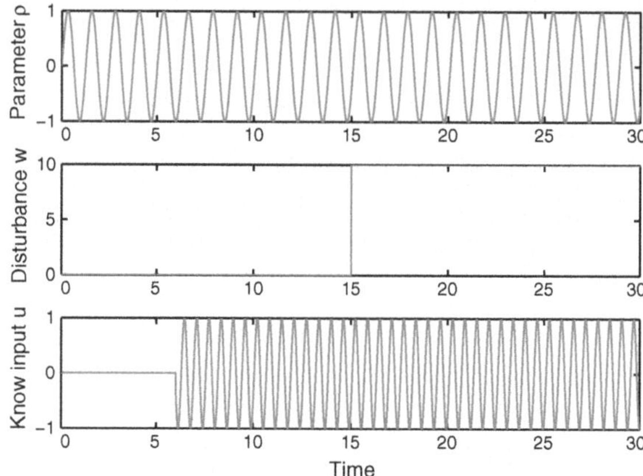

Fig. 7.1 Time evolution of the parameter ρ (*top*), the known input u (*center*) and the disturbance w (*bottom*) for Examples 1 and 2

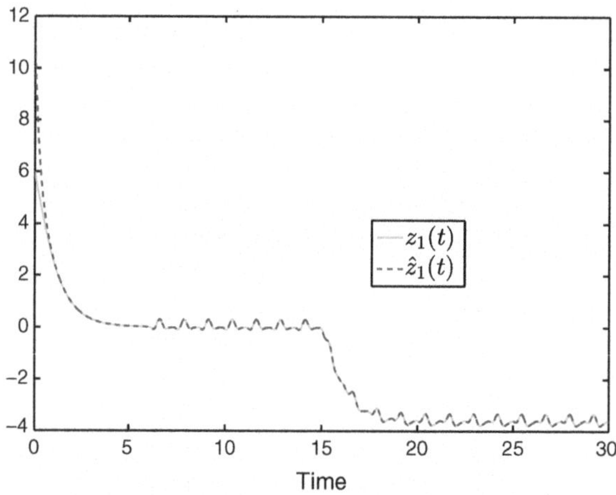

Fig. 7.2 Full-order observer with memory: Observation of the first state of the system (7.17), (7.18)

For simulation purposes, the constant delay is set to $h(t) = 2$, $t \in \mathbb{R}_{\geq 0}$, the parameter is chosen as $\rho(t) = \sin(t)$, the known input $u(t) = \sin(10t)$ is applied at $t = 6$ seconds and the step disturbance $w(t)$ of amplitude 10 is applied at $t = 15$ seconds; see Fig. 7.1. The state trajectories and their respective estimated values are depicted in Figs. 7.2 and 7.3. We can clearly observe that since e and w are decoupled, the observation remains accurate even in the presence of a disturbance of large amplitude.

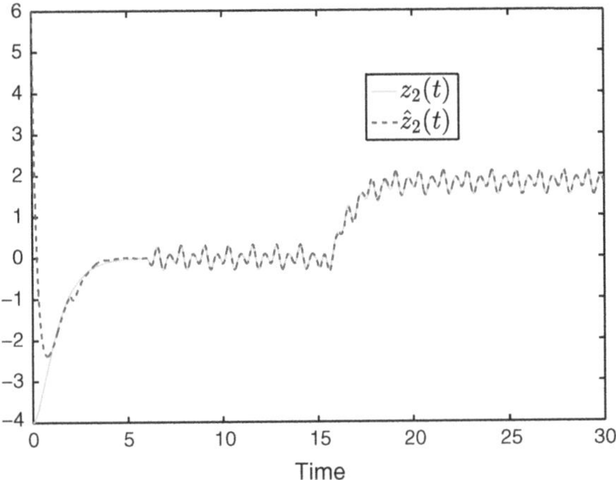

Fig. 7.3 Full-order observer with memory: Observation of the second state of the system (7.17), (7.18)

7.1.2.2 Design of Reduced-Order Observer with Exact Memory

Let us consider now the system (7.17), (7.18) with the difference that the measured and estimated outputs are now given by:

$$y(t) = \begin{bmatrix} 0 & 1 \end{bmatrix} x(t) \text{ and } z(t) = \begin{bmatrix} 1 & 0 \end{bmatrix} x(t). \tag{7.21}$$

So, the goal is to obtain a reduced-order observer estimating the first state from the measurement of the second state. The matrix ϕ is again full-column rank, hence H is also arbitrary in this case. Using Theorem 7.1.4 with $\bar{h} = 2$, we find a minimal $\gamma = 0.4126$, showing that total decoupling does not seem to be possible using the considered stability result. The observer matrices are, in this case, given by $H = -0.4567$, $M_0(\rho) = -0.9133$ and

$$\begin{aligned} M_h(\rho) &= 0.2456\rho - 0.0913, \quad N_0(\rho) = 0.2456\rho + 0.0471, \\ N_h(\rho) &= -0.1121\rho + 0.0047, \quad S(\rho) = 1.4566\rho + 1.9133. \end{aligned} \tag{7.22}$$

The simulation results are depicted in Fig. 7.4 in the same environmental conditions as in the first example. In this case, we can clearly see that the disturbance affects the observation error.

7.1.2.3 Observation of a Milling Process

Let us consider the milling process of Sect. 6.2.1 which is described by the model

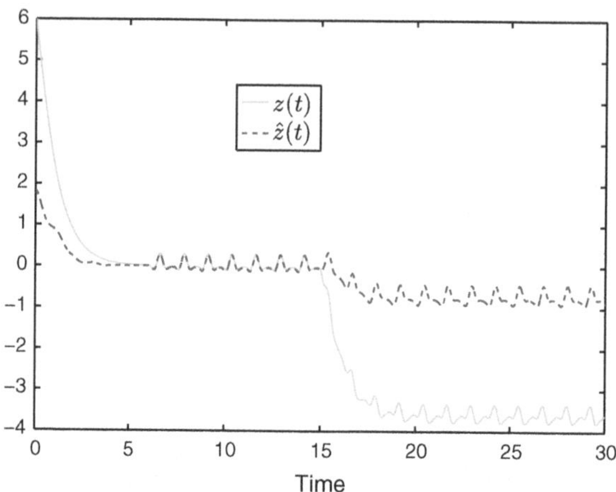

Fig. 7.4 Reduced-order observer with memory: Observation of the first state of the system (7.17)–(7.21)

$$\dot{x}(t) = (A_0 + \rho(t)A_1)x(t) + (A_{h0} + \rho(t)A_{h1})x(t-h) \tag{7.23}$$

where the delay $h = \pi/\omega$ is constant, $\rho(t) = \cos(2\omega t + \beta)$ is the varying parameter, $\beta > 0$ is a constant parameter and $\omega > 0$ is the angular velocity. The system matrices are given by [10]

$$A_0 = \begin{bmatrix} 0 & 0 & 1 & 0 \\ 0 & 0 & 0 & 1 \\ \alpha(k) & 10 & 0 & 0 \\ 5 & -15 & 0 & -0.25 \end{bmatrix}, \quad A_0 = \begin{bmatrix} 0 & 0 & 0 & 0 \\ 0 & 0 & 0 & 0 \\ 0.5k & 0 & 0 & 0 \\ 0 & 0 & 0 & 0 \end{bmatrix},$$
$$A_{h0} = \begin{bmatrix} 0 & 0 & 0 & 0 \\ 0 & 0 & 0 & 0 \\ 0.1710k & 10 & 0 & 0 \\ 0 & 0 & 0 & 0 \end{bmatrix}, \quad A_{h1} = \begin{bmatrix} 0 & 0 & 0 & 0 \\ 0 & 0 & 0 & 0 \\ -0.5k & 0 & 0 & 0 \\ 0 & 0 & 0 & 0 \end{bmatrix} \tag{7.24}$$

where $\alpha(k) = -(10 + 0.1710k)$ and $k > 0$ is the cutting stiffness constant. We assume that the displacements of the blade and the tool are measured (i.e. $x_1(t)$ and $x_2(t)$ are measured) and that the entire state has to be estimated. Hence, we have

$$y(t) = \begin{bmatrix} 1 & 0 & 0 & 0 \\ 0 & 1 & 0 & 0 \end{bmatrix} x(t) \text{ and } z(t) = x(t). \tag{7.25}$$

According to the discussion in [10] (see also Sect. 6.2.1), the parameter needs to be considered as having unbounded derivative. We thus pick

$$P(\rho) = P_0 \text{ and } L(\rho) = L_0 + L_1\rho + L_2\rho^2. \tag{7.26}$$

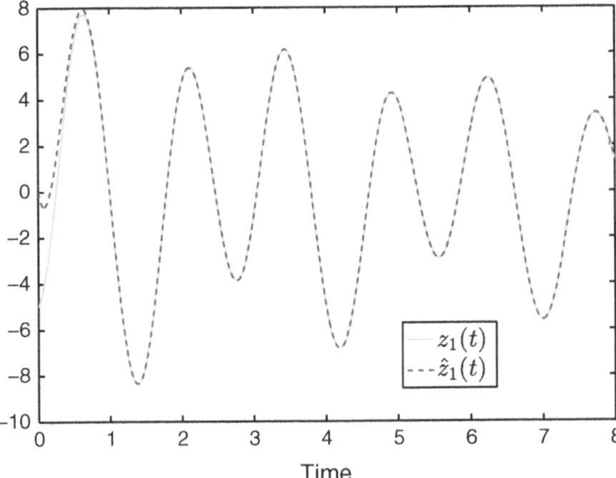

Fig. 7.5 Actual (*plain*) and estimated (*dashed*) outputs for the milling process—State 1

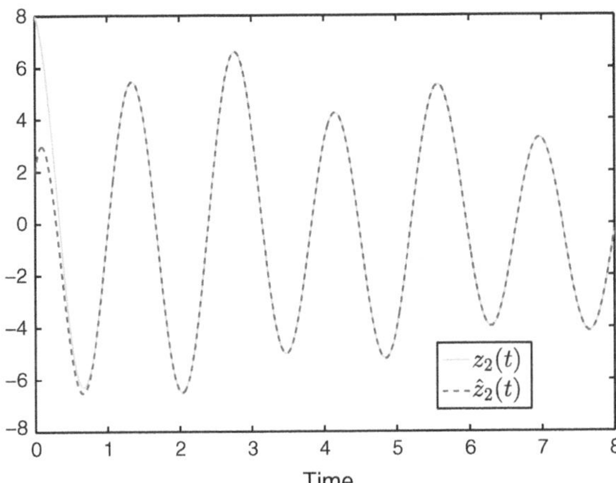

Fig. 7.6 Actual (*plain*) and estimated (*dashed*) outputs for the milling process—State 2

The observer is designed for $\bar{h} = 1$ (hence $h \leq \bar{h}, \omega \geq \pi$), $k = 0.2$. Since ϕ is full-column rank, then H is arbitrary and the observer is designed using Theorem 7.1.4. For simulation purposes, we pick $\omega = 10$ rad/s and $\beta = 7\pi/18$. The simulation results are depicted in Figs. 7.5, 7.6, 7.7 and 7.8 where we can see that the observer is able to track the system state accurately.

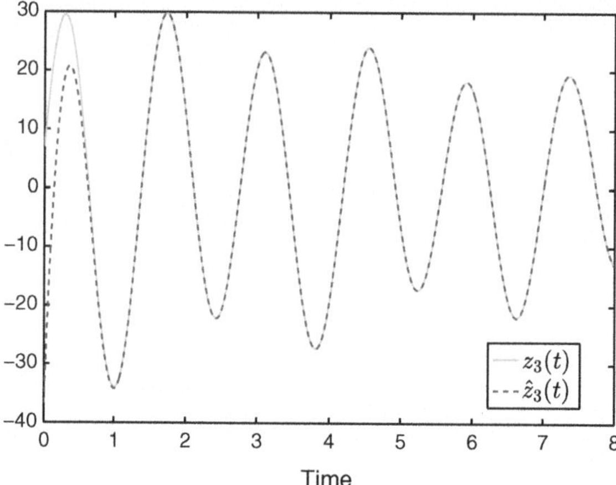

Fig. 7.7 Actual (*plain*) and estimated (*dashed*) outputs for the milling process—State 3

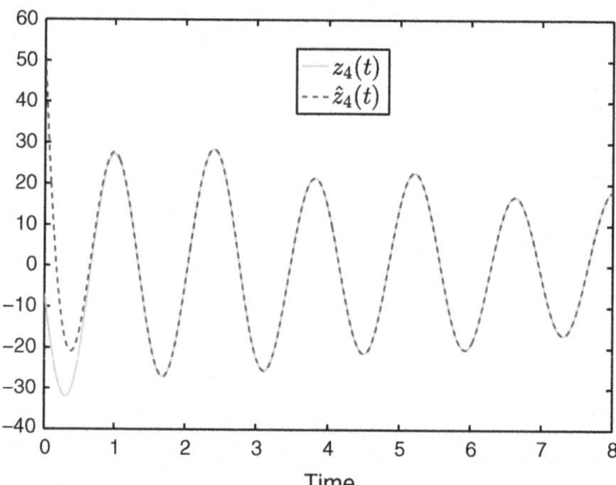

Fig. 7.8 Actual (*plain*) and estimated (*dashed*) outputs for the milling process—State 4

7.1.3 Memoryless Observer

The observers designed in the previous section all assume a perfect knowledge of the delay value. This is, in many cases, very unrealistic since it may be very difficult to measure or estimate the delay value in real time. Memoryless observers can then be used to circumvent this problem, but with a possible reduction of accuracy. This accuracy deterioration will, however, be limited when the delayed part has a low impact on the overall dynamics of the system.

We thus consider the following observer structure:

$$\begin{aligned}\dot{\xi}(t) &= M(\rho)\xi(t) + N(\rho)y(t) + S(\rho(t))u(t)\\ \hat{z}(t) &= \xi(t) + Hy(t)\end{aligned} \tag{7.27}$$

where $\xi \in \mathbb{R}^r$ is the state of the observer. We have the following result:

Proposition 7.1.5 *Assume that there exist matrix functions $M : \boldsymbol{\Delta}_\rho \to \mathbb{R}^{r\times r}$, $N : \boldsymbol{\Delta}_\rho \to \mathbb{R}^{r\times s}$ and a matrix $H \in \mathbb{R}^{r\times s}$ such that the nonlinear matrix equalities*

$$TA(\rho) - M(\rho)F - N(\rho)C - HCA(\rho) = 0 \tag{7.28a}$$

$$S(\rho) - FB(\rho) = 0 \tag{7.28b}$$

hold for all $\rho \in \boldsymbol{\Delta}_\rho$ where $F = T - HC$. Then, the dynamics of the observation error reduces to

$$\dot{e}(t) = M(\rho(t))e(t) + FA_h(\rho(t))x(t - h(t)) + FE(\rho(t))w(t). \tag{7.29}$$

Note that, unlike in the exact memory case, the observation error cannot be made independent of the state of the system unless H can be chosen such that $(T - HC)A_h(\rho) = 0$ for all $\rho \in \boldsymbol{\Delta}_\rho$. The following result is the memoryless counterpart of Lemma 7.1.2:

Lemma 7.1.6 *There exist matrix functions $M : \boldsymbol{\Delta}_\rho \to \mathbb{R}^{n\times n}$, $N : \boldsymbol{\Delta}_\rho \to \mathbb{R}^{r\times s}$ such that the equalities* (7.28a) *hold if and only if the constant matrix $H \in \mathbb{R}^{r\times s}$ is such that the equality*

$$[T - HC]A(\rho)\left[I - \phi^+\phi\right] = 0 \tag{7.30}$$

holds for all $\rho \in \boldsymbol{\Delta}_\rho$ where

$$\phi = \begin{bmatrix} T \\ C \end{bmatrix}. \tag{7.31}$$

The following result is the counterpart of Proposition 7.1.3:

Proposition 7.1.7 *Assume that the conditions of Lemma 7.1.6 are fulfilled, then for all $L : \boldsymbol{\Delta}_\rho \to \mathbb{R}^{r\times(r+s)}$ the matrices*

$$\begin{aligned} M(\rho) &= [TA(\rho) - HCA(\rho) - L(\rho)\Xi]\,\Delta_1 \\ N(\rho) &= [TA(\rho) - HCA(\rho) - L(\rho)\Xi]\,\Delta_2 + M(\rho)H \\ S(\rho) &= FB(\rho) \end{aligned} \tag{7.32}$$

where $F = T - HC$, $\Xi = \left(I - \phi\phi^+\right)$,

$$\Delta_1 = \begin{bmatrix} I_r \\ 0 \end{bmatrix} \text{ and } \Delta_2 = \begin{bmatrix} 0 \\ I_s \end{bmatrix}$$

solve the matrix equations (7.28a) *and* (7.28b).

Assuming now that $w \in L_2$ and that the system to be observed is asymptotically stable, then it is clear that x is also in L_2. Let $\mathscr{D}_h$ be the delay operator defined in (5.107). Then we have $||\mathscr{D}_h(x)||_{L_2} < \infty$ as well since $\dot{h}(t) \le \mu < 1$. Thus, the term $x(t - h(t))$ can be viewed as an L_2 disturbance acting on the observation error.[1] Hence, the model (7.29) can be rewritten as

$$\dot{e}(t) = M(\rho(t))e(t) + F\left[A_h(\rho(t))\ E(\rho(t))\right]\tilde{w}(t) \tag{7.33}$$

where $\tilde{w}(t) := \mathrm{col}(x(t-h(t)), w(t))$. We then apply the usual Bounded-Real Lemma in order to design suitable $M(\rho)$ and H such that the L_2-gain of the transfer $\tilde{w} \to e$ is small. This result is stated below.

Theorem 7.1.8 *Assume that there exist a continuously differentiable matrix function $P : \boldsymbol{\Delta}_\rho \to \mathbb{S}^r_{\succ 0}$, a matrix function $\bar{L} : \boldsymbol{\Delta}_\rho \to \mathbb{R}^{r\times(r+s)}$, $X \in \mathbb{R}^{r\times r}$, $\bar{H} \in \mathbb{R}^{r\times r}$ and scalars $\gamma, \sigma > 0$ such that the matrix inequality:*

$$\begin{bmatrix} -\mathrm{He}[X] & P(\rho) + \mathcal{A}(\rho) & \mathcal{E}(\rho) & X^{\mathrm{T}} & 0 \\ \star & -\sigma P(\rho) + \sum_i \dfrac{\partial P(\rho)}{\partial \rho_i}\nu_i & 0 & 0 & I_r \\ \star & \star & -\gamma I_{n+m} & 0 & 0 \\ \star & \star & \star & -P(\rho)/\sigma & 0 \\ \star & \star & \star & \star & -\gamma I_r \end{bmatrix} \prec 0$$

holds for all $(\rho, \nu) \in \boldsymbol{\Delta}_\rho \times \mathbf{V}_\nu$ where

[1] Note, however, that it would be perhaps more relevant to consider $x(t - h(t))$ as a bounded disturbance, i.e. $x \in L_\infty$.

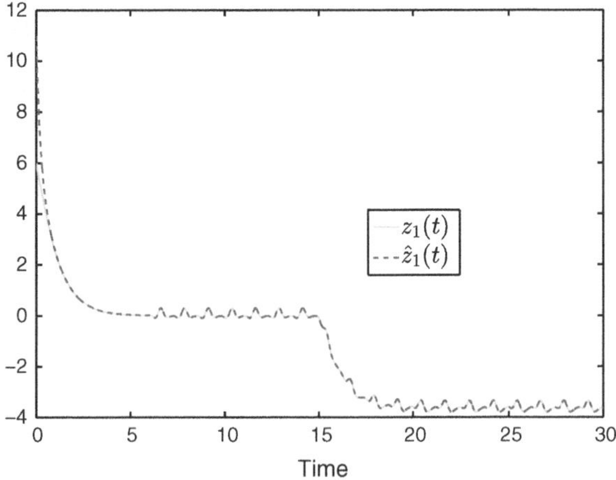

Fig. 7.9 Full-order memoryless observer: Observation of the first state of the system (7.17), (7.18)

$$\begin{aligned}\mathcal{A}(\rho) &= X^{\mathrm{T}}\Theta_0(\rho) - \bar{H}\Theta_H - \bar{L}(\rho)\Xi \\ \mathcal{E}(\rho) &= \left(X^{\mathrm{T}}T - \bar{H}C\right)\left[A_h(\rho)\ E(\rho)\right].\end{aligned} \tag{7.34}$$

Then, there exists an r-order memoryless observer of the form (7.27) for system (7.1) such that, for all $(\rho, h) \in \mathscr{P}^\nu \times \mathscr{H}_{\mu,\bar{h}}$, the dynamics of the error is asymptotically stable and the L_2-gain of the transfer $w \to e$ is less than γ. Moreover, such an observer is given by the matrices of Proposition 7.1.7 where $L(\rho) = X^{-T}\bar{L}(\rho)$ and $H = X^{-T}\bar{H}$.

Proof The proof is based on the use of Theorem 2.5.6 and the same changes of variables as in the exact memory case. ■

7.1.4 Examples

7.1.4.1 Full-Order Memoryless Observer Design

We consider back the system (7.17), (7.18), for which we design a memoryless observer using Theorem 7.1.8. We obtain the minimal value $\gamma = 0.1697$ and the trajectories depicted in Figs. 7.9 and 7.10. Not surprisingly, perfect tracking of the state is not achievable anymore. Note, however, that the estimate is quite close to the trajectory of the second state thanks to the small L_2-gain.

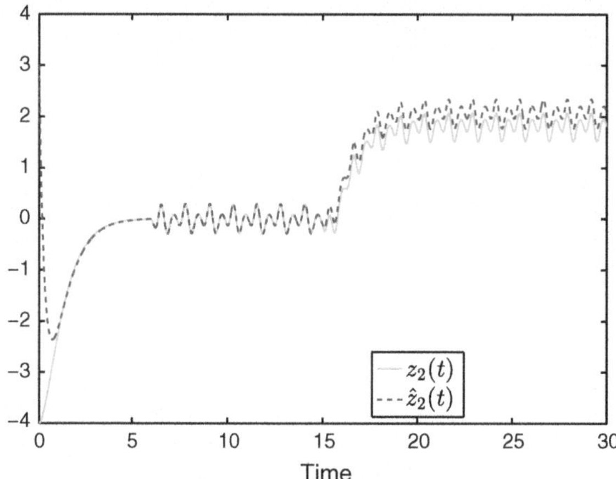

Fig. 7.10 Full-order memoryless observer: Observation of the second state of the system (7.17), (7.18)

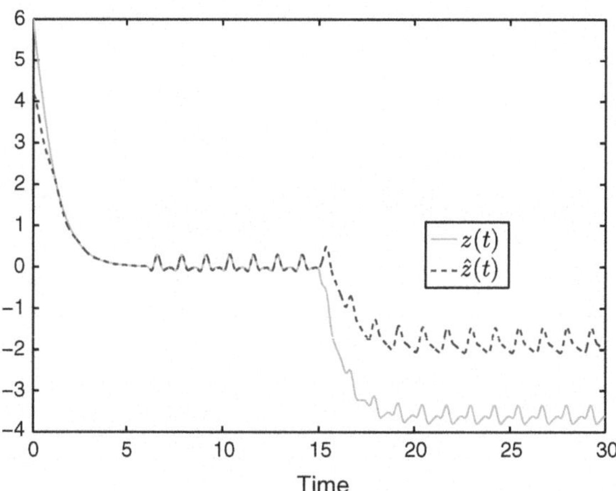

Fig. 7.11 Reduced-order memoryless observer: Observation of the first state of the system

7.1.4.2 Reduced-Order Memoryless Observer Design

Use the same matrices as in (7.21), we get $\gamma = 0.3872$ together with the trajectory depicted in Fig. 7.11. Interestingly, even though the observer is memoryless we get a smaller γ and a better observation error than in the exact memory case. This is certainly due to the conservatism of the result that is used to characterize stability of LPV time-delay systems.

7.2 Simple Observers for LPV Time-Delay Systems

We consider in this section a less general, yet useful enough, class of observers for systems of the form

$$\begin{aligned}\dot{x}(t) &= A(\rho(t))x(t) + A_h(\rho(t))x(t-h(t)) + B(\rho(t))u(t) + E(\rho(t))w(t)\\ y(t) &= C_y(\rho(t))x(t) + C_{yh}(\rho(t))x(t-h(t)) + F_y(\rho(t))w(t)\\ x(\theta) &= \psi(\theta),\ \theta \in \left[-\bar{h}, 0\right]\end{aligned} \tag{7.35}$$

where $x \in \mathbb{R}^n$, $u \in \mathbb{R}^m$, $w \in \mathbb{R}^p$, $y \in \mathbb{R}^q$ and $\psi \in C([-\bar{h}, 0], \mathbb{R}^n)$ are the state of the system, the known input, the exogenous input, the measured output and the functional initial condition, respectively. The delay and parameter vectors belong to $\mathscr{H}_{\mu,\bar{h}}$ and $\mathscr{P}^\nu$, respectively.

7.2.1 Observer with Exact-Memory

Let us consider, in this section, memoryless observers of the form

$$\begin{aligned}\dot{\hat{x}}(t) &= A(\rho(t))\hat{x}(t) + B(\rho(t))u(t) + L(\rho)(y(t) - \hat{y}(t))\\ \hat{y}(t) &= C_y(\rho(t))\hat{x}(t)) - C_{yh}(\rho(t))\hat{x}(t-h(t))\\ \hat{x}(\theta) &= \hat{\psi}(\theta),\ \theta \in [-\bar{h}, 0]\end{aligned} \tag{7.36}$$

where $\hat{x} \in \mathbb{R}^n$ and $\hat{\psi} \in C([-\bar{h}, 0], \mathbb{R}^n)$ are the estimate of the state of the system and the functional initial condition, respectively. The goal is therefore to find an observer such that, for all $(\rho, h) \in \mathscr{P}^\nu \times \mathscr{H}_{\mu,\bar{h}}$,

1. the observation error $e := x - \hat{x}$ is asymptotically stable and
2. the L_2-gain of the transfer $w \to e$ is smaller than $\gamma > 0$.

We then have the following result:

Theorem 7.2.1 *Assume that there exist a continuously differentiable matrix function $P : \boldsymbol{\Delta}_\rho \to \mathbb{S}^n_{\succ 0}$, matrix functions $\bar{L} : \boldsymbol{\Delta}_\rho \to \mathbb{R}^{n\times q}$, $X : \boldsymbol{\Delta}_\rho \to \mathbb{R}^{n\times n}$, constant matrices $Q, R \in \mathbb{S}^n_{\succ 0}$ and a positive scalar γ such that the LMI:*

$$\begin{bmatrix} -\mathrm{He}[X(\rho)] & \Sigma_{12}(\rho) & \Sigma_{13}(\rho) & \Sigma_{14}(\rho) & 0 & X(\rho)^{\mathrm{T}} & \bar{h}R\\ \star & \Sigma_{22}(\rho,\nu) & R & 0 & I_n & 0 & 0\\ \star & \star & \Sigma_{33} & 0 & 0 & 0 & 0\\ \star & \star & \star & -\gamma I_p & 0 & 0 & 0\\ \star & \star & \star & \star & -\gamma I_n & 0 & 0\\ \star & \star & \star & \star & \star & -P(\rho) & -\bar{h}R\\ \star & \star & \star & \star & \star & \star & -R \end{bmatrix} \prec 0 \tag{7.37}$$

holds for all $(\rho, \nu) \in \boldsymbol{\Delta}_\rho \times \mathbf{V}_\nu$ *where*

$$\begin{aligned}
\Sigma_{12}(\rho) =& X(\rho)^{\mathrm{T}} A(\rho) - \bar{L}(\rho) C_y(\rho) + P(\rho) \\
\Sigma_{13}(\rho) =& X(\rho)^{\mathrm{T}} A_h(\rho) - \bar{L}(\rho) C_{yh}(\rho) \\
\Sigma_{14}(\rho) =& X(\rho)^{\mathrm{T}} E(\rho) - \bar{L}(\rho) F_y(\rho) \\
\Sigma_{22}(\rho, \nu) =& \sum_i \frac{\partial P(\rho)}{\partial \rho_i} \nu_i - P(\rho) + Q - R \\
\Sigma_{33} =& -(1-\mu) Q - R.
\end{aligned} \tag{7.38}$$

Then, there exists an observer with exact memory of the form (7.36) *for system* (7.35) *such that, for all* $(\rho, h) \in \mathscr{P}^\nu \times \mathscr{H}_{\mu,\bar{h}}$, *the observation error is asymptotically stable and the* L_2*-gain of the transfer* $w \to e$ *is less than* γ. *Moreover, the gain of the observer is given by*

$$L(\rho) = X(\rho)^{-T} \bar{L}(\rho). \tag{7.39}$$

Proof The dynamical expression of the observation error is given by

$$\begin{aligned}
\dot{e}(t) =& (A(\rho(t)) - L(\rho(t)) C_y(\rho(t))) e(t) \\
&+ (A_h(\rho(t)) - L(\rho(t)) C_{yh}(\rho(t))) e(t - h(t)) \\
&+ (E(\rho(t) - L(\rho(t)) F_y(\rho(t))) w(t).
\end{aligned} \tag{7.40}$$

Substituting then the observation error model in the matrix inequality condition of Theorem 6.3.2 and making the change of variables $\bar{L}(\rho) = X(\rho)^{\mathrm{T}} L(\rho)$ yield the result. ■

7.2.2 Memoryless Observer

Let us consider, in this section, memoryless observers of the form

$$\begin{aligned}
\dot{\hat{x}}(t) &= A(\rho(t))\hat{x}(t) + B(\rho(t))u(t) + L(\rho)(y(t) - C_y(\rho(t))\hat{x}(t)) \\
\hat{x}(\theta) &= \hat{\psi}(\theta), \; \theta \in [-\bar{h}, 0]
\end{aligned} \tag{7.41}$$

where $\hat{x} \in \mathbb{R}^n$ and $\hat{\psi} \in C([-\bar{h}, 0], \mathbb{R}^n)$ are the estimate of the state of the system and the functional initial condition, respectively. The goal is therefore to find an observer such that, for all $(\rho, h) \in \mathscr{P}^\nu \times \mathscr{H}_{\mu,\bar{h}}$,

1. the observation error $e := x - \hat{x}$ is asymptotically stable and

2. the L_2-gain of the transfer $(w, x_h) \to e$ is smaller than $\gamma > 0$

where $x_h(t) := x(t - h(t))$, $t \geq 0$. As in Sect. 7.1.3 and unlike in the case of the observer with exact memory, the delayed state of the system $x(t - h(t))$ acts on the observation error as a disturbance input which we need to assume to be in L_2, which is equivalent to require that the system (7.35) be asymptotically stable. Even if this assumption may seem restrictive at first sight, this is actually not the case since the observer (7.41) cannot be used to observe unstable systems.

We then have the following result:

Theorem 7.2.2 *Assume that the system* (7.41) *is asymptotically stable for all* $(\rho, h) \in \mathscr{P}^\nu \times \mathscr{H}_{\mu,\bar{h}}$. *Assume further that there exist a continuously differentiable matrix function* $P : \boldsymbol{\Delta}_\rho \to \mathbb{S}^n_{\succ 0}$, *a matrix function* $\bar{L} : \boldsymbol{\Delta}_\rho \to \mathbb{R}^{n \times q}$ *and a scalar* $\gamma > 0$ *such that the matrix inequality:*

$$\begin{bmatrix} \Xi_{11}(\rho,\nu) & \Xi_{12}(\rho) & P(\rho)E(\rho) & I_n \\ \star & -\gamma I_n & 0 & 0 \\ \star & \star & -\gamma I_p & 0 \\ \star & \star & \star & -\gamma I_n \end{bmatrix} \prec 0$$

holds for all $(\rho, \nu) \in \boldsymbol{\Delta}_\rho \times \mathbf{V}_\nu$ *where*

$$\begin{aligned} \Xi_{11}(\rho,\nu) &= \mathrm{He}[P(\rho)A(\rho) - \bar{L}(\rho)C_y(\rho)] + \sum_i \frac{\partial P(\rho)}{\partial \rho_i}\nu_i, \\ \Xi_{12}(\rho) &= P(\rho)A_h(\rho) - \bar{L}(\rho)C_{yh}(\rho). \end{aligned} \tag{7.42}$$

Then, there exists a memoryless observer of the form (7.41) *for system* (7.35) *such that, for all* $(\rho, h) \in \mathscr{P}^\nu \times \mathscr{H}_{\mu,\bar{h}}$, *the dynamics of the error is asymptotically stable and the* L_2*-gain of the transfer* $(w, x_h) \to e$ *is less than* γ. *Moreover, the gain of the observer is given by*

$$L(\rho) = P(\rho)^{-1}\bar{L}(\rho). \tag{7.43}$$

Proof The dynamical model of the observation error is given by

$$\begin{aligned} \dot{e}(t) =& (A(\rho(t)) - L(\rho(t))C_y(\rho(t)))e(t) \\ &+ (A_h(\rho(t)) - L(\rho(t))C_{yh}(\rho(t)))x(t - h(t)) \\ &+ (E(\rho(t) - L(\rho(t))F_y(\rho(t)))w(t). \end{aligned} \tag{7.44}$$

Since the observation error behaves as a non-delayed system, we use the bounded-real lemma, i.e. Lemma 2.6.5, where the disturbance vector is set to be $\mathrm{col}(x(t-h(t)), w(t))$. Substituting then the above dynamical model of the observation error in the matrix inequality condition of Lemma 2.6.5 and making the change of variables $\bar{L}(\rho) = P(\rho)L(\rho)$ yield the result. ■

7.3 Filtering of LPV Time-Delay Systems

In this section, we will be interested in filtering, which is another way for estimating the state or more general any output of the system. This problem has been addressed for instance in [9, 11–15].

Let us then consider the following uncertain LPV time-delay system

$$\begin{aligned} \dot{x}(t) &= \left[A(\rho(t)) + A^{\Delta}(\rho(t))\right]x(t) + \left[A_h(\rho(t)) + A_h^{\Delta}(\rho(t))\right]x(t-h(t)) \\ &\quad + \left[E(\rho(t)) + E^{\Delta}(\rho(t))\right]w(t) \\ x(s) &= \psi(s),\ s \in [-\bar{h}, 0] \\ y(t) &= \left[C_y(\rho(t)) + C_y^{\Delta}(\rho(t))\right]x(t) + \left[C_{yh}(\rho(t)) + C_{yh}^{\Delta}(\rho(t))\right]x(t-h(t)) \\ &\quad + \left[F_y(\rho(t)) + F_y^{\Delta}(\rho(t))\right]w(t) \\ z(t) &= C(\rho(t))x(t) + C_h(\rho(t))x(t-h(t)) + F(\rho(t))w(t) \end{aligned} \tag{7.45}$$

where $x \in \mathbb{R}^n, w \in \mathbb{R}^m, y \in \mathbb{R}^p$ and $z \in \mathbb{R}^q$ are the state of the system, the exogenous inputs, the measured output and the output that has to be estimated, respectively. The parameters and delay trajectories belong to $\mathscr{P}^{\nu}$ and $\mathscr{H}_{\mu,\bar{h}}$, respectively. The uncertain part of the matrices of the system is assumed to be described as

$$\begin{bmatrix} A^{\Delta}(\rho) & A_h^{\Delta}(\rho) & E^{\Delta}(\rho) \\ C_y^{\Delta}(\rho) & C_{yh}^{\Delta}(\rho) & F_y^{\Delta}(\rho) \end{bmatrix} = H(\rho)\Delta G(\rho) \tag{7.46}$$

where

$$\begin{aligned} H(\rho) &= \mathrm{diag}(H_0(\rho), H_1(\rho)) \\ G(\rho) &= \begin{bmatrix} G_0(\rho) & G_1(\rho) & G_2(\rho) \\ G_3(\rho) & G_4(\rho) & G_5(\rho) \end{bmatrix} \end{aligned}$$

and Δ is any constant or time-varying matrix belonging to the set

$$\boldsymbol{\Delta} := \left\{ \Delta \in \mathbb{R}^{\delta \times \delta} : \Delta^{\mathrm{T}}\Delta \preceq I \right\} \tag{7.47}$$

where $\delta > 0$ is the dimension of the uncertain matrix. The matrices G_i's and H_i's are assumed to be known.

The objective of the filtering problem is to find a filter of the form

$$\begin{aligned}\dot{x}_F(t) &= A_F(\rho(t))x_F(t) + A_{Fh}(\rho(t))x_F(t-h(t)) + B_F(\rho(t))y(t)\\ x_F(s) &= \psi_F(s),\ s \in \left[-\bar{h}, 0\right]\\ \hat{z}(t) &= C_F(\rho(t))x_F(t) + C_{Fh}(\rho(t))x_F(t-h(t)) + D_F(\rho(t))y(t)\end{aligned} \tag{7.48}$$

such that the L_2-gain of the transfer $w \to z - \hat{z}$ is less than γ for all $\Delta : \mathbb{R}_{\geq 0} \to \boldsymbol{\Delta}$ and for some $\gamma > 0$.

7.3.1 Filter with Exact Memory

The first step towards the derivation of synthesis conditions consists of constructing the extended system:

$$\begin{aligned}\dot{x}_a(t) &= \bar{\mathcal{A}}(\rho(t))x_a(t) + \bar{\mathcal{A}}_h(\rho(t))x_a(t-h(t)) + \bar{\mathcal{E}}(\rho(t))w(t)\\ z_e(t) &= \bar{\mathcal{C}}(\rho(t))x_a(t) + \bar{\mathcal{C}}_h(\rho(t))x_a(t-h(t)) + \bar{\mathcal{F}}(\rho(t))w(t)\end{aligned} \tag{7.49}$$

where $x_a(t) := \mathrm{col}(x(t), e(t))$, $e(t) = x(t) - x_F(t)$, $z_e(t) := z(t) - \hat{z}(t)$ and

$$\begin{aligned}\bar{\mathcal{A}}(\rho) &= \begin{bmatrix} A(\rho) + A^{\Delta}(\rho) & 0\\ A(\rho) + A^{\Delta}(\rho) - B_F(\rho)\left(C_y(\rho) + C_y^{\Delta}(\rho)\right) - A_F(\rho) & A_F(\rho)\end{bmatrix}\\ \bar{\mathcal{A}}_h(\rho) &= \begin{bmatrix} A_h(\rho) + A_h^{\Delta}(\rho) & 0\\ A_h(\rho) + A_h^{\Delta}(\rho) - B_F(\rho)\left(C_{yh}(\rho) + C_{yh}^{\Delta}(\rho)\right) - A_{Fh}(\rho) & A_{Fh}(\rho)\end{bmatrix}\\ \bar{\mathcal{E}} &= \begin{bmatrix} E(\rho) + E^{\Delta}(\rho)\\ E(\rho) + E^{\Delta}(\rho) - B_F(\rho)\left(F_y(\rho) + F_y^{\Delta}(\rho)\right)\end{bmatrix}\\ \bar{\mathcal{C}} &= \begin{bmatrix} C(\rho) - D_F(\rho)\left(C_y(\rho) + C_y^{\Delta}(\rho)\right) - C_F(\rho) & C_F(\rho)\end{bmatrix}\\ \bar{\mathcal{C}}_h(\rho) &= \begin{bmatrix} C_h(\rho) - D_F(\rho)\left(C_{yh}(\rho) + C_{yh}^{\Delta}(\rho)\right) - C_{Fh}(\rho) & C_{Fh}(\rho)\end{bmatrix}\\ \bar{\mathcal{F}}(\rho) &= F(\rho) - D_F(\rho)\left(F_y(\rho) + F_y^{\Delta}(\rho)\right).\end{aligned}$$

The following result provides sufficient constructive conditions for a filter with exact memory of the form (7.48):

Theorem 7.3.1 ([15]) *Assume that there exist matrix functions $X_1, X_2, X_3 : \boldsymbol{\Delta} \to \mathbb{R}^{n\times n}$, a differentiable matrix function $P : \boldsymbol{\Delta}_\rho \to \mathbb{S}^{2n}_{\succ 0}$, constant matrices $Q, R \in \mathbb{S}^{2n}_{\succ 0}$, matrix functions $\tilde{A}_F : \boldsymbol{\Delta}_\rho \to \mathbb{R}^{n\times n}$, $\tilde{A}_{Fh} : \boldsymbol{\Delta}_\rho \to \mathbb{R}^{n\times n}$, $\tilde{B}_F : \boldsymbol{\Delta}_\rho \to \mathbb{R}^{n\times p}$, $C_F : \boldsymbol{\Delta}_\rho \to \mathbb{R}^{q\times n}$, $C_{Fh} : \boldsymbol{\Delta}_\rho \to \mathbb{R}^{q\times n}$, $D_F : \boldsymbol{\Delta}_\rho \to \mathbb{R}^{q\times p}$ and scalars $\gamma, \varepsilon > 0$ such that LMI*

$$\begin{bmatrix} \mathrm{He}[X(\rho)] & \Theta_{12} & \Theta_{13} & \Theta_{14} & 0 & -X(\rho)^{\mathrm{T}} & \bar{h}R & \Theta_{18} \\ \star & \Theta_{22} & \Theta_{23} & \Theta_{24} & \Theta_{25} & 0 & 0 & 0 \\ \star & \star & \Theta_{33} & \Theta_{34} & \Theta_{35} & 0 & 0 & 0 \\ \star & \star & \star & \Theta_{44} & \Theta_{45} & 0 & 0 & 0 \\ \star & \star & \star & \star & -\gamma I_q & 0 & 0 & \Theta_{58} \\ \star & \star & \star & \star & \star & -P(\rho) & -\bar{h}R & 0 \\ \star & \star & \star & \star & \star & \star & -R & 0 \\ \star & \star & \star & \star & \star & \star & \star & -\varepsilon I \end{bmatrix} \prec 0 \quad (7.50)$$

holds for all $(\rho, \nu) \in \boldsymbol{\Delta}_\rho \times \mathbf{V}_\nu$ where $X_{13} = X_1 + X_3$, $X_{23} = X_2 + X_3$, $\Theta_{45} = (F - D_F F_y)^{\mathrm{T}}$, $\Theta_{58} = \begin{bmatrix} 0 & -D_F H_1 \end{bmatrix}$, $\Theta_{44} = -\gamma I_m + \varepsilon(G_2^{\mathrm{T}} G_2 + G_5^{\mathrm{T}} G_5)$, $\Theta_{35} = \begin{bmatrix} C_h - D_F C_{yh} - C_{Fh} C_{Fh} \end{bmatrix}^{\mathrm{T}}$, $\Theta_{25} = \begin{bmatrix} C - D_F C_y - C_F C_F \end{bmatrix}^{\mathrm{T}}$,
$X(\rho) = \begin{bmatrix} X_1(\rho) & X_2(\rho) \\ X_3(\rho) & X_3(\rho) \end{bmatrix}$,

$$\Theta_{22} = \sum_i \frac{\partial P}{\partial \rho_i} \nu_i - P(\rho) + Q - R + \varepsilon \begin{bmatrix} G_0^{\mathrm{T}} G_0 + G_3^{\mathrm{T}} G_3 & 0 \\ 0 & 0 \end{bmatrix},$$

$$\Theta_{12} = P + \begin{bmatrix} X_{13}^{\mathrm{T}} A - \tilde{B}_F C_y - \tilde{A}_F & \tilde{A}_F \\ X_{23}^{\mathrm{T}} A - \tilde{B}_F C_y - \tilde{A}_F & \tilde{A}_F \end{bmatrix}, \quad \Theta_{14} = \begin{bmatrix} X_{13}^{\mathrm{T}} E - \tilde{B}_F F_y \\ X_{23}^{\mathrm{T}} E - \tilde{B}_F F_y \end{bmatrix}$$

$$\Theta_{13} = \begin{bmatrix} X_{13}^{\mathrm{T}} A_h - \tilde{B}_F C_{yh} - \tilde{A}_{Fh} & \tilde{A}_{Fh} \\ X_{23}^{\mathrm{T}} A_h - \tilde{B}_F C_{yh} - \tilde{A}_{Fh} & \tilde{A}_{Fh} \end{bmatrix}, \quad \Theta_{18} = \begin{bmatrix} X_{13}^{\mathrm{T}} H_0 - \tilde{B}_F H_1 \\ X_{23}^{\mathrm{T}} H_0 - \tilde{B}_F H_1 \end{bmatrix}$$

$$\Theta_{23} = R + \varepsilon \begin{bmatrix} G_0^{\mathrm{T}} G_1 + G_3^{\mathrm{T}} G_4 & 0 \\ 0 & 0 \end{bmatrix}, \quad \Theta_{24} = \varepsilon \begin{bmatrix} G_0^{\mathrm{T}} G_2 + G_3^{\mathrm{T}} G_5 \\ 0 \end{bmatrix}$$

$$\Theta_{33S} = -(1-\mu)Q - R + \varepsilon \begin{bmatrix} G_1^{\mathrm{T}} G_1 + G_4^{\mathrm{T}} G_4 & 0 \\ 0 & 0 \end{bmatrix},$$

$$\Theta_{34} = \varepsilon \begin{bmatrix} G_1^{\mathrm{T}} G_2 + G_4^{\mathrm{T}} G_5 \\ 0 \\ \\ \end{bmatrix}.$$

Then, the filter with exact memory (7.48) *with the matrices given by*

$$A_F(\rho) = X_3(\rho)^{-T} \tilde{A}_F(\rho), \quad A_{Fh}(\rho) = X_3(\rho)^{-T} \tilde{A}_{Fh}(\rho)$$
$$\text{and } B_F(\rho) = X_3(\rho)^{-T} \tilde{B}_F(\rho)$$

ensures that the L_2*-gain of the transfer* $w \to z - \hat{z}$ *is less than* γ *for all* $(\rho, h, \Delta) \in \mathscr{P}^{\nu} \times \mathscr{H}_{\mu,\bar{h}} \times \boldsymbol{\Delta}$.

Proof The proof is based on the substitution of the extended system (7.49) into the stability condition (6.30). The changes of variables $\tilde{A}_F(\rho) = X_3(\rho)^T A_F(\rho)$, $\tilde{A}_{Fh}(\rho) = X_3(\rho)^T A_{Fh}(\rho)$ and $\tilde{B}_F(\rho) = X_3(\rho)^T B_F(\rho)$ yields an LMI of the form

$$\Psi + \mathcal{U}^{\mathrm{T}} \Delta \mathcal{V} + \mathcal{V}^{\mathrm{T}} \Delta^{\mathrm{T}} \mathcal{U} \prec 0 \tag{7.51}$$

for some $\Psi, \mathcal{U}, \mathcal{V}$ and where $\Delta \in \boldsymbol{\Delta}$. Applying then Petersen's Lemma (see Appendix C.10) we get the inequality

$$\Psi + \varepsilon^{-1} \mathcal{U}^{\mathrm{T}} \mathcal{U} + \varepsilon \mathcal{V}^{\mathrm{T}} \mathcal{V} \prec 0 \tag{7.52}$$

for some $\varepsilon > 0$. A Schur complement finally yields the result. ■

7.3.2 Memoryless Filter

A memoryless filter can be simply designed by setting the matrices A_{Fh} and C_{Fh} to 0 in the filter model. This leads to the following corollary:

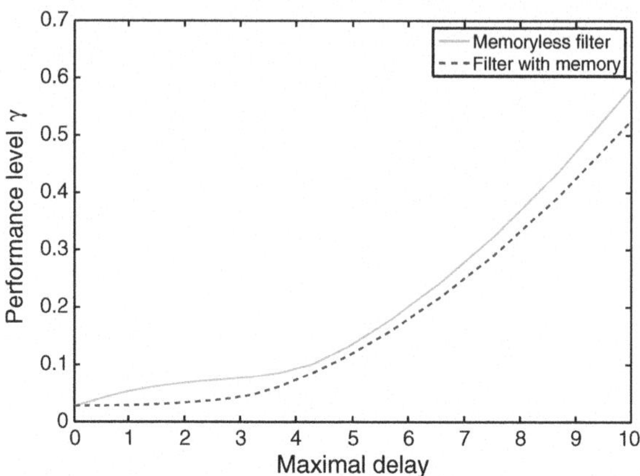

Fig. 7.12 Evolution of the worst case L_2 gain for the filter with memory (*dashed*) and the memoryless filter (*plain*) from Theorem 7.3.1—Nominal case

Corollary 7.3.2 ([15]) *Assume that there exist matrix functions* $X_1, X_2, X_3 : \boldsymbol{\Delta} \to \mathbb{R}^{n\times n}$, *a differentiable matrix function* $P : \boldsymbol{\Delta}_\rho \to \mathbb{S}^{2n}_{\succ 0}$, *constant matrices* $Q, R \in \mathbb{S}^{2n}_{\succ 0}$, *matrix functions* $\tilde{A}_F : \boldsymbol{\Delta}_\rho \to \mathbb{R}^{n\times n}$, $\tilde{B}_F : \boldsymbol{\Delta}_\rho \to \mathbb{R}^{n\times p}$, $C_F : \boldsymbol{\Delta}_\rho \to \mathbb{R}^{q\times n}$, $D_F : \boldsymbol{\Delta}_\rho \to \mathbb{R}^{q\times p}$ *and scalars* $\gamma, \varepsilon > 0$ *such that LMI* (7.50) *holds for all* $(\rho, \nu) \in \boldsymbol{\Delta}_\rho \times \mathbf{V}_\nu$ *with* $\tilde{A}_{Fh} = \tilde{C}_{Fh} = 0$.
Then, the memoryless filter (7.48) *with the matrices given by* $A_{Fh} = C_{Fh} = 0$,

$$A_F(\rho) = X_3(\rho)^{-T}\tilde{A}_F(\rho) \text{ and } B_F(\rho) = X_3(\rho)^{-T}\tilde{B}_F(\rho)$$

ensures that the L_2*-gain of the transfer* $w \to z - \hat{z}$ *is less than* γ *for all* $(\rho, h, \Delta) \in \mathscr{P}^\nu \times \mathscr{H}_{\mu,\bar{h}} \times \boldsymbol{\Delta}$.

7.3.3 Examples

Consider the following LPV time-delay system taken from [9]:

$$\begin{aligned}
\dot{x}(t) &= \begin{bmatrix} 0 & 1+0.2\rho(t) \\ -2 & -3+0.1\rho(t) \end{bmatrix} x(t) + \begin{bmatrix} 0.2\rho(t) & 0.1 \\ -0.2+0.1\rho(t) & -0.3 \end{bmatrix} x(t-h(t)) \\
&\quad + \begin{bmatrix} -0.2 \\ -0.2 \end{bmatrix} w(t) \\
z(t) &= \begin{bmatrix} 0.3 & 1.5 \\ -0.45 & 0.75 \end{bmatrix} x(t) + \begin{bmatrix} 0.5\rho(t) \\ -0.5 \end{bmatrix} w(t)
\end{aligned}$$

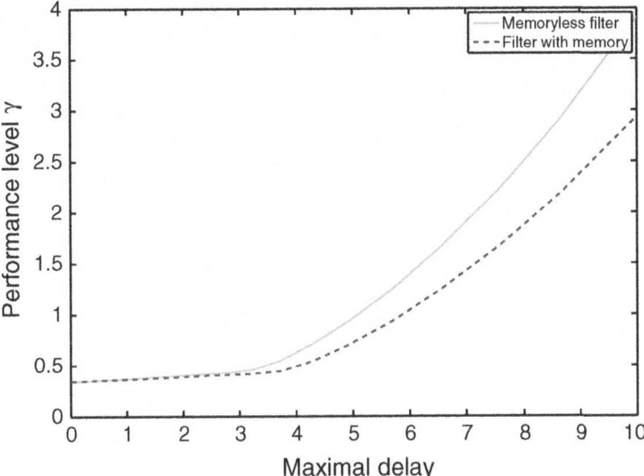

Fig. 7.13 Evolution of the worst case L_2 gain for the filter with memory (*dashed*) and the memoryless filter (*plain*) from Theorem 7.3.1—Robust case

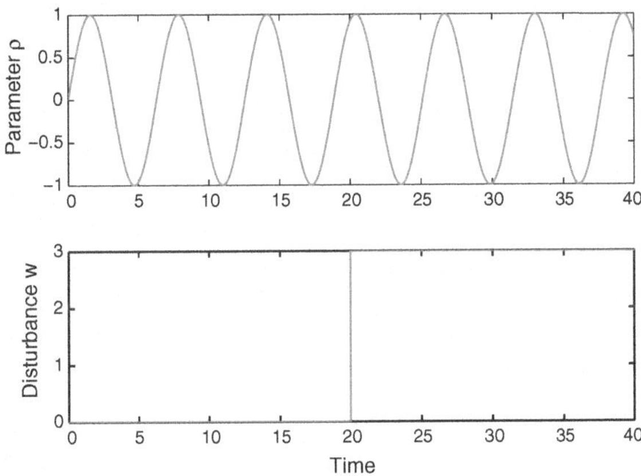

Fig. 7.14 Parameter and disturbance trajectories

$$y(t) = \begin{bmatrix} 0 & 1 \\ 0.5 & 0 \end{bmatrix} x(t) + \begin{bmatrix} 0 \\ 1 + 0.1\rho(t) \end{bmatrix} w(t) \tag{7.53}$$

where $\rho(t) \in [-1, 1]$ and $\dot{\rho}(t) \in [-1, 1]$. We also consider the following matrices driving the uncertain parameters $H_0 = H_1 = 0.1I_2$, $G_0 = G_1 = G_3 = G_4 = I_2$ and $G_2 = G_5 = \begin{bmatrix} 1 & 1 \end{bmatrix}^T$.

In view of applying Theorem 7.3.1, all the parameter-dependent variables are expressed over the basis $\{1, \rho\}$. Using Theorem 7.3.1, the evolution of the worst-case performance level γ as a function of the maximal constant delay $\bar{h}$ is depicted in Figs. 7.12 and 7.13 for the nominal and uncertain cases, respectively.

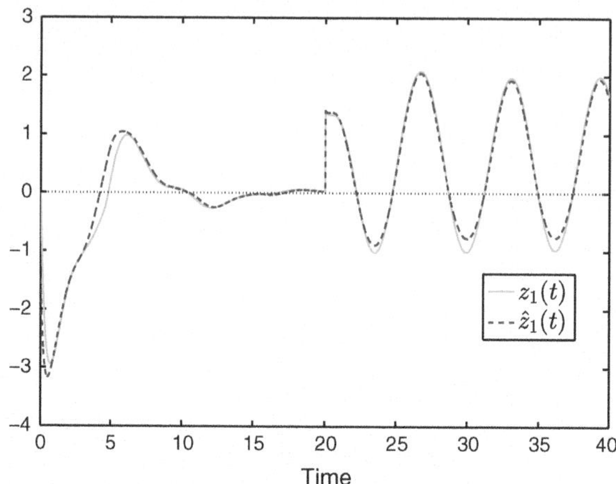

Fig. 7.15 Trajectories of z_1 and $\hat{z}_1$ using a filter with exact memory

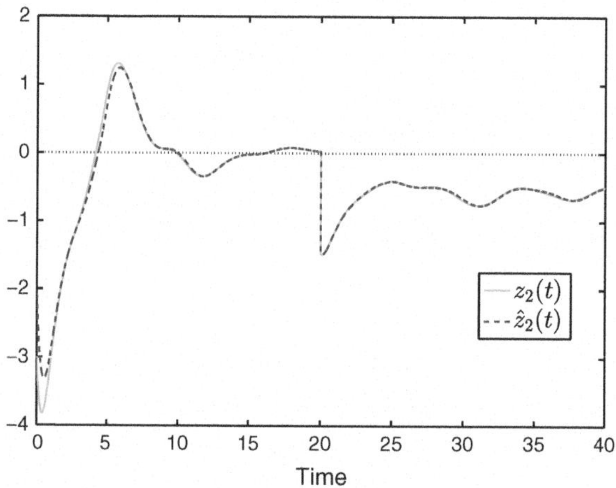

Fig. 7.16 Trajectories of z_2 and $\hat{z}_2$ using a filter with exact memory

For simulation purposes, we consider a constant delay $\bar{h} = 4.5$. The parameter and disturbance trajectories are depicted in Fig. 7.14. We obtain the trajectories depicted in Figs. 7.15 and 7.16 the case of a filter with memory. We can see that the filter is able to track quite accurately the output z.

References

1. L.H. Keel, S.P. Bhattacharyya, Robust, fragile or optimal ? IEEE Trans. Autom. Control. **42**(8), 1098–1105 (1997)
2. P. Dorato, Non-fragile controller design: an overview, in American control conference, pp. 2829–2831, Philadelphia, Pennsylvania, USA, 1998
3. D. Peaucelle, D. Arzelier, Ellipsoidal sets for resilient and robust static output feedback. IEEE Trans. Autom. Control. **50**, 899–904 (2005)
4. C. Briat, J.J. Martinez, Design of $\mathcal{H}_\infty$ bounded non-fragile controllers for discrete-time systems, in 48th IEEE Conference on Decision and Control, pp. 2192–2197, 2009
5. M. Darouach, Linear functional observers for systems with delays in state variables. IEEE Trans. Autom. Control. **46**(3), 491–496 (2001)
6. C. Briat, O. Sename, and J.-F. Lafay, Design of LPV observers for LPV Time-Delay Systems: An Algebraic Approach. Int. J. Control **84**(9), 1533–1542 (2011)
7. C.R. Mitra, S.K. Mitra, *Generalized Inverse of Matrices and its Applications* (Wiley, New York, 1971)
8. R.E. Skelton, T. Iwasaki, K.M. Grigoriadis, *A Unified Algebraic Approach to Linear Control Design* (Taylor & Francis, New York, 1997)
9. J. Mohammadpour, K.M. Grigoriadis, Less conservative results of delay-depdendent $\mathcal{H}_\infty$ filtering for a class of time-delayed LPV systems. Int. J. Control. **80**(2), 281–291 (2007)
10. X. Zhang, P. Tsiotras, C. Knospe, Stability analysis of LPV time-delayed systems. Int. J. Control. **75**, 538–558 (2002)
11. L. Wu, P. Shi, C. Wang, H. Gao, Delay-dependent robust $\mathcal{H}_\infty$ and $\mathcal{L}_2 - \mathcal{L}_\infty$ filtering for LPV systems with both discrete and distributed delays. IEE proc. Control. Theor. Appl. **153**, 483–492 (2006)
12. J. Mohammadpour, K. M. Grigoriadis. Rate-dependent mixed $\mathcal{H}_2/\mathcal{H}_\infty$ filtering for time varying state delayed LPV systems, in Conference on Decision and Control, San Diego, USA, 2006
13. J. Mohammadpour, K. M. Grigoriadis. Delay-dependent $\mathcal{H}_\infty$ filtering for a class of time-delayed LPV systems, in American Control Conference, Minneapolis, USA, 2006
14. J. Mohammadpour, K.M. Grigoriadis, Delay-dependent $\mathcal{H}_\infty$ filtering for time-delayed LPV systems. Syst. Control. Lett. **57**, 290–299 (2008)
15. C. Briat, O. Sename, J.-F. Lafay. $\mathcal{H}_\infty$ filtering of uncertain LPV systems with time-delays, in 10th European Control Conference, Budapest, Hungary, 2009

Chapter 8
Control of LPV Time-Delay Systems

Control! Control! You must learn control !
Master Yoda

Abstract This chapter pertains of the control of linear parameter-varying time-delay systems in the framework of parameter-dependent differential equations and Lyapunov-Krasovskii functionals. State-feedback and output-feedback controllers are considered both in the memoryless and with-memory cases. Controllers with approximate memory, which implement a different delay than the one in the system, are also introduced and shown to generalize the concepts of memoryless controllers and controllers with exact memory. Some examples with simulations are given for illustration.

8.1 State-Feedback Controllers

Let us consider the following class of LPV time-delay systems

$$\begin{aligned} \dot{x}(t) &= A(\rho(t))x(t) + A_h(\rho(t))x(t-h(t)) + B(\rho(t))u(t) + E(\rho(t))w(t) \\ z(t) &= C(\rho(t))x(t) + C_h(\rho(t))x(t-h(t)) + D(\rho(t))u(t) + F(\rho(t))w(t) \\ x(s) &= \phi_x(s),\ s \in [-\bar{h}, 0] \end{aligned} \tag{8.1}$$

where $x \in \mathbb{R}^n$, $u \in \mathbb{R}^m$, $w \in \mathbb{R}^p$, $z \in \mathbb{R}^q$, $\psi_x \in C([-\bar{h}, 0], \mathbb{R}^n)$ are the state of the system, the control input, the exogenous input, the controlled output and the initial condition, respectively. As in the previous chapter, the delay and the parameter vector are assumed to belong to $\mathscr{H}_{\mu,\bar{h}}$ and $\mathscr{P}^\nu$, respectively.

C. Briat, *Linear Parameter-Varying and Time-Delay Systems*,
Advances in Delays and Dynamics 3, DOI 10.1007/978-3-662-44050-6_8

In the following, we will be interested in the design of three different types of gain-scheduled state-feedback controllers. Namely:

1. Memoryless state-feedback controllers:

$$u(t) = K(\rho(t))x(t). \tag{8.2}$$

2. State-feedback controllers with exact memory:

$$u(t) = K(\rho(t))x(t) + K_h(\rho(t))x(t - h(t)). \tag{8.3}$$

3. State-feedback controllers with approximate memory:

$$u(t) = K(\rho(t))x(t) + K_d(\rho(t))x(t - d(t)), \ d \in \mathscr{D}_\delta \tag{8.4}$$

where

$$\mathscr{D}_\delta := \left\{ d : \mathbb{R}_{\geq 0} \to \mathbb{R}_{\geq 0} : \ |d(t) - h(t)| \leq \delta, \ h \in \mathscr{H}_{\mu,\bar{h}}, \ t \geq 0 \right\}. \tag{8.5}$$

As for observers and filters, the state-feedback controller with memory should improve over the memoryless one since the delay information is considered. However, knowing the delay in real time may not be possible and controllers with approximate memory are more realistic since they relax the constraint of exact delay knowledge. Note that estimating the delay or measuring it exactly is not an easy task; see e.g. [1–4]. In this regard, controllers with approximate memory can therefore be seen as a generalization of controllers with memory. It will also be emphasized later that they encompass memoryless controllers in more subtle sense that by simply setting $K_d = 0$, see e.g. [10].

8.1.1 Delay-Independent Stabilization—Generic Case

In this section, the goal is to design state-feedback controllers with exact memory of the form (8.3) in such a way that the closed-loop system is delay-independent stable. Memoryless controllers of the form (8.2) are recovered by simply setting K_h to 0. The closed-loop system obtained from the interconnection of the system (8.1) and the controller (8.3) is given by

$$\begin{aligned} \dot{x}(t) &= A_{cl}(\rho(t))x(t) + A_{hcl}(\rho(t))x_h(t) + E(\rho(t))w(t) \\ z(t) &= C_{cl}(\rho(t))x(t) + C_{hcl}(\rho(t))x_h(t) + F(\rho(t))w(t) \end{aligned} \tag{8.6}$$

where

$$\begin{aligned} A_{cl}(\rho) &= A(\rho) + B(\rho)K(\rho), \quad A_{hcl}(\rho) = A_h(\rho) + B(\rho)K_h(\rho), \\ C_{cl}(\rho) &= C(\rho) + D(\rho)K(\rho), \quad C_{hcl}(\rho) = C_h(\rho) + D(\rho)K_h(\rho). \end{aligned} \tag{8.7}$$

We have the following result on delay-independent stabilization:

Theorem 8.1.1 *Assume that there exist a continuously differentiable matrix function $X : \mathbf{\Delta}_\rho \to \mathbb{S}^n_{\succ 0}$, a constant matrix $\tilde{Q} \in \mathbb{S}^n_{\succ 0}$, matrix functions $Y, Y_h : \mathbf{\Delta}_\rho \to \mathbb{R}^{m\times n}$ and a scalar $\gamma > 0$ such that the LMI*

$$\begin{bmatrix} \Xi_{11}(\rho,\nu) & A_h(\rho)\tilde{Q} + B(\rho)Y_h(\rho) & E(\rho) & [C(\rho)+D(\rho)Y(\rho)]^{\mathrm{T}} & X(\rho) \\ \star & -\tilde{Q}_\mu & 0 & [C_h(\rho)+D(\rho)Y_h(\rho)]^{\mathrm{T}} & 0 \\ \star & \star & -\gamma I_p & F(\rho)^{\mathrm{T}} & 0 \\ \star & \star & \star & -\gamma I_q & 0 \\ \star & \star & \star & \star & -\tilde{Q} \end{bmatrix} \prec 0 \tag{8.8}$$

holds for all $(\rho,\nu) \in \mathbf{\Delta}_\rho \times \mathbf{V}_\nu$ where $\tilde{Q}_\mu = (1-\mu)\tilde{Q}$ and

$$\Xi_{11}(\rho,\nu) = \mathrm{He}[A(\rho)X(\rho) + B(\rho)Y(\rho)] - \sum_i \frac{\partial X(\rho)}{\partial \rho_i}\nu_i.$$

In such a case, a stabilizing state-feedback controller with exact memory is given by (8.3) with gains

$$K(\rho) = Y(\rho)X^{-1} \text{ and } K_h(\rho) = Y_h(\rho)\tilde{Q}^{-1}. \tag{8.9}$$

Moreover, the L_2-gain of the transfer $w \to z$ of the closed-loop system (8.6) is less than γ for all $(h,\rho) \in \mathscr{H}_{\mu,\infty} \times \mathscr{P}^\nu$.

Proof The proof is based on the delay-independent stability result of Sect. 5.6.1, i.e. Theorem 5.6.1, extended to the LPV case. That is, we consider the following Lyapunov-Krasovskii functional

$$V(x_t,\rho_t) = x(t)^{\mathrm{T}}P(\rho(t))x(t) + \int_{t-h(t)}^{t} x(s)^{\mathrm{T}}Qx(s)ds. \tag{8.10}$$

Differentiating the functional along the trajectories of the closed-loop system and considering the supply-rate $-w(t)^{\mathrm{T}}w(t)+\gamma^{-1}z(t)^{\mathrm{T}}z(t)$ yields the matrix inequality

$$\begin{bmatrix} \mathrm{He}[P(\rho)A_{cl}(\rho)] + \dot{P}(\rho(t)) & P(\rho)A_{hcl}(\rho) & PE(\rho) & P(\rho)C_{cl}(\rho) \\ \star & -(1-\dot{h}(t))Q & 0 & P(\rho)C_{hcl}(\rho) \\ \star & \star & -\gamma I_p & F(\rho)^{\mathrm{T}} \\ \star & \star & \star & -\gamma I_q \end{bmatrix} \prec 0. \tag{8.11}$$

A congruence transformation with respect to $\mathrm{diag}(X(\rho), \tilde{Q}, I_p, I_q)$, $X(\rho) := P(\rho)^{-1}$, $\tilde{Q} = Q^{-1}$, the change of variables $Y(\rho) = K(\rho)X(\rho)$, $Y_h(\rho) = K_h(\rho)\tilde{Q}$ and a Schur complement yield the result. Again, we have used the fact that for any differentiable matrix function $Z(t)$, we have that $\frac{d}{dt}Z(t)^{-1} = -Z(t)^{-1}\left[\frac{d}{dt}Z(t)\right]Z(t)^{-1}$. ■

8.1.2 Delay-Dependent Stabilization—Generic Case

Three types of controllers are considered in this section. First, conditions for delay-dependent stabilization using memoryless and exact-memory controllers. The results are then extended to the case of controllers with approximate memory.

8.1.2.1 Memoryless and Exact Memory State-Feedback Controllers

Let us first consider memoryless and exact memory controllers. We then have the following stabilization result:

Theorem 8.1.2 ([13]) *Assume that there exist a continuously differentiable matrix function $P : \mathbf{\Delta}_\rho \to \mathbb{S}^n_{\succ 0}$, constant matrices $Q, R \in \mathbb{S}^n_{\succ 0}$, $X \in \mathbb{R}^{n\times n}$, matrix functions $Y, Y_h : \mathbf{\Delta}_\rho \to \mathbb{R}^{m\times n}$ and a scalar $\gamma > 0$ such that the LMI*

$$\begin{bmatrix} -\mathrm{He}[X] & \Xi_{12} & \Xi_{13} & E(\rho) & 0 & X & \bar{h}R \\ \star & \Xi_{22} & R & 0 & \Xi_{24} & 0 & 0 \\ \star & \star & \Xi_{33} & 0 & \Xi_{34} & 0 & 0 \\ \star & \star & \star & -\gamma I_p & F(\rho)^{\mathrm{T}} & 0 & 0 \\ \star & \star & \star & \star & -\gamma I_q & 0 & 0 \\ \star & \star & \star & \star & \star & -P(\rho) & -\bar{h}R \\ \star & \star & \star & \star & \star & \star & -R \end{bmatrix} \prec 0 \tag{8.12}$$

holds for all $(\rho, \nu) \in \mathbf{\Delta}_\rho \times \mathbf{V}_\nu$ where

$$\begin{aligned} \Xi_{12} &= P(\rho) + A(\rho)X + B(\rho)Y(\rho), & \Xi_{23} &= A_h(\rho)X + B(\rho)Y_h(\rho), \\ \Xi_{22} &= \textstyle\sum_i \frac{\partial P(\rho)}{\partial \rho_i}\nu_i - P(\rho) + Q - R, & \Xi_{33} &= -(1-\mu)Q - R, \\ \Xi_{24} &= [C(\rho)X + D(\rho)Y(\rho)]^{\mathrm{T}}, & \Xi_{34} &= [C_h(\rho)X + D(\rho)Y_h(\rho)]^{\mathrm{T}}. \end{aligned}$$

In such a case, a stabilizing state-feedback controller with exact memory is given by (8.3) with gains

$$K(\rho) = Y(\rho)X^{-1} \text{ and } K_h(\rho) = Y_h(\rho)X^{-1}. \tag{8.13}$$

Moreover, the L_2-gain of the transfer $w \to z$ of the closed-loop system is less than γ for all $(h, \rho) \in \mathscr{H}_{\mu,\bar{h}} \times \mathscr{P}^{\nu}$.

Proof Substitute first the closed-loop system (8.6) into the LMI (6.30). Set X to be a constant matrix and, finally, perform a congruence transformation with respect to

$$\text{diag}(X^{-1}, X^{-1}, X^{-1}, I_{p+q}, X^{-1}, X^{-1}).$$

We obtain the final result by making the following linearizing change of variables

$$\begin{array}{lll} X \leftarrow X^{-1}, & P(\rho) \leftarrow X^{-T}P(\rho)X^{-1}, & Q \leftarrow X^{-T}QX^{-1}, \\ R \leftarrow X^{-T}RX^{-1}, & Y(\rho) \leftarrow K(\rho)X^{-1}, & Y_h(\rho) \leftarrow K_h(\rho)X^{-1}. \end{array}$$

The proof is complete. ∎

We have the following corollary for the design of memoryless controllers:

Corollary 8.1.3 ([13]) *Assume that there exist a continuously differentiable matrix function $P : \mathbf{\Delta}_\rho \to \mathbb{S}^n_{\succ 0}$, constant matrices $Q, R \in \mathbb{S}^n_{\succ 0}$, $X \in \mathbb{R}^{n\times n}$, a matrix function $Y : \mathbf{\Delta}_\rho \to \mathbb{R}^{m\times n}$ and a scalar $\gamma > 0$ such that the parameter dependent LMI*

$$\begin{bmatrix} -\text{He}[X] & \Xi_{12} & \Xi_{13} & E(\rho) & 0 & X & \bar{h}R \\ \star & \Xi_{22} & R & 0 & \Xi_{24} & 0 & 0 \\ \star & \star & \Xi_{33} & 0 & \Xi_{34} & 0 & 0 \\ \star & \star & \star & -\gamma I_p & F(\rho)^{\mathrm{T}} & 0 & 0 \\ \star & \star & \star & \star & -\gamma I_q & 0 & 0 \\ \star & \star & \star & \star & \star & -P(\rho) & -\bar{h}R \\ \star & \star & \star & \star & \star & \star & -R \end{bmatrix} \prec 0 \tag{8.14}$$

holds for all $(\rho, \nu) \in \mathbf{\Delta}_\rho \times \mathbf{V}_\nu$ where

$$\begin{array}{ll} \Xi_{12} = P(\rho) + A(\rho)X + B(\rho)Y(\rho), & \Xi_{23} = A_h(\rho)X, \\ \Xi_{22} = \sum_i \dfrac{\partial P(\rho)}{\partial \rho_i}\nu_i - P(\rho) + Q - R, & \Xi_{33} = -(1-\mu)Q - R, \\ \Xi_{24} = [C(\rho)X + D(\rho)Y(\rho)]^{\mathrm{T}}, & \Xi_{34} = [C_h(\rho)X]^{\mathrm{T}}. \end{array}$$

In such a case, a stabilizing memoryless state-feedback controller is given by (8.2) with gain

$$K(\rho) = Y(\rho)X^{-1}. \tag{8.15}$$

Moreover, the L_2-gain of the transfer $w \to z$ of the closed-loop system is less than γ for all $(h, \rho) \in \mathscr{H}_{\mu,\bar{h}} \times \mathscr{P}^{\nu}$.

8.1.2.2 State-Feedback Controllers with Approximate Memory

Let us consider now the state-feedback controller with approximate memory given by (8.4). The closed-loop system given by the interconnection of the control law (8.4) and system (8.1) is governed by the expressions:

$$\begin{aligned} \dot{x}(t) &= A_{cl}(\rho(t))x(t) + A_h(\rho(t))x(t-h(t)) + E(\rho(t))w(t) \\ &\quad + B(\rho(t))K_h(\rho(t))x(t-d(t)) \\ z(t) &= C_{cl}(\rho(t))x(t) + C_h(\rho(t))x(t-h(t)) + F(\rho(t))w(t) \\ &\quad + D(\rho(t))K_h(\rho(t))x(t-d(t)) \end{aligned} \tag{8.16}$$

where

$$A_{cl}(\rho) = A(\rho) + B(\rho)K(\rho), \quad C_{cl}(\rho) = C(\rho) + D(\rho)K(\rho). \tag{8.17}$$

The main difficulty in the stability analysis of the above system arises from the fact that the delays d and h are not independent since $d(t)$ evolves within a ball of radius $\delta > 0$ centered around $h(t)$. A way for capturing this interdependence relies on the use of operator-based model-transformations.

Proposition 8.1.4 ([17]) *The operator $\Gamma : L_2 \to L_2$ defined as*

$$\Gamma(w)(t) = \frac{1}{\sqrt{2\delta}} \int_{t-d(t)}^{t-h(t)} w(s)ds \quad \text{with } (h, d) \in \mathscr{H}_{\mu,\bar{h}} \times \mathscr{D}_{\delta} \tag{8.18}$$

has L_2-gain smaller than 1.

Proof Let us pose that $d(t) = h(t) + \varepsilon(t)$, then we have that

$$\Gamma(w)(t)^{\mathrm{T}}\Gamma(w)(t) = \frac{1}{2\delta^2}\left(\int_{t-h(t)-\varepsilon(t)}^{t-h(t)} w(s)ds\right)^{\mathrm{T}}\left(\int_{t-h(t)-\varepsilon(t)}^{t-h(t)} w(s)ds\right)$$

$$
\begin{aligned}
&\leq \frac{1}{2\delta^2}|\varepsilon(t)| \int_{t-h(t)-\max\{0,\varepsilon(t)\}}^{t-h(t)-\min\{0,\varepsilon(t)\}} w(s)^{\mathrm{T}} w(s) ds \\
&\leq \frac{1}{2\delta} \int_{t-h(t)-\max\{0,\varepsilon(t)\}}^{t-h(t)-\min\{0,\varepsilon(t)\}} w(s)^{\mathrm{T}} w(s) ds
\end{aligned}
$$

where the second inequality follows from Jensen's inequality (see Sect. 5.6.7) and the last one from the fact that $|\varepsilon(t)| \leq \delta$. The integral in the latter expression can be decomposed as

$$
\begin{aligned}
\int_{t-h(t)\max\{0,\varepsilon(t)\}}^{t-h(t)-\min\{0,\varepsilon(t)\}} w(s)^{\mathrm{T}} w(s) ds &= \mathcal{I}(\varepsilon(t)) \int_{t-h(t)-\varepsilon(t)}^{t-h(t)} w(s)^{\mathrm{T}} w(s) ds \\
&\quad + (1 - \mathcal{I}(\varepsilon(t))) \int_{t-h(t)}^{t-h(t)-\varepsilon(t)} w(s)^{\mathrm{T}} w(s) ds \\
&\leq \mathcal{I}(\varepsilon(t)) \int_{t-h(t)-\delta}^{t-h(t)} w(s)^{\mathrm{T}} w(s) ds \\
&\quad + (1 - \mathcal{I}(\varepsilon(t))) \int_{t-h(t)}^{t-h(t)+\delta} w(s)^{\mathrm{T}} w(s) ds \\
&\leq \int_{t-h(t)-\delta}^{t-h(t)} w(s)^{\mathrm{T}} w(s) ds \\
&\quad + \int_{t-h(t)}^{t-h(t)+\delta} w(s)^{\mathrm{T}} w(s) ds \\
&\leq \int_{t-h(t)-\delta}^{t-h(t)+\delta} w(s)^{\mathrm{T}} w(s) ds
\end{aligned}
$$

where $\mathcal{I}(\varepsilon(t)) = 1$ when $\varepsilon(t) \geq 0$ and 0 otherwise. We then have now that

$$
\int_0^\infty \Gamma(w)(t)^{\mathrm{T}} \Gamma(w)(t) dt \leq \frac{1}{2\delta} \int_0^\infty \int_{t-h(t)-\delta}^{t-h(t)+\delta} w(s)^{\mathrm{T}} w(s) ds dt
$$

As in the proof of Proposition 5.7.2, since $h(t)$ is differentiable and $\dot{h}(t) < 1$, we can exchange the order of integration to get that

$$\int_0^\infty \Gamma(w)(t)^{\mathrm{T}}\Gamma(w)(t)dt \leq \int_0^\infty w(s)^{\mathrm{T}} w(s)ds.$$

The proof is complete. ∎

The following result provides a way for designing a stabilizing controller with approximate memory of the form (8.4):

Theorem 8.1.5 ([17]) *Assume that there a exist continuously differentiable matrix function $P : \mathbf{\Delta}_\rho \to \mathbb{S}^n_{\succ 0}$, matrix functions $S : \mathbf{\Delta}_\rho \to \mathbb{S}^n_{\succ 0}$, $Y, Y_d : \mathbf{\Delta}_\rho \to \mathbb{R}^{m\times n}$, constant matrices $Q, R \in \mathbb{S}^n_{\succ 0}$, $X \in \mathbb{R}^{n\times n}$ and a scalar $\gamma > 0$ such that the parameter dependent LMI*

$$\begin{bmatrix} -\mathrm{He}[X] & \Omega_{12} & \Omega_{13} & \Omega_{14} & E(\rho) & 0 & X & S(\rho) & \bar{h}R \\ \star & \Omega_{22} & R & \zeta\delta R & 0 & \Omega_{26} & 0 & 0 & 0 \\ \star & \star & \Omega_{33} & \zeta\delta\Omega_{33} & 0 & \Omega_{36} & 0 & 0 & 0 \\ \star & \star & \star & \Omega_{44} & 0 & \Omega_{45} & 0 & 0 & 0 \\ \star & \star & \star & \star & -\gamma I_p & F(\rho)^{\mathrm{T}} & 0 & 0 & 0 \\ \star & \star & \star & \star & \star & -\gamma I_q & 0 & 0 & 0 \\ \star & \star & \star & \star & \star & \star & -P(\rho) & -S(\rho) & -\bar{h}R \\ \star & \star & \star & \star & \star & \star & \star & -S(\rho) & 0 \\ \star & \star & \star & \star & \star & \star & \star & \star & -R \end{bmatrix} \prec 0 \quad (8.19)$$

holds for all $(\rho, \nu) \in \mathbf{\Delta}_\rho \times \mathbf{V}_\nu$ with $\zeta = \sqrt{2}$ and

$$\begin{aligned} \Omega_{12} &= P(\rho) + A(\rho)X + B(\rho)Y(\rho), & \Omega_{13} &= A_h(\rho)X + B(\rho)Y_d(\rho), \\ \Omega_{14} &= \zeta\delta A_h(\rho)X, & \Omega_{22} &= \textstyle\sum_i \frac{\partial P(\rho)}{\partial \rho_i}\nu_i - P(\rho) + Q - R, \\ \Omega_{33} &= -(1-\mu)Q - R, & \Omega_{26} &= (C(\rho)X + D(\rho)Y(\rho))^{\mathrm{T}}, \\ \Omega_{36} &= (C_h(\rho)X + D(\rho)Y_d(\rho))^{\mathrm{T}}, & \Omega_{44} &= \zeta^2\delta^2\Omega_{33} - S(\rho), \\ \Omega_{45} &= \zeta\delta C_h(\rho)^{\mathrm{T}}. \end{aligned}$$

Then, the state-feedback control law with approximate memory (8.4) with gains

$$K(\rho) = Y(\rho)X^{-1} \text{ and } K_d(\rho) = Y_d(\rho)X^{-1} \quad (8.20)$$

stabilizes the system (8.1) and the L_2-gain of the transfer $w \to z$ of the closed-loop system is less than γ for all $(\rho, h, d) \in \mathscr{P}^\nu \times \mathscr{H}_{\mu,\bar{h}} \times \mathscr{D}_\delta$.

Proof Let us consider the Lyapunov-Krasovskii functional

$$V(x_t) = x(t)^{\mathrm{T}} P(\rho(t))x(t) + \int_{t-h(t)}^{t} x(s)^{\mathrm{T}} Q x(s) ds$$

$$+ \bar{h} \int_{-\bar{h}}^{0} \int_{t+\theta}^{t} \dot{x}(s)^{\mathrm{T}} R \dot{x}(s) ds d\theta. \tag{8.21}$$

Note that the functional does not depend on the controller delay d. Computing the derivative of the functional along the trajectory of the system (8.16) yields

$$\dot{V} \leq X^{\mathrm{T}} \left(\Pi(\rho, \dot{\rho}) + \bar{h}^2 \Xi(\rho)^{\mathrm{T}} R \Xi(\rho) \right) X \tag{8.22}$$

where $X(t) = \mathrm{col}(x(t), x(t-h(t)), x(t-d(t)), w(t))$ and

$$\Pi(\rho, \dot{\rho}) = \begin{bmatrix} \Pi_{11} & \Pi_{12} & \Pi_{13} & P(\rho)E(\rho) \\ \star & -(1-\mu)Q - R & 0 & 0 \\ \star & \star & 0 & 0 \\ \star & \star & \star & 0 \end{bmatrix}$$

where

$$\begin{aligned} \Pi_{11} &= \mathrm{He}\left[A_{cl}(\rho)^{\mathrm{T}} P(\rho) \right] + \sum_i \frac{\partial P(\rho)}{\partial \rho_i} \dot{\rho}_i + Q - R, \\ \Pi_{12} &= P(\rho) A_h(\rho) + R, \\ \Pi_{13} &= P(\rho) B(\rho) K_d(\rho), \\ \Xi(\rho) &= \begin{bmatrix} A_{cl}(\rho) & A_h(\rho) & B(\rho)K_d(\rho) & E(\rho) \end{bmatrix}. \end{aligned}$$

Using now the relation

$$w_0 = \Delta(\dot{x}) = \frac{1}{\zeta\delta} \left[x(t-h(t)) - x(t-d(t)) \right] \tag{8.23}$$

we can express $X(t)$ as

$$X(t) = \underbrace{\begin{bmatrix} I & 0 & 0 & 0 \\ 0 & I & \zeta\delta I & 0 \\ 0 & I & 0 & 0 \\ 0 & 0 & 0 & I \end{bmatrix}}_{M} \underbrace{\begin{bmatrix} x(t) \\ x(t-d(t)) \\ w_0(t) \\ w(t) \end{bmatrix}}_{Y(t)} \tag{8.24}$$

and thus we get

$$\dot{V} \leq Y(t)^{\mathrm{T}} M^{\mathrm{T}} \left(\Pi(\rho, \dot{\rho}) + \bar{h}^2 \Xi(\rho)^{\mathrm{T}} R \Xi(\rho) \right) M Y(t). \tag{8.25}$$

In order to consider the uncertain operator and characterize the L_2-gain of the transfer $w \rightarrow z$, we add the supply-rate

$$- w_0^{\mathrm{T}} S(\rho) w_0 + z_0^{\mathrm{T}} S(\rho) z_0 - \gamma w^{\mathrm{T}} w + \gamma z^{\mathrm{T}} z \tag{8.26}$$

to (8.25). Above, the signals w_0 and z_0 are defined as

$$z_0(t) = \dot{x}(t), \quad w_0(t) = \Gamma(z_0)(t). \tag{8.27}$$

The matrix function $S : \mathbf{\Delta}_\rho \rightarrow \mathbb{S}^n_{\succ 0}$ therefore plays the role of a parameter-dependent D-scaling for considering the uncertain operator Γ; see Sect. 2.6.3. Applying the same relaxation procedure as for Theorem 6.3.1, we get a matrix inequality that can be linearized by first applying the congruence transformation

$$\mathrm{diag}(X^{-1}, X^{-1}, X^{-1}, X^{-1}, I_{p+q}, X^{-1}, X^{-1}, X^{-1})$$

and by then performing the change of variables

$$\begin{array}{llll} X & \leftarrow X^{-1}, & P(\rho) & \leftarrow X^{-T} P(\rho) X^{-1}, \\ Q & \leftarrow X^{-T} Q X^{-1}, & R & \leftarrow X^{-T} R X^{-1}, \\ S(\rho) & \leftarrow X^{-T} S(\rho) X^{-1}, & Y(\rho) & \leftarrow K(\rho) X^{-1}, \\ Y_d(\rho) & \leftarrow K_d(\rho) X^{-1}. & & \end{array} \tag{8.28}$$

The proof is complete. ■

In the result above, the delay $d(t)$ may have arbitrarily fast variation rate, leading then to a high norm for the operator Γ. The norm of this operator can be, for instance, reduced by assuming a differentiable delay $d(t)$ such that $\dot{d}(t) < 1$. Considering it to be constant, i.e. $d(t) = d$, for instance equal to $\bar{h}/2$, could also be a solution. Note, however, that in the latter case, $h(t)$ and $d(t)$ would be independent and thus there would be no need for considering the operator Γ. Note that, in such a case, design conditions could have been indeed directly derived from a Lyapunov-Krasovskii functional of the form

$$\begin{aligned} V(x_t) = {} & x(t)^{\mathrm{T}} P(\rho(t)) x(t) + \int_{t-h(t)}^{t} x(s)^{\mathrm{T}} Q x(s) ds + \int_{t-d}^{t} x(s)^{\mathrm{T}} Q x(s) ds \\ & + \bar{h} \int_{-\bar{h}}^{0} \int_{t+\theta}^{t} \dot{x}(s)^{\mathrm{T}} R \dot{x}(s) ds d\theta + d \int_{-d}^{0} \int_{t+\theta}^{t} \dot{x}(s)^{\mathrm{T}} R \dot{x}(s) ds d\theta. \end{aligned}$$

When the delay is exactly known, i.e. when $\delta = 0$, then Theorem 8.1.5 reduces to Theorem 8.1.2. This means that Theorem 8.1.5 encompasses the case of exact memory. This is stated in the following result:

Proposition 8.1.6 ([17]) *Assume $\delta = 0$, then condition (8.19) is equivalent to condition (8.12).*

Proof The proof is only sketched since it relies on simple manipulations. By setting first δ to 0 in condition (8.19), the 4th column/row reduces to a 0 column/row except for the diagonal entry which is equal to $-S(\rho)$. Since $S(\rho) \succ 0$, then this row/column can be removed. Then, all we have to do is analyze the behavior of the LMI condition with respect to the terms in $S(\rho)$ located on the 8th row/column. A Schur complement on the block (8, 8) yields the matrix inequality

$$\Upsilon_1(\rho) + \Upsilon_2(\rho) \prec 0$$

where $\Upsilon_1(\rho)$ is exactly the matrix in (8.12) and

$$\Upsilon_2(\rho) = \begin{bmatrix} S(\rho) & 0 & \dots & 0 & -S(\rho) & 0 \\ 0 & 0 & \dots & 0 & 0 & 0 \\ \vdots & \vdots & \ddots & \vdots & \vdots & \vdots \\ 0 & 0 & \dots & 0 & 0 & 0 \\ -S(\rho) & 0 & \dots & 0 & S(\rho) & 0 \\ 0 & 0 & \dots & 0 & 0 & 0 \end{bmatrix}$$

Since $\Upsilon_2(\rho)$ is positive semidefinite, then by choosing $S(\rho) = S_0$ sufficiently small, the condition (8.12) is retrieved. ■

8.1.2.3 Example

Let us consider the LPV time-delay system (8.1) with matrices [5, 6]:

$$\begin{aligned} \dot{x}(t) &= \begin{bmatrix} 0 & 1 + \phi\rho(t) \\ -2 & -3 + \sigma\rho(t) \end{bmatrix} x(t) + \begin{bmatrix} \phi\rho(t) & 0.1 \\ -0.2 + \sigma\rho(t) & -0.3 \end{bmatrix} x(t - h(t)) \\ &\quad + \begin{bmatrix} 0.2 \\ 0.2 \end{bmatrix} w(t) + \begin{bmatrix} \phi\rho(t) \\ 0.1 + \sigma\rho(t) \end{bmatrix} u(t) \\ z(t) &= \begin{bmatrix} 0 & 10 \\ 0 & 0 \end{bmatrix} x(t) + \begin{bmatrix} 0 \\ 0.1 \end{bmatrix} u(t) \end{aligned} \tag{8.29}$$

where $\phi = 0.2$, $\sigma = 0.1$, $\rho(t) := \sin(t)$. In the following, the parameter dependent matrices that have to be determined are considered to be quadratic in ρ.

Design of a Memoryless Controller
Let us consider that $\bar{h} = 10$ and that $\mu = 0.9$. Using Corollary 8.1.3, we obtain the controller (8.2) with the matrix

$$K(\rho) = \begin{bmatrix} 0.5724 - 6.3679\rho - 1.4898\rho^2 \\ -0.7141 - 4.1617\rho - 0.8425\rho^2 \end{bmatrix}^{\mathrm{T}} \tag{8.30}$$

which ensures that the L_2-gain of the transfer $w \to z$ is less that $\gamma = 12.8799$.

Design of a Controller with Exact Memory
Considering the same parameters as for the memoryless case, Theorem 8.1.2 yields the controller matrices

$$\begin{aligned} K(\rho) &= \begin{bmatrix} 1.0524 - 2.8794\rho - 0.4854\rho^2 \\ -0.7731 - 1.8859\rho + 0.1181\rho^2 \end{bmatrix}^{\mathrm{T}} \\ K_h(\rho) &= \begin{bmatrix} -0.6909 + 0.5811\rho + 0.1122\rho^2 \\ -0.0835 + 0.3153\rho + 0.0689\rho^2 \end{bmatrix}^{\mathrm{T}} \end{aligned} \tag{8.31}$$

which ensures that the L_2-gain of the transfer $w \to z$ is less that $\gamma = 4.1641$. This is a clear improvement over the memoryless controller.

For comparison purposes, the minimal achievable L_2-gain for different values of the maximal delay $\bar{h}$ for both types of controllers are depicted in Figs. 8.1–8.4. It is interesting to note that when $\bar{h} \leq 1$, the controllers perform quite the same. The effect of the delayed part in the controller starts to be visible when the maximal delay becomes larger.

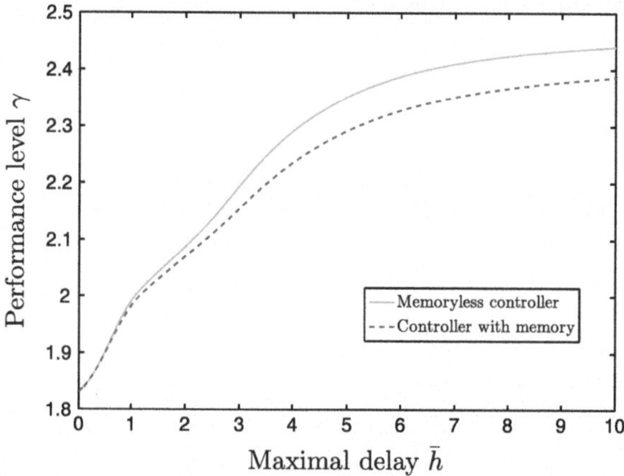

Fig. 8.1 Comparison of the minimal L_2-gains for different maximal delays and controllers—Case $\mu = 0$

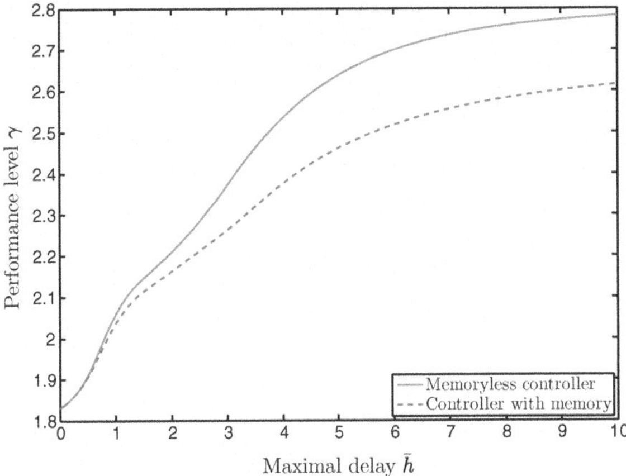

Fig. 8.2 Comparison of the minimal L_2-gains for different maximal delays and controllers—Case $\mu = 0.5$

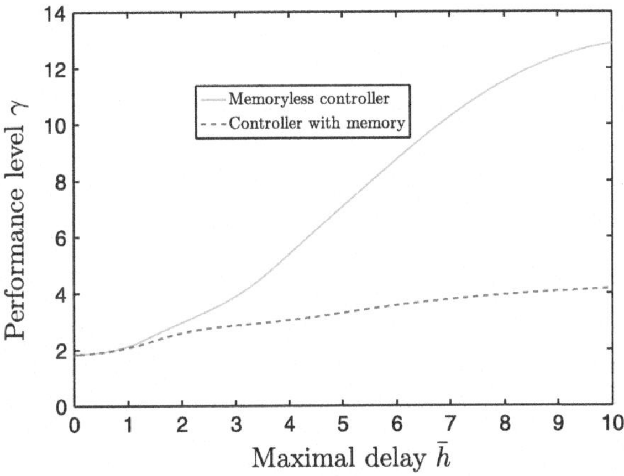

Fig. 8.3 Comparison of the minimal L_2-gains for different maximal delays and controllers—Case $\mu = 0.9$

Design of a Controller with Approximate Memory

Two remarkable values for δ deserve to be pointed out. The first one is $\delta = 0$ since, as stated in Proposition 8.1.6, we should recover the results of the exact memory case. Theorem 8.1.5 with $\bar{h} = 10$, $\mu = 0.9$ and $\delta = 0$ yields the controller (8.4) with the matrices

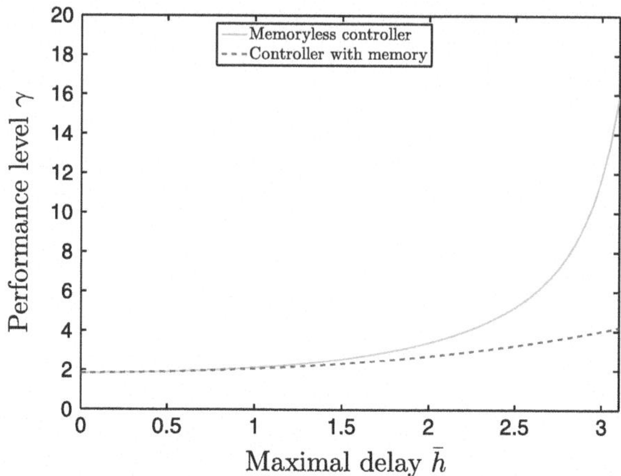

Fig. 8.4 Comparison of the minimal L_2-gains for different maximal delays and controllers—Case $\mu = 0.99$

$$
\begin{aligned}
K(\rho)|_{\delta=0} &= \begin{bmatrix} 1.0542 - 2.8895\rho - 0.4827\rho^2 \\ -0.7714 - 1.8912\rho + 0.1216\rho^2 \end{bmatrix}^{\mathrm{T}} \\
K_d(\rho)|_{\delta=0} &= \begin{bmatrix} -0.6885 + 0.5849\rho + 0.1116\rho^2 \\ -0.0817 + 0.3148\rho + 0.0667\rho^2 \end{bmatrix}^{\mathrm{T}}
\end{aligned} \tag{8.32}
$$

which ensures that the L_2-gain of the transfer $w \to z$ is less that $\gamma|_{\delta=0} = 4.1658$, a value very close to the one obtained in the exact memory case. Note also that the controller matrices are almost equal to the ones obtained in the exact memory case.

The second value for δ is $\delta = \bar{h}$ since, in this case, the variability of $d(t)$ is equal to the maximal delay value. Therefore, this is equivalent to say that $h(t)$ is unknown. Theorem 8.1.5 with $\bar{h} = 10$, $\mu = 0.9$ and $\delta = 10$ yields the controller (8.4) with the matrices

$$
\begin{aligned}
K(\rho)|_{\delta=10} &= \begin{bmatrix} 0.4375 - 6.3445\rho - 1.3576\rho^2 \\ -0.9547 - 4.1173\rho - 0.6064\rho^2 \end{bmatrix}^{\mathrm{T}} \\
K_d(\rho)|_{\delta=10} &= \begin{bmatrix} -0.0160 - 0.0003\rho + 0.0125\rho^2 \\ -0.0006 - 0.0006\rho + 0.0010\rho^2 \end{bmatrix}^{\mathrm{T}}
\end{aligned} \tag{8.33}
$$

which ensures that L_2-gain of the transfer $w \to z$ is less that $\gamma|_{\delta=10} = 13.1165$, a value very close to the one obtained in the memoryless case. More importantly, it is interesting to point out that the norm of K_d is small, which reflects the poor knowledge of $h(t)$ and the poor confidence degree in the delayed part of the controller which

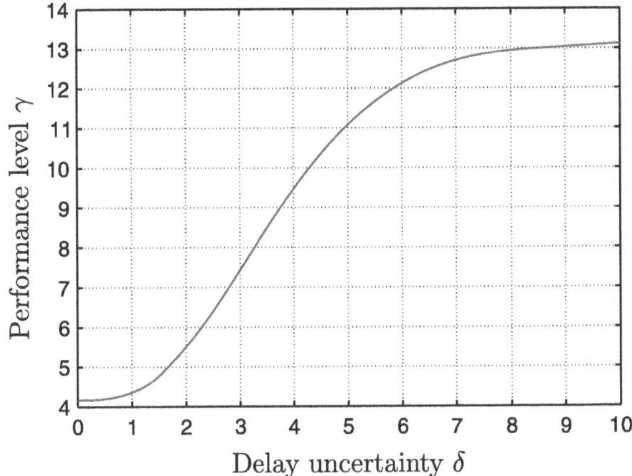

Fig. 8.5 Evolution of the worst-case L_2 gain of the closed-loop system with respect to maximal delay uncertainty δ

result in the design of an almost memoryless controller. Note that the matrix $K(\rho)$ is also very close to the one obtained in the memoryless case.

These facts illustrate well that controllers with approximate memory can be considered as a general and unified formulation for memoryless controllers and controllers with memory. The evolution of the minimal L_2-gain for different values of δ is depicted in Fig. 8.5.

8.1.3 Delay-Independent Stabilization—LFT Case

We illustrate in this section that design conditions for delay-independent stabilization via gain-scheduled state-feedback controllers can be obtained in the LFT framework. To do so, we first rewrite the system (8.1) as an LPV time-delay system in LFT-form

$$\begin{aligned}\dot{x}(t) &= Ax(t) + A_h x(t-h(t)) + E_0 w_0(t) + E_1 w_1(t) + Bu(t)\\ z_0(t) &= C_0 x(t) + C_{0h} x(t-h(t)) + F_{00} w_0(t) + F_{10} w_1(t) + D_0 u(t)\\ z_1(t) &= C_1 x(t) + C_{1h} x(t-h(t)) + F_{11} w_1(t) + F_{11} w_1(t) + D_1 u(t)\\ w_0(t) &= \Theta(\rho(t)) z_0(t)\end{aligned} \tag{8.34}$$

where $w_0, z_0 \in \mathbb{R}^{n_0}$ are the signals involved in the scheduling channel. The matrix $\Theta(\rho)$ is assumed to be a diagonal matrix depending linearly on the parameters, see e.g. Sect. 3.5, and we consider parameter trajectories in $\mathscr{P}_1^{\infty}$. The set of D-scalings associated with the structure of Θ is given by

$$\mathcal{D}(\Theta) = \left\{ L \in \mathbb{S}^{n_0}_{\succ 0} : L^{1/2}\Theta(\rho) = \Theta(\rho)L^{1/2},\ \rho \in [-1, 1]^{N_p} \right\} \tag{8.35}$$

where N_p is the number of parameters and $L^{1/2}$ is the unique positive square root of L.

8.1.3.1 State-Feedback Controller with Exact Memory

Let us first address the case of gain-scheduled state-feedback controllers with exact memory which take the form

$$\begin{aligned} \begin{bmatrix} u(t) \\ z_c(t) \end{bmatrix} &= \bar{K} \begin{bmatrix} x(t) \\ x(t-h(t)) \\ w_c(t) \end{bmatrix} \\ w_c(t) &= \Theta(\rho(t))z_c(t) \end{aligned} \tag{8.36}$$

where z_c/w_c is the scheduling channel of the controller. We then have the following result:

Theorem 8.1.7 *Assume that there exist matrices $\tilde{P}, \tilde{Q} \in \mathbb{S}^n_{\succ 0}$ and $L_1, \tilde{L}_1 \in \mathcal{D}(\Theta)$ such that the LMIs*

$$\begin{bmatrix} L_1 & I \\ \star & \tilde{L}_1 \end{bmatrix} \succ 0$$

$$\begin{bmatrix} -L_1 & 0 \\ \star & -\gamma I_p \end{bmatrix} + \begin{bmatrix} F_{00} & F_{01} \\ F_{10} & F_{11} \end{bmatrix}^{\mathrm{T}} \begin{bmatrix} L_1 & 0 \\ \star & \gamma^{-1} I_q \end{bmatrix} \begin{bmatrix} F_{00} & F_{01} \\ F_{10} & F_{11} \end{bmatrix} \prec 0 \tag{8.37}$$

$$\mathcal{N}^{\mathrm{T}} \left(\begin{bmatrix} \mathrm{He}[A\tilde{P}] & \tilde{P}C_0^{\mathrm{T}} & \tilde{P}C_1^{\mathrm{T}} \\ \star & -\tilde{L}_1 & 0 \\ \star & \star & -\gamma I_q \end{bmatrix} + ZMZ^{\mathrm{T}} \right) \mathcal{N} \prec 0$$

hold where

$$M = \begin{bmatrix} \frac{1}{1-\mu}\tilde{Q} & 0 & 0 \\ \star & \tilde{L}_1 & 0 \\ \star & \star & \gamma^{-1} I_p \end{bmatrix}, \quad Z = \begin{bmatrix} A_h & E_0 & E_1 \\ C_{0h} & F_{00} & F_{01} \\ C_{1h} & F_{10} & F_{11} \end{bmatrix} \tag{8.38}$$

and $\mathcal{N}$ is a basis of the null-space of the matrix $\begin{bmatrix} B^{\mathrm{T}} & D_0^{\mathrm{T}} & D_1^{\mathrm{T}} \end{bmatrix}$.

Then, there exists a gain-scheduled state-feedback controller of the form (8.36) which stabilizes the system (8.34) and ensures that the L_2-gain of the *transfer $w \to z$ of the closed-loop system (8.34)–(8.36) is less than γ for all $(h, \rho) \in \mathscr{H}_{\mu,\infty} \times \mathscr{P}_1^{\infty}$.*

Proof The proof is based on the use of the Lyapunov-Krasovskii functional

$$V(x_t) = x(t)^{\mathrm{T}} P x(t) + \int_{t-h(t)}^{t} x(s)^{\mathrm{T}} Q x(s) ds \tag{8.39}$$

which serves the role of storage function. The following supply-rate combining the scaled small-gain and the L_2-performance

$$s(w, z) = -\gamma w_1^{\mathrm{T}} w_1 + \gamma^{-1} z_1^{\mathrm{T}} z_1 - \begin{bmatrix} w_0 \\ w_c \end{bmatrix}^{\mathrm{T}} L \begin{bmatrix} w_0 \\ w_c \end{bmatrix} + \begin{bmatrix} z_0 \\ z_c \end{bmatrix}^{\mathrm{T}} L \begin{bmatrix} z_0 \\ z_c \end{bmatrix} \tag{8.40}$$

is also considered with

$$L := \begin{bmatrix} L_1 & L_2 \\ \star & L_3 \end{bmatrix} \tag{8.41}$$

and $L \in \mathcal{D}(\mathrm{diag}(\Theta, \Theta))$. The derivative of the storage function is upper-bounded as in Sect. 5.6.7, i.e. using Jensen's inequality, and after adding the supply-rate, we get the matrix inequality

$$\mathcal{S}^{\mathrm{T}} \left[\begin{array}{ccccc|ccccc} 0 & 0 & 0 & 0 & 0 & P & 0 & 0 & 0 & 0 \\ \star & -(1-\mu)Q & 0 & 0 & 0 & 0 & 0 & 0 & 0 & 0 \\ \star & \star & -L_1 & -L_2 & 0 & 0 & 0 & 0 & 0 & 0 \\ \star & \star & \star & -L_3 & 0 & 0 & 0 & 0 & 0 & 0 \\ \star & \star & \star & \star & -\gamma I_p & 0 & 0 & 0 & 0 & 0 \\ \hline \star & \star & \star & \star & \star & 0 & 0 & 0 & 0 & 0 \\ \star & \star & \star & \star & \star & \star & Q & 0 & 0 & 0 \\ \star & \star & \star & \star & \star & \star & \star & L_1 & L_2 & 0 \\ \star & \star & \star & \star & \star & \star & \star & \star & L_3 & 0 \\ \star & \star & \star & \star & \star & \star & \star & \star & \star & \gamma^{-1} I_q \end{array}\right] \mathcal{S} \prec 0 \tag{8.42}$$

where

$$\mathcal{S} := \left[\begin{array}{ccccc} I & 0 & 0 & 0 & 0 \\ 0 & I & 0 & 0 & 0 \\ 0 & 0 & I & 0 & 0 \\ 0 & 0 & 0 & I & 0 \\ 0 & 0 & 0 & 0 & I \\ \hline A & A_h & E_0 & 0 & E_1 \\ I & 0 & 0 & 0 & 0 \\ C_0 & C_{0h} & F_{00} & 0 & F_{01} \\ 0 & 0 & 0 & 0 & 0 \\ C_1 & C_{1h} & F_{10} & 0 & F_{11} \end{array}\right] + \left[\begin{array}{cc} 0 & 0 \\ 0 & 0 \\ 0 & 0 \\ 0 & 0 \\ 0 & 0 \\ \hline B & 0 \\ 0 & 0 \\ D_0 & 0 \\ 0 & I \\ D_1 & 0 \end{array}\right] \bar{K} \begin{bmatrix} I & 0 & 0 & 0 & 0 \\ 0 & I & 0 & 0 & 0 \\ 0 & 0 & 0 & I & 0 \end{bmatrix}.$$

The rank of $\mathcal{S}$ is equal to $2n + p + 2n_0$ and the number of negative eigenvalues of the central matrix in (8.42) is $2n + 2n_0 + p$. The dualijection lemma, i.e. Lemma 3.5.1, can thus be applied. By doing so, we get the conditions of the theorem where we have set $\tilde{P} = P^{-1}$, $\tilde{Q} = Q^{-1}$ and $\tilde{L} = L^{-1} = \begin{bmatrix} \tilde{L}_1 & \tilde{L}_2 \\ \star & \tilde{L}_3 \end{bmatrix}$. ■

The controller can be constructed following the same procedure as in the non-delayed case; see Sect. 3.5. That is, we first construct the matrices L and $\tilde{L}$ and then we solve for $\bar{K}$ in the LMI (8.42). The controller can also be shown to be well-posed using the same arguments as in Sect. 3.5.

8.1.3.2 Memoryless State-Feedback Controller

Let us consider now memoryless gain-scheduled state-feedback controllers of the form

$$\begin{aligned} \begin{bmatrix} u(t) \\ z_c(t) \end{bmatrix} &= \bar{K} \begin{bmatrix} x(t) \\ w_c(t) \end{bmatrix} \\ w_c(t) &= \Theta(\rho(t)) z_c(t). \end{aligned} \tag{8.43}$$

We then have the following stabilization result:

Theorem 8.1.8 *Assume that there exist matrices $\tilde{P}, Q \in \mathbb{S}^n_{\succ 0}$ and $L_1, \tilde{L}_1 \in \mathcal{D}(\Theta)$ such that the LMIs*

$$\begin{bmatrix} L_1 & I \\ \star & \tilde{L}_1 \end{bmatrix} \succ 0$$

$$\begin{bmatrix} -(1-\mu)Q & 0 & 0 \\ \star & -L_1 & 0 \\ \star & \star & -\gamma I_p \end{bmatrix} + \begin{bmatrix} C_{0h}^{\mathrm{T}} & C_{1h}^{\mathrm{T}} \\ F_{00}^{\mathrm{T}} & F_{10}^{\mathrm{T}} \\ F_{01}^{\mathrm{T}} & F_{11}^{\mathrm{T}} \end{bmatrix} \begin{bmatrix} L_1 & 0 \\ \star & \gamma^{-1} I_q \end{bmatrix} \begin{bmatrix} C_{0h}^{\mathrm{T}} & C_{1h}^{\mathrm{T}} \\ F_{00}^{\mathrm{T}} & F_{10}^{\mathrm{T}} \\ F_{01}^{\mathrm{T}} & F_{11}^{\mathrm{T}} \end{bmatrix}^{\mathrm{T}} \prec 0$$

$$\begin{bmatrix} \mathcal{N} & 0 \\ 0 & I_n \end{bmatrix}^{\mathrm{T}} \begin{bmatrix} \mathcal{M}_1 & \mathcal{M}_2 \\ \star & -(1-\mu)Q \end{bmatrix} \begin{bmatrix} \mathcal{N} & 0 \\ 0 & I_n \end{bmatrix} \prec 0 \tag{8.44}$$

hold where

$$\mathcal{M}_1 = \begin{bmatrix} \mathrm{He}[A\tilde{P}] & \tilde{P}C_0^{\mathrm{T}} & \tilde{P}C_1^{\mathrm{T}} \\ \star & -\tilde{L}_1 & 0 \\ \star & \star & -\gamma I_q \end{bmatrix} + \begin{bmatrix} E_0 & E_1 \\ F_{00} & F_{01} \\ F_{10} & F_{11} \end{bmatrix} \begin{bmatrix} \tilde{L}_1 & 0 \\ \star & \gamma^{1} I_p \end{bmatrix} \begin{bmatrix} E_0 & E_1 \\ F_{00} & F_{01} \\ F_{10} & F_{11} \end{bmatrix}^{\mathrm{T}}$$

$$\mathcal{M}_2 = \begin{bmatrix} A_h^{\mathrm{T}} & C_{0h}^{\mathrm{T}} & C_{1h}^{\mathrm{T}} \end{bmatrix}^{\mathrm{T}}$$

and $\mathcal{N}$ is a basis of the null-space of the matrix $\begin{bmatrix} B^{\mathrm{T}} & D_0^{\mathrm{T}} & D_1^{\mathrm{T}} \end{bmatrix}$.

Then, there exists a gain-scheduled state-feedback controller of the form (8.43) which stabilizes the system (8.34) and ensures that the L_2-gain of the transfer $w \to z$ of the closed-loop system (8.34)–(8.43) is less than γ for all $(h, \rho) \in \mathscr{H}_{\mu,\infty} \times \mathscr{P}_1^{\infty}$.

Proof The proof follows the same lines as the one of Theorem 8.1.7. ■

The controller can be constructed following the same procedure as in the non-delayed case; see Sect. 3.5. That is, we first construct the matrices L and $\tilde{L}$ and then we solve for $\bar{K}$ in the LMI (8.42). The controller can also be shown to be well-posed using the same arguments as in Sect. 3.5.

8.1.4 Delay-Dependent Stabilization—LFT Case

We address in this section the problem of delay-dependent stabilization. Both memoryless and exact memory gain-scheduled state-feedback controllers are considered.

8.1.4.1 State-Feedback Controller with Exact Memory

The following result is the delay-dependent counterpart of Theorem 8.1.7:

Theorem 8.1.9 *Assume that there exist matrices $\tilde{P}, \tilde{R} \in \mathbb{S}^n_{\succ 0}$ and $L_1, \tilde{L}_1 \in \mathcal{D}(\Theta)$ such that the LMIs*

$$\begin{bmatrix} L_1 & I \\ \star & \tilde{L}_1 \end{bmatrix} \succ 0$$

$$\begin{bmatrix} \mathcal{M}_1^1 & \mathcal{M}_2^1 \\ \star & -\bar{h}^{-2}\tilde{R} \end{bmatrix} \prec 0 \tag{8.45}$$

$$\begin{bmatrix} \mathcal{N} & 0 \\ 0 & I_n \end{bmatrix}^{\mathrm{T}} \begin{bmatrix} \mathcal{M}_1^2 & \mathcal{M}_2^2 \\ \star & -\bar{h}^{-2}\tilde{R} \end{bmatrix} \begin{bmatrix} \mathcal{N} & 0 \\ 0 & I_n \end{bmatrix} \prec 0$$

hold where

$$\mathcal{M}_1^1 = \begin{bmatrix} -L_1 & 0 \\ \star & -\gamma I_p \end{bmatrix} + \begin{bmatrix} F_{00} & F_{01} \\ F_{10} & F_{11} \end{bmatrix}^{\mathrm{T}} \begin{bmatrix} L_1 & 0 \\ \star & \gamma^{-1} I_q \end{bmatrix} \begin{bmatrix} F_{00} & F_{01} \\ F_{10} & F_{11} \end{bmatrix} \prec 0$$

$$\mathcal{M}_1^2 = \begin{bmatrix} E_0^{\mathrm{T}} & E_1^{\mathrm{T}} \end{bmatrix}^{\mathrm{T}}$$

$$\mathcal{M}_1^2 = \begin{bmatrix} \mathrm{He}[(A+A_h)\tilde{P} & C_0^{\mathrm{T}}\tilde{P} & C_1^{\mathrm{T}}\tilde{P} \\ \star & \tilde{L}_1 & 0 \\ \star & \star & -\gamma I_q \end{bmatrix}$$

$$+ \begin{bmatrix} A_h & E_0 & E_1 \\ C_{0h} & F_{00} & F_{01} \\ C_{1h} & F_{10} & F_{11} \end{bmatrix} \begin{bmatrix} \tilde{R} & 0 & 0 \\ \star & \tilde{L}_1 & 0 \\ \star & \star & \gamma^{-1} I_p \end{bmatrix} \begin{bmatrix} A_h & E_0 & E_1 \\ C_{0h} & F_{00} & F_{01} \\ C_{1h} & F_{10} & F_{11} \end{bmatrix}^{\mathrm{T}}$$

$$\mathcal{M}_2^2 = \left[\tilde{P}(A+A_h)^{\mathrm{T}} \quad \tilde{P}(C_0 + C_{0h})^{\mathrm{T}} \quad \tilde{P}(C_1 + C_{1h})^{\mathrm{T}}\right]^{\mathrm{T}} \tag{8.46}$$

and $\mathcal{N}$ is a basis of the null-space of the matrix $\begin{bmatrix} B^{\mathrm{T}} & D_0^{\mathrm{T}} & D_1^{\mathrm{T}} \end{bmatrix}$.

Then, there exists a gain-scheduled state-feedback controller of the form (8.36) which stabilizes the system (8.34) and ensures that the L_2-gain of the transfer $w \to z$ of the closed-loop system (8.34)–(8.36) is less than γ for all $(h, \rho) \in \mathscr{H}_{\infty,\bar{h}} \times \mathscr{P}_1^{\infty}$.

Proof The proof follows the same lines as the proof of Theorem 8.1.7 with the difference that the following Lyapunov-Krasovskii functional (storage function) is considered

$$V(x_t) = x(t)^{\mathrm{T}} P x(t) + \bar{h} \int_{-\bar{h}}^{0} \int_{t+\theta}^{t} x(s)^{\mathrm{T}} R x(s) ds d\theta. \tag{8.47}$$

Note that, in this case, we have

$$\mathcal{S}^{\mathrm{T}} \left[\begin{array}{ccccc|ccccc} 0 & 0 & 0 & 0 & 0 & P & 0 & 0 & 0 & 0 \\ \star & -R & 0 & 0 & 0 & 0 & 0 & 0 & 0 & 0 \\ \star & \star & -L_1 & -L_2 & 0 & 0 & 0 & 0 & 0 & 0 \\ \star & \star & \star & -L_3 & 0 & 0 & 0 & 0 & 0 & 0 \\ \star & \star & \star & \star & -\gamma I_p & 0 & 0 & 0 & 0 & 0 \\ \hline \star & \star & \star & \star & \star & 0 & 0 & 0 & 0 & 0 \\ \star & \star & \star & \star & \star & \star & \bar{h}^2 R & 0 & 0 & 0 \\ \star & \star & \star & \star & \star & \star & \star & L_1 & L_2 & 0 \\ \star & \star & \star & \star & \star & \star & \star & \star & L_3 & 0 \\ \star & \star & \star & \star & \star & \star & \star & \star & \star & \gamma^{-1} I_q \end{array}\right] \mathcal{S} \prec 0 \tag{8.48}$$

where

$$\mathcal{S} := \left[\begin{array}{ccccc} I & 0 & 0 & 0 & 0 \\ -I & I & 0 & 0 & 0 \\ 0 & 0 & I & 0 & 0 \\ 0 & 0 & 0 & I & 0 \\ 0 & 0 & 0 & 0 & I \\ \hline A & A_h & E_0 & 0 & E_1 \\ A & A_h & E_0 & 0 & E_1 \\ C_0 & C_{0h} & F_{00} & 0 & F_{01} \\ 0 & 0 & 0 & 0 & 0 \\ C_1 & C_{1h} & F_{10} & 0 & F_{11} \end{array}\right] + \left[\begin{array}{cc} 0 & 0 \\ 0 & 0 \\ 0 & 0 \\ 0 & 0 \\ 0 & 0 \\ B & 0 \\ B & 0 \\ D_0 & 0 \\ 0 & I \\ D_1 & 0 \end{array}\right] K \begin{bmatrix} I & 0 & 0 & 0 & 0 \\ 0 & I & 0 & 0 & 0 \\ 0 & 0 & 0 & I & 0 \end{bmatrix}.$$

The rank of $\mathcal{S}$ is equal to $2n + 2n_0 + p$ whereas the number of negative eigenvalues of the central matrix in (8.48) is equal to $2n + 2n_0 + p$. We can therefore apply the dualijection lemma. We need, however, to consider the generalized dualijection lemma, i.e. Lemma 3.5.2, since the upper-block of $\mathcal{S}$ is not equal to the identity matrix. By doing so, we obtain the conditions stated in the theorem where we have set $\tilde{R} = R^{-1}$, $\tilde{P} = P^{-1}$, $\tilde{L} = \begin{bmatrix} \tilde{L}_1 & \tilde{L}_2 \\ \star & \tilde{L}_3 \end{bmatrix} := L^{-1} = \begin{bmatrix} L_1 & L_2 \\ \star & L_3 \end{bmatrix}^{-1}$. ■

The controller can be constructed following the same procedure as in the non-delayed case; see Sect. 3.5. That is, we first construct the matrices L and $\tilde{L}$ and then we solve for $\tilde{K}$ in the LMI (8.42). The controller can also be shown to be well-posed using the same arguments as in Sect. 3.5.

8.1.4.2 Memoryless State-Feedback Controller

This result is the delay-dependent counterpart of Theorem 8.1.8:

Theorem 8.1.10 *Assume that there exist matrices $\tilde{P}$, R, $\tilde{R} \in \mathbb{S}^n_{\succ 0}$ and $L_1, \tilde{L}_1 \in \mathcal{D}(\Theta)$ verifying $R\tilde{R} = I$ such that the LMIs*

$$\begin{bmatrix} L_1 & I \\ \star & \tilde{L}_1 \end{bmatrix} \succ 0$$

$$\begin{bmatrix} -R & 0 & 0 \\ \star & -L_1 & 0 \\ \star & \star & -\gamma I_p \end{bmatrix} + Z^{\mathrm{T}} \begin{bmatrix} \bar{h}^2 R & 0 & 0 \\ \star & L_1 & 0 \\ \star & \star & \gamma^{-1} I_q \end{bmatrix} Z \prec 0 \tag{8.49}$$

$$\begin{bmatrix} \mathcal{N} & 0 \\ 0 & I_n \end{bmatrix}^{\mathrm{T}} \begin{bmatrix} \mathcal{M}_{11} & \mathcal{M}_{12} \\ \star & -\bar{h}^{-2}\tilde{R} \end{bmatrix} \begin{bmatrix} \mathcal{N} & 0 \\ 0 & I_n \end{bmatrix} \prec 0$$

hold where

$$Z = \begin{bmatrix} A_h & E_0 & E_1 \\ C_{0h} & F_{00} & F_{01} \\ C_{1h} & F_{10} & F_{11} \end{bmatrix}$$

$$\mathcal{M}_{11} = \begin{bmatrix} \mathrm{He}[(A + A_h)\tilde{P} & C_0^{\mathrm{T}}\tilde{P} & C_1^{\mathrm{T}}\tilde{P} \\ \star & \tilde{L}_1 & 0 \\ \star & \star & -\gamma I_q \end{bmatrix} + Z \begin{bmatrix} \tilde{R} & 0 & 0 \\ \star & \tilde{L}_1 & 0 \\ \star & \star & \gamma^{-1} I_p \end{bmatrix} Z^{\mathrm{T}}$$

$$\mathcal{M}_{12}^{\mathrm{T}} = \left[\tilde{P}(A + A_h)^{\mathrm{T}} \;\; \tilde{P}(C_0 + C_{0h})^{\mathrm{T}} \;\; \tilde{P}(C_1 + C_{1h})^{\mathrm{T}}\right] \tag{8.50}$$

and $\mathcal{N}$ is a basis of the null-space of the matrix $\begin{bmatrix} B^{\mathrm{T}} & D_0^{\mathrm{T}} & D_1^{\mathrm{T}} \end{bmatrix}$.

Then, there exists a gain-scheduled state-feedback controller of the form (8.43) which stabilizes the system (8.34) and ensures that the L_2-gain of the transfer $w \to z$ of the closed-loop system (8.34)–(8.43) is less than γ for all $(h, \rho) \in \mathscr{H}_{\infty,\bar{h}} \times \mathscr{P}_1^{\infty}$.

The controller can be constructed following the same procedure as in the non-delayed case; see Sect. 3.5. That is, we first construct the matrices L and $\tilde{L}$ and then we solve for $\bar{K}$ in the LMI (8.42). The controller can also be shown to be well-posed using the same arguments as in Sect. 3.5.

It seems important to stress that the stabilization conditions of the theorem above are not convex due to the presence of the constraint $R\tilde{R} = I$. The conditions can, however, be solved in an iterative and effective way using the so-called *cone complementary algorithm* derived in [7]; see Appendix C.14.

8.1.5 Further Remarks on LFT-Based Approaches

As stated in the introductory statements of Chap. 1, in Sect. 6.1, the design of gain-scheduled controllers using the LFT-based LPV formulation very often leads to complications. The results obtained above seem to, however, go against this claim. This is actually not true because the considered functionals have been carefully chosen such that the resulting conditions are tractable.

If we were interested in deriving stabilization results using the Lyapunov-Krasovskii functional

$$V(x_t) = x(t)^{\mathrm{T}} P x(t) + \int_{t-h(t)}^{t} x(s)^{\mathrm{T}} Q x(s) ds \bar{h} \int_{-\bar{h}}^{0} \int_{t+\theta}^{t} x(s)^{\mathrm{T}} R x(s) ds d\theta, \quad (8.51)$$

we would have run into troubles. The dualijection lemma would not indeed apply to the LMI condition corresponding to this functional since the rank of the outer-factor matrix S would be smaller than the number of negative eigenvalues of the central matrix. Note that, moreover, the projection lemma would also fail to produce tractable synthesis conditions due to the presence of non-linearizable nonlinear terms. These difficulties are certainly the reasons why so few results based on this framework have been reported in the literature.

8.2 Observer-Based Output Feedback Controllers

This section is devoted to the design of both memoryless and exact-memory observer-based controllers, that is, controllers consisting of two blocks:

1. an observer-block that estimates the state of the system from the knowledge of the known input and measured output, and
2. a state-feedback-block that computes the control input from the estimated state value.

Two approaches can be used to design such controllers. The first one, which is the simplest, consists of first designing an observer according to some observation performance measure, and then designing the state-feedback controller according to a control performance measure. The second approach consists of designing the observer and the controller at the same time, either according to a joint performance measure, or simply a control performance measure. The first approach is the simplest since the two design problems can be easily cast as convex problems with some degree of conservatism. The second one is, however, more difficult to deal with and several approximations are in general needed before coming up with tractable design conditions.

An interesting question then is: are these approaches equivalent or, in other words, is one better or more suitable than the other? When the system is perfectly known, i.e. no uncertainty is involved, the first approach should be considered since (1) it is simpler and (2) there is no advantage in considering the simultaneous design approach. The latter point can be justified from the fact that when the observer error is decoupled from the state and the known input, the observer gain can then be designed such that the L_2-gain of the transfer of the disturbance to the observation error is less than γ_o, independently of the values of the state of the system and the known input (which contains the control input). Then, the controller can be designed such that the L_2-gain of the transfer of the disturbance to the controlled output is less than γ_c, without impacting the performance of the observer: this is the so-called *separation principle*. Therefore, the simultaneous design is not relevant in this case. Note that the same arguments still hold if another performance measure is considered.

When the system is, however, subject to uncertainties, the above principle does not usually apply since the observation error depends on the control input and the state of the system. This is notably the case when memoryless-observers are considered. In such a case, the simultaneous design seems to be the only applicable since the extended system, describing the dynamics of the system and the observer, has to be considered; see e.g. [8–11]. There is, however, a workaround to this and the separate design can still be applied in a rigorous way. The idea is simply to augment the disturbance vector in the observation error model to contain all the terms that do not depend on the error itself, i.e. the new disturbance vector therefore gathers the state of the system, the initial disturbances, etc. In this case, separate design is made possible and the computed performance observation level will be still valid even after the introduction of the controller. It is, nevertheless, unclear which design will yield the best performance.

In the following, we will be interested in designing several observer-based control laws. The inherent difficulty of the simultaneous design approach will be notably emphasized.

8.2.1 Memoryless Observer-Based Output Feedback

Let us consider here LPV time-delay systems of the form

$$
\begin{aligned}
\dot{x}(t) &= A(\rho(t))x(t) + A_h(\rho(t))x(t-h(t)) + B(\rho(t))u(t) + E(\rho(t))w(t) \\
z(t) &= C(\rho(t))x(t) + C_h(\rho(t))x(t-h(t)) + D(\rho(t))u(t) + F(\rho(t))w(t) \\
y(t) &= C_y(\rho(t))x(t) + C_{yh}(\rho(t))x(t) + F_y(\rho(t))w(t) \\
x(s) &= \psi_x(s),\ s \in [-\bar{h}, 0]
\end{aligned} \tag{8.52}
$$

where $x \in \mathbb{R}^n$, $u \in \mathbb{R}^m$, $y \in \mathbb{R}^p$, $w \in \mathbb{R}^q$, $z \in \mathbb{R}^r$ and $\psi_x \in C([-\bar{h}, 0], \mathbb{R}^n)$ are respectively the state of the system, the control input, the measured output, the exogenous input, the controlled output and the initial condition, respectively. The delay and parameters are assumed to satisfy $h \in \mathscr{H}_{\mu,\bar{h}}$ and $\rho \in \mathscr{P}^\nu$.

8.2.1.1 Memoryless Controller—Simultaneous Design

This section aims at developing sufficient conditions to the existence of a memoryless observer-based control law of the form

$$
\begin{aligned}
\dot{\hat{x}}(t) &= A(\rho(t))\hat{x}(t) + B(\rho(t))u(t) + L(\rho(t))(y(t) - C_y(\rho(t))\hat{x}(t)) \\
u(t) &= K(\rho(t))\hat{x}(t) \\
\xi(s) &= \psi_\xi(s),\ s \in [-\bar{h}, 0]
\end{aligned} \tag{8.53}
$$

where $\hat{x} \in \mathbb{R}^n$ and $\psi_\xi \in C([-\bar{h}, 0], \mathbb{R}^n)$ are the state and the initial condition of the observer, respectively. Above, the matrix functions $L : \mathbf{\Delta}_\rho \to \mathbb{R}^{n\times p}$ and $K : \mathbf{\Delta}_\rho \to \mathbb{R}^{m\times n}$ have to be determined such that:

1. the error $e := x - \hat{x}$ is asymptotically stable;
2. the closed-loop system is asymptotically stable;
3. the L_2-gain of the transfer $w \to z$ is smaller than $\gamma > 0$.

We have the following result:

Theorem 8.2.1 *Assume that there exist a continuously differentiable matrix function $P : \mathbf{\Delta}_\rho \to \mathbb{S}^{2n}_{\succ 0}$, matrix functions $X_0, X_c : \mathbf{\Delta}_\rho \to \mathbb{R}^{n\times n}$, $K : \mathbf{\Delta}_\rho \to \mathbb{R}^{m\times n}$, $L_o : \mathbf{\Delta}_\rho \to \mathbb{R}^{n\times p}$, constant matrices $Q, R \in \mathbb{S}^{2n}_{\succ 0}$, $Z_1, Z_2 \in \mathbb{S}^m_{\succ 0}$ and a constant scalar $\gamma > 0$ such that the matrix inequality*

$$
\left[\begin{array}{cccc|cc}
-\mathrm{He}[X] & \Omega_2(\rho) & \Omega_3(\rho) & \Omega_5(\rho) & \Omega_c^1(\rho) & 0 \\
\star & \Omega_4(\rho,\nu) & \Omega_6(\rho) & 0 & 0 & \Omega_c^2(\rho) \\
\star & \star & \Omega_8(\rho) & 0 & 0 & 0 \\
\star & \star & \star & \Omega_{10}(\rho) & 0 & 0 \\
\hline
\star & \star & \star & \star & -\Omega_d & 0 \\
\star & \star & \star & \star & \star & -\Omega_d^{-1}
\end{array}\right] \prec 0
$$

holds for all $(\rho, \nu) \in \mathbf{\Delta}_\rho \times \mathbf{V}_\nu$ *where* $X(\rho) = diag(X_o(\rho), X_c(\rho))$, $\Omega_d = diag(Z_1, Z_2)$ *and*

$$\Omega_2(\rho) = \left[\begin{array}{cc} A(\rho)^{\mathrm{T}}X_o(\rho) - C_y(\rho)^{\mathrm{T}}L_o(\rho)^{\mathrm{T}} & 0 \\ 0 & A(\rho)^{\mathrm{T}}X_c \\ \hline 0 & 0 \\ A_h(\rho)^{\mathrm{T}}X_o(\rho) - C_{yh}(\rho)^{\mathrm{T}}L_o(\rho)^{\mathrm{T}} & A_h(\rho)^{\mathrm{T}}X_c(\rho) \end{array}\right]^{\mathrm{T}} + \left[P(\rho) \mid 0\right],$$

$$\Omega_3(\rho) = \begin{bmatrix} X_o(\rho)^{\mathrm{T}}E(\rho) - L_o(\rho)F_y(\rho) & 0 \\ X_c(\rho)^{\mathrm{T}}E(\rho) & 0 \end{bmatrix}, \quad \Omega_8(\rho) = \begin{bmatrix} -\gamma(\rho)I_q & F(\rho)^{\mathrm{T}} \\ \star & -\gamma(\rho)I_r \end{bmatrix}$$

$$\Omega_4(\rho, \nu) = \left[\begin{array}{c|c} \sum_i \dfrac{\partial P(\rho)}{\partial \rho}\nu_i - P(\rho) + Q - R & R \\ \hline \star & -(1-\mu)Q - R \end{array}\right],$$

$$\Omega_6(\rho) = \left[\begin{array}{cc} 0 & 0 \\ 0 & C(\rho)^{\mathrm{T}} \\ \hline 0 & 0 \\ 0 & C_h(\rho)^{\mathrm{T}} \end{array}\right], \quad \Omega_c^1(\rho) = \begin{bmatrix} 0 & 0 \\ X_c(\rho)^{\mathrm{T}}B(\rho) & X_c(\rho)^{\mathrm{T}}B(\rho) \end{bmatrix},$$

$$\Omega_{10}(\rho) = \begin{bmatrix} -P(\rho) & -\bar{h}R \\ \star & -R \end{bmatrix}, \quad \Omega_c^2(\rho) = \begin{bmatrix} K(\rho)^{\mathrm{T}} & 0 \\ 0 & K(\rho)^{\mathrm{T}} \end{bmatrix}, \quad \Omega_5(\rho) = \begin{bmatrix} X \\ \bar{h}R \end{bmatrix}^{\mathrm{T}}.$$

Then, the observer-based control law (8.53) with the gains $K(\rho)$ *and* $L(\rho) = X_o(\rho)^{-T}L_o(\rho)$ *asymptotically stabilizes the system (8.52) and ensures that the* L_2*-gain of the transfer* $w \to z$ *of the closed-loop system is less than* γ *for all* $(h, \rho) \in \mathscr{H}_{\mu,\bar{h}} \times \mathscr{P}^\nu$.

Proof The closed-loop system is given by

$$\begin{aligned} \begin{bmatrix} \dot{e}(t) \\ \dot{x}(t) \end{bmatrix} &= \begin{bmatrix} A(\rho(t)) - L(\rho(t))C_y(\rho(t)) & 0 \\ -B(\rho(t))K(\rho(t)) & A(\rho(t)) + B(\rho(t))K(\rho(t)) \end{bmatrix} \begin{bmatrix} e(t) \\ x(t) \end{bmatrix} \\ &\quad + \begin{bmatrix} 0 & A_h(\rho(t)) - L(\rho(t))C_{yh}(\rho(t)) \\ 0 & A_h(\rho(t)) \end{bmatrix} \begin{bmatrix} e(t-h(t)) \\ x(t-h(t)) \end{bmatrix} \\ &\quad + \begin{bmatrix} E(\rho(t)) - L(\rho(t))F_y(\rho(t)) \\ E(\rho(t)) \end{bmatrix} w(t) \\ z(t) &= \begin{bmatrix} -D(\rho)K(\rho) & C(\rho) + D(\rho)K(\rho) \end{bmatrix} \begin{bmatrix} e(t) \\ x(t) \end{bmatrix} \\ &\quad + \begin{bmatrix} 0 & C_h(\rho) \end{bmatrix} \begin{bmatrix} e(t-h(t)) \\ x(t-h(t)) \end{bmatrix} + F(\rho)w(t) \end{aligned} \tag{8.54}$$

where $e(t) := x(t) - \hat{x}(t)$ is the observation error. Let then $X := \text{diag}(X_o, X_c)$ and substitute the closed-loop system expression (8.54) in the LMI (6.30). The main difficulty lies in the presence the bilinear terms $X_c(\rho)^{\mathrm{T}}B(\rho)K(\rho)$. Note that they cannot be linearized using a congruence transformation. The idea is then to upper-bound them as

$$
\begin{aligned}
2x_3^{\mathrm{T}}X_c(\rho)^{\mathrm{T}}B(\rho)K(\rho)x_2 &\leq x_3^{\mathrm{T}}X_c(\rho)^{\mathrm{T}}B(\rho)Z_1B(\rho)^{\mathrm{T}}X_c(\rho)x_3 \\
&\quad + x_2^{\mathrm{T}}K(\rho)^{\mathrm{T}}Z_1^{-1}K(\rho)x_2 \\
2x_4^{\mathrm{T}}X_c(\rho)^{\mathrm{T}}B(\rho)K(\rho)x_2 &\leq x_4^{\mathrm{T}}X_c(\rho)^{\mathrm{T}}B(\rho)Z_2B(\rho)^{\mathrm{T}}X_c(\rho)x_4 \\
&\quad + x_2^{\mathrm{T}}K(\rho)^{\mathrm{T}}Z_2^{-1}K(\rho)x_2
\end{aligned}
$$

where these inequalities hold true for all real valued vectors x_2, x_3, x_4 and for any positive definite matrices Z_1 and Z_2. Substituting these expressions into the conditions, performing several Schur complements and making the change of variables $L_o(\rho) := X_o(\rho)^T L(\rho)$ yield the result. ■

The above result deserves few remarks. First of all, in order to derive (more or less) tractable conditions, we had to make several simplifications and approximations:

1. the matrix X had to be set to be block-diagonal, i.e. $X = \text{diag}(X_o, X_c)$.
2. the cross-product term $X_c(\rho)B(\rho)K(\rho)$ had been approximated by quadratic terms in the same spirit as in time-delay systems; see the discussion on the conservatism of bounds on cross-terms in Sect. 5.6.4.

These manipulations are obviously conservative and deteriorate the efficiency of the considered Lyapunov-Krasovskii functional. Note, moreover, that the matrix inequality condition is not an LMI due to the presence of the matrices Z_1, Z_1^{-1}, Z_2 and Z_2^{-1}. The cone complementary algorithm can, however, be used to deal with such a matrix inequality problem; see Appendix C.14. An LMI can be easily obtained by setting $Z_1 = Z_2 = I$, a simplification increasing even more the conservatism of the approach.

8.2.1.2 Memoryless Controller—Separate Design

Whereas the one-step procedure yields design conditions that may be difficult to check, the two-step procedure, consisting of designing first the observer and then the controller, turns out to be more tractable, as shown in the following.

8.2.1.3 General Observer Case

We consider now the quite general observer structure considered in Sect. 7.1.3. Note that in this case, we need to assume that C_y is independent of ρ and that $C_{yh} = 0$,

$F_y(\rho) = 0$ in the system (8.52). A memoryless observer-based control law can therefore be designed using the following procedure:

1. First design a memoryless observer using the results of Sect. 7.1.3 that makes the L_2-gain of the transfer $\text{col}(w, x_h) \to e$ smaller than some $\gamma_o > 0$ where $x_h(t) := x(t - h(t))$.
2. Then consider the control law

$$u(t) = K(\rho(t))\hat{x}(t) \tag{8.55}$$

   and build the augmented system

$$\begin{aligned}
\begin{bmatrix} \dot{x}(t) \\ \dot{e}(t) \end{bmatrix} &= \begin{bmatrix} A(\rho) + B(\rho)K(\rho) & -B(\rho)K(\rho) \\ 0 & M(\rho) \end{bmatrix} \begin{bmatrix} x(t) \\ e(t) \end{bmatrix} \\
&\quad + \begin{bmatrix} A_h(\rho) & 0 \\ (I - HC_y)A_h(\rho) & 0 \end{bmatrix} \begin{bmatrix} x(t - h(t)) \\ e(t - h(t)) \end{bmatrix} + \begin{bmatrix} E(\rho) \\ (I - HC_y)E(\rho) \end{bmatrix} w(t) \\
z(t) &= \begin{bmatrix} C(\rho) + D(\rho)K(\rho) & -D(\rho)K(\rho) \end{bmatrix} \begin{bmatrix} x(t) \\ e(t) \end{bmatrix} \\
&\quad + \begin{bmatrix} C_h(\rho) & 0 \end{bmatrix} \begin{bmatrix} x(t - h(t)) \\ e(t - h(t)) \end{bmatrix} + F(\rho)w(t)
\end{aligned} \tag{8.56}$$

   where $M(\rho)$ and H are the matrices of the observer that have been computed in the first step.
3. Find the controller gain K such that the transfer $w \to z$ of the closed-loop system has L_2-gain smaller than some $\gamma_c > 0$.

The last step of the above procedure can be performed using the following result:

Theorem 8.2.2 *Assume that there exist a continuously differentiable matrix function $P : \mathbf{\Delta}_\rho \to \mathbb{S}^{2n}_{\succ 0}$, constant matrices $Q, R \in \mathbb{S}^{2n}_{\succ 0}$, $X \in \mathbb{R}^{2n \times 2n}$, a matrix function $Y : \mathbf{\Delta}_\rho \to \mathbb{R}^{m \times n}$ and a scalar $\gamma_c > 0$ where*

$$X := \begin{bmatrix} X_1 & X_2 \\ X_3 & X_4 \end{bmatrix}, \ X_i \in \mathbb{R}^{n \times n}$$

verifies $X_1 - X_3 = X_2 - X_4$ and such that the parameter dependent LMI

$$\begin{bmatrix}
-\text{He}[X] & \Xi_{12}(\rho) & \Xi_{13}(\rho) & \Xi_{14}(\rho) & 0 & X & \bar{h}R \\
\star & \Xi_{22}(\rho, \nu) & R & 0 & \Xi_{24}(\rho) & 0 & 0 \\
\star & \star & \Xi_{33} & 0 & \Xi_{34}(\rho) & 0 & 0 \\
\star & \star & \star & -\gamma_c I_p & F(\rho)^{\mathrm{T}} & 0 & 0 \\
\star & \star & \star & \star & -\gamma_c I_q & 0 & 0 \\
\star & \star & \star & \star & \star & -P(\rho) & -\bar{h}R \\
\star & \star & \star & \star & \star & \star & -R
\end{bmatrix} \prec 0 \tag{8.57}$$

holds for all $(\rho, \nu) \in \boldsymbol{\Delta}_\rho \times \mathbf{V}_\nu$ *where*

$$\begin{aligned}
\Xi_{12}(\rho) &= P(\rho) + \begin{bmatrix} A(\rho)X_1 + B(\rho)Y(\rho) & A(\rho)X_2 + B(\rho)Y(\rho) \\ M(\rho)X_3 & M(\rho)X_4 \end{bmatrix}, \\
\Xi_{14}(\rho) &= \left[E(\rho)^{\mathrm{T}} \ [(I - HC_y)E(\rho)]^{\mathrm{T}}\right]^{\mathrm{T}}, \\
\Xi_{13}(\rho) &= \begin{bmatrix} A_h(\rho)X_1 & A_h(\rho)X_2 \\ (I - HC_y)A_h(\rho)X_1 & (I - HC_y)A_h(\rho)X_2 \end{bmatrix}, \\
\Xi_{22}(\rho, \nu) &= \sum_i \frac{\partial P(\rho)}{\partial \rho_i}\nu_i - P(\rho) + Q - R, \\
\Xi_{33} &= -(1 - \mu)Q - R, \\
\Xi_{24}(\rho) &= \left[C(\rho)X_1 + D(\rho)Y(\rho) \ \ C(\rho)X_2 + D(\rho)Y(\rho)\right]^{\mathrm{T}}, \\
\Xi_{34}(\rho) &= \left[C_h(\rho)X_1 \ \ C_h(\rho)X_2\right]^{\mathrm{T}}.
\end{aligned}$$

In such a case, a stabilizing memoryless state-feedback controller is given by (8.55) with

$$K(\rho) = Y(\rho)(X_1 - X_3)^{-1}. \tag{8.58}$$

Moreover, the L_2*-gain of the transfer* $w \to z$ *of the closed-loop system (8.56) is less than* γ_c *for all* $(h, \rho) \in \mathscr{H}_{\mu,\bar{h}} \times \mathscr{P}^\nu$.

Proof The proof follows the same lines as the one of Theorem 8.1.2. ■

8.2.1.4 Simple Observer Case

Let us consider now the simple memoryless observer (7.41) of Sect. 7.2.2. Such a memoryless observer-based control law can be designed using the following procedure:

1. First design a memoryless observer using the results of Sect. 7.2.2 that makes the L_2-gain of the transfer $\mathrm{col}(w, x_h) \to e$ smaller than some $\gamma_o > 0$ where $x_h(t) := x(t - h(t))$.
2. Then consider the control law

$$u(t) = K(\rho(t))\hat{x}(t) \tag{8.59}$$

and build the augmented system

$$\begin{bmatrix} \dot{x}(t) \\ \dot{e}(t) \end{bmatrix} = \begin{bmatrix} A(\rho) + B(\rho)K(\rho) & -B(\rho)K(\rho) \\ 0 & A(\rho) - L(\rho)C_y(\rho) \end{bmatrix} \begin{bmatrix} x(t) \\ e(t) \end{bmatrix}$$

$$
\begin{aligned}
&+\begin{bmatrix} A_h(\rho) & 0 \\ A_h(\rho) - L(\rho)C_{yh}(\rho) & 0 \end{bmatrix}\begin{bmatrix} x(t-h(t)) \\ e(t-h(t)) \end{bmatrix} \\
&+\begin{bmatrix} E(\rho) \\ E(\rho) - L(\rho)F_y(\rho) \end{bmatrix} w(t) \\
z(t) &= \begin{bmatrix} C(\rho) + D(\rho)K(\rho) & -D(\rho)K(\rho) \end{bmatrix}\begin{bmatrix} x(t) \\ e(t) \end{bmatrix} \\
&+\begin{bmatrix} C_h(\rho) & 0 \end{bmatrix}\begin{bmatrix} x(t-h(t)) \\ e(t-h(t)) \end{bmatrix} \\
&+F(\rho)w(t)
\end{aligned}
\tag{8.60}
$$

where $L(\rho)$ is the gain of the observer that has been computed in the first step of this procedure.

3. Find the controller gain K such that the transfer $w \to z$ of the closed-loop system has L_2-gain smaller than some $\gamma_c > 0$.

The last step of the above procedure is addressed using the following result:

Theorem 8.2.3 *Assume that there exist a continuously differentiable matrix function $P : \mathbf{\Delta}_\rho \to \mathbb{S}^{2n}_{\succ 0}$, constant matrices $Q, R \in \mathbb{S}^{2n}_{\succ 0}$, $X \in \mathbb{R}^{2n \times 2n}$, a matrix function $Y : \mathbf{\Delta}_\rho \to \mathbb{R}^{m \times n}$ and a scalar $\gamma_c > 0$ where*

$$
X := \begin{bmatrix} X_1 & X_2 \\ X_3 & X_4 \end{bmatrix}, \quad X_i \in \mathbb{R}^{n \times n}
$$

verifies $X_1 - X_3 = X_2 - X_4$ and such that the parameter dependent LMI

$$
\begin{bmatrix}
-\mathrm{He}[X] & \Xi_{12}(\rho) & \Xi_{13}(\rho) & \Xi_{14}(\rho) & 0 & X & \bar{h}R \\
\star & \Xi_{22}(\rho,\nu) & R & 0 & \Xi_{24}(\rho) & 0 & 0 \\
\star & \star & \Xi_{33} & 0 & \Xi_{34}(\rho) & 0 & 0 \\
\star & \star & \star & -\gamma_c I_p & F(\rho)^{\mathrm{T}} & 0 & 0 \\
\star & \star & \star & \star & -\gamma_c I_q & 0 & 0 \\
\star & \star & \star & \star & \star & -P(\rho) & -\bar{h}R \\
\star & \star & \star & \star & \star & \star & -R
\end{bmatrix} \prec 0 \tag{8.61}
$$

holds for all $(\rho, \nu) \in \mathbf{\Delta}_\rho \times \mathbf{V}_\nu$ where

$$
\Xi_{12}(\rho) = P(\rho) + \begin{bmatrix} A(\rho)X_1 + B(\rho)Y(\rho) & A(\rho)X_2 + B(\rho)Y(\rho) \\ (A(\rho) - L(\rho)C_y(\rho))X_3 & (A(\rho) - L(\rho)C_y(\rho))X_4 \end{bmatrix},
$$

$$
\Xi_{13}(\rho) = \begin{bmatrix} A_h(\rho)X_1 & A_h(\rho)X_2 \\ [A_h(\rho) - L(\rho)C_{yh}(\rho)]X_1 & [A_h(\rho) - L(\rho)C_{yh}(\rho)]X_2 \end{bmatrix},
$$

$$
\Xi_{14}(\rho) = \begin{bmatrix} E(\rho)^{\mathrm{T}} & [E(\rho) - L(\rho)F_y(\rho)]^{\mathrm{T}} \end{bmatrix}^{\mathrm{T}},
$$

$$\Xi_{22}(\rho,\nu) = \sum_i \frac{\partial P(\rho)}{\partial \rho_i}\nu_i - P(\rho) + Q - R,$$
$$\Xi_{33} = -(1-\mu)Q - R,$$
$$\Xi_{24}(\rho) = \left[C(\rho)X_1 + D(\rho)Y(\rho) \;\; C(\rho)X_2 + D(\rho)Y(\rho)\right]^{\mathrm{T}},$$
$$\Xi_{34}(\rho) = \left[C_h(\rho)X_1 \;\; C_h(\rho)X_2\right]^{\mathrm{T}}.$$

In such a case, a stabilizing memoryless state-feedback controller is given by (8.59) with

$$K(\rho) = Y(\rho)(X_1 - X_3)^{-1}. \tag{8.62}$$

Moreover, the L_2-gain of the transfer $w \to z$ of the closed-loop system (8.60) is less than γ_c for all $(h,\rho) \in \mathscr{H}_{\mu,\bar{h}} \times \mathscr{P}^{\nu}$.

Proof The proof follows the same lines as the one of Theorem 8.1.2. ■

8.2.2 Observer-Based Output Feedback with Exact Memory

We derive in this section, the exact memory counterparts of the results of the previous sections addressing memoryless control.

8.2.2.1 General Observer Case

Let us consider first the general observer structure considered in Sect. 7.1.1. As in the memoryless case, we need to assume that C_y is independent of ρ and that $C_{yh} = 0$, $F_y(\rho) = 0$ in the system (8.52). An observer-based control law with exact memory can be designed by following the procedure:

1. First design an observer with exact memory using the results of Sect. 7.1.1 that makes the L_2-gain of the transfer $w \to e$ smaller than some $\gamma_o > 0$.
2. Then consider the control law with exact memory

$$u(t) = K_0(\rho(t))\hat{x}(t) + K_h(\rho(t))\hat{x}(t-h(t)) \tag{8.63}$$

   and build the augmented system

$$\begin{bmatrix} \dot{x}(t) \\ \dot{e}(t) \end{bmatrix} = \begin{bmatrix} A(\rho) + B(\rho)K_0(\rho) & -B(\rho)K_0(\rho) \\ 0 & M_0(\rho) \end{bmatrix} \begin{bmatrix} x(t) \\ e(t) \end{bmatrix}$$

$$
\begin{aligned}
&+\begin{bmatrix} A_h(\rho)+B(\rho)K_h(\rho) & -B(\rho)K_h(\rho) \\ 0 & M_h(\rho) \end{bmatrix}\begin{bmatrix} x(t-h(t)) \\ e(t-h(t)) \end{bmatrix} \\
&+\begin{bmatrix} E(\rho) \\ (I-HC_y)E(\rho) \end{bmatrix} w(t) \\
z(t) = &\begin{bmatrix} C(\rho)+D(\rho)K_0(\rho) & -D(\rho)K_0(\rho) \end{bmatrix}\begin{bmatrix} x(t) \\ e(t) \end{bmatrix} \\
&+\begin{bmatrix} C_h(\rho)+D(\rho)K_h(\rho) & -D(\rho)K_h(\rho) \end{bmatrix}\begin{bmatrix} x(t-h(t)) \\ e(t-h(t)) \end{bmatrix} + F(\rho)w(t)
\end{aligned} \tag{8.64}
$$

where $M_0(\rho)$, $M_h(\rho)$ and H are the matrices of the observer that have been computed in the first step.

3. Find the controller gain $K_0(\rho)$ and $K_h(\rho)$ such that the transfer $w \to z$ of the closed-loop system has L_2-gain smaller than some $\gamma_c > 0$.

The last step can be performed using the following result:

Theorem 8.2.4 *Assume that there exist a continuously differentiable matrix function $P : \mathbf{\Delta}_\rho \to \mathbb{S}^{2n}_{\succ 0}$, constant matrices $Q, R \in \mathbb{S}^{2n}_{\succ 0}$, $X \in \mathbb{R}^{2n\times 2n}$, a matrix function $Y : \mathbf{\Delta}_\rho \to \mathbb{R}^{m\times n}$ and a scalar $\gamma_c > 0$ where*

$$
X := \begin{bmatrix} X_1 & X_2 \\ X_3 & X_4 \end{bmatrix}, \quad X_i \in \mathbb{R}^{n\times n}
$$

verifies $X_1 - X_3 = X_2 - X_4$ and such that the parameter dependent LMI

$$
\begin{bmatrix}
-\mathrm{He}[X] & \Xi_{12}(\rho) & \Xi_{13}(\rho) & \Xi_{14}(\rho) & 0 & X & \bar{h}R \\
\star & \Xi_{22}(\rho) & R & 0 & \Xi_{24}(\rho) & 0 & 0 \\
\star & \star & \Xi_{33}(\rho,\nu) & 0 & \Xi_{34}(\rho) & 0 & 0 \\
\star & \star & \star & -\gamma I_p & F(\rho)^{\mathrm{T}} & 0 & 0 \\
\star & \star & \star & \star & -\gamma_c I_q & 0 & 0 \\
\star & \star & \star & \star & \star & -P(\rho) & -\bar{h}R \\
\star & \star & \star & \star & \star & \star & -R
\end{bmatrix} \prec 0 \tag{8.65}
$$

holds for all $(\rho,\nu) \in \mathbf{\Delta}_\rho \times \mathbf{V}_\nu$ where

$$
\Xi_{12}(\rho) = P(\rho) + \begin{bmatrix} A(\rho)X_1 + B(\rho)Y_0(\rho) & A(\rho)X_2 + B(\rho)Y_0(\rho) \\ M_0(\rho)X_3 & M_0(\rho)X_4 \end{bmatrix},
$$

$$
\Xi_{14}(\rho) = \begin{bmatrix} E(\rho)^{\mathrm{T}} & [(I-HC_y)E(\rho)]^{\mathrm{T}} \end{bmatrix}^{\mathrm{T}},
$$

$$\Xi_{23}(\rho) = \begin{bmatrix} A_h(\rho)X_1 + B(\rho)Y_h(\rho) & A_hX_2 + B(\rho)Y_h(\rho) \\ M_h(\rho)X_3 & M_h(\rho)X_4 \end{bmatrix},$$

$$\Xi_{22}(\rho) = \sum_i \frac{\partial P(\rho)}{\partial \rho_i} \nu_i - P(\rho) + Q - R,$$

$$\Xi_{33} = -(1-\mu)Q - R,$$

$$\Xi_{24}(\rho) = \begin{bmatrix} C(\rho)X_1 + D(\rho)Y_0(\rho) & C(\rho)X_2 + D(\rho)Y_0(\rho) \end{bmatrix}^{\mathrm{T}},$$

$$\Xi_{34}(\rho) = \begin{bmatrix} C_h(\rho)X_1 + D(\rho)Y_h(\rho) & C_h(\rho)X_2 + D(\rho)Y_h(\rho) \end{bmatrix}^{\mathrm{T}}.$$

In such a case, a stabilizing memoryless state-feedback controller is given by (8.63) with

$$K_0(\rho) = Y_0(\rho)(X_1 - X_3)^{-1} \text{ and } K_h(\rho) = Y_h(\rho)(X_1 - X_3)^{-1}. \tag{8.66}$$

Moreover, the L_2-gain of the transfer $w \to z$ of the closed-loop system (8.64) is less than γ_c for all $(h, \rho) \in \mathscr{H}_{\mu,\bar{h}} \times \mathscr{P}^{\nu}$.

Proof The proof follows the same lines as the one of Theorem 8.1.2. ■

8.2.2.2 Simple Observer Case

Let us consider now the simple memoryless observer (7.36) of Sect. 7.2.1. A stabilizing observer-based control law with exact memory can be obtained by following the procedure:

1. First design an observer with exact memory using the result of Sect. 7.2.1 that makes the L_2-gain of the transfer $w \to e$ smaller than some $\gamma_o > 0$.
2. Then consider the control law with exact memory

$$u(t) = K_0(\rho(t))\hat{x}(t) + K_h(\rho(t))\hat{x}(t - h(t)) \tag{8.67}$$

and build the augmented system

$$\begin{bmatrix} \dot{x}(t) \\ \dot{e}(t) \end{bmatrix} = \begin{bmatrix} A(\rho) + B(\rho)K_0(\rho) & -B(\rho)K_0(\rho) \\ 0 & A(\rho) - L(\rho)C_y(\rho) \end{bmatrix} \begin{bmatrix} x(t) \\ e(t) \end{bmatrix}$$
$$+ \begin{bmatrix} A_h(\rho) + B(\rho)K_h(\rho) & -B(\rho)K_h(\rho) \\ 0 & A_h(\rho) - L(\rho)C_{yh}(\rho) \end{bmatrix} \begin{bmatrix} x(t-h(t)) \\ e(t-h(t)) \end{bmatrix}$$

$$+\begin{bmatrix} E(\rho) \\ E(\rho)-L(\rho)F_y(\rho) \end{bmatrix} w(t)$$

$$\begin{aligned} z(t) &= \begin{bmatrix} C(\rho)+D(\rho)K_0(\rho) & -D(\rho)K_0(\rho) \end{bmatrix} \begin{bmatrix} x(t) \\ e(t) \end{bmatrix} \\ &+ \begin{bmatrix} C_h(\rho)+D(\rho)K_h(\rho) & -D(\rho)K_h(\rho) \end{bmatrix} \begin{bmatrix} x(t-h(t)) \\ e(t-h(t)) \end{bmatrix} \\ &+ F(\rho)w(t) \end{aligned} \tag{8.68}$$

where $L(\rho)$ is the gain of the observer that has been computed in the first step of this procedure.

3. Find the controller gain $K_0(\rho)$ and $K_h(\rho)$ such that the transfer $w \to z$ of the closed-loop system has L_2-gain smaller than some $\gamma_c > 0$.

The last step of the above procedure can be performed using the following result:

Lemma 8.2.5 *Assume that there exist a continuously differentiable matrix function $P : \mathbf{\Delta}_\rho \to \mathbb{S}^{2n}_{\succ 0}$, constant matrices $Q, R \in \mathbb{S}^{2n}_{\succ 0}$, $X \in \mathbb{R}^{2n\times 2n}$, a matrix function $Y : \mathbf{\Delta}_\rho \to \mathbb{R}^{m\times n}$ and a scalar $\gamma_c > 0$ where*

$$X := \begin{bmatrix} X_1 & X_2 \\ X_3 & X_4 \end{bmatrix},\ X_i \in \mathbb{R}^{n\times n}$$

verifies $X_1 - X_3 = X_2 - X_4$ and such that the parameter dependent LMI

$$\begin{bmatrix} -\mathrm{He}[X] & \Xi_{12}(\rho) & \Xi_{13}(\rho) & \Xi_{14}(\rho) & 0 & X & \bar{h}R \\ \star & \Xi_{22}(\rho,\nu) & R & 0 & \Xi_{24}(\rho) & 0 & 0 \\ \star & \star & \Xi_{33} & 0 & \Xi_{34}(\rho) & 0 & 0 \\ \star & \star & \star & -\gamma_c I_p & F(\rho)^{\mathrm{T}} & 0 & 0 \\ \star & \star & \star & \star & -\gamma_c I_q & 0 & 0 \\ \star & \star & \star & \star & \star & -P(\rho) & -\bar{h}R \\ \star & \star & \star & \star & \star & \star & -R \end{bmatrix} \prec 0 \tag{8.69}$$

holds for all $(\rho,\nu) \in \mathbf{\Delta}_\rho \times \mathbf{V}_\nu$ where

$$\Xi_{12}(\rho) = P(\rho) + \begin{bmatrix} A(\rho)X_1 + B(\rho)Y_0(\rho) & A(\rho)X_2 + B(\rho)Y_0(\rho) \\ (A(\rho)-L(\rho)C_y(\rho))X_3 & (A(\rho)-L(\rho)C_y(\rho))X_4 \end{bmatrix},$$

$$\Xi_{13}(\rho) = \begin{bmatrix} A_h(\rho)X_1 + B(\rho)Y_h(\rho) & A_h(\rho)X_2 + B(\rho)Y_h(\rho) \\ (A_h(\rho)-L(\rho)C_{yh}(\rho))X_3 & (A_h(\rho)-L(\rho)C_{yh}(\rho))X_4 \end{bmatrix},$$

$$\Xi_{14}(\rho) = \begin{bmatrix} E(\rho)^{\mathrm{T}} & [E(\rho)-L(\rho)F_y(\rho)]^{\mathrm{T}} \end{bmatrix}^{\mathrm{T}},$$

$$\begin{aligned}
\Xi_{22}(\rho,\nu) &= \sum_i \frac{\partial P(\rho)}{\partial \rho_i}\nu_i - P(\rho) + Q - R,\\
\Xi_{33} &= -(1-\mu)Q - R,\\
\Xi_{24}(\rho) &= \left[C(\rho)X_1 + D(\rho)Y_0(\rho) \;\; C(\rho)X_2 + D(\rho)Y_0(\rho)\right]^{\mathrm{T}},\\
\Xi_{34}(\rho) &= \left[C_h(\rho)X_1 + D(\rho)Y_h(\rho) \;\; C_h(\rho)X_2 + D(\rho)Y_h(\rho)\right]^{\mathrm{T}}.
\end{aligned}$$

In such a case, a stabilizing memoryless state-feedback controller is given by (8.55) with

$$K_0(\rho) = Y_0(\rho)(X_1 - X_3)^{-1} \text{ and } K_h(\rho) = Y_h(\rho)(X_1 - X_3)^{-1}. \tag{8.70}$$

Moreover, the L_2-gain of the transfer $w \to z$ of the closed-loop system (8.68) is less than γ_c for all $(h,\rho) \in \mathscr{H}_{\mu,\bar{h}} \times \mathscr{P}^{\nu}$.

Proof The proof follows the same lines as the one of Theorem 8.1.2. ∎

8.3 Dynamic Output-Feedback Controllers

The main difference between observer-based and dynamic output feedback controllers is that, for the latter, we are not interested in estimating the state of the system. We are simply interested in dynamically computing a stabilizing control input from the measurements. The structure of such controllers is usually "full", in the sense that no internal structure is considered, except in some special design problems such as decentralized control or structured control (like PID control). The goal of this section is to show that the design conditions of exact-memory controllers are convex whereas they are nonconvex for memoryless controllers. Design results have been obtained e.g. in [12, 13].

8.3.1 Dynamic Output Feedback with Exact Memory

In this section, we will be interested in designing a dynamic output feedback controller with exact memory of the form

$$\begin{aligned}
\dot{x}_c(t) &= A_c(\rho)x_c(t) + A_{hc}(\rho)x_c(t-h(t)) + B_c(\rho)y(t)\\
u(t) &= C_c(\rho)x_c(t) + C_{hc}(\rho)x_c(t-h(t)) + D_c(\rho)y(t)\\
x_c(s) &= \phi_c(s),\;\; s \in [-\bar{h}, 0]
\end{aligned} \tag{8.71}$$

which stabilizes the system (8.52) and ensures that the L_2-gain of the transfer $w \to z$ is smaller than some $\gamma > 0$. Above, $x_c \in \mathbb{R}^n$ and $\phi_c \in C([-\bar{h}, 0], \mathbb{R}^n)$ denote the state and the initial condition of the controller, respectively. The closed-loop system is given in this case by[1]

$$\begin{bmatrix} \dot{x}(t) \\ \dot{x}_c(t) \end{bmatrix} = \underbrace{\begin{bmatrix} A + BD_cC_y & BC_c \\ B_cC_y & A_c \end{bmatrix}}_{A_{cl}} \begin{bmatrix} x(t) \\ x_c(t) \end{bmatrix} + \underbrace{\begin{bmatrix} A_h + BD_cC_{yh} & BC_{hc} \\ B_cC_{yh} & A_{hc} \end{bmatrix}}_{A_{hcl}} \begin{bmatrix} x(t-h(t)) \\ x_c(t-h(t)) \end{bmatrix},$$
$$+ \underbrace{\begin{bmatrix} E + BD_cF_y \\ B_cF_y \end{bmatrix}}_{E_{cl}} w(t)$$

$$z(t) = \underbrace{\begin{bmatrix} C + DD_cC_y & DC_c \end{bmatrix}}_{C_{cl}} \begin{bmatrix} x(t) \\ x_c(t) \end{bmatrix} + \underbrace{\begin{bmatrix} C_h + DD_cC_{yh} & DC_{hc} \end{bmatrix}}_{C_{hcl}} \begin{bmatrix} x(t-h(t)) \\ x_c(t-h(t)) \end{bmatrix}$$
$$+ \underbrace{(F + DD_cF_y)}_{F_{cl}} w(t).$$

The derivation of tractable conditions for the design of such controllers relies on the ideas proposed in [14, 15], and in Sects. 3.3.2 and 3.3.4. Following the same procedure, we obtain the result below:

Theorem 8.3.1 *Assume that there exist a continuously differentiable matrix function $\tilde{P} : \mathbf{\Delta}_\rho \to \mathbb{S}^{2n}_{\succ 0}$, constant matrices $W_1, X_1 \in \mathbb{S}^n_{\succ 0}$, $\tilde{Q}, \tilde{R} \in \mathbb{S}^{2n}_{\succ 0}$ and a scalar $\gamma > 0$ such that the LMIs*

$$\begin{bmatrix} W_1 & I \\ I & X_1 \end{bmatrix} \succ 0$$

and

$$\begin{bmatrix} -2\tilde{X} & \Xi_{12}(\rho) & \Xi_{13}(\rho) & \Xi_{14}(\rho) & 0 & \tilde{X} & \bar{h}\tilde{R} \\ \star & \Xi_{22}(\rho,\nu) & \tilde{R} & 0 & \Xi_{25}(\rho) & 0 & 0 \\ \star & \star & \Xi_{33} & 0 & \Xi_{35}(\rho) & 0 & 0 \\ \star & \star & \star & -\gamma I & \mathcal{F}(\rho)^{\mathrm{T}} & 0 & 0 \\ \star & \star & \star & \star & -\gamma I & 0 & 0 \\ \star & \star & \star & \star & \star & -\tilde{P}(\rho) & -\bar{h}\tilde{R} \\ \star & \star & \star & \star & \star & \star & -\tilde{R} \end{bmatrix} \prec 0$$

[1] We drop the dependence on the parameters to improve clarity.

hold for all $(\rho, \nu) \in \boldsymbol{\Delta}_\rho \times \mathbf{V}_\nu$ *where* $\Xi_{33} = -(1-\mu)\tilde{Q} - \tilde{R}$,

$$\Xi_{22}(\rho, \nu) = -\tilde{P}(\rho) + \tilde{Q} - \tilde{R} + \sum_i \frac{\partial \tilde{P}(\rho)}{\partial \rho_i} \nu_i,$$

$$\tilde{X} = \begin{bmatrix} W_1 & I \\ I & X_1 \end{bmatrix},$$

$$\Xi_{12}(\rho) = \begin{bmatrix} A(\rho)W_1 + B(\rho)\mathcal{C}_c(\rho) & A(\rho) + B(\rho)\mathcal{D}_c(\rho)C_y(\rho) \\ \mathcal{A}_c(\rho) & X_1 A(\rho) + \mathcal{B}_c(\rho)C_y(\rho) \end{bmatrix} + P(\rho),$$

$$\Xi_{13}(\rho) = \begin{bmatrix} A_h(\rho)W_1 + B(\rho)\mathcal{C}_c(\rho) & A(\rho) + B(\rho)\mathcal{D}_c(\rho)C_{yh}(\rho) \\ \mathcal{A}_{hc}(\rho) & X_1 A_h(\rho) + \mathcal{B}_c(\rho)C_{yh}(\rho) \end{bmatrix},$$

$$\Xi_{14}(\rho) = \begin{bmatrix} E(\rho) + B(\rho)\mathcal{D}_c(\rho)F_y(\rho) \\ X_1 E(\rho) + \mathcal{B}_c(\rho)F_y(\rho) \end{bmatrix},$$

$$\Xi_{25}(\rho) = \left[C(\rho)W_1 + D(\rho)\mathcal{C}_c(\rho) \;\; C(\rho) + D(\rho)\mathcal{D}_c(\rho)C_y(\rho)\right]^{\mathrm{T}},$$

$$\Xi_{35}(\rho) = \left[C_h(\rho)W_1 + D(\rho)\mathcal{C}_{yh}(\rho) \;\; C_h(\rho) + D(\rho)\mathcal{D}_c(\rho)C_{yh}(\rho)\right]^{\mathrm{T}},$$

$$\Xi_{45}(\rho) = \left[F(\rho) + D(\rho)\mathcal{D}_c(\rho)F_y(\rho)\right]^{\mathrm{T}}.$$

In such a case, the dynamic output feedback (8.71) with the matrices given by

$$\begin{bmatrix} A_c(\rho) & A_{hc}(\rho) & B_c(\rho) \\ C_c(\rho) & C_{hc}(\rho) & D_c(\rho) \end{bmatrix} = \mathcal{M}_1(\rho)^{-1}\left(\mathcal{K}(\rho) - \mathcal{M}_2(\rho)\right)\mathcal{M}_3(\rho)^{-1}$$

$$\mathcal{M}_1(\rho) = \begin{bmatrix} X_2 & X_1 B(\rho) \\ 0 & I \end{bmatrix},$$

$$\mathcal{K}(\rho) = \begin{bmatrix} \mathcal{A}_c(\rho) & \mathcal{A}_{hc}(\rho) & \mathcal{B}_c(\rho) \\ \mathcal{C}_c(\rho) & \mathcal{C}_{hc}(\rho) & \mathcal{D}_c(\rho) \end{bmatrix},$$

$$\mathcal{M}_2(\rho) = \begin{bmatrix} X_1 A(\rho)W_1 & X_1 A_h(\rho)W_1 & 0 \\ 0 & 0 & 0 \end{bmatrix},$$

$$\mathcal{M}_3(\rho) = \begin{bmatrix} W_2^{\mathrm{T}} & 0 & 0 \\ 0 & W_2^{\mathrm{T}} & 0 \\ C_y(\rho)W_1 & C_{yh}(\rho)W_1 & I \end{bmatrix},$$

$$X^{-1} = \begin{bmatrix} X_1 & X_2 \\ \star & X_3 \end{bmatrix}^{-1} = \begin{bmatrix} W_1 & W_2 \\ \star & W_3 \end{bmatrix}$$

stabilizes the system (8.52) and makes the L_2*-gain of the transfer* $w \to z$ *of the closed-loop system smaller than* γ *for all* $(h, \rho) \in \mathscr{H}_{\mu,\bar{h}} \times \mathscr{P}^\nu$.

Proof First of all, rewrite the closed-loop system as

$$\left[\begin{array}{c|c|c} A_{cl} & A_{hcl} & E_{cl} \\ \hline C_{cl} & C_{hcl} & F_{cl} \end{array}\right] = \Theta + \left[\begin{array}{cc} 0 & B \\ I & 0 \\ \hline 0 & D \end{array}\right] \Omega \left[\begin{array}{cc|cc|c} 0 & I & 0 & 0 & 0 \\ 0 & 0 & 0 & I & 0 \\ C_y & 0 & C_{yh} & 0 & F_y \end{array}\right]$$

where

$$\Theta := \left[\begin{array}{cc|cc|c} A & 0 & A_h & 0 & E \\ 0 & 0 & 0 & 0 & 0 \\ \hline C & 0 & C_h & 0 & F \end{array}\right] \text{ and } \Omega := \begin{bmatrix} A_c & A_{hc} & B_c \\ C_c & C_{hc} & D_c \end{bmatrix}.$$

Then, restrict X to be a symmetric positive definite matrix[2] of the form

$$X = \begin{bmatrix} X_1 & X_2 \\ \star & X_3 \end{bmatrix}.$$

and define

$$W := X^{-1} = \begin{bmatrix} W_1 & W_2 \\ \star & W_3 \end{bmatrix}.$$

Substituting the closed-loop system matrices in the LMI (6.30) yields

$$\begin{bmatrix} -2X & \Upsilon_{12}(\rho) & X^{\mathrm{T}} A_{hcl}(\rho) & X^{\mathrm{T}} E_{cl}(\rho) & 0 & X & \bar{h}R \\ \star & \Upsilon_{22}(\rho, \nu) & R & 0 & C_{cl}(\rho)^{\mathrm{T}} & 0 & 0 \\ \star & \star & \Upsilon_{33} & 0 & C_{hcl}(\rho)^{\mathrm{T}} & 0 & 0 \\ \star & \star & \star & -\gamma I & F_cl(\rho)^{\mathrm{T}} & 0 & 0 \\ \star & \star & \star & \star & -\gamma I & 0 & 0 \\ \star & \star & \star & \star & \star & -P(\rho) & -\bar{h}R \\ \star & \star & \star & \star & \star & \star & -R \end{bmatrix} \prec 0$$

where $\Upsilon_{33} = -(1-\mu)Q - R$ and

$$\begin{aligned} \Upsilon_{12}(\rho) &= P(\rho) + X^{\mathrm{T}} A_{cl}(\rho), \\ \Upsilon_{22}(\rho, \nu) &= -P(\rho) + Q - R + \textstyle\sum_i \frac{\partial P(\rho)}{\partial \rho_i} \nu_i. \end{aligned}$$

To linearize this inequality, a congruence transformation is first performed with respect to the matrix $\mathrm{diag}(Z, Z, Z, I, I, Z, Z)$ where

$$Z := \begin{bmatrix} W_1 & I \\ W_2^{\mathrm{T}} & 0 \end{bmatrix}.$$

[2] Avoiding this simplification is possible; see e.g. [16].

This leads to

$$\begin{bmatrix} \tilde{\Upsilon}_{11} & \tilde{\Upsilon}_{12}(\rho) & \tilde{\Upsilon}_{13}(\rho) & \tilde{\Upsilon}_{14}(\rho) & 0 & Z^{\mathrm{T}}XZ & \bar{h}\tilde{R} \\ \star & \tilde{\Upsilon}_{22}(\rho,\nu) & \tilde{R} & 0 & Z^{\mathrm{T}}C_{cl}(\rho)^{\mathrm{T}} & 0 & 0 \\ \star & \star & \tilde{\Upsilon}_{33} & 0 & Z^{\mathrm{T}}C_{hcl}(\rho)^{\mathrm{T}} & 0 & 0 \\ \star & \star & \star & -\gamma I & F_{c}l(\rho)^{\mathrm{T}} & 0 & 0 \\ \star & \star & \star & \star & -\gamma I & 0 & 0 \\ \star & \star & \star & \star & \star & -\tilde{P}(\rho) & -\bar{h}\tilde{R} \\ \star & \star & \star & \star & \star & \star & -\tilde{R} \end{bmatrix} \prec 0 \tag{8.72}$$

with $\tilde{\Upsilon}_{11} = -2Z^{\mathrm{T}}XZ$, $\tilde{\Upsilon}_{14}(\rho) = Z^{\mathrm{T}}X^{\mathrm{T}}E_{cl}(\rho)$, $\tilde{\Upsilon}_{33} = -(1-\mu)\tilde{Q} + \tilde{R}$, $\tilde{P}(\rho) = Z^{\mathrm{T}}P(\rho)Z$, $\tilde{Q} = Z^{\mathrm{T}}QZ$, $\tilde{R} = Z^{\mathrm{T}}RZ$,

$$\tilde{\Upsilon}_{12} = \tilde{P}(\rho) + Z^{\mathrm{T}}X^{\mathrm{T}}A_{cl}(\rho)Z, \quad \tilde{\Upsilon}_{13} = Z^{\mathrm{T}}X^{\mathrm{T}}A_{hcl}(\rho)Z \text{ and}$$

$$\tilde{\Upsilon}_{22} = -\tilde{P}(\rho) + \tilde{Q} - \tilde{R} + \sum_{i} \frac{\partial \tilde{P}(\rho)}{\partial \rho_i}\nu_i.$$

Noting that

$$Z^{\mathrm{T}}X = \begin{bmatrix} I & 0 \\ X_1 & X_2 \end{bmatrix} \text{ and } Z^{\mathrm{T}}XZ = \begin{bmatrix} W_1 & I \\ I & X_1 \end{bmatrix}$$

we obtain

$$\begin{aligned} \mathcal{Z} &:= \left[\begin{array}{c|c|c} Z^{\mathrm{T}}XA_{cl}Z & Z^{\mathrm{T}}XA_{hcl}Z & Z^{\mathrm{T}}XE_{cl} \\ \hline C_{cl}Z & C_{hcl}Z & F_{cl} \end{array}\right] \\ &= \left[\begin{array}{cc|cc|c} AW_1 & A & A_hW_1 & A & E \\ 0 & X_1A & 0 & X_1A_h & X_1E \\ \hline CW_1 & C & C_hW_1 & C_h & F \end{array}\right] + \Theta_1 \begin{bmatrix} \mathcal{A}_c & \mathcal{A}_{hc} & \mathcal{B}_c \\ \mathcal{C}_c & \mathcal{C}_{hc} & \mathcal{D}_c \end{bmatrix} \Theta_2 \end{aligned}$$

where

$$\Theta_1 = \left[\begin{array}{cc} 0 & B \\ I & 0 \\ \hline 0 & D \end{array}\right], \quad \Theta_2 = \left[\begin{array}{cc|cc|c} I & 0 & 0 & 0 & 0 \\ 0 & 0 & I & 0 & 0 \\ 0 & C_y & 0 & C_{yh} & F_y \end{array}\right],$$

$$\begin{bmatrix} \mathcal{A}_c & \mathcal{A}_{hc} & \mathcal{B}_c \\ \mathcal{C}_c & \mathcal{C}_{hc} & \mathcal{D}_c \end{bmatrix} = \begin{bmatrix} X_1AW_1 & X_1A_hW_1 & 0 \\ 0 & 0 & 0 \end{bmatrix} + \Theta_3 \begin{bmatrix} A_c & A_{hc} & B_c \\ C_c & C_{hc} & D_c \end{bmatrix} \Theta_4,$$

$$\Theta_3 = \begin{bmatrix} X_2 & X_1B \\ 0 & I \end{bmatrix} \text{ and } \Theta_4 = \begin{bmatrix} W_2^{\mathrm{T}} & 0 & 0 \\ 0 & W_2^{\mathrm{T}} & 0 \\ C_yW_1 & C_{yh}W_1 & I \end{bmatrix}.$$

We get finally that $F_{cl} = F + D\mathcal{D}_c F_y$ and

$$\begin{aligned}
Z^{\mathrm{T}} X A_{cl} Z &= \begin{bmatrix} AW_1 + B\mathcal{C}_c & A + B\mathcal{D}_c C_y \\ \mathcal{A}_c & X_1 A + \mathcal{B}_c C_y \end{bmatrix}, \\
Z^{\mathrm{T}} X A_{hcl} Z &= \begin{bmatrix} A_h W_1 + B\mathcal{C}_c & A + B\mathcal{D}_c C_{yh} \\ \mathcal{A}_{hc} & X_1 A_h + \mathcal{B}_c C_{yh} \end{bmatrix} \\
C_{cl} Z &= \begin{bmatrix} CW_1 + D\mathcal{C}_c & C + D\mathcal{D}_c C_y \end{bmatrix}, \\
C_{hcl} Z &= \begin{bmatrix} C_h W_1 + D\mathcal{C}_{yh} & C_h + D\mathcal{D}_c C_{yh} \end{bmatrix} \text{ and} \\
Z^{\mathrm{T}} X E_{cl} &= \begin{bmatrix} E + B\mathcal{D}_c F_y \\ X_1 E + \mathcal{B}_c F_y \end{bmatrix}
\end{aligned}$$

from which it is obvious that the expressions are affine in the variables X_1, W_1, $\mathcal{A}_c$, $\mathcal{A}_{hc}$, $\mathcal{B}_c$, $\mathcal{C}_c$, $\mathcal{C}_{ch}$ and $\mathcal{D}_c$. Substitution in the inequality (8.72) yields the result. ■

8.3.2 Memoryless Dynamic Output Feedback

Whereas conditions for designing dynamic output-feedback controllers with exact memory take the form of LMIs, the design conditions for memoryless controllers are nonconvex. To illustrate this, let us consider a memoryless controller of the form

$$\begin{aligned}
x_c(t) &= A_c(\rho(t))x_c(t) + B_c(\rho(t))y(t) \\
u(t) &= C_c(\rho(t))x_c(t) + D_c(\rho(t))y(t).
\end{aligned} \tag{8.73}$$

In such a case, the matrix $\mathcal{Z}$ is given by

$$\begin{aligned}
\mathcal{Z} = &\left[\begin{array}{cc|cc|c} AW_1 & A & A_h W_1 & A_h & E \\ 0 & X_1 A & X_1 A_h W_1 & X_1 A_h & X_1 E \\ \hline CW_1 & C & C_h W_1 & C_h & F \end{array}\right] \\
&+ \left[\begin{array}{cc} 0 & B \\ I & 0 \\ \hline 0 & D \end{array}\right] \begin{bmatrix} \mathcal{A}_c & \mathcal{B}_c \\ \mathcal{C}_c & \mathcal{D}_c \end{bmatrix} \left[\begin{array}{cc|cc|c} I & 0 & 0 & 0 & 0 \\ 0 & C_y & 0 & 0 & D_y \end{array}\right]
\end{aligned} \tag{8.74}$$

where

$$\begin{bmatrix} \mathcal{A}_c & \mathcal{B}_c \\ \mathcal{C}_c & \mathcal{D}_c \end{bmatrix} = \begin{bmatrix} X_1 A W_1 & 0 \\ 0 & 0 \end{bmatrix} + \begin{bmatrix} X_2 & X_1 B \\ 0 & I \end{bmatrix} \begin{bmatrix} A_c & B_c \\ C_c & D_c \end{bmatrix} \begin{bmatrix} W_2^{\mathrm{T}} & 0 \\ C_y W_1 & I \end{bmatrix} \tag{8.75}$$

and is a nonlinear function of the decision variables due to the presence of the bilinear term $X_1 A_h W_1$. This nonlinear term can be upper-bounded in the same way as for the design of the observer-based output feedback of Sect. 8.2.1 to yield more tractable synthesis conditions.

References

1. Y. Orlov, L. Belkoura, J.-P. Richard, M. Dambrine, Adaptive identification of linear time-delay systems. International Journal of Robust and Nonlinear Control **13**(9), 857–872 (2003)
2. S. Drakunov, S. Perruquetti, J.P. Richard, L. Belkoura, Delay identification in time-delay systems using variable structure control. Annual Reviews in Control **30**(2), 143–158 (2006)
3. L. Belkoura, J.P. Richard, M. Fliess, *Real time identification of time-delay systems* (In IFAC Workshop on Time-Delay Systems, Nantes, France, 2007)
4. L. Belkoura, J.P. Richard, M. Fliess, *A convolution approach for delay systems identification* (In IFAC World Congress, Seoul, South Korea, 2008)
5. F. Wu, K.M. Grigoriadis, LPV systems with parameter-varying time delays: analysis and control. Automatica **37**, 221–229 (2001)
6. F. Zhang and K. M. Grigoriadis. Delay-dependent stability analysis and $\mathcal{H}_\infty$ control for state-delayed LPV system. In Mediterranean Conference on Control and Automation, pages 1532–1537, 2005.
7. L. El-Ghaoui, F. Oustry, M. Ait, Rami. A Cone Complementary linearization algorithm for static output-feedback and related problems. IEEE Transactions on Automatic Control **42**, 1171–1176 (1997)
8. H.H. Choi, M.J. Chung, Observer-based $\mathcal{H}_\infty$ controller design for state delayed linear systems. Automatica **32**(7), 1073–1075 (1996)
9. H.H. Choi, M.J. Chung, Robust observer-based $\mathcal{H}_\infty$ controller design for linear uncertain time-delay systems. Automatica **33**(9), 1749–1752 (1997)
10. O. Sename and C. Briat. Observer-based $\mathcal{H}_\infty$ control for time-delay systems: a new LMI solution. In 6th IFAC Workshop on Time Delay Systems, L'Aquila, Italy, 2006.
11. O. Sename, Is a mixed design of observer-controllers for time-delay systems interesting ? Asian Journal of Control **9**(2), 180–189 (2007)
12. K. Tan, K.M. Grigoriadis, F. Wu, $\mathcal{H}_\infty$ and $\mathcal{L}_2$-to-$\mathcal{L}_\infty$ gain control of linear parameter varying systems with parameter varying delays. Control Theory and Applications **150**, 509–517 (2003)
13. C. Briat. Robust Control and Observation of LPV Time-Delay Systems. Grenoble Institute of Technology (2008) http://www.briat.info/thesis/PhDThesis.pdf
14. C.W. Scherer, P. Gahinet, M. Chilali, Multiobjective output-feedback control via LMI optimization. IEEE Transaction on Automatic Control **42**(7), 896–911 (1997)
15. C. W. Scherer and S. Weiland. Linear matrix inequalities in control. Lecture Notes; Available online http://www.imng.uni-stuttgart.de/simtech/Scherer/lmi/, 2005
16. J. de Caigny, J.F. Camino, R.C.L.F. Oliveira, P. L.D. Peres, and J. Swevers. Gain-scheduled dynamic output feedback control for discrete-time LPV systems. International Journal of Robust and Nonlinear Control **22**, 535–558 (2012)
17. C. Briat, O. Sename, J.-F. Lafay, Memory resilient gain-scheduled state-feedback control of uncertain LTI/LPV systems with time-varying delays. Syst. Control Lett. **59**, 451–459 (2010)

Appendix A
Technical Results in Linear Algebra

When I consider what people generally want in calculating,
I found that it always is a number.
Abū ‘Abdallāh Muḥammad ibn Mūsā al-Khwārizmī

A.1 Determinant Formulas

Let $A \in \mathbb{C}^{n \times n}$ be a given matrix, its determinant is denoted by $\det(A)$. If A and B are both square matrices of same dimensions, we have that

$$\det(AB) = \det(A)\det(B) = \det(BA).$$

Another well-known fact is

$$\det\left(\begin{bmatrix} A & B \\ 0 & D \end{bmatrix}\right) = \det(A)\det(D) \tag{A.1}$$

where both A and D are square. If A is square and nonsingular, then we can use the latter relations and the equality

$$\begin{bmatrix} A & B \\ C & D \end{bmatrix} = \begin{bmatrix} I & 0 \\ CA^{-1} & I \end{bmatrix} \begin{bmatrix} A & B \\ 0 & D - CA^{-1}B \end{bmatrix}$$

to get that

$$\det\left(\begin{bmatrix} A & B \\ C & D \end{bmatrix}\right) = \det(A)\det(D - CA^{-1}B),$$

a relation known as the *Schur (determinant) complement* or the *Schur formula*. This formula has been introduced in [1, 2] which have been later translated into English in [3]. For more details, see [4].

C. Briat, *Linear Parameter-Varying and Time-Delay Systems*,
Advances in Delays and Dynamics 3, DOI 10.1007/978-3-662-44050-6

Symmetrically, when D is nonsingular, we have that

$$\det\left(\begin{bmatrix} A & B \\ C & D \end{bmatrix}\right) = \det(D)\det(A - BD^{-1}C).$$

If $A = I$ and $D = I$ and BC is a square matrix, we arrive at the following very useful identity

$$\det(I - BC) = \det(I - CB).$$

A.2 Block-Matrices

Let us consider the matrix

$$M := \begin{bmatrix} A & B \\ C & D \end{bmatrix} \tag{A.2}$$

that we assume to be square and invertible. We then have the following result [4, 5]:

Proposition A.2.1 (Banachiewicz inversion formulas) *The inverse of M is given by*

$$\begin{aligned} M^{-1} &= \begin{bmatrix} A^{-1} + A^{-1}B(D - CA^{-1}B)^{-1}CA^{-1} & -A^{-1}B(D - CA^{-1}B)^{-1} \\ -(D - CA^{-1}B)^{-1}CA^{-1} & (D - CA^{-1}B)^{-1} \end{bmatrix} \\ &= \begin{bmatrix} (A - BD^{-1}C)^{-1} & -(A - BD^{-1}C)^{-1}BD^{-1} \\ -D^{-1}C(A - BD^{-1}C)^{-1} & D^{-1} + D^{-1}C(A - BD^{-1}C)^{-1}BD^{-1} \end{bmatrix}. \end{aligned}$$

The first formula is well-defined whenever A is invertible whereas the second one is when D is invertible. By identification of the blocks we get the well-known *matrix inversion lemma* which has been first introduced in [6]:

Lemma A.2.2 (Duncan inversion formulas) *We have the following identities*

$$\left(A - BD^{-1}C\right)^{-1} = A^{-1} + A^{-1}B\left(D - CA^{-1}B\right)^{-1}CA^{-1}$$

and

$$(A - BDC)^{-1} = A^{-1} + A^{-1}B\left(D^{-1} - CA^{-1}B\right)^{-1}CA^{-1}.$$

The following identity also holds:

$$A^{-1}B\left(D - CA^{-1}B\right)^{-1} = \left(A - BD^{-1}C\right)^{-1} BD^{-1}. \tag{A.3}$$

For more details about these formulas, see [4].

A.3 Singular-Values Decomposition

The singular value decomposition is a specific way of factorizing a rectangular matrix. It plays important roles in robust analysis and control, see e.g [7].

Theorem A.3.1 *Let $M \in \mathbb{C}^{k\times n}$ be a matrix of rank r. Then, there exist unitary matrices[a] U and V such that*

$$M = U\Sigma V^*$$

where U and V satisfy

$$MM^*U = U\Sigma\Sigma^* \qquad M^*MV = V\Sigma^*\Sigma$$

and Σ has the canonical structure

$$\Sigma := \begin{bmatrix} \Sigma_0 & 0 \\ 0 & 0 \end{bmatrix} \text{ where } \Sigma_0 := diag(\sigma_1, \ldots, \sigma_r) \prec 0$$

The numbers $\sigma_i > 0$, $i = 1, \ldots, r$ are called the nonzero singular values of M.

[a] A matrix U is unitary if $U^*U = UU^* = I$.

Proof The proof is given, for instance, in [8]. For more details on the singular value decomposition, see [9] or any other book on linear algebra. ■

A.4 Moore-Penrose Pseudoinverse

When a matrix is not invertible, it may still be possible to define an inverse referred to as the *generalized inverse* or the *Moore-Penrose pseudoinverse*.

Theorem A.4.1 *For every matrix $M \in \mathbb{R}^{n\times m}$, there exists a unique matrix $M^+ \in \mathbb{R}^{m\times n}$, the Moore-Penrose pseudoinverse of M, which satisfies the following identities*

$$MM^+M = M, \qquad M^+MM^+ = M^+,$$
$$(MM^+)^* = MM^+ \text{ and } (M^+M)^* = M^+M.$$

The explicit form of M^+ is given by

$$M^+ := V \begin{bmatrix} \Sigma_0^{-1} & 0 \\ 0 & 0 \end{bmatrix} U^*$$

where the matrices U, V and Σ_0 are obtained from the singular values decomposition of M.

Moreover, when

- M is full-row rank n, then $M^+ = M^*(MM^*)^{-1}$,
- M is full-column rank m, then $M^+ = (M^*M)^{-1}M^*$.

A.5 Solving $AX = B$

The solution X of the matrix equation $AX = B$ is trivial when A is square and nonsingular. The question is less easy when A is a rectangular or a singular square matrix. The following result, proved in [8], addresses these latter cases.

Theorem A.5.1 *Let $A \in \mathbb{R}^{n_1\times n_2}$, $X \in \mathbb{R}^{n_2\times n_3}$ and $B \in \mathbb{R}^{n_1\times n_3}$. Then, the following statements are equivalent:*

1. *The equation $AX = B$ has at least one solution X.*
2. *A and B satisfy $(I - AA^+)B = 0$.*

In such a case, all the solutions are given by

$$X = A^+B + (I - A^+A)\, Z$$

where $Z \in \mathbb{R}^{n_2\times n_3}$ is arbitrary and A^+ is the Moore-Penrose pseudoinverse of A.

A.6 Solving $BXC + (BXC)^{\mathrm{T}} + Q \prec 0$

This inequality arises, for instance, in the design of dynamic output feedback controllers for linear systems (see also the projection lemma in Appendix C.12). The proof of the following result can be found in [8].

Theorem A.6.1 *Let the matrices $B \in \mathbb{R}^{n\times m}$, $C \in \mathbb{R}^{k\times n}$ and $Q \in \mathbb{S}^n$ be given. Then, the following statements are equivalent:*

1. *There exists a $X \in \mathbb{R}^{m\times k}$ satisfying*

$$BXC + (BXC)^{\mathrm{T}} + Q \prec 0 \tag{A.4}$$

2. *The conditions*

$$\begin{array}{l} \mathcal{N}_B Q \mathcal{N}_B^{\mathrm{T}} \prec 0 \text{ or } BB^{\mathrm{T}} \succ 0 \\ \mathcal{N}_C^{\mathrm{T}} Q \mathcal{N}_C \prec 0 \text{ or } C^{\mathrm{T}} C \succ 0 \end{array}$$

hold where $\mathcal{N}_B$ and $\mathcal{N}_C$ are bases of the left and right null-spaces of B and C, respectively.
Suppose that the above statements hold. Let r_b and r_c be the ranks of B and C, respectively. Let, furthermore, (B_ℓ, B_r) and (C_ℓ, C_r) be any full rank factors of B and C (i.e. $B = B_\ell B_r$ and $C = C_\ell C_r$). Then, all the matrices X solutions of (A.4) *are given by:*

$$X = B_r^+ K C_\ell^+ Z - B_r^+ B_r Z C_\ell C_\ell^+$$

where Z is any arbitrary matrix and

$$\begin{aligned} K &:= -R^{-1} B_\ell^{\mathrm{T}} \Phi C_r^{\mathrm{T}} \left(C_r \Phi C_r^{\mathrm{T}}\right)^{-1} + S^{1/2} L \left(C_r \Phi C_r^{\mathrm{T}}\right)^{-1/2} \\ S &:= R^{-1} - R^{-1} B_\ell^{\mathrm{T}} - R^{-1} B_\ell^{\mathrm{T}} \left[\Phi - \Phi C_r^{\mathrm{T}} \left(C_r^{\Phi} C_r^{\mathrm{T}}\right)^{-1} C_r \Phi\right] B_\ell R^{-1} \end{aligned}$$

where L is any arbitrary matrix such that $||L|| < 1$ (i.e. $\bar{\sigma}(L) < 1$) and R is any arbitrary positive definite matrix such that

$$\Phi := \left(B_\ell R^{-1} B_\ell^{\mathrm{T}} - Q\right)^{-1} \succ 0.$$

The solution for X above is quite intricate can be made simpler [10]. Two alternative solutions denoted by X_1 and X_2 have been obtained in [10, 11] and are given by

$$X_1 := -\tau_1 B^{\mathrm{T}} \Psi_1 C^{\mathrm{T}} \left(C \Psi_1 C^{\mathrm{T}}\right)^{-1} \text{ and } X_2 := -\tau_2 \left(B^{\mathrm{T}} \Psi_2 B\right)^{-1} B^{\mathrm{T}} \Psi_2 C^{\mathrm{T}}$$

where $\tau_1, \tau_2 > 0$ are sufficiently large scalars such that

$$\Psi_1 := \left(\tau_1 BB^{\mathrm{T}} - Q\right)^{-1} \succ 0 \text{ and } \Psi_2 := \left(\tau_2 C^{\mathrm{T}} C - Q\right)^{-1} \succ 0.$$

Appendix B
Linear Matrix Inequalities

...the "dragon" of optimization is multiheaded and it takes a special sword to cut-off each head.

V. F. Dem'yanov and L. V. Vasil'ev

B.1 Preliminaries

An LMI problem is the problem of finding $x \in \mathbb{R}^m$ such that the matrix inequality

$$\mathcal{L}(x) := \mathcal{L}_0 + \sum_{i=1}^{m} \mathcal{L}_i x_i \succ 0 \tag{B.1}$$

holds where the inequality is understood in terms of the location of the eigenvalues, i.e. the eigenvalues are positive, and the matrices $\mathcal{L}_i \in \mathbb{S}^n$, $i = 1, \ldots, n$, are symmetric and known. It turns out that this problem is convex since the set

$$\mathcal{S} = \left\{x \in \mathbb{R}^m : \mathcal{L}(x) \succ 0\right\} \tag{B.2}$$

is convex.

Example B.1.1 The LMI

$$\begin{bmatrix} R^2 & x_1 & x_2 \\ x_1 & 1 & 0 \\ x_2 & 0 & 1 \end{bmatrix} \succ 0 \tag{B.3}$$

describes the set $\mathcal{S} = \left\{x \in \mathbb{R}^2 : x_1^2 + x_2^2 < R^2\right\}$ which is the ball of radius R.

C. Briat, *Linear Parameter-Varying and Time-Delay Systems*, Advances in Delays and Dynamics 3, DOI 10.1007/978-3-662-44050-6

Several LMIs $\mathcal{L}^i(x) \succ 0$, $i = 1, \ldots, q$ can be expressed into a single one as $\text{diag}_{i=1}^q\{\mathcal{L}^i(x)\} \succ 0$, which is consistent with the fact that an intersection of convex sets is also a convex set.

Example B.1.2 For instance, the LMI condition

$$A^{\mathrm{T}}P + PA \prec 0 \tag{B.4}$$

for some $P \in \mathbb{S}^n_{\succ 0}$ can be rewritten as

$$\mathcal{L} := \begin{bmatrix} A^{\mathrm{T}}P + PA & 0 \\ \star & -P \end{bmatrix} \prec 0. \tag{B.5}$$

Let $\{P_i\}$ be a basis of the set of symmetric matrices of dimension n, then P can be decomposed as

$$P = \sum_{i=1}^{\frac{n(n+1)}{2}} P_i x_i. \tag{B.6}$$

We thus have that

$$\mathcal{L} = \sum_{i=1}^{\frac{n(n+1)}{2}} x_i \mathcal{L}_i \tag{B.7}$$

where

$$\mathcal{L}_i = \begin{bmatrix} A^{\mathrm{T}}P_i + P_i A & 0 \\ \star & -P_i \end{bmatrix} \tag{B.8}$$

and $x \in \mathbb{R}^{\frac{n(n+1)}{2}}$.

Optimization problems involving LMIs arise in many problems in systems and control theory, e.g. in the computation of the H_∞-norm of linear systems. Such problems are formally expressed as

$$\begin{aligned} &\min c^{\mathrm{T}}x \\ &\quad \text{s.t. } x \in \mathbb{R}^m \\ &\qquad \mathcal{L}(x) \succ 0 \end{aligned} \tag{B.9}$$

where $c \in \mathbb{R}^m$. Algorithms have been proposed to solve the above problem in an efficient manner, for instance using interior point methods; see e.g. [12, 13]. Sophisticated solvers such as SeDuMi [14] or SDPT3 [15] can be used together with the Yalmip interface [16] or the CVX interface [17, 18] to solve LMI problems.

Semi-Infinite LMI Problems
Semi-infinite LMI optimization problems take the form

$$\begin{aligned} &\min c^{\mathrm{T}} x \\ &\quad \text{s.t. } x \in \mathbb{R}^m \\ &\quad \mathcal{L}(x,\delta) := \mathcal{L}_0(\delta) + \sum_{i=1}^{m} \mathcal{L}_i(\delta) x_i \succ 0,\ \delta \in \Delta \end{aligned} \tag{B.10}$$

where Δ is a compact set. It is meant, above, that a single $x \in \mathbb{R}^m$, i.e. independent of δ, must be determined such that the LMI condition holds for all $\delta \in \Delta$. We are hence in presence of an LMI-problem involving an infinite number of LMI constraints parametrized by δ.

Example B.1.3 The quadratic stability of the uncertain linear system

$$\dot{x} = A(\delta)x$$

with $\delta \in \Delta$ is characterized by the semi-infinite dimensional LMI problem:

Find $P \in \mathbb{S}^n$ such that the LMIs

$$\begin{aligned} A(\delta)^{\mathrm{T}} P + PA(\delta) &\prec 0 \\ P &\succ 0 \end{aligned} \tag{B.11}$$

hold for all $\delta \in \Delta$.

Infinite-Dimensional LMI Problems
On the other hand, infinite-dimensional LMI optimization problems take the form

$$\begin{aligned} &\min c^{\mathrm{T}} x(\delta) \\ &\quad \text{s.t. } x : \Delta \to \mathbb{R}^m \\ &\quad \mathcal{L}_0(\delta) + \sum_{i=1}^{m} \mathcal{L}_i(\delta) x_i(\delta) \succ 0,\ \delta \in \Delta \end{aligned} \tag{B.12}$$

where Δ is compact. The main difference lies in the fact that the decision variable $x : \Delta \to \mathbb{R}^m$ is now a *function*, whence the name *infinite-dimensional LMI problem*.

Example B.1.4 The robust stability of an LTI uncertain linear system $\dot{x} = A(\delta)x$ with $\delta \in \Delta$, Δ compact, is characterized in terms of the following infinite-dimensional LMI problem:

Find $P : \Delta \to \mathbb{S}^n$ such that the LMIs

$$\begin{aligned} A(\delta)^{\mathrm{T}} P(\delta) + P(\delta) A(\delta) &\prec 0 \\ P(\delta) &\succ 0 \end{aligned} \tag{B.13}$$

hold for all $\delta \in \Delta$.

B.2 Solving Infinite-Dimensional LMI Problems

A method for converting infinite-dimensional variables into finite-dimensional ones is based on the projection of the infinite-dimensional decision variables onto a finite-dimensional basis of functions, e.g. a polynomial basis, as exemplified below[1]:

$$f_{\alpha_i}(\rho) = \rho^{\alpha_i}, \; i = 1, \ldots, N_b.$$

The matrix $P(\rho)$ in (B.13) and Theorem 2.4.1 can then be expressed over this basis as

$$P(\rho) = \sum_{i=1}^{N_b} P_i f_{\alpha_i}(\rho)$$

where the matrices $P_i \in \mathbb{S}^n$, $i = 1, \ldots, N_b$, are our new finite-dimensional decision variables. The following corollary follows from Theorem 2.4.1 where we have used the projection method described above:

Corollary B.2.1 *The system (2.36) is robustly stable if there exist matrices $P_i \in \mathbb{S}^n_{\succ 0}$, $i = 1, \ldots, N_b$, such that the LMIs*

$$\sum_{i=1}^{N_b} P_i f_{\alpha_i}(\rho) \succ 0$$

and

$$\mathrm{He}\left[\left(\sum_{i=1}^{N_b} P_i f_{\alpha_i}(\rho)\right) A(\rho)\right] + \sum_{i=1}^{N} \nu_i \left(\sum_{i=1}^{N_b} P_i \frac{\partial f_{\alpha_i}(\rho)}{\partial \rho_i}\right) \prec 0$$

hold for all $(\rho, \nu) \in \boldsymbol{\Delta}_\rho \times \mathbf{V}_\nu$.

[1] We use here a multi-index notation.

The main difficulty in this procedure lies in the fact that it is unclear how to decide what basis should be used, i.e. the type of basis functions and their number. An informal agreement between practioners consists in selecting a basis of functions that are close to the parameter dependence of the system, e.g. a polynomial basis when the system depends polynomially on the parameters.

B.3 Solving Semi-Infinite LMI Problems

This section aims at introducing different relaxation schemes for solving semi-infinite LMI problems. These relaxation schemes mainly consist of converting the initial problem into a tractable finite-dimensional problem. Three methods are presented: the first one addresses the case of parameter-dependent LMIs that are affine in the parameters. By exploiting this structure, it is possible to obtain interesting finite-dimensional results. The second method, generally referred to as the *gridding approach*, can be applied to parameter-dependent LMIs with any parameter dependence. Finally, the third and last one, is based on *sum-of-squares programming* and can be applied to polynomially parameter-dependent LMIs. Some other methods are also briefly mentioned for completeness in the last part of this section.

B.3.1 Relaxation of Affine Parameter-Dependent LMIs

We consider in this section LMIs that are affine in the parameters, that is, LMIs taking the form

$$\mathcal{M}(x,\delta) := \mathcal{M}_0(x) + \sum_{i=1}^{N} \delta_i \mathcal{M}_i(x) \prec 0 \tag{B.14}$$

where the $\mathcal{M}_i$'s are symmetric matrices, $x \in \mathbb{R}^n$ is the vector of decision variables and $\delta \in [-1, 1]^N$ is the vector of parameters, N being the number of parameters. Note that a change of variables can always change the domain of values of the parameters to $[-1, 1]^N$. For instance, when the initial domain is given by $\delta_i \in [\delta_i^-, \delta_i^+]$, $\delta_i^- < \delta_i^+$, the change of variables

$$\delta_i = \frac{\delta_i^+ - \delta_i^-}{2}\tilde{\delta}_i + \frac{\delta_i^+ + \delta_i^-}{2}, \quad \delta_i \in [-1, 1] \tag{B.15}$$

gives the LMI condition

$$\mathcal{M}(x,\tilde{\delta}) := \widetilde{\mathcal{M}}_0(x) + \sum_{i=1}^{N} \tilde{\delta}_i \widetilde{\mathcal{M}}_i(x) \prec 0 \tag{B.16}$$

where $\delta \in [-1, 1]^N$ and

$$\begin{aligned}\widetilde{\mathcal{M}}_0(x) &:= \mathcal{M}_0(x) + \textstyle\sum_{i=1}^N \frac{\delta_i^+ + \delta_i^-}{2}\mathcal{M}_i(x)\\ \widetilde{\mathcal{M}}_i(x) &:= \frac{\delta_i^+ - \delta_i^-}{2}\mathcal{M}_i(x).\end{aligned} \tag{B.17}$$

Vertex LMIs

The next result has been proved in Sect. 2.4.1 in the particular case of quadratic stability of generic LPV systems. We state here the general case:

Theorem B.3.1 *Let us consider the affine parameter-dependent LMI* (B.14). *Then, the following statements are equivalent:*

1. *There exists $x \in \mathbb{R}^n$ such that the LMI*

$$\mathcal{M}_0(x) + \sum_{i=1}^N \delta_i \mathcal{M}_i(x) \prec 0 \tag{B.18}$$

holds for all $\delta \in [-1, 1]^N$.

2. *There exists $x \in \mathbb{R}^n$ such that the LMI*

$$\mathcal{M}_0(x) + \sum_{i=1}^N v_i \mathcal{M}_i(x) \prec 0 \tag{B.19}$$

holds for all $v \in \{-1, 1\}^N$.

Proof The proof exploits the convexity of the polytope $[-1, 1]^N$, from which we can state that for any $\delta \in [-1, 1]^N$, there exists $\lambda \in \Lambda_{2^N}$ such that

$$\delta = \sum_{i=1}^{2^N} \lambda_i v_i, \;\; v_i \in \{-1, 1\}^N. \tag{B.20}$$

Proof of 2 ⇒ 1: Assume that the LMIs (B.19) hold for all $v \in \{-1, 1\}^N$. Then, considering $\mathcal{M}_0(x) + v_i\mathcal{M}_i(x)$,multiplying it by λ_i and summing over i yields

$$\sum_{i=1}^{2^N} \lambda_i \left[\mathcal{M}_0(x) + v_i \mathcal{M}_i(x)\right]. \tag{B.21}$$

Since by assumption the LMIs (B.19) hold for all $v \in \{-1, 1\}^N$ and using the facts that (1) a sum of negative definite matrices is negative definite; and (2) for any

$\delta \in [-1, 1]^N$, there is $\lambda \in \Lambda_{2^N}$ such that (B.20) holds, then we can conclude that (B.21) holds for all $\lambda \in \Lambda_{2^N}$. Therefore, (B.18) holds for all $\delta \in [-1, 1]^N$.

Proof of 1 ⇒ 2: Assume that the LMI (B.18) holds for all $\delta \in [-1, 1]^N$, then it must also hold on all the vertices of $[-1, 1]^N$, and therefore for all $v \in \{-1, 1\}^N$. ■

The advantage of the above result lies in the equivalence between the statements. We have indeed been able to convert an infinite set of LMIs into a finite set. The compensation, however, lies on the level of tractability of the finite-dimensional representation since the number of LMIs we have to check is an exponential number of the parameters, i.e. we exactly have 2^N LMIs, which may be prohibitive when N is "large".

Matrix Cube Theorem

The matrix cube theorem has been proposed in [19–21] and extended in [22]. The idea behind this theorem is to find a finite dimensional LMI condition that approximates the semi-infinite LMI (B.14). This result is stated below

Theorem B.3.2 (Matrix cube theorem) *Assume that there exist symmetric matrices X_i, $i = 1, \ldots, N$, and a vector $x \in \mathbb{R}^n$ such that the LMIs*

$$-X_i \pm \mathcal{M}_i(x) \preceq 0, \; i = 1, \ldots, N \tag{B.22}$$

and

$$\mathcal{M}_0(x) + \sum_{i=1}^{N} X_i \prec 0 \tag{B.23}$$

hold. Then, the LMI (B.14) *holds for all $\delta \in [-1, 1]^N$ with the same x.*

Proof Assume that the statements of the theorem holds. Then, from (B.22), we have that the X_i's are positive semidefinite.

Note that since $\sum_{i=1}^{N} |\delta_i| [-X_i \pm \mathcal{M}_i(x)] \preceq 0$, then we have

$$\sum_{i=1}^{N} -X_i \pm |\delta_i| \mathcal{M}_i(x) \preceq \sum_{i=1}^{N} |\delta_i| [-X_i \pm \mathcal{M}_i(x)] \preceq 0 \tag{B.24}$$

and thus

$$\sum_{i=1}^{N} X_i \succeq \sum_{i=1}^{N} \delta_i \mathcal{M}_i(x) \tag{B.25}$$

where we have used the fact that $\delta_i \in [-1, 1]$. Combining this with (B.23), we get that the matrix inequalities

$$\mathcal{M}_0(x) + \sum_{i=1}^{N} \delta_i \mathcal{M}_i(x) \preceq \mathcal{M}_0(x) + \sum_{i=1}^{N} X_i \prec 0 \quad \text{(B.26)}$$

hold for all $\delta \in [-1, 1]^N$, which is equivalent to the feasibility of the LMI (B.14). The proof is complete. ■

Even though the above result is only sufficient, the number of LMIs to solve is equal to $2N + 1$. Hence, the finite-dimensional approximation scales linearly with respect to the number of parameters. This a great advantage over the approach based on the vertices of the polytope which scales exponentially. A conservatism analysis carried out in [19] indicates that when the matrices $\mathcal{M}_i$'s are of small-ranks, then the approach is not too conservative in the sense that when the above theorem does not hold, then it is possible to slightly increase the size of the box containing the parameters to make the original problem involving the LMI (B.14) infeasible. This is stated in the following result:

Theorem B.3.3 (Conservatism of the matrix cube theorem) *Assume that the conditions of Theorem B.3.2 are not fulfilled and let*

$$\zeta_x = \max_{i=1,\ldots,N} rank[\mathcal{M}_i(x)].$$

Then, the LMI (B.14) *with fixed x is not feasible for the enlarged parameter domain* $\delta \in [-\vartheta(\zeta_x), \vartheta(\zeta_x)]^N$ *where* $\vartheta(\zeta)$ *is a universal function such that*

$$\vartheta(1) = 1, \ \vartheta(2) = \frac{\pi}{2} \simeq 1.57, \ \vartheta(3) \simeq 1.73, \ \vartheta(4) = 2 \quad \text{(B.27)}$$

and

$$\vartheta(\zeta) \leq \frac{\pi\sqrt{\zeta}}{2} \quad \text{(B.28)}$$

for all positive $\zeta \in \mathbb{N}$.

Proof The proof can be found in [19]. ■

B.3.2 Relaxation of General Parameter-Dependent LMIs by Gridding

Gridding is certainly the most straightforward way for dealing with semi-infinite constraints. This procedure can be applied to any LMI with any parameter dependence. It is thus very general. The idea is to approximate the semi-infinite constraint LMI by a finite number of LMIs, each one of them corresponding to a specific point in the parameter space. To illustrate this, let us assume that we have the following feasibility problem:

Problem B.3.4 Find $x \in \mathbb{R}^n$ such that the LMI

$$\mathcal{L}(x, \delta) \prec 0 \tag{B.29}$$

holds for all $\delta \in \Delta$ where Δ is compact.

The gridding approach simply proposes to substitute this problem by the following one:

Problem B.3.5 Find $x \in \mathbb{R}^n$ such that the LMI $\mathcal{L}(x, \delta) \prec 0$ holds for all $\delta \in \Delta_g$ where Δ_g is a finite collection of points in Δ.

The rationale behind this approach is that assuming that the initial problem is unfeasible, then by choosing a sufficiently dense set of points, a critical point will finally be sampled and infeasibility will be inferred from the gridded conditions. The main problem lies in the fact that it is not really known how to sample the parameter space, i.e. how the points should be distributed and how many points should be considered.[2] Even though the computational complexity of the gridded conditions grows linearly with the number of samples, the curse of dimensionality makes the method impractical when the number of parameters is large. Assuming for instance that the number of parameters is given by N_p and that the number of samples for each parameter is given by N, the number of LMIs to consider is then equal to N^{N_p}.

It is finally very important to stress that, even if for a large number of samples the gridded problem is feasible, one cannot conclude on the feasibility of the original problem since we may still miss critical points located between samples. The gridded problem feasibility is then *a necessary condition only* for the feasibility of the original problem. In spite of this, the gridding approach is still very useful when dealing with problems with very few parameters for which a very thin grid can be considered. In certain problems having a particular structure, it is possible to consider balls around each sample so that the complete parameter space is covered. In such cases, the inaccuracy problem of the gridding method is resolved, see e.g. [26, 27]. In all the other cases, it is possible to have an a posteriori certificate of accuracy, as shown in [28]. The latter one is discussed below on few examples:

---

[2] Probabilistic approaches can be employed for dealing with such a problem and obtain probabilistic certificates of feasibility, see e.g. [23–25].

Proposition B.3.6 *Let us consider the semi-infinite dimensional LMI* (B.11) *characterizing quadratic stability of an LPV system. Let us consider that N parameters are involved and that the LMI is shown to be feasible over the grid*

$$\boldsymbol{\Delta}_g := \{\delta_{1,1}, \ldots, \delta_{1,n_1}\} \times \ldots \times \{\delta_{N,1}, \ldots, \delta_{N,n_N}\} \tag{B.30}$$

and let $h_j > 0$ be defined as $\delta_{j,i+1} - \delta_{j,i} \leq h_j$ for all $i = 1, \ldots, n_j - 1$, $j = 1, \ldots, N$. Assume further that

1. $||P||_F \leq T$, *for some $T > 0$ and where $|| \cdot ||_F$ denotes the Frobenius norm*[a], *and*
2. $A(\delta)^{\mathrm{T}}P + PA(\delta) \prec -\varphi I$ *holds for some $\varphi > 0$ and for all $\delta \in \boldsymbol{\Delta}_g$.*

Then, the LMI (B.11) *also holds for all $\delta \in \boldsymbol{\Delta}$ provided that the condition*

$$h_j \leq \frac{\varphi}{2TN}\left(\max_{\delta \in \Delta}\left|\left|\frac{\partial A(\delta)}{\partial \delta_j}\right|\right|_F\right)^{-1} \tag{B.31}$$

holds for all $i = 1, \ldots, N$.

[a] *The Frobenius norm of a matrix M is defined as*

$$||M||_F := \left(\sum_{i,j} |m_{i,j}|^2\right)^{1/2} = \sqrt{trace(M^*M)}.$$

It also coincides with the Euclidian norm of the vector containing the singular values of M.

Proof First note that for any $\delta \in \Delta$, there exist some integers $k_1, \ldots, k_N$ such that

$$\delta \in \boldsymbol{\Delta}_k := [\delta_{1,k_1}, \delta_{1,k_1+1}] \times \ldots \times [\delta_{N,k_N}, \delta_{N,k_N+1}]$$

and let $\bar{\delta} = \left[\delta_{1,k_1}, \ldots, \delta_{N,k_N}\right]$. Define then

$$D(\delta, \bar{\delta}) := \left[A(\delta) - A\left(\bar{\delta}\right)\right]^{\mathrm{T}} P + P\left[A(\delta) - A\left(\bar{\delta}\right)\right]. \tag{B.32}$$

Thus, we have

$$\begin{aligned}||D\left(\delta,\bar{\delta}\right)||_F &= 2\left|\left|P\left[A(\delta)-A(\bar{\delta})\right]\right|\right|_F\\ &\le 2||P||_F\sum_{j=1}^N|\delta_i-\delta_{i,k_j}|\cdot\left|\left|\frac{\partial A}{\partial\delta_j}(\xi_j)\right|\right|_F \quad \text{for some } \xi_j\in\boldsymbol{\Delta}_k\\ &\le 2T\sum_{j=1}^N h_j\max_{\delta\in\boldsymbol{\Delta}}\left|\left|\frac{\partial A(\delta)}{\partial\delta_j}\right|\right|_F\\ &\le\varphi\end{aligned}$$

where we have used the conditions of the proposition. Using then the fact that for any real matrix M, we have $||M||_2 \le ||M||_F$ and hence $||D(\delta,\bar{\delta})||_2 \le \varphi$. This thus implies that if $A(\delta)^{\mathrm{T}}P + PA(\delta) \prec -\varphi I$ holds for some $\varphi > 0$ and for all $\delta \in \boldsymbol{\Delta}_g$, then $A(\delta)^{\mathrm{T}}P + PA(\delta) \prec 0$ holds for all $\delta \in \boldsymbol{\Delta}$. The proof is complete. ■

The application of this idea to more complex matrix inequalities is straightforward due to the decomposability property of the Frobenius norm. For instance, we have that

$$\begin{aligned}\left|\left|\begin{bmatrix}A(\rho)^{\mathrm{T}}P+PA(\rho) & PE(\rho) & C(\rho)^{\mathrm{T}}\\ \star & -\gamma I_m & F(\rho)^{\mathrm{T}}\\ \star & \star & -\gamma I_p\end{bmatrix}\right|\right|_F &= 2||PA(\rho)||_F+2||PE(\rho)||_F\\ &\quad +2||C(\rho)||_F+2||F(\rho)||_F\\ &\quad +(m+p)\gamma.\end{aligned} \tag{B.33}$$

Note that considering directly the 2-norm would be much more difficult.

B.3.3 Relaxation of Polynomially Parameter-Dependent LMIs Using Sum of Squares Programming

Sum of squares (SOS) programming is a powerful tool that can be used for dealing with infinite-dimensional and semi-infinite semidefinite optimization problems. The key idea behind sum of squares programming is to merge tools from algebraic geometry and optimization theory for characterizing positivity of polynomials over compact semialgebraic sets. Although, the next results are mostly stated for scalar polynomials, they can be easily extended to the matrix case. The toolboxes SOSTOOLS [29, 30], Yalmip [16] and SOSOPT [31] can be used to handle sum of squares programming problems.

Preliminary Results on Sum of Squares and Positive Polynomials

The following result considers the case of univariate polynomials:

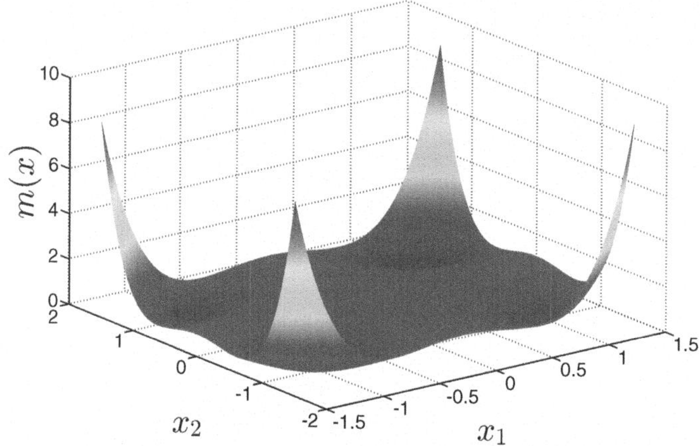

Fig. B.1 Motzkin's polynomial

Theorem B.3.7 *Let $p(x) = \sum_{i=0}^{n} p_i x^i$ be a real univariate polynomial of degree n. Then, the following statements are equivalent:*

1. *The polynomial $p(x)$ is nonnegative over $x \in \mathbb{R}$.*
2. *There exist polynomials $h_i(x)$, $i = 1, \ldots, n_0$, $n_0 \leq n$, such that*

$$p(x) = \sum_{i=1}^{n_0} h_i(x)^2. \tag{B.34}$$

Moreover, the degree n of $p(x)$ is even and $p_n > 0$.

Proof **Proof of 1 ⇒ 2**: Assume that $p(x) \geq 0$ for all $x \in \mathbb{R}$. Since $p(x)$ is univariate, it can then be factorized as

$$\begin{aligned} p(x) &= p_n \prod_i (x - r_i)^{n_i} \prod_k (x - \alpha_k + j\beta_k)^{m_k} (x - \alpha_k - j\beta_k)^{m_k} \\ &= p_n \prod_i (x - r_i)^{n_i} \prod_k \left[(x - \alpha_k)^2 + \beta_k^2\right]^{m_k} \end{aligned}$$

where r_i denotes the i-th real root with multiplicity n_i, and $\alpha_k \pm j\beta_k$ denote the k-th complex conjugate roots with multiplicity m_k. Since the second product in the expression above is always positive, then nonnegativity only depends on the product $p_n \prod_i (x - r_i)^{n_i}$. Clearly, if n_i is odd, the polynomial cannot be nonnegative regardless of the sign of p_n. Therefore, n_i is an even number and p_n is positive. Defining then $n_i := 2n_i'$, we get that

$$p(x) = p_n \prod_i (x - r_i)^{2n_i'} \prod_k \left[(x - \alpha_k)^2 + \beta_k^2\right]^{m_k}.$$

Using the fact that the product of sums of squares polynomials is also a sum of squares polynomial,[3] we can conclude that $p(x)$ is a sum of squares.

Proof of 2 ⇒ 1: It is straightforward to see that if the polynomial is a sum of squares, then it is nonnegative. ∎

The above result unfortunately does not extend to the multivariate case since a nonnegative multivariate polynomial may not be a sum of squares. A well-known example is Motzkin's polynomial (Fig. B.1)

$$m(x) = 1 + x_1^2 x_2^2 \left(x_1^2 + x_2^2 - 3\right) \tag{B.35}$$

which is globally nonnegative[4] but cannot be written as a sum of squares [34]. The problem of writing positive polynomials as a sum of squares is related to the celebrated 27th Hilbert's problem, solved by Artin in 1927.

Theorem B.3.8 (Artin's solution [35]) *Any multivariate polynomial $p(x)$ that is nonnegative for all $x \in \mathbb{R}^n$ can be expressed as a sum of squares of rational functions.*

An immediate corollary of this result is given below:

Corollary B.3.9 *Any multivariate polynomial $p(x)$ that is nonnegative for all $x \in \mathbb{R}^n$ can be expressed as*

$$p(x) = \frac{n(x)}{d(x)} \tag{B.36}$$

where $n(x)$ and $d(x)$ are sum of squares polynomials.

Proof From Artin's Theorem, the polynomial can be written as

$$p(x) = \sum_i \frac{n_i(x)^2}{d_i(x)^2} \tag{B.37}$$

for some polynomials $n_i(x)$, $d_i(x)$. This sum of squares of quotient can be expressed with the same denominator $d(x) = d_1(x)^2 d_2(x)^2 \dots$ which is product of squares and the numerator $n(x)$ is a sum of products of squares. Since the set of sum of squares polynomials is closed under multiplication and addition, the numerator and the denominator are sums of squares. The proof is complete. ∎

---

[3] The set of sum of squares polynomials is closed under multiplication.

[4] This can be proved from the arithmetic-geometric mean inequality, i.e. the arithmetic mean is greater or equal to the geometric mean [32, 33].

The latter result is very useful in the context of determining whether a polynomial is nonnegative in the multivariate case since if we can find sum of squares polynomials $n(x)$ and $d(x)$ such that $p(x)d(x) = n(x)$ then the polynomial $p(x)$ is nonnegative for all $x \in \mathbb{R}^n$. This condition is, moreover, necessary.

Example B.3.10 As an example, Motzkin's polynomial $m(x)$ defined in (B.35), known to be not sum of squares representable, can be expressed as [34, 36]:

$$m(x) = \left(\frac{x_1^2 - x_2^2}{x_1^2 + x_2^2}\right)^2 + \left(\frac{x_1 x_2(x_1^2 + x_2^2 - 2)}{x_1^2 + x_2^2}\right)^2 + \left(\frac{x_1^2 x_2(x_1^2 + x_2^2 - 2)}{x_1^2 + x_2^2}\right)^2 + \left(\frac{x_1 x_2^2(x_1^2 + x_2^2 - 2)}{x_1^2 + x_2^2}\right)^2. \tag{B.38}$$

Let us now consider a compact semi-algebraic set $\mathcal{S}$ given by

$$\mathcal{S} := \left\{x \in \mathbb{R}^n : f_1(x) \geq 0, \ldots, f_m(x) \geq 0\right\} \tag{B.39}$$

where the $f_i(x)$'s are polynomials. Let us also define the quadratic module of the polynomials $f_1, \ldots, f_m$ as

$$M(f_1, \ldots, f_m) := \{\sigma_0(x) + \sigma_1(x) f_1(x) + \ldots + \sigma_m(x) f_m(x) : \sigma_i(x) \text{ SOS}\}. \tag{B.40}$$

We have the following very important result:

Theorem B.3.11 (Putinar's Positivstellensatz, [37]) *Assume that there exists* $N > 0$ *such that* $N - ||x||_2^2 \in M(f_1, \ldots, f_m)$*. If* $f(x)$ *is positive on* $\mathcal{S}$*, then* $f(x) \in M(f_1, \ldots, f_m)$*.*

The first condition implies that the module $M(f_1, \ldots, f_m)$ is Archimedean. Under this condition, the above result states that if $f(x)$ is positive on $\mathcal{S}$, then it can be written as a linear combination of the polynomials $1, f_1, \ldots, f_m$ where the coefficients of the linear combination are taken as sum of squares polynomials. Note that even though the positivstellensatz provides a necessary condition, it is easily seen that if $f(x)$ can be written as an element of $M(f_1, \ldots, f_m)$ or even admits a lower bound in $M(f_1, \ldots, f_m)$, then it is positive on $\mathcal{S}$. It is interesting to point out that the positivstellensatz is akin to the S-procedure (see Appendix C.8), the latter being however more specialized to quadratic forms.

We show now that checking whether a polynomial is sum of squares can be cast as an SDP problem:

Proposition B.3.12 *Assume that $p(x)$ is a multivariate polynomial of degree $2d$ in $x \in \mathbb{R}^n$. Then, the following statements are equivalent:*

1. *The polynomial $p(x)$ is a sum of squares.*
2. *The polynomial admits a representation of the form*

$$p(x) = z(x)^{\mathrm{T}} Q z(x) \tag{B.41}$$

where $Q \succeq 0$ and $z(x)$ is a vector containing monomials up to degree d.

Moreover, the number of squares is equal to rank$[Q]$ and the maximal dimension of $z(x)$ is given by $(n+d)!/n!d!$.

Proof **Proof of 1 ⇒ 2**: Assume that $p(x)$ is a sum of squares, then there exist polynomials $h_i(x)$ such that $p(x) = \sum_i h_i(x)^2$. Letting $h_i(x) = q_i^{\mathrm{T}} z(x)$ for some vector $z(x)$ of monomials of degrees up to d and some vector q_i, we can write that

$$\begin{aligned} p(x) &= \textstyle\sum_i z(x)^{\mathrm{T}} q_i q_i^{\mathrm{T}} z(x) \geq 0, & \\ &= \textstyle\sum_i z(x)^{\mathrm{T}} Q_i z(x) \geq 0 & \text{where } Q_i = q_i q_i^{\mathrm{T}} \\ &= z(x)^{\mathrm{T}} Q z(x) \geq 0 & \text{where } Q = \textstyle\sum_i Q_i. \end{aligned}$$

The number of squares obviously equals rank$[Q]$.

Proof of 2 ⇒ 1: Conversely, assume that $p(x)$ can be written as in (B.41). Since Q is positive-semidefinite, it therefore admits a Cholesky decomposition $Q = L^{\mathrm{T}} L$ where L is upper triangular. Defining $\mathrm{col}_i\{h_i(x)\} := Lz(x)$, we finally obtain that $p(x) = \sum_i h_i(x)^2$. Note, however, that the sum of squares decomposition may not be unique when the polynomial $p(x)$ is nonnegative since the Cholesky decomposition of Q may not be unique. ■

The following definition will be useful in the sequel:

Definition B.3.13 Let $z(x)$ be a vector of dimension d containing monomials in $x \in \mathbb{R}^n$. We then define the set

$$\mathcal{K}(z) := \left\{ Q \in \mathbb{S}^{d \times d} \ : \ z(x)^{\mathrm{T}} Q z(x) = 0, \ \text{for all } x \in \mathbb{R}^n \right\}. \tag{B.42}$$

The above set allows us to characterize the dependency between the different entries of $z(x)$. Note that the above set can also be extended to cope with matrices of the form $z(x) \otimes I$.

Example B.3.14 Let $z(x) = \text{col}(1, x, x^2)$, then we have

$$\mathcal{K}(z) := \text{span}\left\{\begin{bmatrix} 0 & 0 & -1 \\ 0 & 2 & 0 \\ -1 & 0 & 0 \end{bmatrix}\right\}. \tag{B.43}$$

Example B.3.15 Let $z(x) = \text{col}(1, x)$ and define $Z(x) = z(x) \otimes I_2$, then we have

$$\mathcal{K}(Z) := \text{span}\left\{\begin{bmatrix} 0 & S \\ S^{\mathrm{T}} & 0 \end{bmatrix}\right\} \tag{B.44}$$

where $S + S^{\mathrm{T}} = 0$.

Proposition B.3.16 *Let $p(x)$ be a polynomial of degree 2d in the variable $x \in \mathbb{R}^n$. Let also $z_f(x)$ be the vector containing all the monomials in x up to degree d. Then, there exists $M \in \mathbb{S}^{\frac{(n+d)!}{n!d!}}$ such that we have*

$$p(x) = z_f(x)^{\mathrm{T}} M z_f(x).$$

Furthermore, $p(x)$ also admits the following representation

$$p(x) = z_f(x)^{\mathrm{T}} (M + N) z_f(x)$$

which is valid for any $N \in \mathcal{K}(z_f)$.

The above matrix representation is referred to as the *Gram matrix method* [38] or the *square matricial representation* (SMR) [39, 40]. Using the above representation, we can state the following result:

Theorem B.3.17 ([39–41]) *Let $p(x)$ be a homogeneous polynomial of degree 2d in the variable $x \in \mathbb{R}^n$. Then, p is SOS if and only if*

$$M + N \succeq 0 \tag{B.45}$$

where M is such that $p(x) = z_f(x)^{\mathrm{T}} M z_f(x)$ and $N \in \mathcal{K}(z_f)$.

Example B.3.18 Let us consider the polynomial $p(x) = x^4 + 2x^2 + 2x + 1$ and let us prove that it is positive over $\mathbb{R}$. We have $z_f(x) = \mathrm{col}(1, x, x^2)$ together with the following representation

$$p(x) = z_f(x)^{\mathrm{T}} \begin{bmatrix} 1 & 1 & -1 \\ 1 & 4 & 0 \\ -1 & 0 & 1 \end{bmatrix} z_f(x). \tag{B.46}$$

But, the central matrix is not positive definite. For the chosen $z_f(x)$, we have

$$\mathcal{K}(z_f) = \mathrm{span} \begin{bmatrix} 0 & 0 & -1 \\ 0 & 2 & 0 \\ -1 & 0 & 0 \end{bmatrix}$$

and

$$\begin{bmatrix} 1 & 1 & -1 \\ 1 & 4 & 0 \\ -1 & 0 & 1 \end{bmatrix} - \begin{bmatrix} 0 & 0 & -1 \\ 0 & 2 & 0 \\ -1 & 0 & 0 \end{bmatrix} = \begin{bmatrix} 1 & 1 & 0 \\ 1 & 2 & 0 \\ 0 & 0 & 1 \end{bmatrix} \succ 0. \tag{B.47}$$

This illustrates the importance of considering the interdependency of the entries of $z_f(x)$.

Sums of Squares as an Optimization Tool

We illustrate in this part how to apply the concepts and results described above to solve Problem B.3.4. Let us assume first that Δ admits the representation

$$\Delta = \left\{ \delta \in \mathbb{R}^{N_p} : \ f_i(\delta) \geq 0, \ i = 1, \ldots, N_f \right\} \tag{B.48}$$

where the functions f_i's are polynomial functions. Note that this representation is not unique. In this case, we are not looking for an exact representation of $\mathcal{L}(x, \delta)$ expressed in terms of an element in the quadratic module, but for a lower-bound for $-\mathcal{L}(x, \delta)$ in the quadratic module. In the spirit of Putinar's Positivstellensatz, this can be formalized into the following LMI problem:

Problem B.3.19 Find $x \in \mathbb{R}^n$ (or $\min c^{\mathrm{T}} x$) such that the LMI

$$\mathcal{L}(x, \delta) + \sum_{i=1}^{N_p} f_i(\delta) M_i(\delta) + \varepsilon I \preceq 0 \tag{B.49}$$

holds for all $\delta \in \mathbb{R}^{N_p}$ where $\varepsilon > 0$ and $M_i(\delta)$ is a SOS matrix polynomial, i.e. $M_i(\delta)$ can be written as $M(\delta) = \sum_j Q_j^i(\delta)^{\mathrm{T}} Q_j^i(\delta)$.

The above inequality can be made independent of δ by factorizing it as

$$\mathcal{L}(x,\delta) + \sum_{i=1}^{N_p} f_i(\delta) M_i(\delta) + \varepsilon I = Z(\delta)^{\mathrm{T}} \mathcal{M}(x) Z(\delta) \tag{B.50}$$

where $\mathcal{M}(x)$ is symmetric and $Z(\delta) := z(\delta) \otimes I$ for some vector of monomials $z(\delta)$. Considering then the square matricial representation, we obtain the sufficient SDP condition

$$\mathcal{M}(x) + \mathcal{N}(y) \preceq 0 \tag{B.51}$$

where $\mathcal{N}(y)$ belongs to $\mathcal{K}(Z)$ and y emphasizes the fact that $\mathcal{N}(y)$ contains decision variables.

B.3.4 Other Methods

Many other methods exist for dealing with parameter dependent LMIs. When the LMI depends quadratically on parameters belonging to the unit-simplex, relaxation methods as in [42–44] can be used. Pólya's Theorem can be used as well in this setting (and for parameters belonging to the positive orthant), see e.g. [45, 46]. Slack variables approaches are considered in [47]. Scalings and the S-procedure are used in [48]. Relaxations based on the full-block S-procedure are considered in [49]. When the parameters belong to a compact polyhedron, Handelman's Theorem generalized to the matrix case can be used [50–53]. For polynomial techniques based on sums of squares, see [39–41, 51, 54, 55]. Finally, Lasserre's moments method can also be used to address polynomial problems; see e.g. [56–58] and references therein.

Appendix C
Technical Results in Robust Analysis, Control and LMIs

One cannot really argue with a mathematical theorem.
Stephen Hawking

C.1 Kalman-Yakubovich-Popov Lemma

The Kalman-Yakubovich-Popov Lemma is a very general and important result that relates conditions in the frequency domain to conditions in the time-domain. It is for instance central in IQC theory, H_∞-control, positive-real analysis, etc. Several versions of the Kalman-Yakubovich-Popov Lemma (KYP Lemma) can be stated. The original ones can be found in [59–61]. A "modern" version, taken from [62], is the one which is considered in this monograph. It is given below:

Lemma C.1.1 (Kalman-Yakubovich-Popov Lemma) *Let us consider matrices $A \in \mathbb{R}^{n\times n}$, $E \in \mathbb{R}^{n\times m}$ and*

$$M = \begin{bmatrix} M_{11} & M_{12} \\ M_{12}^{\mathrm{T}} & M_{22} \end{bmatrix} \in \mathbb{S}^{n+m}.$$

Then, the following statements are equivalent:

1. *There exists a matrix $P \in \mathbb{S}^n$ such that*

$$M + \begin{bmatrix} I & 0 \\ A & E \end{bmatrix}^{\mathrm{T}} \begin{bmatrix} 0 & P \\ P & 0 \end{bmatrix} \begin{bmatrix} I & 0 \\ A & E \end{bmatrix} \prec 0$$

holds.

C. Briat, *Linear Parameter-Varying and Time-Delay Systems*,
Advances in Delays and Dynamics 3, DOI 10.1007/978-3-662-44050-6

2. *$M_{22} \prec 0$ and for all $\omega \in \mathbb{R}$ and all complex vectors $col(x, w) \neq 0$*

$$\begin{bmatrix} A - j\omega I & E \end{bmatrix} \begin{bmatrix} x \\ w \end{bmatrix} = 0 \quad \textit{implies that} \quad \begin{bmatrix} x \\ w \end{bmatrix}^* M \begin{bmatrix} x \\ w \end{bmatrix} < 0.$$

Moreover, if the pair (A, E) is controllable, the corresponding equivalence also holds for non-strict inequalities. Finally, if M is given by

$$M = \begin{bmatrix} 0 & I \\ C & F \end{bmatrix}^{\mathrm{T}} \begin{bmatrix} Q & S \\ S^{\mathrm{T}} & R \end{bmatrix} \begin{bmatrix} 0 & I \\ C & F \end{bmatrix}$$

then statement 2 is equivalent to the condition that for all $\omega \in \mathbb{R}$ such that $\det(j\omega I - A) \neq 0$, we have that

$$\begin{bmatrix} I \\ H(j\omega) \end{bmatrix}^* \begin{bmatrix} Q & S \\ S^{\mathrm{T}} & R \end{bmatrix} \begin{bmatrix} I \\ H(j\omega) \end{bmatrix} \prec 0$$

where $H(s) := C(sI - A)^{-1}E + F$.

This lemma tells us that geometric conditions in the frequency-domain admit time-domain interpretations in terms of the state-space data. Interestingly, the time-domain condition is expressed as a tractable LMI condition in the matrix $P \in \mathbb{S}^n$ whereas the frequency-domain one is a semi-infinite condition in ω, which is much less tractable due to the rational structure of $H(s)$. Even more importantly, the scope of the time-domain condition is way broader since it remains valid for linear time-varying systems, such as LPV systems.

Extensions, re-derivations and specializations of the KYP Lemma have also been obtained, for instance, in [63–71].

C.2 Facts on Dissipativity Theory

For more details on dissipativity theory, readers should refer, for instance, to [62, 72, 73].

C.2.1 Dissipative Systems, Storage Functions and Supply Rates

Dissipativity theory is devoted to the analysis of dynamical systems. It has been initiated by J. Willems in [72]. The main concept behind this theory is the notion of energy, and how this energy changes when the dynamical system evolves over time.

The key remark consists of noticing that if the system stores less energy that it is supplied to it, then the difference of energy must have been dissipated by the system. To illustrate this, let us consider the general system Σ governed by the equations

$$\Sigma : \begin{cases} \dot{x} = f(x, w) \\ z = h(x, w) \end{cases} \tag{C.1}$$

where $x \in \mathcal{X} \subset \mathbb{R}^n$, $w \in \mathcal{W} \subset \mathbb{R}^p$ and $z \in \mathcal{Z} \subset \mathbb{R}^q$ are the state, the input and the output of the system, respectively.

Definition C.2.1 (*Supply function*) A function

$$\begin{aligned} s : \mathcal{W} \times \mathcal{Z} &\rightarrow \mathbb{R} \\ (w, z) &\rightarrow s(w, z) \end{aligned} \tag{C.2}$$

that is absolutely integrable over any time interval is referred to as a **supply function** or **supply rate**.

The supply rate $s(\cdot, \cdot)$ defined above should be interpreted as the supply delivered to the system. The value $s(w, z)$ represents the rate at which the supply circulates into the system whenever the pair (w, z) is generated. When the integral

$$\int_0^T s(w(t), z(t))dt$$

is positive, then work is done **on** the system.When, on the other hand, the integral is negative, then the work is done **by** the system.

Definition C.2.2 The system (C.1) with supply function $s(\cdot, \cdot)$ is said to be **dissipative** if there exists a function $V : \mathcal{X} \rightarrow \mathbb{R}$ such that

$$V(x(t_0)) + \int_{t_0}^{t_1} s(w(t), z(t))dt \geq V(x(t_1)) \tag{C.3}$$

holds for all $t_0 \leq t_1$ and all signals (w, x, z) that satisfy (C.1). The pair (Σ, s) is said to be **conservative** if the equality holds for all $t_0 \leq t_1$ and all signals (w, x, z) that satisfy (C.1). In any of these cases, the function V is called a **storage function.**[a]

[a]Note that the function V is not necessarily a positive definite function, unlike a Lyapunov function.

The inequality (C.3) says that, for any interval $[t_0, t_1]$, the change of internal storage $V(x(t_1)) - V(x(t_0))$ will never exceed the amount of supply that flows into the system. This means then that part of what is supplied is stored, while the remaining part is dissipated. When the function V is differentiable, the condition (C.3) is equivalent to the condition that

$$\dot{V}(x(t)) - s(w(t), z(t)) \leq 0 \tag{C.4}$$

holds for all $t \geq 0$.

C.2.2 Linear Dissipative Systems, Quadratic Storage Functions and Quadratic Supply Rates

Let us consider the linear system Σ_ℓ governed by the equations

$$\Sigma_\ell : \begin{cases} \dot{x} = Ax + Ew \\ z = Cx + Fw \end{cases} \tag{C.5}$$

where $x \in \mathcal{X} \subset \mathbb{R}^n$, $w \in \mathcal{W} \subset \mathbb{R}^p$ and $z \in \mathcal{Z} \subset \mathbb{R}^q$ are the state, the input and the output of the system, respectively. Let $\widehat{G}_\ell(s) := C(sI - A)^{-1}E + F$ be the corresponding transfer function.

For the particular case of linear systems, we assume that $x^* = 0$ is the point of neutral storage, i.e. $V(x^*) = 0$, and we consider quadratic supply rates of the form

$$s(w, z) = \begin{bmatrix} w \\ z \end{bmatrix}^{\mathrm{T}} \begin{bmatrix} Q & S \\ S^{\mathrm{T}} & R \end{bmatrix} \begin{bmatrix} w \\ z \end{bmatrix} \tag{C.6}$$

where $Q \in \mathbb{S}^p$, $R \in \mathbb{S}^q$ and $S \in \mathbb{R}^{p \times q}$.

The following result is central in the dissipativity analysis of linear systems:

Theorem C.2.3 (Dissipativity theorem) *Suppose that system Σ_ℓ defined by* (C.5) *is controllable and let the supply function be given by* (C.6). *Then, the following statements are equivalent:*

1. *(Σ_ℓ, s) is dissipative.*
2. *(Σ_ℓ, s) admits a quadratic storage function $V(x) = x^{\mathrm{T}}Px$ with $P \in \mathbb{S}^n$.*
3. *There exists $P \in \mathbb{S}^n$ such that*

$$\begin{bmatrix} A^{\mathrm{T}}P + PA & PE \\ \star & 0 \end{bmatrix} - \begin{bmatrix} 0 & I \\ C & F \end{bmatrix}^{\mathrm{T}} \begin{bmatrix} Q & S \\ S^{\mathrm{T}} & R \end{bmatrix} \begin{bmatrix} 0 & I \\ C & F \end{bmatrix} \preceq 0.$$

4. *For all $\omega \in \mathbb{R}$ with* $\det(j\omega I - A) \neq 0$, *we have that*

$$\begin{bmatrix} I \\ \widehat{G}_\ell(j\omega) \end{bmatrix}^* \begin{bmatrix} Q & S \\ S^{\mathrm{T}} & R \end{bmatrix} \begin{bmatrix} I \\ \widehat{G}_\ell(j\omega) \end{bmatrix} \succeq 0.$$

Proof The proof can be found in [62]. ■

It seems important to stress that the equivalence between the two last statements follows from the Kalman-Yakubovich-Popov Lemma.

C.3 Schur Complement

The term *Schur complement* has been introduced by Emilie Virginia Haynsworth in [74]. In the same article, she also proved the *inertia additivity formula*[5] which is, nowadays, called the *Haynsworth inertia additivity formula*. In some words, she proved that the inertia of some matrices is additive on the Schur complement, and is a direct consequence of the *Guttman rank additivity formula* [75]. See [4] for more details.

In the context of LMIs, a certain form of the inertia additivity formula can be written into the form stated in the following result [12]:

Lemma C.3.1 (Schur complement formula) *Let us consider the matrices $M_{11} \in \mathbb{S}^{n_1}$, $M_{22} \in \mathbb{S}^{n_2}$ and $M_{12} \in \mathbb{R}^{n_1 \times n_2}$. Then, the following statements are equivalent:*

1. *The matrix inequality*

$$\begin{bmatrix} M_{11} & M_{12} \\ M_{12}^{\mathrm{T}} & M_{22} \end{bmatrix} \prec 0 \tag{C.7}$$

holds.

2. *The matrix inequalities*

$$M_{11} \prec 0 \text{ and } M_{22} - M_{12}^{\mathrm{T}} M_{11}^{-1} M_{12} \prec 0 \tag{C.8}$$

hold.

3. *The matrix inequalities*

$$M_{22} \prec 0 \text{ and } M_{11} - M_{12} M_{22}^{-1} M_{12}^{\mathrm{T}} \prec 0 \tag{C.9}$$

hold.

[5] The inertia of a symmetric matrix is the triplet (n_-, n_0, n_+) where n_-, n_0 and n_+ are the numbers of negative, zero and positive eigenvalues, respectively.

This lemma is very useful since it may convert the nonlinear matrix inequalities of statements 2 and 3, into the linear matrix inequality of statement 1.

This result also admits the following non-strict version [12]:

Lemma C.3.2 *Let us consider the matrices* $M_{11} \in \mathbb{S}^{n_1}$, $M_{22} \in \mathbb{S}^{n_2}$ *and* $M_{12} \in \mathbb{R}^{n_1 \times n_2}$. *Then, the following statements are equivalent:*

1. *The matrix inequality*
$$\begin{bmatrix} M_{11} & M_{12} \\ M_{12}^{\mathrm{T}} & M_{22} \end{bmatrix} \preceq 0 \tag{C.10}$$
holds.
2. *The relations*
$$M_{22} \preceq 0, \; M_{11} - M_{12}M_{22}^{+}M_{12}^{\mathrm{T}} \preceq 0 \text{ and } M_{12}(I - M_{22}M_{22}^{+}) = 0$$
hold where M_{22}^{+} *is the Moore-Penrose pseudoinverse of* M_{22}.

C.4 Generalized Relaxation Lemma

In the Relaxation Lemma 5.6.23, the matrix M_{22} is assumed to be positive definite. The result below relaxes this assumption:

Lemma C.4.1 *Let the matrices* $M_{11} \in \mathbb{S}^{n}$, $M_{22} \in \mathbb{S}^{m}$ *and* $M_{12} \in \mathbb{R}^{n \times m}$ *be given. Then, the following statements are equivalent:*

1. *The matrix inequality*
$$M_{11} - M_{12}M_{22}M_{12}^{\mathrm{T}} \prec 0 \tag{C.11}$$
holds.
2. *There exist a matrix* $N \in \mathbb{R}^{m \times n}$ *and a scalar* $\tau \geq 0$ *such that the matrix inequality*
$$M_{11} + N^{\mathrm{T}}M_{12}^{\mathrm{T}} + M_{12}N + N^{\mathrm{T}}\left(\tau I + M_{22}\right)^{-1}N + \tau M_{12}M_{12}^{\mathrm{T}} \prec 0 \tag{C.12}$$
holds with $\tau I + M_{22} \succ 0$.
3. *There exist a matrix* $N \in \mathbb{R}^{m \times n}$ *and a scalar* $\tau \geq 0$ *such that the matrix inequality*

$$\begin{bmatrix} M_{11} + N^{\mathrm{T}} M_{12}^{\mathrm{T}} + M_{12} N & N^{\mathrm{T}} & \tau M_{12} \\ \star & -(\tau I + M_{22}) & 0 \\ \star & \star & -\tau I \end{bmatrix} \prec 0 \tag{C.13}$$

holds.

Proof **Proof of 1 ⇔ 2**: Define first

$$f^* := \min_{N \in \mathbb{R}^{m \times n}} \left\{ f(N) := \mathrm{He}\left[N^{\mathrm{T}} M_{12}^{\mathrm{T}} \right] + N^{\mathrm{T}} (\tau I + M_{22})^{-1} N + \tau M_{12} M_{12}^{\mathrm{T}} \right\} \tag{C.14}$$

where the minimum is considered as in Lemma 5.6.23. To see the equivalence, it is enough to show that we have $f^* = -M_{12} M_{22} M_{12}^{\mathrm{T}}$. Completing the squares yields

$$\begin{aligned} f(N) = {} & \left[N + (\tau I + M_{22}) M_{12}^{\mathrm{T}} \right]^{\mathrm{T}} (\tau I + M_{22})^{-1} \left[N + (\tau I + M_{22}) M_{12}^{\mathrm{T}} \right] \\ & - M_{12} M_{22} M_{12}^{\mathrm{T}}. \end{aligned} \tag{C.15}$$

Since $\tau I + M_{22} \succ 0$, the above quadratic term in N is convex, positive semidefinite and lower bounded by 0. The lower bound is attained for $N = -(\tau I + M_{22}) M_{12}^{\mathrm{T}}$ and thus we have

$$f^* = -M_{12} M_{22} M_{12}^{\mathrm{T}}. \tag{C.16}$$

Therefore the two first statements are equivalent.

Proof of 2 ⇔ 3: This follows from the Schur complement formula.. ■

An alternative proof can be derived using the projection lemma; see Appendix C.12.

C.5 Positive Real Lemma

The positive-real lemma allows one to check whether a system is strictly positive-real. The definition of a strictly positive-real transfer function is given below for completeness:

Definition C.5.1 A $p \times p$ transfer function $G(s)$ is said to have the strict positive-real property if

1. $G(s)$ is asymptotically stable.
2. $G(j\omega) + G(-j\omega)^{\mathrm{T}} \succ 0$ for all $\omega \in \mathbb{R}$.
3. $G(\infty) + G(\infty)^{\mathrm{T}} \succ 0$ or

$$\lim_{\omega \to \infty} \omega^{2(p-q)} \det\left(G(j\omega) + G(-j\omega)^{\mathrm{T}} \right) \succ 0$$

where $q = \mathrm{rank}\left[G(\infty) + G(\infty)^{\mathrm{T}} \right]$.

By virtue of the Kalman-Yakubovich-Popov Lemma, the following alternative time-domain condition is obtained:

Theorem C.5.2 *The system* (C.5) *is strictly positive real if and only if there exists a matrix* $P \in \mathbb{S}^n_{\succ 0}$ *such that the LMI*

$$\begin{bmatrix} A^{\mathrm{T}}P + PA & PE - C^{\mathrm{T}} \\ \star & -(F + F^{\mathrm{T}}) \end{bmatrix} \prec 0 \tag{C.17}$$

holds.
Moreover, $V(x) = x^{\mathrm{T}}Px$ *defines a quadratic storage function for the system* (A, E, C, F).

Proof The proof is an application of the Kalman-Yakubovich-Popov Lemma with the quadratic supply-rate $s(w, z) = w^{\mathrm{T}}z + z^{\mathrm{T}}w$. ■

C.6 H_2 Performance

The H_2-norm of a system measures the energy of the impulse responses of the system. The H_2-norm of the system (C.5) is therefore given by

$$\begin{aligned} ||\widehat{G}||^2_{H_2} &= \int_0^{+\infty} \mathrm{trace}[g(t)^{\mathrm{T}}g(t)] \\ &= \frac{1}{2\pi}\mathrm{trace}\left(\int_{-\infty}^{+\infty} \widehat{G}(j\omega)^*\widehat{G}(j\omega)d\omega\right) \end{aligned}$$

where $g(t)$ is the impulse response of (C.5). Note that if $F \neq 0$, then the H_2-norm cannot be finite.

Theorem C.6.1 *Assume that* $F = 0$. *Then, the system* (C.5) *is asymptotically stable and satisfies* $||\widehat{G}||_{H_2} < \nu$ *if and only if there exist* $P \in \mathbb{S}^n_{\succ 0}$ *and* $Z \in \mathbb{S}^q_{\succ 0}$ *such that the conditions*

$$\begin{bmatrix} A^{\mathrm{T}}P + PA & PE \\ E^{\mathrm{T}}P & -I_p \end{bmatrix} \prec 0, \quad \begin{bmatrix} P & C^{\mathrm{T}} \\ C & Z \end{bmatrix} \succ 0, \quad \mathrm{trace}(Z) < \nu^2$$

hold.

Proof From the definition of the H_2-norm, we have that

$$\begin{aligned} ||\widehat{G}||^2_{H_2} &= \text{trace}\left(\int_0^{+\infty} E^{\mathrm{T}} e^{A^{\mathrm{T}}t} C^{\mathrm{T}} C e^{At} E dt\right) \\ &= \text{trace}\left(\int_0^{+\infty} C e^{At} E E^{\mathrm{T}} e^{A^{\mathrm{T}}t} C^{\mathrm{T}} dt\right) \end{aligned} \tag{C.18}$$

where we have used the fact that for two matrices $M_1 \in \mathbb{R}^{n_1 \times n_2}$ and $M_2 \in \mathbb{R}^{n_2 \times n_1}$, we have $\text{trace}(M_1 M_2) = \text{trace}(M_2 M_1)$. Let the controllability Gramian be denoted by

$$W := \int_0^{+\infty} e^{At} E E^{\mathrm{T}} e^{A^{\mathrm{T}}t} dt.$$

It is well-known that it verifies the Lyapunov equation

$$AW + WA^{\mathrm{T}} + EE^{\mathrm{T}} = 0.$$

Substituting then W into (C.18), we get that

$$||\widehat{G}||^2_{H_2} = \text{trace}\left[CWC^{\mathrm{T}}\right] < \nu^2.$$

Since $\text{rank}[E] = p \leq n$, hence $EE^{\mathrm{T}} \succeq 0$ and $W \succeq 0$. Since A is Hurwitz, this means that there exists $X \succeq W$ such that

$$AX + XA^{\mathrm{T}} + EE^{\mathrm{T}} \prec 0, \; CXC^{\mathrm{T}} \prec Z \text{ and } \text{trace}[Z] < \nu^2.$$

Pre- and post-multiplying by $P := X^{-1}$, we get

$$PA + A^{\mathrm{T}}P + PEE^{\mathrm{T}}P \prec 0, \; CP^{-1}C^{\mathrm{T}} \prec Z \text{ and } \text{trace}[Z] < \nu^2.$$

A Schur complement yields the LMIs

$$\begin{bmatrix} PA + A^{\mathrm{T}}P & PE \\ \star & -I_p \end{bmatrix} \prec 0, \; \begin{bmatrix} Z & C \\ \star & P \end{bmatrix} \succ 0 \text{ and } \text{trace}[Z] < \nu^2$$

The proof is complete. ∎

C.7 Generalized H_2 Performance

The generalized H_2-norm is defined as the L_2-QL_∞-gain of a system, i.e. the gain from the set of input signals of bounded energy to the set of output signals of finite-amplitude.[6] In the scalar case, the L_2-QL_∞ induced-norm coincides with the H_2-norm. It can be notably proved that

$$||\Sigma_\ell||_{L_2-QL_\infty} = \frac{1}{2\pi}\lambda_{max}\left(\int_{-\infty}^{+\infty} \widehat{G}(j\omega)\widehat{G}(j\omega)^* d\omega\right) \tag{C.19}$$

We have the following result:

Theorem C.7.1 *Assume that $F = 0$. Then, the system (C.5) is asymptotically stable and satisfies $||\Sigma_\ell||_{L_2-QL_\infty} < \nu$ if and only if there exists $P \in \mathbb{S}^n_{\succ 0}$ such that the LMIs*

$$\begin{bmatrix} A^{\mathrm{T}}P + PA & PE \\ \star & -I_p \end{bmatrix} \prec 0 \text{ and } \begin{bmatrix} P & C^{\mathrm{T}} \\ \star & \nu^2 I_q \end{bmatrix} \succ 0 \tag{C.20}$$

hold.

Proof It is readily verified that the first LMI in (C.20) is equivalent to the dissipativity of the system (C.5) with quadratic storage function $V(x) = x^{\mathrm{T}}Px$ with $P \in \mathbb{S}^n_{\succ 0}$ and supply-rate $s(w, z) = w^{\mathrm{T}}w$. Equivalently, we have that (assuming zero initial conditions)

$$x(t)^{\mathrm{T}}Px(t) \le \int_0^t w(s)^{\mathrm{T}}w(s)ds$$

for all $t \ge 0$ and all $w \in L_2$. From the second LMI of (C.20), we have that $C^{\mathrm{T}}C \prec \nu^2 P$ and thus $z(t)^{\mathrm{T}}z(t) = x(t)^{\mathrm{T}}C^{\mathrm{T}}Cx(t) \le \nu^2 x(t)^{\mathrm{T}}Px(t)$. Hence, we have

$$\begin{aligned} z(t)^{\mathrm{T}}z(t) &\le \nu^2 x(t)^{\mathrm{T}}Px(t) \\ &\le \nu^2 \int_0^t w(s)^{\mathrm{T}}w(s)ds \\ &\le \nu^2 ||w||^2_{L_2}. \end{aligned} \tag{C.21}$$

Taking finally the supremum on the left-hand side of (C.21), we obtain

[6] The QL_∞-norm is defined here as $||w||_{QL_\infty} = \sup_{t\ge 0}\sqrt{w(t)^{\mathrm{T}}w(t)}$.

$$||z||^2_{QL_\infty} := \sup_{t \geq 0} z(t)^{\mathrm{T}} z(t) \leq \nu^2 ||w||^2_{L_2}.$$

The proof is complete. ∎

C.8 *S*-Procedure

The S-procedure allows one to deal easily with conditional feasibility problems in the LMI framework. Let us start with the following set

$$\mathcal{M} := \left\{ \xi \in \mathbb{R}^n : \xi^{\mathrm{T}} M_i \xi \leq 0, \ i = 1, \ldots, N \right\} \tag{C.22}$$

where the given matrices $M_1, \ldots, M_N$ are symmetric. Given a symmetric matrix M_0, we are then interested in checking whether

$$\xi^{\mathrm{T}} M_0 \xi < 0 \text{ for all } \xi \in \mathcal{M}, \ \xi \neq 0. \tag{C.23}$$

A sufficient condition for this problem is given below:

Lemma C.8.1 (S- procedure) *If there exist scalars $\tau_1, \ldots, \tau_N \geq 0$ such that*

$$M_0 - \sum_{i=1}^{N} \tau_i M_i \prec 0 \tag{C.24}$$

then $\xi^{\mathrm{T}} M_0 \xi < 0$ for all $\xi \in \mathcal{M}$, $\xi \neq 0$. The converse is not true in general unless $N = 1$ for real valued problems, or $N = 2$ for complex valued problems.

Despite of its conservatism, it is a very useful tool in robust analysis and control. It indeed plays a crucial role in the derivation of the full-block S-procedure [76], IQC techniques [77], Lur'e systems [78]...

Historically, the first result of this kind was obtained by Finsler in [79] and was later generalized by Hestenes and McShane in [80]. In the field of automatic control, the idea was certainly first used by Lur'e and Postnikov in [78]. In [81, 82], Yakubovich proved a theorem that is now referred to as the S-lemma. Megretski and Treil extended the results in [83] to infinite dimensional spaces. For a complete survey of the S-lemma, see [84].

C.9 Dualization Lemma

The dualization lemma has been introduced independently in [85, 86]. It has been proven to be very useful for deriving synthesis conditions for LPV controllers in the LFT setting; see Sect. 3.5. We need first the following preliminary result:

Proposition C.9.1 *Let $P \in \mathbb{S}^n$ be a nonsingular matrix and define two matrices $S_r \in \mathbb{R}^{n\times r}$ and $S_\ell \in \mathbb{R}^{n\times \ell}$ with $r+\ell=n$ such that $S := \begin{bmatrix} S_\ell & S_r \end{bmatrix}$ is a basis of $\mathbb{R}^n$. Assume that $S_r^{\mathrm{T}} P S_r \prec 0$ and $S_\ell^{\mathrm{T}} P S_\ell \succ 0$, then P has exactly r negative and ℓ positive eigenvalues.*

Conversely, if P has exactly r negative and ℓ positive eigenvalues, then there exist matrices $R_r \in \mathbb{R}^{n\times r}$ and $R_\ell \in \mathbb{R}^{n\times \ell}$ with $r+\ell=n$ such that $R := \begin{bmatrix} R_\ell & R_r \end{bmatrix}$ is a basis of $\mathbb{R}^n$ and such that the inequalities $R_r^{\mathrm{T}} P R_r \prec 0$ and $R_\ell^{\mathrm{T}} P R_\ell \succ 0$ hold.

Proof A proof can be found in [62]. ■

We can now state the dualization lemma:

Lemma C.9.2 (Dualization Lemma) *Let $M \in \mathbb{S}^n$ be such that it has q negative eigenvalues and $n-q$ positive eigenvalues. Let $S \in \mathbb{R}^{n\times q}$ be a full rank matrix, i.e. of rank p. Then, the following statements are equivalent:*

1. *The LMI $S^{\mathrm{T}} M S \prec 0$ holds.*
2. *The LMI $S^{\perp T} M^{-1} S^{\perp} \prec 0$ holds where $S^{\perp}$ is a basis of the orthogonal complement of span(S), i.e. $S^{\mathrm{T}} S^{\perp} = 0$.*

Proof The proof can be found in [62, 86]. ■

C.10 Petersen's Lemma

Petersen's Lemma [87–90] is a result dealing with uncertain matrices in LMIs. Both real-valued and complex-valued uncertainties can be considered with this result. The real version is provided below:

Lemma C.10.1 (Petersen's Lemma) *Let the matrices* $\Psi \in \mathbb{S}^n$, $P \in \mathbb{R}^{\delta\times n}$, $Q \in \mathbb{R}^{\delta\times n}$ *be given and let* $\Delta \in \boldsymbol{\Delta}_f$ *be a (possibly time-varying) uncertain matrix where*

$$\boldsymbol{\Delta}_f := \left\{\Delta \in \mathbb{R}^{\delta\times\delta} : \Delta^{\mathrm{T}}\Delta \leq R,\ R \in \mathbb{S}^{\delta}_{\succ 0}\right\}.$$

Then, the following statements are equivalent:

1. *The matrix inequality*

$$\Psi + P^{\mathrm{T}}\Delta Q + Q^{\mathrm{T}}\Delta^{\mathrm{T}}P \prec 0 \tag{C.25}$$

   holds for all $\Delta \in \boldsymbol{\Delta}_f$.
2. *There exists a scalar* $\varepsilon > 0$ *such that the matrix inequality*

$$\Psi + \varepsilon^{-1}P^{\mathrm{T}}P + \varepsilon Q^{\mathrm{T}}RQ \prec 0$$

   holds.

Proof The original proof of this result in [87] is quite involved and relies on several intermediary results on quadratic forms. We provide here an alternative one based on the S-procedure which is much simpler than the original one.

The LMI (C.25) is equivalent to saying that $x^{\mathrm{T}}\Psi x + 2x^{\mathrm{T}}P^{\mathrm{T}}\Delta Qx < 0$ for all $x \neq 0$. Let $y := \Delta Qx$, therefore (C.25) rewrites

$$\begin{bmatrix} x \\ y \end{bmatrix}^{\mathrm{T}} \begin{bmatrix} \Psi & P^{\mathrm{T}} \\ P & 0 \end{bmatrix} \begin{bmatrix} x \\ y \end{bmatrix} < 0 \text{ for all } (x, y) \neq 0 \tag{C.26}$$

with the additional constraint that $y = \Delta Qx$. Note that we have

$$\begin{aligned} y^{\mathrm{T}}y &= x^{\mathrm{T}}Q^{\mathrm{T}}\Delta^{\mathrm{T}}\Delta Qx \\ &\leq x^{\mathrm{T}}Q^{\mathrm{T}}RQx \end{aligned} \tag{C.27}$$

where the last inequality completely characterizes the set $\boldsymbol{\Delta}_f$. So, we want to check that the inequality (C.26) holds for all pairs (x, y) verifying $y^{\mathrm{T}}y \leq x^{\mathrm{T}}Q^{\mathrm{T}}RQx$. Invoking then the S-procedure, we obtain the condition

$$\begin{bmatrix} \Psi & P^{\mathrm{T}} \\ P & 0 \end{bmatrix} + \varepsilon \begin{bmatrix} Q^{\mathrm{T}}RQ & 0 \\ 0 & -I \end{bmatrix} \prec 0 \tag{C.28}$$

where $\varepsilon > 0$ is the scalar term introduced by the S-procedure. Since, the S-procedure is lossless in the single constraint case, the equivalence with (C.25) follows. A Schur complement yields the final result. ■

It is interesting to note that the above result, when specialized to the quadratic stability analysis of uncertain systems of the form $\dot{x} = (A + E\Delta C)x$, is identical to the stability condition obtained using the scaled small-gain condition with scaling εI for full-block uncertainty.

Rational Version of Petersen's Lemma

In [91], the above result is generalized to the case of LMIs depending rationally on Δ. This result is stated below:

Lemma C.10.2 (Rational Petersen's Lemma) *Let the matrices $\Psi \in \mathbb{S}^n$, $P \in \mathbb{R}^{\delta\times n}$, $Q \in \mathbb{R}^{\delta\times n}$ and $S \in \mathbb{R}^{\delta\times\delta}$ be given and let $\Delta \in \boldsymbol{\Delta}_f$ be a (possibly time-varying) uncertain matrix where*

$$\boldsymbol{\Delta}_f := \left\{\Delta \in \mathbb{R}^{\delta\times\delta} : \Delta^{\mathrm{T}}\Delta \le R,\ R \in \mathbb{S}^{\delta}_{\succ 0}\right\}.$$

Assume further that $I - S^{\mathrm{T}}RS \succ 0$. Then, the following statements are equivalent:

1. *The matrix inequality*[a]
$$\Psi + P^{\mathrm{T}}\Delta(I - S\Delta)^{-1}Q + Q^{\mathrm{T}}(I - S\Delta)^{-\mathrm{T}}\Delta^{\mathrm{T}}P \prec 0 \qquad \text{(C.29)}$$
holds for all $\Delta \in \boldsymbol{\Delta}_f$.
2. *There exists a scalar $\varepsilon > 0$ such that the matrix inequality*
$$\Psi + \begin{bmatrix} \varepsilon^{1/2}Q \\ \varepsilon^{-1/2}P \end{bmatrix}^{\mathrm{T}} \begin{bmatrix} R^{-1} & -S \\ -S^{\mathrm{T}} & I \end{bmatrix}^{-1} \begin{bmatrix} \varepsilon^{1/2}Q \\ \varepsilon^{-1/2}P \end{bmatrix} \prec 0 \qquad \text{(C.30)}$$
holds.
3. *There exists a scalar $\varepsilon > 0$ such that the matrix inequality*
$$\begin{bmatrix} \Psi & \varepsilon^{1/2}Q^{\mathrm{T}} & \varepsilon^{-1/2}P^{\mathrm{T}} \\ \star & -R^{-1} & S \\ \star & \star & -I \end{bmatrix} \prec 0 \qquad \text{(C.31)}$$
holds.

[a] *Note that the LMI is well-posed, i.e. $(I - S\Delta)$ invertible for all $\Delta \in \boldsymbol{\Delta}_f$ since $I - S^{\mathrm{T}}RS \succ 0$.*

Proof The proof in [91] is quite involved. As for Lemma C.10.1, we provide here a simpler one based on the S-procedure.

Proof of 1 ⇔ 2: Pre- and post-multiplying (C.29) by x^{T} and x, and letting $y = \Delta(I - S\Delta)^{-1}x$ yield

$$x^{\mathrm{T}}\Psi x + 2x^{\mathrm{T}}P^{\mathrm{T}}y < 0 \tag{C.32}$$

for all $(x, y) \neq 0$ verifying $y = \Delta(I - S\Delta)^{-1}Qx$, $\Delta \in \boldsymbol{\Delta}_f$. Invoking the equality (A.3) and reorganizing the terms yield that $y = \Delta(Qx + Sy)$. Thus, we have that

$$\begin{aligned} y^{\mathrm{T}}y &= (Qx + Sy)^{\mathrm{T}}\Delta^{\mathrm{T}}\Delta(Qx + Sy) \\ &\leq (Qx + Sy)^{\mathrm{T}}R(Qx + Sy) \end{aligned} \tag{C.33}$$

and therefore

$$\begin{bmatrix} x \\ y \end{bmatrix}^{\mathrm{T}} \begin{bmatrix} Q^{\mathrm{T}}RQ & Q^{\mathrm{T}}RS \\ \star & S^{\mathrm{T}}RS - I \end{bmatrix} \begin{bmatrix} x \\ y \end{bmatrix} \geq 0. \tag{C.34}$$

Hence, we end up with the problem of checking whether inequality (C.32) holds for all $(x, y) \neq 0$ such that (C.34) holds. Applying then the S-procedure, we obtain the matrix inequality

$$\begin{bmatrix} \Psi & P^{\mathrm{T}} \\ \star & 0 \end{bmatrix} + \varepsilon \begin{bmatrix} Q^{\mathrm{T}}RQ & Q^{\mathrm{T}}RS \\ \star & S^{\mathrm{T}}RS - I \end{bmatrix} \prec 0 \tag{C.35}$$

where $\varepsilon > 0$. Since $S^{\mathrm{T}}RS - I \prec 0$, this is equivalent to say that

$$\Psi + \varepsilon Q^{\mathrm{T}}RQ + \left(\varepsilon^{-1/2}P + \varepsilon^{1/2}S^{\mathrm{T}}RQ\right)^{\mathrm{T}}\left(I - S^{\mathrm{T}}RS\right)^{-1}\left(\varepsilon^{-1/2}P + \varepsilon^{1/2}S^{\mathrm{T}}RQ\right) \prec 0 \tag{C.36}$$

where we have used the Schur complement formula. This expression can be reformulated as

$$\Psi + \varepsilon Q^{\mathrm{T}}RQ + \begin{bmatrix} \varepsilon^{1/2}Q \\ \varepsilon^{-1/2}P \end{bmatrix}^{\mathrm{T}} \begin{bmatrix} RS \\ I \end{bmatrix} (I - S^{\mathrm{T}}RS)^{-1} \begin{bmatrix} RS \\ I \end{bmatrix}^{\mathrm{T}} \begin{bmatrix} \varepsilon^{1/2}Q \\ \varepsilon^{-1/2}P \end{bmatrix} \prec 0. \tag{C.37}$$

Using the Banachiewicz inversion formulas of Appendix A.2, we can prove that

$$\begin{bmatrix} RS \\ I \end{bmatrix} (I - S^{\mathrm{T}}RS)^{-1} \begin{bmatrix} RS \\ I \end{bmatrix}^{\mathrm{T}} = \begin{bmatrix} R^{-1} & -S \\ -S^{\mathrm{T}} & I \end{bmatrix}^{-1} - \begin{bmatrix} R & 0 \\ 0 & 0 \end{bmatrix}. \tag{C.38}$$

Substituting the right-hand side of the above expression (C.37) yields (C.34). From the losslessness of the S-procedure in the single constraint case, the equivalence between the first two statements follows.

Proof of 2 ⇔ 3: Since $I - S^{\mathrm{T}}RS \succ 0$, then the matrix

$$\begin{bmatrix} R^{-1} & -S \\ -S^{\mathrm{T}} & I \end{bmatrix} \tag{C.39}$$

is positive definite. A Schur complement on (C.30) yields C.31. The proof is complete. ■

Consideration of D-scalings—Scaled Petersen's Lemma
We aim at showing now that Petersen's Lemma can be strengthened to incorporate constant D-scalings. To this aim, let us consider uncertain matrices Δ belonging to the set

$$\boldsymbol{\Delta}_d := \left\{\Delta \in \mathbb{R}^{\delta\times\delta} : \ \Delta \text{ block diagonal}, \ ||\Delta||_2 \le 1\right\} \tag{C.40}$$

to which we associate the set of D-scalings $\mathcal{D}(\Delta)$ defined as

$$\mathcal{D}(\Delta) := \left\{L \in \mathbb{S}^{\delta}_{\succ 0} : \ L^{1/2}\Delta = \Delta L^{1/2}, \ \Delta \in \boldsymbol{\Delta}_d\right\} \tag{C.41}$$

where $L^{1/2}$ is the unique positive square root of L. We then have the following result:

Lemma C.10.3 (Scaled Petersen's Lemma) *Assume that the matrices $\Psi \in \mathbb{S}^n$, $P \in \mathbb{R}^{\delta\times n}$, $Q \in \mathbb{R}^{\delta\times n}$ are given and let $\Delta \in \boldsymbol{\Delta}_d$. Assume moreover that there exists a matrix $\tilde{L} \in \mathcal{D}(\Delta)$ such that the matrix inequality*

$$\Psi + P^{\mathrm{T}}\tilde{L}^{-1}P + Q^{\mathrm{T}}\tilde{L}Q \prec 0 \tag{C.42}$$

holds. Then, the matrix inequality

$$\Psi + P^{\mathrm{T}}\Delta Q + Q^{\mathrm{T}}\Delta^{\mathrm{T}}P \prec 0 \tag{C.43}$$

holds for all $\Delta \in \boldsymbol{\Delta}_d$.

Proof Let us consider first the matrix inequality (C.43). Since $L^{1/2}\Delta = \Delta L^{1/2}$ for all $\Delta \in \boldsymbol{\Delta}_d$, we then have that $\Delta = L^{-1/2}\Delta L^{1/2}$. Substituting this expression for Δ into (C.43), we get that

$$\Psi + P^{\mathrm{T}}L^{-1/2}\Delta L^{1/2}Q + Q^{\mathrm{T}}L^{1/2}\Delta^{\mathrm{T}}L^{-1/2}P \prec 0 \tag{C.44}$$

for all $\Delta \in \boldsymbol{\Delta}_d$. Now invoking Petersen's Lemma, we get the inequality (C.42) where we have set $\tilde{L} := \varepsilon L$, where $\varepsilon > 0$ is the parameter introduced by Petersen's lemma. ■

It is important to stress that, unlike for the original robustness result, the above one is only sufficient. Necessity is indeed lost since D-scalings are usually conservative unless Δ meets specific structural conditions; see Proposition 2.6.14.

Consideration of Full-Block Scalings

The consideration of full-block scalings is a mixing of Petersen's result and the full-block S-procedure. The employed linearization procedure is notably taken from [85]. Let us consider the following set of uncertainty

$$\boldsymbol{\Delta}_q := \left\{ \Delta \in \mathbb{R}^{m\times p} : \begin{bmatrix} \Delta \\ I \end{bmatrix}^{\mathrm{T}} \begin{bmatrix} U & V \\ \star & W \end{bmatrix} \begin{bmatrix} \Delta \\ I \end{bmatrix} \preceq 0 \right\} \tag{C.45}$$

where $U \in \mathbb{S}^m_{\succ 0}$, $W \in \mathbb{S}^p_{\prec 0}$ and $V \in \mathbb{R}^{m\times p}$ are given matrices. We then have the following result:

Lemma C.10.4 (Full-Block Petersen's Lemma) *Let the matrices $\Psi \in \mathbb{S}^n$, $P \in \mathbb{R}^{m\times n}$, $Q \in \mathbb{R}^{p\times n}$ be given and let $\Delta \in \boldsymbol{\Delta}_q$. The following statements are equivalent:*

1. *The matrix inequality*

$$\Psi + P^{\mathrm{T}}\Delta Q + Q^{\mathrm{T}}\Delta^{\mathrm{T}}P \prec 0 \tag{C.46}$$

holds for all $\Delta \in \boldsymbol{\Delta}_q$.

2. *The matrix inequality*

$$\Psi + \begin{bmatrix} P \\ Q \end{bmatrix}^{\mathrm{T}} \begin{bmatrix} U^{-1} & -U^{-1}V \\ \star & -W + V^{\mathrm{T}}U^{-1}V \end{bmatrix} \begin{bmatrix} P \\ Q \end{bmatrix} \prec 0$$

holds.

Proof First note that the quadratic form in (C.45) rewrites

$$\begin{bmatrix} \Delta \\ I \end{bmatrix}^{\mathrm{T}} \begin{bmatrix} U & V \\ \star & W \end{bmatrix} \begin{bmatrix} \Delta \\ I \end{bmatrix} = (\Delta + U^{-1}V)^{\mathrm{T}}U(\Delta + U^{-1}V) + W - V^{\mathrm{T}}U^{-1}V.$$

Since $U \in \mathbb{S}^m_{\succ 0}$ and $W \in \mathbb{S}^p_{\prec 0}$, then we have that $W - V^{\mathrm{T}}U^{-1}V \prec 0$. A Schur complement yields

$$\begin{bmatrix} -U^{-1} & \Delta + U^{-1}V \\ \star & W - V^{\mathrm{T}}U^{-1}V \end{bmatrix} \preceq 0. \tag{C.47}$$

This inequality alternatively characterizes the set $\boldsymbol{\Delta}_q$ in an affine way, i.e. we have that

$$\boldsymbol{\Delta}_q := \left\{ \Delta \in \mathbb{R}^{m\times p} : \begin{bmatrix} 0 & \Delta \\ \star & 0 \end{bmatrix} \preceq \begin{bmatrix} U^{-1} & -U^{-1}V \\ \star & -W + V^{\mathrm{T}}U^{-1}V \end{bmatrix} \right\}. \tag{C.48}$$

Considering now the inequality (C.46) and rewriting it as

$$\Psi + \begin{bmatrix} P^{\mathrm{T}} & Q^{\mathrm{T}} \end{bmatrix} \begin{bmatrix} 0 & \Delta \\ \Delta^{\mathrm{T}} & 0 \end{bmatrix} \begin{bmatrix} P \\ Q \end{bmatrix} \prec 0, \tag{C.49}$$

we can observe that the Δ-terms are located in the off-diagonal entries as in (C.48). Substituting then the Δ-dependent matrix in (C.49) by the upper-bound defined in (C.48) yields the condition

$$\Psi + \begin{bmatrix} P^{\mathrm{T}} & Q^{\mathrm{T}} \end{bmatrix} \begin{bmatrix} U^{-1} & -U^{-1}V \\ \star & -W + V^{\mathrm{T}}U^{-1}V \end{bmatrix} \begin{bmatrix} P \\ Q \end{bmatrix} \prec 0.$$

The proof is complete. ■

Whenever the matrices U, V and W contain decision variables, the condition of statement 2 is clearly nonlinear whereas the condition in statement 1 actually is. The changes of variables $\tilde{U} = U^{-1} \succ 0$, $\tilde{V} = -U^{-1}V$ and $\tilde{W} = -W + V^{\mathrm{T}}U^{-1}V \succ 0$ linearizes the expression and we get the matrix inequalities

$$\begin{bmatrix} -\tilde{U} & \Delta - \tilde{V} \\ \star & -\tilde{W} \end{bmatrix} \preceq 0 \tag{C.50}$$

and

$$\Psi + \begin{bmatrix} P^{\mathrm{T}} & Q^{\mathrm{T}} \end{bmatrix} \begin{bmatrix} \tilde{U} & \tilde{V} \\ \star & \tilde{W} \end{bmatrix} \begin{bmatrix} P \\ Q \end{bmatrix} \prec 0. \tag{C.51}$$

If the structures of the matrices $\tilde{U}$, $\tilde{V}$ and $\tilde{W}$ are chosen such that (C.50) structurally holds, then it can be removed from the conditions and only remains the inequality (C.51).

C.11 Finsler's Lemma

Finsler's lemma [8, 79, 84, 92] is a very useful tool in robust control to deal with LMIs conditions coupled with equality constraints. This result is highly connected to the S-procedure discussed in Appendix C.8. Initially provided in [79], the lemma was stated as follows

Lemma C.11.1 (Original Finsler's Lemma) *Let S_1 and S_2 by symmetric matrices of the same dimension. Assume then that for all $x \neq 0$ such that $x^{\mathrm{T}}S_2x = 0$, we have that $x^{\mathrm{T}}S_1x > 0$. Then, there exists $y \in \mathbb{R}$ such that $S_1 + yS_2$ is positive definite.*

In control theory, the following more general version of this result is very often considered:

Lemma C.11.2 (Finsler's Lemma) *Let us consider a symmetric matrix $M \in \mathbb{S}^n$ and a full-rank matrix $B \in \mathbb{R}^{m \times n}$, $m < n$. Then, the following statements are equivalent:*

1. *The inequality $x^{\mathrm{T}} M x < 0$ holds for all $x \in \mathcal{X}$ where*

$$\mathcal{X} := \{x \in \mathbb{R}^n : \; Bx = 0, \; x \neq 0\}.$$

2. *There exists a scalar $\tau \in \mathbb{R}$ such that the inequality*

$$M - \tau B^{\mathrm{T}} B \prec 0$$

holds. Moreover, when such a τ exists, it must satisfy the inequality

$$\tau > \tau_{min} := \lambda_{max} \left[D^{\mathrm{T}} \left(M - M B_{\perp} \left(B_{\perp}^{\mathrm{T}} M B_{\perp} \right)^{-1} B_{\perp} M \right) D \right]$$

where $D := (B_r B_l^{\mathrm{T}})^{-1/2} B_l^{+}$, (B_r, B_l) is any full rank factor of B (i.e. $B = B_l B_r$) and $B_{\perp}$ is any basis of the right null space of B.

3. *There exists a symmetric matrix $X \in \mathbb{S}^m$ such that the inequality*

$$M - B^{\mathrm{T}} X B \prec 0$$

holds.

4. *There exists a matrix $N \in \mathbb{R}^{m \times n}$ such that the inequality*

$$M + N^{\mathrm{T}} B + B^{\mathrm{T}} N \prec 0$$

holds.

5. *The inequality*

$$B_{\perp}^{\mathrm{T}} M B_{\perp} \prec 0$$

holds where $B_{\perp}$ is any basis of the right-null-space of B.

6. *There exist a matrix $W \in \mathbb{S}_{\succeq 0}^{n+m}$ and a scalar $\tau > 0$ such that the conditions*

$$\begin{bmatrix} M & B^{\mathrm{T}} \\ B & \tau I_m \end{bmatrix} \prec W \text{ and } \operatorname{rank}(W) = m$$

are satisfied.

In statements 1 and 2, we can recognize the original Finsler's lemma where $S_1 = M$ and $S_2 = B^{\mathrm{T}} B$. Statement 3 is the 'matrix version' of Finsler's lemma as found in [8]. Statements 1 and 5 can be shown to be equivalent using elementary algebra. Statement 5 can be retrieved from statement 4 using the projection lemma (see Appendix C.12) or conversely, statement 4 can be obtained from statement 5 through the use of the creation lemma (inverse procedure of the elimination/projection lemma). Finally, statement 6 which is not part of the initial definition of Finsler's Lemma has been obtained in [93].

The Finsler's lemma can be robustified in order to account for uncertainties in the matrix B; this generalization has been provided in [94].

Lemma C.11.3 (Robust Finsler's Lemma [94]) *Let us consider a symmetric matrix $M \in \mathbb{S}^n$ and a matrix $B \in \mathbb{R}^{m\times n}$, $m < n$, and a compact subset of real matrices $\mathcal{K} \subset \mathbb{R}^{p\times m}$, $p \leq m$. The following statements are equivalent:*

1. *The inequality*
$$x^{\mathrm{T}} M x < 0 \tag{C.52}$$
holds for all $x \in \mathcal{X}_{\mathcal{K}}$ where
$$\mathcal{X}_{\mathcal{K}} := \left\{x \in \mathbb{R}^n : \ KBx = 0,\ x \neq 0,\ K \in \mathcal{K}\right\}. \tag{C.53}$$
2. *There exists a matrix $Z \in \mathbb{S}^m$ such that the matrix inequalities*
$$\begin{aligned} M + B^{\mathrm{T}} Z B &\prec 0, \\ K_{\perp}^{\mathrm{T}} Z K_{\perp} &\succeq 0 \end{aligned} \tag{C.54}$$
hold for all $K \in \mathcal{K}$ where $K_{\perp}$ is a basis of the null-space of K.

Proof The proof can be found in [94]. ■

C.12 Projection/Elimination Lemma

The projection lemma is useful for eliminating decision variables from LMIs, allowing us then to reduce the computational complexity of the problem. It also has a convexifying effect on certain nonlinear matrix inequalities, see e.g. [85, 95, 96] and Sect. 3.5.

Lemma C.12.1 (Projection Lemma) *Let $\Psi \in \mathbb{S}^n$ be a symmetric matrix and $P \in \mathbb{R}^{p\times n}$, $Q \in \mathbb{R}^{m\times n}$ be given matrices. Then, the following statements are equivalent:*

1. *There exists a matrix $\Omega \in \mathbb{R}^{p\times m}$ such that*

$$\Psi + P^{\mathrm{T}}\Omega Q + Q^{\mathrm{T}}\Omega^{\mathrm{T}}P \prec 0. \tag{C.55}$$

2. *The LMIs*

$$\begin{aligned} P_\perp^{\mathrm{T}}\Psi P_\perp &\prec 0\\ Q_\perp^{\mathrm{T}}\Psi Q_\perp &\prec 0 \end{aligned} \tag{C.56}$$

hold where $P_\perp$ and $Q_\perp$ are bases of the null-space of P and Q, respectively.

3. *There exist scalars $\tau_1, \tau_2 \in \mathbb{R}$ such that the LMIs*

$$\begin{aligned} \Psi - \tau_1 P^{\mathrm{T}}P &\prec 0\\ \Psi - \tau_2 Q^{\mathrm{T}}Q &\prec 0 \end{aligned}$$

hold.

Proof The proof is based on the one provided in [97]. We just have to show that statement 1 is equivalent to statement 2 The equivalence between statements 2 and 3 is a direct consequence of Finsler's Lemma; see Appendix C.11.

Proof of 1 ⇒ 2: This implication is straightforward. It is enough to pre and post-multiply (C.55) by $P_\perp^{\mathrm{T}}$ and $P_\perp$, respectively, to get the first inequality of (C.56). The second one is obtained by considered $Q_\perp$ instead.

Proof of 2 ⇒ 1: This part of the proof is more involved but is based on elementary linear algebra and matrix analysis. It is taken from [97]. Let us first consider the matrices $\mathcal{K}_{PQ}$, $\mathcal{K}_P$, $\mathcal{K}_Q$ and $\mathcal{K}_r$ where $\mathcal{K}_{PQ}$ is a basis of the null-space of $\begin{bmatrix} P & Q\end{bmatrix}$, $\begin{bmatrix}\mathcal{K}_{PQ} & \mathcal{K}_P\end{bmatrix}$ is a basis of the null-space of P, $\begin{bmatrix}\mathcal{K}_{PQ} & \mathcal{K}_Q\end{bmatrix}$ is a basis of the null-space of Q and such that the matrix

$$\mathcal{K} := \begin{bmatrix}\mathcal{K}_{PQ} & \mathcal{K}_P & \mathcal{K}_Q & \mathcal{K}_r\end{bmatrix}$$

is invertible. In this case, the inertia of the matrix

$$\Phi := \Psi + P^{\mathrm{T}}\Omega Q + Q^{\mathrm{T}}\Omega^{\mathrm{T}}P$$

is identical to the inertia of the matrix $\mathcal{K}^{\mathrm{T}}\Phi\mathcal{K}$. Define $P\mathcal{K} =: \begin{bmatrix}0 & 0 & P_1 & P_2\end{bmatrix}$ and $Q\mathcal{K} =: \begin{bmatrix}0 & Q_1 & 0 & Q_2\end{bmatrix}$, we then have

$$\Phi = \begin{bmatrix} \Psi_{11} & \Psi_{12} & \Psi_{13} & \Psi_{14} \\ \star & \Psi_{22} & \Psi_{23} + \Upsilon_{11}^{\mathrm{T}} & \Psi_{24} + \Upsilon_{21}^{\mathrm{T}} \\ \star & \star & \Psi_{33} & \Psi_{34} + \Upsilon_{12} \\ \star & \star & \star & \Psi_{44} + \Upsilon_{22} + \Upsilon_{22}^{\mathrm{T}} \end{bmatrix} \tag{C.57}$$

where

$$\mathcal{K}^{\mathrm{T}} \Psi \mathcal{K} =: \begin{bmatrix} \Psi_{11} & \Psi_{12} & \Psi_{13} & \Psi_{14} \\ \star & \Psi_{22} & \Psi_{23} & \Psi_{24} \\ \star & \star & \Psi_{33} & \Psi_{34} \\ \star & \star & \star & \Psi_{44} \end{bmatrix}$$

and

$$\Upsilon = \begin{bmatrix} \Upsilon_{11} & \Upsilon_{12} \\ \Upsilon_{21} & \Upsilon_{22} \end{bmatrix} := \begin{bmatrix} P_1^{\mathrm{T}} \\ P_2^{\mathrm{T}} \end{bmatrix} \Omega \begin{bmatrix} Q_1 & Q_2 \end{bmatrix}.$$

The goal now is to show that under the conditions (C.56), we can build a matrix Ω such that the inequality (C.55) holds. First of all, note that for any Ψ_{44}, there exists Υ_{22} such that $\Psi_{44} + \Upsilon_{22} + \Upsilon_{22}^{\mathrm{T}} \prec 0$. On the other hand, the matrix

$$\Gamma := \begin{bmatrix} \Psi_{11} & \Psi_{12} & \Psi_{13} \\ \star & \Psi_{22} & \Psi_{23} + \Upsilon_{11}^{\mathrm{T}} \\ \star & \star & \Psi_{33} \end{bmatrix} \tag{C.58}$$

is negative definite if and only if the matrix

$$\begin{aligned} \Gamma' &:= \begin{bmatrix} I & 0 & 0 \\ -\Psi_{12}^{\mathrm{T}} \Psi_{11}^{-1} & I & 0 \\ -\Psi_{13}^{\mathrm{T}} \Psi_{11}^{-1} & 0 & I \end{bmatrix} \Gamma \begin{bmatrix} I & 0 & 0 \\ -\Psi_{12}^{\mathrm{T}} \Psi_{11}^{-1} & I & 0 \\ -\Psi_{13}^{\mathrm{T}} \Psi_{11}^{-1} & 0 & I \end{bmatrix}^{\mathrm{T}} \\ &= \begin{bmatrix} \Psi_{11} & 0 & 0 \\ \star & \Psi_{22} - \Psi_{12}^{\mathrm{T}} \Psi_{11}^{-1} \Psi_{12} & \Upsilon_{11}^{\mathrm{T}} + \Theta^{\mathrm{T}} \\ \star & \star & \Psi_{33} - \Psi_{13}^{\mathrm{T}} \Psi_{11}^{-1} \Psi_{13} \end{bmatrix} \end{aligned} \tag{C.59}$$

is negative definite as well with $\Theta := \Psi_{23}^{\mathrm{T}} - \Psi_{13}^{\mathrm{T}} \Psi_{11}^{-1} \Psi_{12}$. Since Υ_{11} is arbitrary, Γ' is negative definite if and only if the inequalities

$$\begin{aligned} \Psi_{11} &\prec 0 \\ \Psi_{22} - \Psi_{12}^{\mathrm{T}} \Psi_{11}^{-1} \Psi_{12} &\prec 0 \\ \Psi_{33} - \Psi_{13}^{\mathrm{T}} \Psi_{11}^{-1} \Psi_{13} &\prec 0 \end{aligned} \tag{C.60}$$

hold. These conditions are equivalent to the LMIs

$$\begin{bmatrix} \Psi_{11} & \Psi_{12} \\ \star & \Psi_{22} \end{bmatrix} \prec 0 \text{ and } \begin{bmatrix} \Psi_{11} & \Psi_{13} \\ \star & \Psi_{33} \end{bmatrix} \prec 0, \tag{C.61}$$

which are identical, in turn, to the LMI conditions in (C.56). Hence, if the conditions (C.56) hold, then it is possible to find Υ_{11} such that $\Gamma \prec 0$. A Schur complement on (C.57) yields the inequality

$$\Psi_{44} + \Upsilon_{22} + \Upsilon_{22}^{\mathrm{T}} - \begin{bmatrix} \Psi_{14} \\ \Psi_{24} + \Upsilon_{21}^{\mathrm{T}} \\ \Psi_{34} + \Upsilon_{12} \end{bmatrix}^{\mathrm{T}} \Gamma^{-1} \begin{bmatrix} \Psi_{14} \\ \Psi_{24} + \Upsilon_{21}^{\mathrm{T}} \\ \Psi_{34} + \Upsilon_{12} \end{bmatrix} \prec 0$$

which is obviously satisfied by choosing a sufficiently small $\Upsilon_{22} + \Upsilon_{22}^{\mathrm{T}} \prec 0$. Hence, when the conditions (C.61) hold, we can build a matrix Υ such that the matrix (C.57) is negative definite, and hence it is possible to build a matrix

$$\Omega = \begin{bmatrix} P_1^{\mathrm{T}} \\ P_2^{\mathrm{T}} \end{bmatrix}^{+} \Upsilon \left[Q_1 \; Q_2 \right]^{+}$$

such that the inequality (C.55) holds. The proof is complete. ■

The *elimination lemma* is a corollary of the projection lemma where P or Q is equal to the identity matrix. It is therefore related to Finsler's lemma.

Lemma C.12.2 (Elimination/Creation Lemma) *Let $\Psi \in \mathbb{S}^n$ be a symmetric matrix and $P \in \mathbb{R}^{p \times n}$ be a given matrix. Then, the following statements are equivalent:*

1. *There exists a matrix $\Omega \in \mathbb{R}^{p \times n}$ such that*

$$\Psi + P^{\mathrm{T}}\Omega + \Omega^{\mathrm{T}} P \prec 0. \tag{C.62}$$

2. *The LMI*

$$P_{\perp}^{\mathrm{T}} \Psi P_{\perp} \prec 0 \tag{C.63}$$

holds where $P_{\perp}$ is a basis of the null-space of P.

3. *There exist a scalars $\tau \in \mathbb{R}$ such that the LMI*

$$\Psi - \tau P^{\mathrm{T}} P \prec 0$$

holds.

The elimination lemma consists of passing from the condition (C.62) to the condition (C.63) whereas the *creation lemma* is the opposite direction.

C.13 Completion Lemma

This theorem shows that it is possible to construct a matrix and its inverse from only one block of each only. It has consequences in the construction of Lyapunov matrices in the dynamic output feedback synthesis problem; see e.g. [96, 98] and Sects. 3.3.2, 3.5.1, and 3.5.2.

Theorem C.13.1 (Completion Lemma) *Let $X \in \mathbb{S}^n_{\succ 0}$ and $Y \in \mathbb{S}^n_{\succ 0}$ be given matrices. Then, the following statements are equivalent:*

1. *There exist $X_2 \in \mathbb{R}^{n\times r}$, $X_3 \in \mathbb{R}^{r\times r}$, $Y_2 \in \mathbb{R}^{n\times r}$ and $Y_3 \in \mathbb{R}^{r\times r}$ such that*

$$\begin{bmatrix} X & X_2 \\ X_2^{\mathrm{T}} & X_3 \end{bmatrix} \succ 0 \quad \text{and} \quad \begin{bmatrix} X & X_2 \\ X_2^{\mathrm{T}} & X_3 \end{bmatrix}^{-1} = \begin{bmatrix} Y & Y_2 \\ Y_2^{\mathrm{T}} & Y_3 \end{bmatrix}.$$

2. *The following conditions*

$$\begin{bmatrix} X & I_n \\ I_n & Y \end{bmatrix} \succeq 0 \quad \text{and} \quad rank \begin{bmatrix} X & I_n \\ I_n & Y \end{bmatrix} \le n + r$$

hold.

Proof **Proof of 2 ⇒ 1**: From the LMI

$$\begin{bmatrix} X & I_n \\ \star & Y \end{bmatrix} \succeq 0,$$

we can state that $X - Y^{-1} \succeq 0$. It is hence possible to compute a matrix $\tilde{X}_2 \in \mathbb{R}^{n\times r}$ satisfying $\tilde{X}_2\tilde{X}_2^{\mathrm{T}} = X - Y^{-1}$. Thus, we have that $X - \tilde{X}_2\tilde{X}_2^{\mathrm{T}} \succ 0$ which is equivalent to

$$\begin{bmatrix} X & \tilde{X}_2 \\ \star & I_r \end{bmatrix} \succ 0.$$

A congruence transformation with respect to $\mathrm{diag}(I_n, \tilde{X}_3^{\mathrm{T}})$ where $\tilde{X}_3$ is nonsingular yields

$$\begin{bmatrix} X & X_2 \\ \star & X_3 \end{bmatrix} \succ 0 \qquad \text{(C.64)}$$

where $X_2 := \tilde{X}_2\tilde{X}_3$ and $X_3 := \tilde{X}_3^{\mathrm{T}}\tilde{X}_3$. This proves that it is possible to complete the matrix with X_2 and X_3 such that the completed matrix is positive definite. For completeness, we can check back whether the (1, 1) block of the inverse of the matrix in (C.64) is equal to Y. Applying the Banachiewicz inversion formula (see

[5] or Appendix A.2), we obtain that the (1, 1) block of the inverse of the matrix in (C.64) is equal to $(X - X_2 X_3^{-1} X_2)^{-1}$. Substituting the expressions $X_2 = \tilde{X}_2 \tilde{X}_3$ and $X_3 = \tilde{X}_3^{\mathrm{T}} \tilde{X}_3$ into $(X - X_2 X_3^{-1} X_2)^{-1}$ yields that $(X - X_2 X_3^{-1} X_2)^{-1} = Y$. The proof is complete.

Proof of 1 ⇒ 2: Using the Banachiewicz inversion formula, we can state that

$$Y = X^{-1} + X^{-1} X_2 \left(X_3 - X_2^{\mathrm{T}} X^{-1} X_2 \right)^{-1} X_2^{\mathrm{T}} X^{-1}.$$

Moreover, since $\operatorname{rank}[X_3 - X_2^{\mathrm{T}} X^{-1} X_2] = r$, we then have

$$X^{-1} X_2 \left(X_3 - X_2^{\mathrm{T}} X^{-1} X_2 \right)^{-1} X_2^{\mathrm{T}} X^{-1} \succeq 0$$

which implies that $Y \succeq X^{-1}$ and $\operatorname{rank}[Y - X^{-1}] \leq r$. This implies that the conditions

$$\begin{bmatrix} X & I_n \\ \star & Y \end{bmatrix} \succeq 0$$

and $\operatorname{rank}[X] + \operatorname{rank}[Y - X^{-1}] \leq n + r$ hold. According to the Guttman rank additivity formula [75], the rank condition is equivalent to

$$\operatorname{rank} \begin{bmatrix} X & I_n \\ \star & Y \end{bmatrix} \leq n + r.$$

The proof is complete. ■

We have the following corollary when $r = n$:

Corollary C.13.2 *Let $X \in \mathbb{S}^n_{\succ 0}$ and $Y \in \mathbb{S}^n_{\succ 0}$ be given matrices. Then, the following statements are equivalent:*

1. *There exist matrices $X_2, Y_2 \in \mathbb{R}^{n \times n}$ and $X_3, Y_3 \in \mathbb{S}^n$ such that the relations*

$$\begin{bmatrix} X & X_2 \\ X_2^{\mathrm{T}} & X_3 \end{bmatrix} \succ 0 \quad \text{and} \quad \begin{bmatrix} X & X_2 \\ X_2^{\mathrm{T}} & X_3 \end{bmatrix}^{-1} = \begin{bmatrix} Y & Y_2 \\ Y_2^{\mathrm{T}} & Y_3 \end{bmatrix}$$

   hold.

2. *The LMI*

$$\begin{bmatrix} X & I_n \\ I_n & Y \end{bmatrix} \succ 0$$

   holds.

C.14 Cone Complementary Algorithm

The cone complementary algorithm is a powerful algorithm allowing one to solve, in an iterative way, certain nonconvex semidefinite programs. This algorithm has been proposed in [99] for solving the static output feedback and fixed-order dynamic output feedback problems, which are problems of high interest in linear control design. Let us consider the following matrix inequality problem:

Problem C.14.1 Find matrices $Q, \tilde{Q} \succ 0$ and R verifying $Q\tilde{Q} = I_n$ such that the LMI

$$\mathcal{M}(Q, \tilde{Q}R) \prec 0 \tag{C.65}$$

holds where $\mathcal{M}$ is indeed an affine function of the matrices Q, $\tilde{Q}$ and R. The dimension of the matrix Q is equal to n.

This problem is nonconvex due to the product constraint $Q\tilde{Q} = I$. The idea is to associate the above feasibility problem with the following optimization problem

$$\min_{Q \succ 0, \tilde{Q} \succ 0, R} \quad \text{trace}(Q\tilde{Q}) \text{ s.t. } \mathcal{M}(Q, \tilde{Q}, R) \prec 0.$$

The above optimization problem is the so-called *cone complementary problem* since it can be understood as an extension of linear complementarity problems to the cone of positive semidefinite matrices. Linear complementarity problems have been introduced by Cottle and Dantzig in [100] in 1968; see also the monographs [101, 102]. Several efficient algorithms have been provided to solve this problem; see e.g. [103–105]. The cone complementary algorithm is the generalization of these algorithms to the case of positive semidefinite matrices. It is described below:

Algorithm 1 Cone complementary algorithm [99]

1: Find a feasible point Q_0, $\tilde{Q}_0$, R such that $\mathcal{M}(Q_0, \tilde{Q}_0, R_0) \prec 0$. If there is none then exit; else set $k = 0$.
2: **loop**
3: $\quad S_k \leftarrow Q_k$, $\tilde{S}_k \leftarrow \tilde{Q}_k$.
4: $\quad$ Find $Q_{k+1} \succ 0$ and $\tilde{Q}_{k+1} \succ 0$ that solves the problem

$$\alpha_k := \min_{Q \succ 0, \tilde{Q} \succ 0, R} \quad \text{trace}(\tilde{S}_k Q + S_k \tilde{Q}) \text{ s.t. } \mathcal{M}(Q, \tilde{Q}, R) \prec 0$$

5: $\quad$ **if** α_k close enough to $2n$ **then**
6: $\quad\quad$ **print** "Solution found".
7: $\quad\quad$ **return** Q_{k+1}.
8: $\quad$ **else if** Maximum iteration reached **then**
9: $\quad\quad$ **print** "Maximum number of iterations reached".
10: $\quad\quad$ Exit.
11: $\quad$ **else**
12: $\quad\quad$ $k \leftarrow k + 1$.
13: $\quad$ **end if**
14: **end loop**

This algorithm has been shown to be locally converging to an optimal value α^* that is greater or equal to $2n$, n being the dimension of Q. When the optimal value is equal to $2n$, then the algorithm has converged and the last computed values for Q and $\tilde{Q}$ verify $Q\tilde{Q} = I$.

This algorithm has been applied successfully to the design of controllers for linear systems in [106–110] and in [111, 112] for linear time-delay systems. In the latter references, however, the cone complementary algorithm is used in order to solve a matrix inequality feasibility problem which contains a concave term. This procedure is formulated in the following result:

Lemma C.14.2 *Let $\mathcal{M}(P, Q, x)$ be a symmetric matrix depending affinely on $x \in \mathbb{R}^n$ and on symmetric matrices P, Q. Then, the following statements are equivalent:*

1. *There exist a vector $x \in \mathbb{R}^n$ and positive definite matrices P, Q such that the matrix inequality*
$$\mathcal{M}(P, Q, x) - PQ^{-1}P \prec 0 \tag{C.66}$$
holds.
2. *There exist a vector $x \in \mathbb{R}^n$ and symmetric positive definite matrices P, $\tilde{P}$, Q, Z, $\tilde{Z}$ and $\tilde{Q}$ verifying $P\tilde{P} = I$, $Q\tilde{Q} = I$, $Z\tilde{Z} = I$ such that the matrix inequalities*
$$\begin{gathered}\mathcal{M}(P, Q, x) - Z \prec 0 \\ \begin{bmatrix} \tilde{Z} & \tilde{P} \\ \star & \tilde{Q} \end{bmatrix} \succeq 0\end{gathered} \tag{C.67}$$
hold.

Proof Define Z as $Z \preceq PQ^{-1}P$. Then, the inequality (C.66) holds if and only if the matrix inequalities $\mathcal{M}(x) - Z$ and $Z \preceq PQ^{-1}P$ hold. The latter inequality is equivalent to $\tilde{Z} \preceq P^{-1}QP^{-1}$ which is equivalent to the condition that

$$\begin{bmatrix} Z^{-1} & P^{-1} \\ \star & Q^{-1} \end{bmatrix} \preceq 0. \tag{C.68}$$

The result is finally obtained by setting $\tilde{Z} = Z^{-1}$, $\tilde{P} = P^{-1}$ and $\tilde{Q} = Q^{-1}$. ■

Whereas the first statement is in general very difficult to solve due to the strong nonlinear nature of the conditions, the second one exhibits a much nicer structure and can be solved using the cone complementary algorithm.

References

1. J. Schur, Uber potenzreihen, die im innern des einheitskreises beschrankt sind [i]. Journal fur die reine und angewandte Mathematik **147**, 205–232 (1917)
2. J. Schur, Uber potenzreihen, die im innern des einheitskreises beschrankt sind [ii]. Journal fur die reine und angewandte Mathematik **148**, 122–145 (1917)
3. I. Gohberg, Schur Methods in Operator Theory and Signal Processing. Operator Theory: Advances and Applications. (Birkhauser Verlag, Basel, 1986).
4. F. Zhang, *The Schur Complement and Its Applications* (Springer, New York, 2005)
5. T. Banachiewicz, Zur berechnung der determinanten, wie auch der inversen, und zur darauf basierten auflosung der systeme linearer gleichungen. Acta Astronomica, Serie C **3**, 41–67 (1937)
6. W.J. Duncan, Some devices for the solution of large sets of simultaneous linear equations. The Lond. Edinb. Dublin Philos. Mag. J. Sci. Seventh Ser. 35, 660–670 (1917).
7. K. Zhou, J.C. Doyle, K. Glover, *Robust and Optimal Control* (Prentice Hall, Upper Saddle River, 1996)
8. R.E. Skelton, T. Iwasaki, K.M. Grigoriadis, *A Unified Algebraic Approach to Linear Control Design* (Taylor & Francis, London, 1997)
9. R.A. Horn, C.R. Johnson, *Matrix Analysis* (Cambridge University Press, Cambridge, 1990)
10. T. Iwasaki, R.E. Skelton, A unified approach to fixed order controller design via linear matrix inequalities. Math. Probl. Eng. **1**, 59–75 (1995)
11. T. Iwasaki, R.E. Skelton, All controllers for the general $\mathcal{H}_\infty$ control probblems: LMI existence conditions and state-space formula. Automatica **30**(8), 1307–1317 (1994)
12. S. Boyd, L. El-Ghaoui, E. Feron, V. Balakrishnan, *Linear Matrix Inequalities in Systems and Control Theory* (SIAM, Philadelphia, 1994)
13. Y. Nesterov, A. Nemirovskii, *Interior-Point Polynomial Algorithm in Convex Programming* (SIAM, Philadelphia, 1994)
14. J.F. Sturm, Using SEDUMI 1.02, a Matlab Toolbox for Optimization Over Symmetric Cones. Optim. Methods. Softw. 11(12), 625–653 (2001).
15. R.H. Tütüncü, K.C. Toh, M.J. Todd, Solving semidefinite-quadratic-linear programs using SDPT3. Math. Program. Ser. B **95**, 189–217 (2003)
16. J. Löfberg, Yalmip: a toolbox for modeling and optimization in MATLAB, in Proceedings of the CACSD Conference, Taipei, Taiwan, 2004.
17. M. Grant, S. Boyd, CVX: matlab software for disciplined convex programming, version 2.0 beta (2013) http://cvxr.com/cvx
18. M. Grant, S. Boyd, in Graph implementations for nonsmooth convex programs, ed. by V. Blondel, S. Boyd, H. Kimura. Recent Advances in Learning and Control, Lecture Notes in Control and Information Sciences, (Springer, London, 2008) pp. 95–110. http://stanford.edu/boyd/graph_dcp.html
19. A. Ben-Tal, A. Nemirovski, On tractable approximations of uncertain linear matrix inequalities affected by interval uncertainty. SIAM J. Optim. **12**(3), 811–833 (2002)
20. A. Ben-Tal, A. Nemirovski, C. Roos, Extended matrix cube theorems with applications to μ-theory in control. Math. Oper. Res. **28**(3), 497–523 (2003)
21. A. Bental, L. El Ghaoui, A. Nemirovski, *Robust Optimization* (Princeton University Press, Princeton, 2009)
22. B.-D. Chen, S. Lall, Degree bounds for polynomial verification of the matrix cube problem, in 45th IEEE Conference on Decision and Control, San Diego, California, USA, pp. 4405–4410 2006.
23. R. Tempo, E.W. Bai, D. Dabbene, Probabilistic robustness analysis: explicit bounds for the minimum number of samples. Syst. Control Lett. **30**, 237–242 (1997)
24. G.C. Calafiore, F. Dabbene, R. Tempo, Randomized algorithms for probabilistic robustness with real and complex structured uncertainty. Syst. Control Lett. **45**(12), 2218–2235 (2000)
25. G. Calafiore, B.T. Polyak, Stochastic algorithms for exact and approximate feasibility of robust LMIs. IEEE Trans. Autom. Control **46**(11), 1755–1759 (2001)

26. H. Fujioka, A discrete-time approach to stability analysis of systems with aperiodic sample-and-hold devices. IEEE Trans. Autom. Control **54**(10), 2440–2445 (2009)
27. H. Fujioka, T. Nakai, L. Hetel, A switched Lyapunov function approach to stability analysis of non-uniformly sampled-data systems, in American Control Conference, Marriott Waterfront, Baltimore, MD, USA, pp. 1803–1804 2010.
28. F. Wu, Control of linear parameter varying systems. Ph.D thesis, University of California, Berkeley, 1995.
29. S. Prajna, A. Papachristodoulou, P. Seiler, P.A. Parrilo, Sum of squares optimization toolbox for MATLAB, SOSTOOLS, 2004.
30. S. Prajna, A. Papachristodoulou, P. Seiler, P.A. Parrilo, in SOSTOOLS and its control applications, ed. by D. Henrion and A. Garulli. Positive Polynomials in Control, Lecture Notes in Control and Information Science, (Springer, Heidelberg, 2005) pp. 273–292.
31. P. Seiler, SOSOPT: a toolbox for polynomial optimization, 2013.
32. A-L. Cauchy, Analyse Algébrique (French) Algebraic Analysis. Edn J. Gabay, France, 1821.
33. J.L.W.V. Jensen, Sur les fonctions convexes et les inégalités entre les valeurs moyennes. Acta Mathematica **30**, 175–193 (1905)
34. M. Marshall, *Positive Polynomials and Sums of Squares* (American Mathematical Society, Providence, 2008)
35. E. Artin, Uber die Zerlegung definiter Funktionen in Quadrate. Abh. Math. Sem. Univ. Hamburg **5**, 85–99 (1927)
36. J.A. Scott, On sums of squares of multivariate polynomials. Appl. Probab. Trust (Mathematical Spectrum) **39**(3), 123–124 (2007)
37. M. Putinar, Positive polynomials on compact semi-algebraic sets. Indiana Univ. Math. J. **42**(3), 969–984 (1993)
38. M.D. Choi, T.Y. Lam, B. Reznick, Sums of squares of real polynomials. Proceedings on Symposia in Pure Mathematics **58**(2), 103–125 (1995)
39. G. Chesi, A. Tesi, A. Vicino, R. Genesio, *On convexification of some minimum distance problems, in 5th European Control Conference* (Karlsruhe, Germany, 1999)
40. G. Chesi, A. Garulli, A. Tesi, A. Vicino, Solving quadratic distance problems: an LMI-based approach. IEEE Trans. Autom. Control **48**(2), 200–212 (2003)
41. P. Parrilo, Structured Semidefinite Programs and Semialgebraic Geometry Methods in Robustness and Optimization. Ph.D thesis, California Institute of Technology, Pasadena, California, 2000.
42. P. Apkarian, H. D. Tuan, Parametrized LMIs in control theory. in *Conference on Decision and Control*, Tampa, Florida, 1998.
43. H.D. Tuan, P. Apkarian, Relaxations of parametrized LMIs with control applications, in *Conference on Decision and Control* (Tampa, Florida, 1998)
44. H.D. Tuan, P. Apkarian, T. Narikiyo, Y. Yamamoto, Parameterized linear matrix inequality techniques in fuzzy control system design. IEEE Trans. Fuzzy Syst. **9**(2), 324–332 (2001)
45. R.C.L.F. Oliveira, V.F. Montagner, P.L.D. Peres, P.A. Bliman, *LMI relaxations for $\mathcal{H}_\infty$ control of time-varying polytopic systems by means of parameter dependent quadratically stabilizing gains, in 3rd IFAC Symposium on System* (Structure and Control, Foz do Iguassu, Brazil, 2007)
46. R.C.L.F. Oliveira, P.L.D. Peres, Parameter-dependent LMIs in robust analysis: Characterization of homogeneous polynomially parameter-dependent solutions via LMI relaxations. IEEE Trans. Autom. Control **57**(7), 1334–1340 (2002)
47. M. Sato, D. Peaucelle, Robust stability/peformance analysis for uncertain linear systems via multiple slack variable approach: polynomial LTIPD systems, in 46rd IEEE Conference on Decision and Control, New Orleans, LA, USA, 2007.
48. E. Féron, P. Apkarian, P. Gahinet, Analysis and synthesis of robust control systems via parameter-dependent Lyapunov functions. IEEE Trans. Autom. Control **41**, 1041–1046 (1996)
49. C.W. Scherer, Relaxations for robust linear matrix inequality problems with verifications for exactness. SIAM J. Matrix Anal. Appl. **27**(2), 365–395 (2005)

50. D. Handelman, Representing polynomials by positive linear functions on compact convex polyhedra. Pac. J. Math. **132**(1), 35–62 (1988)
51. C.W. Schere, C.W.J. Hol, Matrix sum-of-squares relaxations for robust semi-definite programs. Math. Program. Ser. B, 107 189–211 2006 http://www.dcsc.tudelft.nl/cscherer/pub.html
52. C. Briat, Robust stability analysis of uncertain linear positive systems via integral linear constraints - L_1- and L_∞-gains characterizations, in 50th IEEE Conference on Decision and Control, Orlando, Florida, USA, pp. 3122–3129 2011.
53. C. Briat, Robust stability and stabilization of uncertain linear positive systems via integral linear constraints - L_1- and L_∞-gains characterizations. Int. J. Robust Nonlin. Control **23**(17), 1932–1954 (2013)
54. G. Chesi, A. Garulli, A. Tesi, A. Vicino, *Homogeneous Polynomial Forms for Robustness Analysis of Uncertain Systems* (Lecture Notes in Control and Information Sciences. Springer, Berlin, 2009)
55. G. Chesi, LMI techniques for optimization over polynomials in control: a survey. IEEE Trans. Autom. Control **55**(11), 2500–2510 (2010)
56. J.B. Lasserre, Global optimization with polynomials and the problem of moments. SIAM J. Optim. **11**(3), 796–817 (2001)
57. D. Henrion, J.B. Lasserre, Convergent relaxations of polynomial matrix inequalities and static output feedback. IEEE Trans. Autom. Control **51**(2), 192–202 (2006)
58. J.-B. Lasserre, *Moments, Positive Polynomials and Their Applications* (Imperial College Press, London, 2010)
59. V.M. Popov, Absolute stability of nonlinear systems of automatic control. Autom. Remote Control (Trans. from Automatica i Telemekhanika) 22(8), 857–875 (1961).
60. V.A. Yakubovitch, The solution to certain matrix inequalities in automatic control. Soviet Math. Dokl. **3**, 620–623 (1962)
61. R.E. Kalman, Lyapunov functions for the problem of Lur'e in automatic control. Proceedings of the National Academy of Sciences of the United States of America **49**, 201–205 (1963)
62. C. W. Scherer, S. Weiland. Linear matrix inequalities in control. Lecture Notes, 2005 http://www.imng.uni-stuttgart.de/simtech/Scherer/lmi/
63. A. Rantzer, On the Kalman-Yakubovicth-Popov lemma. Syst. Control Lett. **28**(1), 7–10 (1996)
64. T. Iwasaki, G. Meinsma, M. Fu, Generalized $\mathcal{S}$-procedure and finite frequency KYP lemma. Math. Probl. Eng. **6**, 305–320 (1998)
65. T. Paré, A. Hassibi, J. How, *A KYP lemma and invariance principle for systems with multiple hysteresis nonlinearities* (Standford University, California, USA, Technical report , 1999)
66. B. Hencey, A.G. Alleyne, A KYP lemma for LMI regions. IEEE Trans. Autom. Control **52**(10), 1926–1930 (2007)
67. D. Z. Arov, O. J. Staffans, in The infinite-dimensional continuous time Kalman-Yakubovich-Popov inequality, ed. by M.A. Dritschel. The extended field of operator theory, (Springer, Berlin, 2007) pp. 37–72.
68. T. Tanaka, C. Langbort, *KYP Lemma for internally positive systems and a tractable class of distributed H-infinity control problems, in American Control Conference* (Baltimore, Maryland, USA, 2010), pp. 6238–6243
69. T. Tanaka, C. Langbort, Symmetric formulation of the Kalman-Yakubovich-Popov Lemma and exact losslessness condition, in 50th IEEE Conference on Decision and Control, Orlando, Florida, USA, pp. 5645–5652 2011.
70. G. Pipeleers, L. Vandenberghe, Generalized KYP lemma with real data. IEEE Trans. Autom. Control **56**(12), 2942–2946 (2011)
71. A. Ranter, On the Kalman-Yakubovich-Popov Lemma for positive systems, in 51st IEEE Conference on Decision and Control, Maui, Hawaii, USA, pp. 7482–7484 2012.
72. J.C. Willems, Dissipative dynamical systems i and ii. Ration. Mech. Anal. **45**(5), 321–393 (1972)
73. B. Brogliato, R. Lozano, B. Maschke, O. Egeland, *Dissipative Systems Analysis and Control* (Springer, London, 2007)

74. E.V. Haynsworth, Determination of the inertia of a partitioned Hermitian matrix. Linear Algebra Appl. **1**(1), 73–81 (1968)
75. L. Guttman, Enlargement methods for computing the inverse matrix. Ann. Math. Stat. **17**, 336–343 (1946)
76. C.W. Scherer, LPV control and full-block multipliers. Automatica **37**, 361–375 (2001)
77. A. Megretski, A. Rantzer, System analysis via Integral Quadratic Constraints. IEEE Trans. Autom. Control **42**(6), 819–830 (1997)
78. A.I. Lur'e, V.N. Postnikov, On the theory of stability of controlled systems. Prikl. Mat. i Mekh (Applied mathematics and mechanics) **8**(3), 3–13 (1944)
79. P. Finsler, Über das vorkommen definiter und semi-definiter formen in scharen quadratischer formen. Commentarii Mathematici Helvetici **9**, 188–192 (1937)
80. M.R. Hestenes, E.J. MacShane, A theorem on quadratic forms and its applications in the calculus of variations. Trans. Am. Math. Soc. **47**, 501–512 (1940)
81. V.A. Yakubovich, S-procedure in nonlinear control theory (in russian). Vestnik Leningrad Univ. **1**, 62–77 (1971)
82. V.A. Yakubovich, S-procedure in nonlinear control theory (english translation). Vestnik Leningrad Univ. **4**, 73–93 (1977)
83. A. Megretski, S. Treil, Power distribution in optimization and robustness of uncertain systems. J. Math. Syst. Estimation Control **3**, 301–319 (1993)
84. I. Pólik, T. Terlaky, A survey of the s-lemma. SIAM Rev. **49**(3), 371–418 (2007)
85. C.W. Scherer, Robust generalized $\mathcal{H}_2$ control for uncertain and LPV systems with general scalings, in Conference on Decision and Control, pp. 3970–3975 1996.
86. T. Iwasaki, S. Hara, Well-posedness of feedback systems: insight into exact robustness analysis and approximate computations. IEEE Trans. Autom. Control **43**, 619–630 (1998)
87. I.A. Petersen, A stabilization algorithm for a class of uncertain linear systems. Syst. Control Lett. **8**, 351–357 (1987)
88. P.P. Khargonekar, I.A. Petersen, K. Zhou, Robust stabilization of uncertain linear systems: quadratic stabilizability and $\mathcal{H}_\infty$ control theory. J. Math. Anal. Appl. **254**, 410–432 (2001)
89. L. Xie, M. Fu, C.E. de Souza, $\mathcal{H}_\infty$ control and quadratic stabilization of systems with parameter uncertainty via output feedback. IEEE Trans. Autom. Control **37**(8), 1253–1256 (1992)
90. C.E. de Souza, X. Li, Delay-dependent robust $\mathcal{H}_\infty$ control of uncertain linear state-delayed systems. Automatica **35**, 1313–1321 (1999)
91. L. Xie, Output feedback H_∞ control of systems with parameter uncertainty. Int. J. Control **63**, 741–750 (1996)
92. D.H. Jacobson, *Extensions of Linear-Quadratic Control, Optimization and Matrix Theory* (Academic Press, New York, 1977)
93. S.J. Kim, Y.H. Moon, Structurally constrained $\mathcal{H}_2$ and $\mathcal{H}_\infty$ control: a rank-constrained LMI approach. Automatica **42**, 1583–1588 (2006)
94. T. Iwasaki, G. Shibata, LPV system analysis via quadratic separator for uncertain implicit systems. IEEE Trans. Autom. Control **46**, 1195–1208 (2001)
95. A. Packard, Gain scheduling via linear fractional transformations. Syst. Control Lett. **22**, 79–92 (1994)
96. P. Apkarian, P. Gahinet, A convex characterization of gain-scheduled $\mathcal{H}_\infty$ controllers. IEEE Trans. Autom. Control **5**, 853–864 (1995)
97. P. Gahinet, P. Apkarian, A linear matrix inequality approach to $\mathcal{H}_\infty$ control. Int. J Robust Nonlin. Control **4**, 421–448 (1994)
98. A. Packard, K. Zhou, P. Pandey, G. Becker, A collection of robust control problems leading to LMI's, in 30th conference on Decision and Control, Brighton, England, pp. 1245–1250, December 1991.
99. L. El-Ghaoui, F. Oustry, M.A. Rami, A Cone Complementary linearization algorithm for static output-feedback and related problems. IEEE Trans. Autom. Control **42**, 1171–1176 (1997)
100. R.W. Cottle, G.B. Dantzig, Complementary pivot theory of mathematical programming. Linear Algebra Appl. **1**(1), 103–125 (1968)

101. R.W. Cottle, J.-S. Pang, *The Linear Complementarity Problem* (Academic Press, New York, 1992)
102. K.G. Murty, F.-T. Yu, Linear Complementarity, Linear and Nonlinear Programming. Sigma Series in Applied Mathematics 3 (Heldermann Verlag, Berlin, 1992).
103. Y. Ye, A fully polynomial-time algorithm for computing a stationary point for the general linear complementarity problem. Math. Methods Oper. Res. **18**, 334–345 (1993)
104. Y. Zhang, On the convergence of a class of infeasible interior-point methods for the horizontal linear complementarity problem. SIAM J. Optim. **4**(1), 208–227 (1994)
105. O.L. Mangasarian, J.S. Pang, The extended linear complementarity problem. SIAM J. Matrix Anal. Appl. **2**, 359–368 (1995)
106. L. El-Ghaoui, F. Oustry, H. Lebret, Robust solutions to uncertain semidefinite programs. SIAM J. Control Optim. **9**(1), 33–52 (1998)
107. D. Arzelier, D. Henrion, D. Peaucelle, *Robust state-feedback $\mathcal{D}$-stabilization via a cone complementarity algorithm, in 6th European Control Conference* (Porto, Portugal, 2001)
108. X. Song, S. Zhou, B. Zhang, A cone complementarity linearization approach to robust H_∞ controller design for continuous-time piecewise linear systems with linear fractional uncertainties. Nonlin Anal: Hybrid Syst. **2**(4), 1264–1274 (2008)
109. C. Briat, J.J. Martinez Design of $\mathcal{H}_\infty$ bounded non-fragile controllers for discrete-time systems, in 48th IEEE Conference on Decision and Control, pp. 2192–2197 2009.
110. C. Briat, J.J. Martinez, *$\mathcal{H}_\infty$ bounded resilient state-feedback design for linear continuous-time systems - a robust control approach, in 18th IFAC World Congress* (Milano, Italy, 2011), pp. 9307–9312
111. C. Briat, O. Sename, J.-F. Lafay, *A full-block $\mathcal{S}$-procedure application to delay-dependent $\mathcal{H}_\infty$ state-feedback control of uncertain time-delay systems, in 17th IFAC World Congress* (Seoul, South Korea, 2008), pp. 12342–12347
112. W.H. Chen, W.X. Zheng, On improved robust stabilization of uncertain systems with unknown input delays. Automatica **42**, 1067–1072 (2006)

Index

C. Briat, *Linear Parameter-Varying and Time-Delay Systems*,
Advances in Delays and Dynamics 3, DOI 10.1007/978-3-662-44050-6

The manufacturer's authorised representative in the EU is Springer Nature Customer Service Centre GmbH, Europaplatz 3, 69115 Heidelberg, Germany. If you have any concerns regarding our products, please contact ProductSafety@springernature.com

Printed and bound by CPI Group (UK) Ltd, Croydon, CR0 4YY
15/07/2026
02167619-0006